MOLLUSCAN SHELLFISH AQUACULTURE

MOLLUSCAN SHELLFISH AQUACULTURE

A Practical Guide

Edited by Sandra E. Shumway

5m Books

First published 2021

Published by
5M Books Ltd,
Lings, Great Easton,
Essex CM6 2HH, UK,
www.5mbooks.com

A Catalogue record for this book is available from the British Library

ISBN 9781789180107

Book layout by Servis Filmsetting Ltd, Stockport, Cheshire
Printed by Replika Press Pvt Ltd, India
Photos by the authors unless otherwise indicated

For all those who spawned and failed and kept trying ...
and for those who continue the efforts

Contents

Contributors

Brian Beal, University of Maine at Machias, Machias, Maine, USA & Downeast Institute, Beals, Maine, USA

Stephan Bullard, University of Hartford, West Hartford, Connecticut, USA

Juan Carlos Castilla, Departamento de Ecología y Estación Costera de Investigaciones Marinas (ECIM), Las Cruces, Chile & Pontificia Universidad Católica de Chile, Santiago, Chile

Daniel P. Cheney, Pacific Shellfish Institute, Olympia, Washington, USA

Peter Cook, University of Western Australia, Albany, Western Australia, Australia

R. LeRoy Creswell, Florida Sea Grant Extension Program, Fort Pierce, Florida, USA

Michael Dadswell, Acadia University, Wolfville, Nova Scotia, Canada

Leslie-Anne Davidson, Fisheries and Oceans Canada, Moncton, New Brunswick, Canada

Christopher V. Davis, Pemaquid Oyster Company, Inc., Waldoboro, Maine, USA

Jonathan P. Davis, Baywater Shellfish Farms, Bainbridge Island, Washington, USA

Megan Davis, Florida Atlantic University, Harbor Branch Oceanographic Institution, Ft. Pierce, Florida, USA

Carole R. Engle, Engle-Stone Aquatic$ LLC, Strasburg, VA, USA

Jorge Diogène Fadini, Institute for Food and Agricultural Research and Technology, Sant Carles de la Ràpita, Tarragona, Spain

Lynne Falconer, University of Stirling, Stirling, United Kingdom

Tessa Getchis, Connecticut Sea Grant, Groton, Connecticut, USA

Ximing Guo, Rutgers University, Haskin Shellfish Research Laboratory, Port Norris, New Jersey, USA

Kevin Heasman, Cawthron Institute, East Nelson, New Zealand

Zenjie Jiang, Yellow Sea Fisheries Research Institute, Chinese Academy of Fishery Sciences, Qingdao, China

Richard C. Karney, Martha's Vineyard Shellfish Group, Inc., Vineyard Haven, Massachusetts, USA

John Kraeuter, University of New England, Biddeford, Maine, USA & Haskin Shellfish Research Laboratory, Port Norris, New Jersey, USA

Gregg W. Langlois, California Department of Public Health, Richmond, California, USA

César Lodeiros, Universidad Técnica de Manabí, Bahía de Caráquez, Ecuador & Instituto Oceanográfico de Venezuela, Cumaná, Venezuela

Patricio H. Manríquez, Centro de Estudios Avanzados en Zonas Áridas (CEAZA) and Laboratorio de Ecología y Conducta de la Ontogenia Temprana (LECOT), Coquimbo, Chile

José Manuel Mazón-Suástegui, Centro de Investigaciones Biológicas del Noroeste, La Paz, B.C.S., Mexico
Thane A. Militz, University of the Sunshine Coast, Maroochydore, Queensland, Australia
Steve L. Morton, NOAA/NOS, Charleston, South Carolina, USA
Carter Newell, Pemaquid Mussel Farms, Bucksport, Maine, USA
Guilherme Sabino Rupp, Centro de Desenvolvimento em Aquicultura e Pesca, Florianópolis, Santa Catarina, Brazil
Samia Sarkis, Living Reefs Foundation, Bermuda
Sandra E. Shumway, University of Connecticut, Groton, Connecticut, USA
Aad Smaal, Wageningen Marine Research, Zierikzee, the Netherlands
Roxanna Smolowitz, Roger Williams University, Bristol, Rhode Island, USA
Paul C. Southgate, University of the Sunshine Coast, Maroochydore, Queensland, Australia
Trevor C. Telfer, University of Stirling, Stirling, United Kingdom
Alex Walsh, smartPAINT, Inc., East Falmouth, Massachusetts, USA
William Walton, Auburn University Shellfish Lab, Dauphin Island, Alabama, USA

Preface

Culture of bivalve molluscs has been carried out for centuries, but only in the past few decades have technological advances allowed the efforts to prosper based upon land-based hatchery systems coupled with coastal grow-out.

Until R.R.L. Guillard (1957) isolated and cultured phytoplankton as a food source for bivalve larvae, it was not possible to consider growing shellfish in captivity. His success opened the door for future efforts and the establishment of hatcheries (see Shumway et al., 2017). Technology has progressed considerably since that time, hatchery production of bivalve larvae is now routine, genetic and molecular techniques are used routinely to improve broodstock and battle diseases, and sophisticated models are used to assist growers in site selection and carrying capacity. Governmental regulations have taken hold, global marketing has flourished, and environmental issues have given way to the quest for best management practices and sustainability. Shellfish aquaculture is a successful and sustainable global venture.

Pioneers in the field of shellfish aquaculture faced enormous challenges, including practical, biological and social problems, yet they persevered. Among them, in no particular order, Peter Walne, Peter Korringa, Neil Bourne, Ken Chew, Paul Galtsoff, Victor Loosanoff, Carl Medcof, Fushi Zhang, James Mason, Michael Castagna, Paul Chanley, Ravena Ukeles, Robert Guillard, Herb Hidu, Hal Haskin, Cary Matthiessen, and all those who followed laid the groundwork for what is today a global industry worth billions of dollars.

Walne (1954) published a book based on his Buckland lectures entitled *Culture of Bivalve Molluscs*, which focused primarily on the European flat oyster (*Ostrea edulis*). It was one of the first efforts to summarize knowledge in the field of shellfish aquaculture and it is ironic that he chose to focus on one of the species still most challenging to culture today. Spencer (2002) presented a more general text entitled *Molluscan Shellfish Farming*, also based partially on Buckland lectures, which acknowledged advances in the field, and expanded to include chapters covering issues of importance to the newly developed industry – bivalve predators and their control, criteria for site selection for bivalve aquaculture, and processing for human consumption. During these past several decades there have been numerous publications based on specific species and geographic regions, but no comprehensive effort covering the key species and concerns of the shellfish aquaculture industry.

This book is not meant to be a treatise on all aspects of the biology, ecology, and aquaculture efforts for all species of bivalves, there have been numerous books published that cover all of these aspects and the reader is directed to those volumes for more in-depth and specific information (see the Further Reading list). Rather, it is meant to be a useable text for those interested in an up-to-date introduction to the field and to guide those who decide to join the shellfish aquaculture industry.

This book would not have been possible without the contributions of many experts and they are all thanked for their generous sharing of time and knowledge. Nigel Balmforth, 5m Books, made the work a reality. A great debt of gratitude and thanks goes to Noreen Blaschik Favreau who battled through various drafts and got them all ready for publication. Her efforts are intangible and truly appreciated. Thanks to Eric Heupel who designed the cover, and finally, thanks to Gus for his unwavering support through it all.

In Walne's epilogue, he noted that 'once hatcheries can be shown to be commercially profitable, it is clear that, with continual advance in knowledge, their potential for producing very large quantities of seed stock is considerable', and that 'It is the writer's hope that this book will help to spread information on one way in which this might come about.' His vision was fulfilled and it is hoped that the current volume will provide a useful manual and provide newcomers with the information necessary to spark their imagination and continue the effort to grow and utilize shellfish to their fullest potential.

Sandra E. Shumway
2021

Further reading

(see individual chapters for more specific references)

Andersen, R. (ed.) (2005) *Algal Culturing Techniques*. Elsevier Academic Press, London.

Ankenman Granata L., Flick, G.J., and Martin, R.E. (eds.) (2012) *The Seafood Industry: Species, Products, Processing, and Safety*. Second Edition. Blackwell, Oxford.

Bayne, B.L. (ed.) (1976) *Marine Mussels: their Ecology and Physiology*. Cambridge University Press, Cambridge.

Bayne, B.L. (2017) *Biology of Oysters*. Developments in Aquaculture and Fisheries Science, 41. Elsevier Scientific, Amsterdam.

Berg, C.J. (ed.) (1983) *Culture of Marine Invertebrates – Selected Readings*. Hutchinson Ross, Stroudburg, PA.

Boyle, P. and Rodhouse, P. (2005) *Cephalopods – Ecology and Fisheries*. Blackwell, Oxford.

Burnell, G. and Allan, G. (eds.) (2009) *New Technologies in Aquaculture – Improving production efficiency, quality and environmental management*. CRC Press/Woodhead Publishing, Oxford.

Engle, C.R. and Quagrainie, K. (2006) *Aquaculture Marketing Handbook*. Blackwell, Oxford.

Fretter, V. and Graham, A. (1994) *British Prosobranch Molluscs: Their Functional Anatomy and Ecology*. Second Edition. The Ray Society, Andover.

Gosling, E. (ed.) (1992) *The Mussel Mytilus: Ecology, Physiology, Genetics, and Culture*. Developments in Aquaculture and Fisheries Science, 25. Elsevier Scientific, Amsterdam.

Gosling, E. (2015) *Marine Bivalve Molluscs*. Second Edition. Wiley Blackwell, Oxford.

Guillard, R.R.L. (1975) Culture of phytoplankton for feeding marine invertebrates. In: Smith, W.L., and M.H. Chanley (eds.) *Culture of Marine Invertebrate Animals*. Plenum Press, New York, pp 29-60.

Hackney, C. and Pierson, M.D. (eds.) (1994) *Environmental Indicators and Shellfish Safety*. Chapman & Hall, New York.

Hahn, K.O. (ed.) (1989) *Handbook of Culture of Abalone and Other Marine Gastropods*. CRC Press, Boca Raton, FL.

Helm, M.M., Bourne, N., and Lovatelli, A. (ed.) (2004) *Hatchery Culture of Bivalves – a Practical Manual*. FAO Fisheries Technical Paper. No. 471. FAO, Rome.

Huguenin, J.E. and Colt, J. (1989) *Design and Operating Guide for Aquaculture Seawater Systems*. Developments in Aquaculture and Fisheries Science, 20. Elsevier Scientific, Amsterdam.

Ikenoue, H. and Kafuku, T. (eds.) (1992) *Modern Methods of Aquaculture in Japan*. Second revised edition. Developments in Aquaculture and Fisheries Science, 24. Elsevier Scientific, Amsterdam.

Journal of Shellfish Research (1991–present) Publication of the National Shellfisheries Association. www.shellfish.org.

Kennedy, V.S., Newell, R.I.E., and Eble, A.F. (eds.) (1996) *The Eastern Oyster* Crassostrea virginica. Maryland Sea Grant, College Park, MD.

Kraeuter, J.N. and Castagna, M. (eds.) (2001) *The Biology of the Hard Clam*. Developments in Aquaculture and Fisheries Science, 31. Elsevier Scientific, Amsterdam.

Korringa, P. (1976a) *Farming the cupped Oysters of the Genus* Crassostrea. Developments in Aquaculture and Fisheries Science, 2. Elsevier Scientific, Amsterdam.

Korringa, P. (1976b). *Farming the Flat Oysters of the Genus* Ostrea. Developments in Aquaculture and Fisheries Science, 3. Elsevier Scientific, Amsterdam.

Lawrence, J., Loreal, H., Toyofuku, H., Hess, P., Karunasagar, I., and Ababouch, L. (eds.) (2011) *Assessment and Management of Biotoxin Risks in Bivalve Molluscs*. FAO Fisheries and Aquaculture Technical Paper NO. 551. FAO, Rome.

Lutz, R.A. (1980) *Mussel Culture and Harvest: A North American Perspective*. Developments in Aquaculture and Fisheries Science, 7. Elsevier Scientific, Amsterdam.

Manzi, J.J. and Castagna, M. (eds.) (1989) *Clam Mariculture in North America*. Developments in Aquaculture and Fisheries Science, 19. Elsevier Scientific, Amsterdam.

Menzel, W. (ed.) (1991) *Estuarine and Marine Bivalve Mollusk Culture*. CRC Press, Boca Raton, FL.

Morse, D.E., Chew, K.K., and Mann, R. (eds.) (1984) *Recent Innovations in Cultivation of Pacific Molluscs*. Developments in Aquaculture and Fisheries Science, 14. Elsevier Scientific, Amsterdam.

Newell, R.C. (1979) *Biology of Intertidal Animals*. Marine Ecological Surveys Ltd. Faversham.

Otwell, W.S., Rodrick, G.E., and Martin, R.E. (eds.) (1991) *Molluscan Shellfish Depuration*. CRC Press, Boca Raton, FL.

Quayle, D.B. and Newkirk, G.F. (1989) *Farming Bivalve Molluscs: Methods for Study and Development. Advances in World Aquaculture*, Volume 1. World Aquaculture Society and International Development Research Centre, Ottawa.

Sarkis, S. and Lovatelli, A. (2007) *Installation and Operation of a Modular Bivalve Hatchery*. FAO Fisheries Technical Paper. No. 492. FAO, Rome.

Shepherd, S.A., Tegner, M.J., and Guzman del Proo, S.A. (1992) *Abalone of the World: Biology, Fisheries, and Culture*. Fishing News Books. Blackwell Scientific, Oxford.

Shumway, S.E. (ed.) (2011) *Shellfish Aquaculture and the Environment*. Wiley-Blackwell, Chichester.

Shumway, S.E. and Parsons, G.J. (2016) *Scallops: Biology, Ecology, Aquaculture and Fisheries*. Developments in Aquaculture and Fisheries Science, 40. Elsevier Scientific, Amsterdam.

Shumway, S.E. and Rodrick, G.E. (eds.) (2009) *Shellfish Safety and Quality*. CRC Press, Woodhead Publishing Limited, Cambridge.

Shumway, S.E., Burkholder, J.M., and Morton, S.L. (eds.) (2018) *Harmful Algal Blooms*. John Wiley & Sons, Chichester.

Smaal, A.C., Ferriera, J.G., Grant, J., Petersen, J.K., and Strand, O. (eds.) (2019) *Goods and Services of Marine Bivalves*. Springer, Cham.

Southgate, P.C. and Lucas, J.S. (eds.) (2008) *The Pearl Oyster*. Elsevier Science, Amsterdam.

Spencer, B.E. (2002) *Molluscan Shellfish Farming*. Blackwell, Oxford.

Stickney, R.R. and McVey, J.P. (2002) *Responsible Marine Aquaculture*. CABI Publishing, New York.

Tidwell, J.H. (ed.) (2012) *Aquaculture Production Systems*. Wiley-Blackwell, Chichester.

Tucker, C.S. and Hargreaves, J.A. (eds.) (2008) *Environmental Best Management Practices for Aquaculture*. Wiley-Blackwell, Oxford.

Walne, P.R. (1974) *Culture of Bivalve Molluscs: 50 Years Experience at Conwy*. Fishing News Books Ltd, Farnham.

Ward, T. and Phillips, B. (2008) *Seafood Ecolabelling Principles and Practice*. Wiley- Blackwell, Chichester.

CHAPTER 1

Farming oysters for food and profit

Jonathan P. Davis, Christopher V. Davis, and William Walton

Introduction

Edible oysters have been near the center of human folklore and gastronomic enjoyment for millennia. As the preeminent marine and evolutionary biologist, C.M. Yonge (1960) wrote, "There is much to say about oysters," and indeed, the voluminous literature on oysters attests to the many kinds of resources available to assist the aspiring shellfish farmer. Oysters grow quickly, are full of protein and nutrients, are flavorful, and easily acquired. While a rich fossil record exists for oysters beginning in the Jurassic period about 200 million years B.C., oysters were clearly an important part of native people's diets as evidenced by the presence of shell middens in many parts of the world dominated by oyster shells. Globally and historically, oyster fisheries have been notoriously difficult to manage. This may be due, in part, to the requirement, at least in the wild, for oyster larvae to attach to hard surfaces (such as adult oyster shell). Harvesting large oysters in the wild often results in the loss of the attached juveniles (called 'spat') as well as the loss of available substrate for future recruitment. These simple characteristics of the oyster life history, combined with its amenability to culture probably encouraged humans to initiate their cultivation, once local and easily accessible wild stocks were depleted. The Roman knight, Sergius Orata, is credited by many sources to have farmed what is known today as the European flat oyster (*Ostrea edulis*) in coastal lagoons, or *ostrearia* in the Bay of Naples as early as the 1st century B.C., apparently quite successfully according to the account of Carazzi (1893). Similarly, oyster culture in China can be traced back to over 2400 years ago according to the writer Fan Li and his book *Pisciculture* as briefly described by Wijsman et al. (2019). Quite simply, oysters were, and remain, an easily acquired food that has been greatly desired by humans for centuries. Their ecological role as both reef builders and ecosystem engineers in temperate estuaries has also been broadly recognized. For volumes focused exclusively on this commercially and ecologically important bivalve see Kennedy et al. (1996) and Bayne (2017).

Oysters are by their nature easily exploited and the demise of oyster reefs has been well documented on a worldwide basis (Beck et al., 2011). Serial depletion of cultured oysters is similar to an observed pattern for many species grown for aquaculture (You and Hedgecock, 2019). Often a shellfish disease first induces widespread mortality and the population collapses. Once that species is depleted, another is often introduced. W.K. Brooks (1891) wrote an important volume on the demise of wild oysters (*Crassostrea virginica*) in the Maryland portion of the Chesapeake Bay in the late 1800s, while a recent volume describes the vast ecological changes associated with the loss of eastern oysters from the Chesapeake Bay estuary, including the proposed, but ultimately rejected proposal to introduce the Asian oyster (*C. ariakensis*) to the Chesapeake Bay estuary (Kennedy, 2018). Similarly, the European flat oyster (*O. edulis*) was depleted in France in the 1920s and replaced first with the what was referred to then as the Portuguese oyster (*C. angulata*), and then by *C. gigas*, the Pacific or Japanese oyster (Buestel et al., 2009). Not to be outdone, the commercial fishery for the native oyster on the North American west coast (*O. lurida*) was commercially extinct by the late 1800s in much of its range though extensive efforts, particularly in southern Puget Sound, to rear this species in intertidal claires similar to those used for the culture of *O. edulis* in France, ultimately failed (Goulletquer, 2004; White et al., 2009). Since 1920, commercial sea farming of oysters on the US west coast has focused on the Pacific oyster, once this oyster was successfully transplanted from Japan to Washington State between 1905–1919 (Steele, 1964). The long history of the relationship of oysters to mankind clearly suggests that the demand for oysters has historically outstripped the wild supply and that oyster farmers continually adapted methods and species to meet the demand. In short, methods for cultivating oysters have changed and they will continue to evolve. Successful oyster farmers tend to be resilient and innovative. To be successful, growers must embrace innovation and have the capacity to adapt continuously to a climate in rapid flux, shortages of seed, and other fundamentals and changing business conditions. In short, a can-do attitude is perhaps the fundamental determinant for success.

Because oysters are grown throughout the world, methods of cultivation vary by place and species from simple farming approaches, which rely on natural recruitment, to sophisticated approaches, which include large-scale hatchery production and precision farming from seed to market. The focus of this chapter is to provide practical information to the aspiring oyster farmer to assist in the development of sustainable production of oysters for artisanal use or profit. Much goes into the development of a shellfish farm and, while oysters are cultured in abundance around the world, the ultimate success of an operation will require careful attention to site considerations (see Chapter 10), the species cultivated, and the many details associated with developing farm-based production. While pearl-producing oysters are commonly farmed in subtropical and tropical parts of the world (see Chapter 9), the general focus of this chapter is to provide information on culture techniques for edible oysters.

Biology and ecology of oysters

A thorough understanding of the general biology and ecology of oysters is essential for the aspiring oyster farmer because the success of any operation ultimately depends upon both recognizing and responding to problems in the culture system (hatchery, nursery, or grow-out) before they become uncontrolled calamities resulting in mortality and loss of the crop.

As bivalves, oysters belong to a large and diverse group of molluscs that inhabit both hard substrates and firm sediments as surface dwelling (epibenthic) animals in mainly intertidal estuarine and marine waters on all continental coastlines except Antarctica. Historically, oysters were exceedingly common, occurring in large intertidal and subtidal assemblages, or beds occupying many hectares of intertidal or bottom lands. Oysters are generally classified as belonging to taxonomic families that are characterized by a cupped left (lower) valve and flatter right (upper) valve (for example, *Crassostrea* spp., *Saccostrea* spp.), or two, relatively flatter valves (for example, *Ostrea* spp., *Tiostrea* spp.). In all oysters, the two valves are joined by an elastic ligament at the anterior end. A single adductor muscle, attached to the interior surface of both valves, when contracted serves to hold the valves together, especially when oysters are exposed to air during low tide. When under water the adductor muscle is generally not contracted, and the flexible hinge serves to prop the valves a few millimeters to enable essential bivalve function. Utilizing a highly efficient, cilia-driven feeding and respiratory current over specially modified tissues referred to as ctenidia (gills), microscopic phytoplankton and detrital material are retained and may be ingested. The inhalant current brings suspended materials to the ctenidial surface where highly evolved particle selection and sorting of particles may occur. These behaviors have been extensively studied in bivalves generally including oysters and demonstrate highly evolved mechanisms for particle capture, handling, and nutrient extraction (Ward and Shumway, 2004; for a recent review see Rosa et al., 2018). Oysters generally process large quantities of seawater, but feed discontinuously depending on the availability and quality of seston among other factors (Cranford et al., 2009). Feeding behavior in oysters was also comprehensively reviewed by Bayne (2017). A feature of bivalves associated with feeding on microscopic materials in the water column and certainly shared by oysters is the production of biodeposits. These may be produced as either pseudofeces or feces (fecal pellets). Pseudofeces consist of organic and inorganic materials collected on the ctenidia and labial palps prior to ingestion and rejected by the oyster as unconsolidated materials in mucus. True feces are produced following ingestion and absorption by the animal and are recognized as fecal pellets similar in features to other suspension feeding bivalves.

The sex life and life history of oysters is similar to other bivalves. Oysters may be predominantly male or female, sequentially male and female on an annual basis, or protandric – tending to be male as younger oysters and then changing sex as they get older and larger to become female. Suffice to say that sex and sexuality in oysters is complicated and has been a subject of significant interest for many investigators since the late 1800s. Generally, oysters undergo a cycle of reproduction characterized by the seasonal allocation of resources to germinal tissue development, generally in response to increasing temperature and food availability followed by the spawning of gametes stimulated by a number of environmental cues. Cupped and rock oysters shed gametes directly into the water column where fertilization and development to the larval stages occurs. Flat oysters diverge significantly from this life history pattern as females deposit their eggs directly into their mantle cavity where the eggs are fertilized by sperm (contained in sperm balls) inhaled by the female from surrounding waters. Eggs are fertilized within the mantle cavity and embryos are brooded for a period of time prior to their release into the water column as well-developed veliger larvae. For all oysters, spawning is generally followed by a period of nutrient rebuilding/reallocation of nutritional resources to growth.

Oyster larvae are microscopic and begin life as a fertilized egg only ~50 μm in diameter for *Crassostrea* species. Following fertilization, oyster larvae pass through several developmental stages, all the while feeding on microscopic materials once they become veligers. This stage is characterized by a unique, ciliated organ supremely adapted to enable the oyster larvae to swim and simultaneously feed on microscopic materials in the water column. Depending on

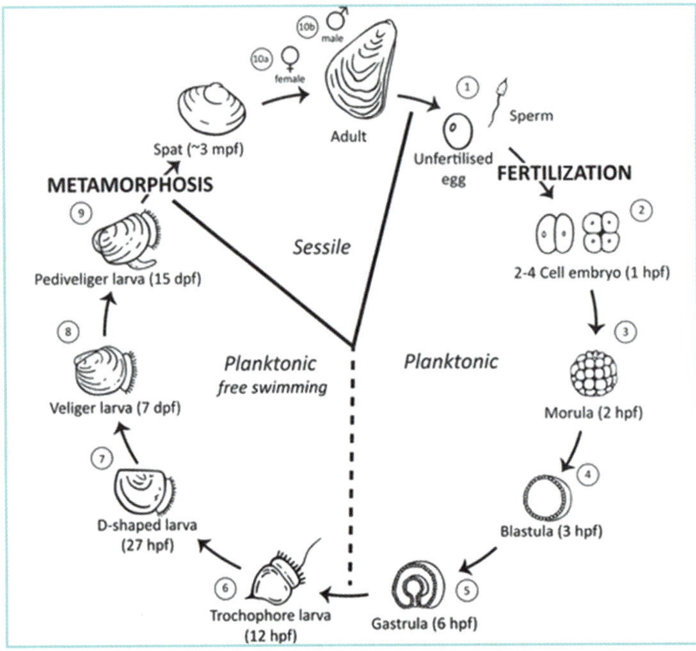

Figure 1.1 Diagram of the oyster life history (hpf = hours post fertilization, mpf = months post fertilization, dpf = days post fertilization). Oyster farmers utilize the cementation stage at the end of the larval period to facilitate subsequent juvenile production. Source: https://commons.wikimedia.org/wiki/File:Oyster_life_cycle.png.

the species and environmental conditions, oyster larvae live in the plankton for days to weeks prior to settling to the bottom as pediveligers. At this stage the typical *Crassostrea* oyster larvae is ~275–300 µm in diameter (Fig. 1.1). Oyster larvae, as is the case for monomyarian bivalves, generally settle and cement themselves via the left valve onto a hard substrate and quickly initiate metamorphosis to the juvenile stage, often in the company of other oysters. There they remain and grow to maturity and remain sedentary for the rest of their lives.

The role of feeding and feeding behavior in oysters is highly relevant to the siting of farms specific to the availability of adequate nutrition and overall carrying capacity to support multiple farms on the same bay or estuary. Oysters respond to temperature, seston quality, and quantity, and can exquisitely modulate their feeding behavior relative to food availability and other physiographic parameters (see extensive review by Bayne, 2017). Oysters are also known for their prodigious filtration capacity that en masse, often in reefs of many millions of individuals, can process huge volumes of seawater on a daily basis. Overstocking of farms and induced trophic risk can be an important source of mortality of oysters. Water filtration behaviors for oysters have been assessed for many species and, as is similar for other bivalves, filtration rates for individual oysters are highly variable and dependent on body size, seawater temperature, physiological state, and the availability of suspended food. Because oyster reefs provide significant habitat for other aquatic organisms for all or part of their lives, oysters are often referred to as ecosystem engineers. Oysters living in either natural reefs or intensive aquaculture have the capacity to impact water filtration dynamics, nutrient cycling, provision of habitat, and influence food web dynamics. Oysters are increasingly recognized for the ecosystem goods and services they can provide to the marine environment (Smaal et al., 2019).

The role of feeding, biodeposition, and growth in bivalves, including their ecological role in estuarine nutrient cycling, has been extensively modeled (Hawkins et al., 2002; Grant et al.,

2005). More recently, modeling efforts have been employed to help site shellfish farms for optimized production (Ferreira et al., 2007; Silva et al., 2011; Hawkins et al., 2013; Newell et al., 2019, and see Chapter 10), including modeling ecosystem benefits specific to molluscan aquaculture (Saurel et al., 2012; Grant and Pastres, 2019; Rose et al., 2019). The stocking density of shellfish, ambient physiographic information, and flux of suspended seston all influence feeding in shellfish and therefore require detailed resolution of the variability in feeding for site specific application of modeling. These data currently are generally lacking for most shellfish sea farm locations and model validation for sites worldwide is urgently needed. Environmental data and especially the application of farm models is important to optimize shellfish production, especially for larger farms where social and ecological carrying capacities are important considerations.

The impacts on farming due to a changing climate are also important to considerations. Growers of Pacific oysters in the Pacific northwest of North America have been significantly impacted by changing ocean and estuarine water chemistry related to continental upwelling, resulting in increased dissolved carbon dioxide and reduced aragonite saturation state, leading to decreased availability of carbonate in the seawater to supply oyster embryos, especially during the first 24 hours of development (Barton et al., 2015). A reduced energetic capacity to biomineralize shell efficiently under low aragonite saturation conditions was proposed as a primary cause for reduced hatchery production of oysters (Waldbusser et al., 2013; Waldbusser and Salisbury, 2014), though other investigations on Pacific oysters found that the supply of endogenous reserves was not a limiting factor during early shell formation in larvae. Rather, constraints on the kinetics regulating rate processes associated with calcification in larvae under low aragonite saturation conditions were proposed to explain commonly observed hatchery failures associated with corrosive seawater conditions (Frieder et al., 2017). The oyster industry in the US Pacific northwest adapted quickly to increasingly corrosive seawater conditions by buffering incoming hatchery seawater with sodium carbonate to simply increase aragonite saturation and enabled a resumption in oyster production for most hatcheries. Looking ahead, impacts to oysters due to a changing climate may not be so simply resolved, as significant challenges will emerge that require continued applied research and implementation of adaptation strategies to help ensure a viable oyster industry into the future.

Oyster aquaculture

On a worldwide basis, oyster aquaculture is generally practiced in near-shore waters between the mid-intertidal to subtidal depths of generally less than 5 m. By a wide margin, cupped oysters (*Crassostrea* spp. and *Saccostrea* spp.) are the focus of interest. China, Japan, and Korea lead the world in production of farmed oysters by a considerable amount, though oysters of a variety of species are reared on all continents except Antarctica. China is the major producer far and away of cupped oysters accounting in 2017 for almost 4.9 million metric tons (MT) produced (FAO, 2017). There are about 15 species of oysters grown or harvested commercially around the world. While most oysters are produced in China, there were an additional 978,918 MT of oysters produced by a further 22 countries (Table 1.1). See Box 1.1 for a brief description of the major oyster species cultivated worldwide. There are other cultured edible oysters, including tropical mangrove oysters (*C. rhizophorae*, *C. corteziensis*, *C. gasar*, and *C. tulipa*);

Table 1.1 Production (metric tons) of the major species of oysters harvested or cultivated worldwide. Adapted from Food and Agriculture Organization of the United Nations, Fisheries and Aquaculture Department, statistical collections: global aquaculture production (2017).

Common name	Scientific name	Country	Metric tons produced, 2017
Chilean oyster	*Crassostrea chiliensis*	Chile	377
Cortez oyster	*Crassostrea corteziensis*	Mexico	7,358
Cupped oyster	*Crassostrea* spp.	Thailand	14,981
		Netherlands	2,900
		Brazil	2,700
		USA	2,639
Eastern oyster	*Crassostrea virginica*	USA	112,408
		Canada	5,788
European oyster	*Ostrea edulis*	France	710
		Netherlands	350
		Spain	287
		Ireland	237
Flat and cupped oysters nei	*Ostrea and Crassostrea* spp.	Australia	8,925
		Portugal	1,455
Gasar cupped oyster	*Crassostrea gasar*	Senegal	194
Hong Kong oyster	*Crassostrea hongkongensis*	China (2018)	*1,800,000*
Indian backwater oyster	*Crassostrea madrasensis*	India	4,000
Mangrove cupped oyster	*Crassostrea rhizophorae*	Cuba	1,089
Pacific oyster	*Crassostrea gigas*	China (2018)	*1,240,000*
		Korea, Republic of	315,255
		Japan	173,800
		France	64,200
		USA	25,845
		Taiwan Province of China	23,179
		Ireland	9,822
		Canada	8,012
		Mexico	7,534
		Australia	3,004
		United Kingdom	2,294
		New Zealand	1,834
		Channel Islands	1,312
		Spain	806
		Portugal	600
		Morocco	411
		Namibia	400
		South Africa	336
		Italy	145
Portugese oyster	*Crassostrea angulata*	China (2018)	*2,100,000*
Slipper cupped oyster	*Crassostrea iredalei*	Philippines	22,944

Cuban production of mangrove oysters, for example, was 1089 MT in 2017 (FAO, 2017). Culture operations for other mangrove species, however, have historically been small in scale with total production of mangrove oysters outside of Cuban production being less than 19 MT (FAO, 2017).

Box 1.1 A brief description of the major species of oysters produced in the world (with Ximing Guo, personal communication, 2020).

Crassostrea gigas – Pacific (Japanese) oyster

Pacific oysters are the most widely produced oyster in the world (introduced to over 60 countries) with major hatchery production in Asia where this species is also native (northern China, Japan and Korea). Pacific oysters have been widely introduced for aquaculture in Australia, New Zealand, France, the UK, Germany, the Netherlands, Spain, North America (west coast), Brazil, Morocco, Algeria, Namibia and South Africa among other locations. Devastating losses due to Ostreid herpesvirus 1-μVar (Pacific Oyster Mortality Syndrome [POMS]) have impacted production, mainly in Asia, Australia, New Zealand and France but breeding for resistance has proved to be an effective strategy for reducing mortalities.

Crassostrea virginica – eastern oyster

Native to eastern North America and long harvested on North American East and Gulf coasts from Chaleur Bay, New Brunswick, Canada to western Texas, eastern oysters are widely farmed and popular in half-shell markets. This oyster is also susceptible to diseases that severely limited wild production over the last 50 years, though breeding for disease resistance and more recently the use of triploids has been effective on the US east coast to boost production. Production in the Gulf of Mexico (western Florida, Louisiana, Mississippi, and Texas) is mainly from wild set though farming of hatchery produced seed is increasing. Changes in freshwater inputs have recently impacted wild production in the eastern Gulf of Mexico. Limited culture of *C. virginica* also occurs in California and Washington State with a small naturalized population in western North America located in southern British Columbia.

Crassostrea angulata – Portuguese or Fujian oyster

This small, thin-shelled and deeply cupped species is found intertidally along the coast of China south of the Yangtze River and is utilized for aquaculture production in southern China (primarily Fujian Province). Seed are primarily obtained via natural recruitment, although hatchery production, especially of triploids, is on the rise. In China, Portuguese oysters are favored for their reported superior flavor compared to *C. gigas*.

Crassostrea hongkongensis

Widely cultured in southern China primarily in Guangdong and Guangxi Provinces, the "white meat" oyster has been important for southern Chinese oyster production for the last 700 years. Much of the production was historically based on on-bottom culture due to the prevalence of typhoons but raft culture has proliferated in recent years.

Crassostrea sikamea – Kumamoto oyster

Native to Ariake Bay in southern Japan, southern Korea and southern China, Kumamoto oysters were originally transplanted to the US West Coast from Japan in the 1940s. *C. sikamea* broodstocks were re-introduced from the Ariake Sea in 2006 to retain the genetic identity of naturalized stocks in North America. Today, Kumamoto oysters remain a premium oyster cultured on the US West Coast for domestic and international half-shell markets.

Crassostrea iredalei – Philippine cupped oyster

Widely cultured at a small scale in the Philippines and Malaysia, the Philippine oyster is important in local diets and has attracted increased attention in recent years for aquaculture production.

Crassostrea madrasensis – Indian backwater oyster

Culture methods are well developed for this widespread, low-salinity tolerant and fast-growing intertidal oyster native to the Bay of Bengal. Culture and harvest of wild populations is widespread in eastern India (particularly in the state of Tamil Nadu) and Bangladesh.

Ostrea edulis – European flat oyster

Native to western Europe from Norway to Morocco, European flat oysters are farmed today, mainly in France but production is limited due to disease. Flat oyster production has been largely supplanted by the Pacific oyster. Only *O. edulis* grown in the Belon estuary in France are legally recognized as Belon oysters. Introduced and naturalized in North America, European flat oysters are harvested in small quantities, primarily in Maine.

Ostrea lurida – Olympia oyster

The flat oyster native to western North America was formerly abundant in estuaries along the west coast, particularly in the coastal estuaries from Northern California to Washington State and Puget Sound. Decimated by over-harvest in the late 1800s, Olympia oysters are being restored into former habitats but the species was largely displaced for aquaculture production by the Pacific oyster, beginning in the 1920s. Olympia oysters today are enjoying a comeback with small scale but active sea farming for this species mainly in Washington State.

Ostrea angasi – Australian mud oyster

Native populations of this flat oyster in southern Australia were severely overharvested with new commercial production currently reported to be augmenting Pacific oyster culture operations impacted by POMS in the last few years.

Saccostrea glomerata – Sydney rock oyster

Native to eastern Australia the Sydney rock oyster is extensively cultured in estuaries on both coasts using intensive off bottom tray-based production. Devastating losses due to QX, the protozoan parasite, *Marteilia sydneyi*, has resulted in contraction of the industry in New South Wales in recent years; however, breeding efforts initiated since 1997 have resulted in increased resistance to mortalities.

Hatchery and nursery techniques

Sourcing oyster seed is one of the earliest decisions a farmer has to make in establishing a successful farming operation. Multiple factors come into play with the farmer essentially having three choices: (1) collecting juvenile oysters, referred to as spat, from the wild; (2) remotely setting larvae acquired from a shellfish hatchery; or (3) acquiring juvenile oysters through the hatchery and nursery process. Collection of wild spat remains an economical and common method for acquiring juveniles for grow-out and was formerly the basis for most production in Asia, including China. In the last few decades, however, a significant portion of the vast Chinese oyster production is now hatchery produced (Guo et al., 1999), as is the case for most world production. Oyster hatcheries come in many different forms, from small, seasonal operations that produce 1 million seed annually for one farm operation, to large industrial hatcheries that produce billions of oyster seed annually, which supply significant segments of the industry. Hatchery production offers the oyster farmer numerous advantages over the collection of wild spat as a source of seed. Reliability of production and timing of seed availability alone will likely offset any additional costs to the farmer, particularly in growing areas that have a short optimal growth season. For example, in northern New England, oyster farmers generally acquire hatchery-reared eastern oyster seed in late May or early June after water temperatures have seasonally warmed to at least 15°C (Fig. 1.2). Attempts to rear small (< 5 mm shell height [SH]) *C. virginica* seed in Maine waters below 15°C typically results in minimal growth and is often accompanied by high rates of mortality

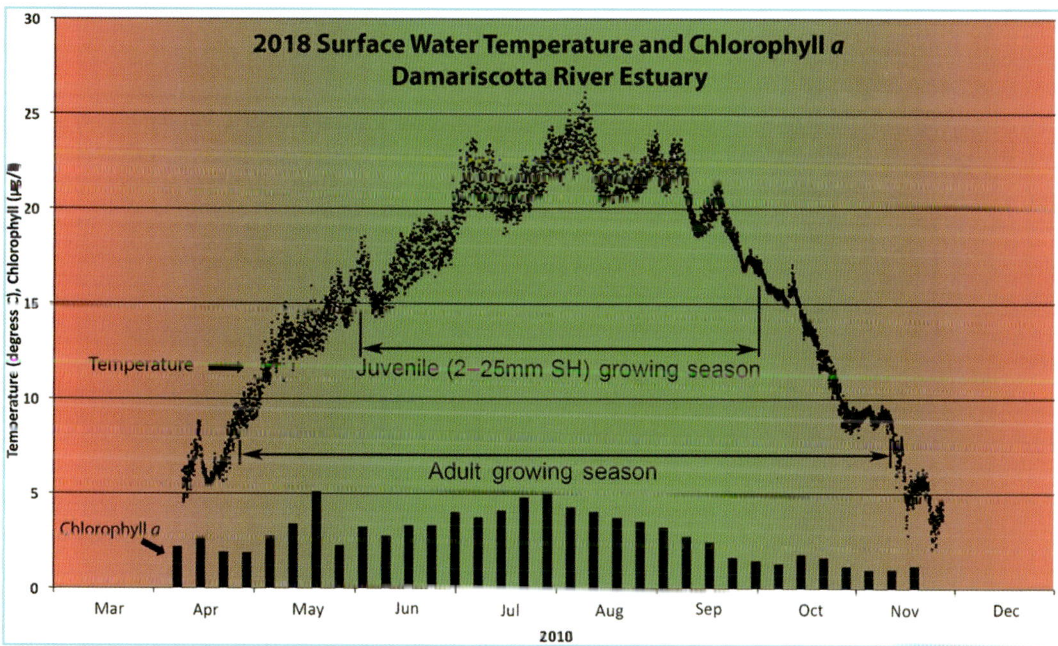

Figure 1.2 Damariscotta River, Maine surface water temperature and seasonal chlorophyll content set the optimal growing season for eastern oysters in this location. Image courtesy of Chris Davis. Data obtained from University of Maine Land/Ocean Biogeochemical Observatory buoy.

should less optimal water temperatures exceed 1 to 2 weeks. Given the relatively short 4 to 5 month growing season for juvenile oysters in northern New England waters, most farmers find it essential that hatchery seed availability coincides with the onset of suitable environmental growing conditions. In contrast, in warmer regions, such as the Chesapeake Bay or Gulf of Mexico, growers may start seed at different times through the year to ensure year-round harvest. Hatchery production of seed also affords the opportunity to take advantage of genetic improvements (for example, disease resistance, higher yield, optimal market characteristics, and polyploidy) that may be extremely important to the grower. In southern New England and the mid-Atlantic US, for example, the availability of hatchery-produced, disease-resistant eastern oyster seed enables the industry to exist in the face of two highly virulent protozoan diseases. Australian, New Zealand, and French oyster industries similarly rely today on disease-resistant strains of Pacific oysters that are only available through hatchery production. While the details of hatchery operations are similar for many bivalves and beyond the scope of this chapter, the fundamentals of hatchery culture specific to oysters are discussed below. These include broodstock management, methods of larval rearing, and methods for setting and rearing juveniles to a size suitable for out-planting to farms.

Broodstock management – applying breeding science to oysters

Sourcing broodstock

Most oyster hatcheries take care to ensure that the broodstock parental lines they select are well-suited to the growing environment in which the offspring will be cultivated. Often, sourcing oysters for broodstock from the same general locale where oysters are intended to be cultivated is a good, fundamental practice. The existence of distinct physiological/genetic races for oysters has been long-established with traits of interest to oyster farmers varying according to locality. Hatchery operators should never introduce oysters for broodstock from outside the general region, unless specific guidance is attained and where warranted disease testing and permitting conditions (for example, holding in quarantine) rigorously applied. There have been numerous examples of oysters harboring pests and diseases and introducing them to an oyster growing region with catastrophic consequences. As an example, a massive die off of eastern oysters in the 1950s due to the MSX pathogen (multinucleated sphere unknown although the causative agent later identified as *Haplosporidium nelsoni*) in Delaware and Chesapeake Bays most likely arose following an illegal transplant of Pacific oysters. In fact, today shellfish producers increasingly use oyster broodstocks bred for some level of disease resistance. The emergence of the MSX pathogen resulted in one of the most important selective breeding programs in the US being established by researchers at Rutgers University. Following four decades of line selection of survivors exposed to waters infected by MSX, the lines from Rutgers University have demonstrated remarkable resistance (although not immunity) to *H. nelsoni* infections, so much so that oysters can be commercially reared to market size prior to succumbing to the disease (Guo et al., 2008). Line breeding programs targeting other oyster pathogens, such as Dermo (*Perkensis marinis*), roseovarius oyster disease (ROD) (*Roseovarius crassostreae*), Denman Island Disease (*Mikrocytos mackini*), and Bonomia (*Bonomia ostreae*), have similarly yielded resistant broodstock lines made available to hatcheries (Proestou et al., 2016). Similar efforts have focused on enhancing growth and yield for various species of oysters including *C. virginica*, *O. edulis* (Newkirk, 1988), *O. chiliensis* (Toro and Newkirk, 1991), and *C. gigas*. A public breeding program focused on Pacific oysters (Molluscan Broodstock Program) housed at Oregon State University on the US west coast was established in 1996 with the intent to increase the average yield in farmed Pacific oysters (de Melo et al., 2016). Long-term breeding in Pacific oysters has been an integral part of the industry in France, Australia, and New Zealand as well. Oysters are also amenable to production improvement through a combined approach that takes advantage of selection in combination with crossbreeding. Higher yields in Pacific oysters are observed following crossing of unrelated, partially inbred oyster lines to induce heterosis or hybrid vigor (Hedgecock and Davis, 2007). Advantages to a combined selection/crossbreeding approach include a reduced risk of inbreeding, a problem inherent to shellfish breeding program, generally. The advantage of using broodstock resistant to diseases of oyster stocks will become increasingly important in the years ahead as increased disease incidence impacts oyster production areas around the world. Recently, the emergence of *Ostreid herpesvirus* 1-μVar (Pacific Oyster Mortality Syndrome [POMS]) has refocused growers around the world on the importance of genetics and breeding in oysters (Dégremont et al., 2015). Emerging diseases combined with the increasing emphasis on marketing oysters having specific traits (for example, survival, meat yield, shell color) coupled with market driven, increased pressure on oyster growing regions to produce more oysters, will likely increase the emphasis on optimizing the management of oysters for their genetic resources.

Conditioning, spawning, and larval culture

The interplay among availability of nutritional resources and endogenous changes due to environmental influences (temperature, food supplies, and light/dark cycles) that serve to initiate reproduction should be controlled by the operator to the extent possible to ensure success. Oysters used for broodstock are more often selected for convenience than for their qualitative value for gamete production. Hatchery operators routinely condition oysters for reproduction outside of the normal window for natural production and much depends upon the nutritional status of the oysters in the field, especially when gamete maturation is attempted outside of the natural cycle of reproduction. The physiological state of broodstock oysters specific to storage metabolism and resources, is therefore important to assess. Control over the nutrition of oysters in the field is difficult to achieve, but having oysters in good condition relative to stored reserves (for example, glycogen content) is nevertheless important. Adequate stored glycogen reserves are used during the conditioning phase to invest eggs, for example, with adequate energy stores (mainly lipids) through the process of vitellogenesis. Both early embryogenesis and healthy larval activity, once feeding stages are reached, are enhanced by having adequate lipid reserves in the egg.

Hatchery operators do not routinely screen oyster broodstocks for glycogen content nor are gametes regularly assessed for lipid content, but the operator should attempt to introduce oysters that have adequate energy stores whenever possible. Using Pacific oysters as an example, the movement of oysters to 'fattening ground' or areas that are richer in natural phytoplankton abundance at certain times of the year relative to other locations has been a feature that growers in the Pacific northwest of North America have long used to ensure that oysters subsequently brought into the hatchery environment for broodstock are in the best condition possible. While regulating storage cycles in oysters has been a minor component of hatchery management for oyster production, recent work suggests that better control over these stored energy reserves in oyster broodstocks is important to consider.

Once in the hatchery, it is important to optimize the feeding protocols that supply both daily energy needs and support gamete production. Oysters are typically fed a mix of hatchery cultured algae over a period of 4 to 6 weeks at an elevated seawater temperature (typically 18–24°C) to induce a shift from storage metabolism to gametogenesis. Cultured species used for conditioning oysters typically include the Prymnesiophytes *Tisochrysis lutea* (formally *Isochrysi* aff. *galbana* clone Tahiti T-Iso) and *Pavlova lutheri* (National Marine Fisheries Service Milford Shellfish Laboratory clone 459), the diatoms *Chaetoceros calcitrans* and *C. gracilis*, and the green alga, *Tetraselmis suecica*, among others. A basic guideline for broodstock oyster feeding is to supply an equivalent mix of live algae (based on cell volume) totaling at least 2×10^9 (T-iso equivalent) cells per oyster per day, or about 5% of their tissue mass per day (see Chapters 12 and 13).

A number of research investigations conducted in the early 1980s (Lannan et al., 1980; Muranaka and Lannan, 1984) documented the relationships between temperature, supplemental feeding, and the rate of gametogenesis in Pacific oysters. Managers generally introduce broodstock into the hatchery for conditioning in late fall and early winter, depending on the schedule for first spawns of the season. Protocols described in Muranaka and Lannan (1984), for example, prescribe an approach based on the use of degree days. Oysters are generally placed into broodstock tanks with either running seawater or seawater that is changed out every day or so, at the ambient temperature they were experiencing in the field. The temperature is ramped 1°C per day until the conditioning temperature is reached (usually between 18–24°C). Oysters are maintained at the prescribed temperature for the number of days required to produce ripe (viable) gametes, based on a prescribed feeding regime and degree days. Following these simple protocols will assist hatchery managers in producing broodstock oysters that are ready to spawn well provisioned gametes as needed to produce regular larval cohorts.

Broodstock use, spawning, and rearing larvae to the pediveliger stage

One of the goals of hatchery technicians is to produce large numbers of viable larvae while at the same time maintaining a high degree of genetic variability in the offspring. Given the high fecundity in cupped oysters (*Crassostrea* spp. and *Saccostrea* spp.), conceivably, very low numbers of parents can satisfy the larval needs of the hatchery, but if low parental numbers are used, issues related to low genetic variability in the offspring may come into play, which can result in poor overall fitness and/or viability with the potential of inbreeding depression occurring in subsequent generations should offspring be used as future broodstock. Thus, it is important to bring into reproductive condition as many broodstock as possible for each spawning event in order to optimize genetic diversity. It is similarly important to use as many male and female oysters as possible. Ideally, the hatchery operator will collect gametes from male and females and conduct pairmatings (use sperm from one male to fertilize the eggs from a single female) and then combine multiple pairmatings into one culture for rearing.

Cupped oysters of the genus *Crassostrea* are oviparous, broadcast spawners that release ova or sperm directly into the water column where fertilization occurs. Spawning is typically achieved using thermal ramps (3–8°C) of slowly increasing seawater temperature. Shallow baths containing broodstock oysters may be warmed by heat lamps, immersion heaters, or the addition of warmed seawater. Male oysters generally initiate shedding of gametes first. Spawning males are immediately removed from the water bath and placed into individual containers and allowed to continue to spawn. Similarly, once females commence spawning they are similarly rinsed and separated into individual spawning dishes and allowed to continue to shed gametes. Eggs from each parent are counted and generally pooled with each parent contributing an equal number of gametes. Similarly, sperm from males is pooled prior to use. Ideally, within 30–60 minutes of the initial spawning event, the egg suspension is fertilized by the addition of sperm at a ratio of approximately 4–10 sperm cells per egg, as confirmed by microscopic examination. As mentioned earlier, hatcheries are generally interested in maximizing genetic diversity of the offspring, in which case, having an equal number of males and females will provide the greatest effective population size.

Oysters are among a small group of bivalves that do not require that gametes be shed naturally, as both eggs and sperm are viable once exposed to seawater. Under some mating designs, especially when gamete contamination must be avoided, there are significant advantages to this approach. The process entails a thorough cleaning of the exterior shells and then removal of the left valve. A scalpel is used to lightly penetrate the ripe gonad to release either the eggs or sperm. A squeeze bottle filled with 1-μm filtered sea water is useful to wash excised gametes into a clean container. The gametes are then sexed by microscopy and eggs and sperm set aside until all broodstock have been stripped. Pooling of eggs and sperm will depend on the mating scheme, but ought to be accomplished within a 1 hour timeframe.

Spawning in flat oysters within the genus *Ostrea* may also be initiated using thermal ramps by inducing males to release sperm (as accumulations of sperm in balls). Sperm balls or individual sperm are inhaled by ripe females and individual sperm fertilize eggs retained in the female mantle cavity. Brooding of fertilized embryos occurs for 1 week or more prior to the release of larvae into the water column. European flat oysters, for example typically release larvae of 170–190 μm shell length between 6 and 8 days post spawn.

The culturing of larvae can be accomplished using either batch or flow-through culture systems. In the batch culture system, larvae suspensions are typically held in conical-bottom tanks or flat-bottom tanks when larger culture volumes (> 1000–2000 L) are required. Culture water (filtered to 1–5 μm) is changed every 48–72 hours to ensure that accumulation of metabolites does not create stressful conditions (Creswell et al., 1990). Both continuous or batch feeding of microalgae of the appropriate species and cell concentration depends upon the size, age, and concentration of larvae in the culture tank. In contrast, flow-through culture systems allow for considerably higher larvae concentrations, thus allowing for a significantly smaller footprint in the hatchery (Fig. 1.3). Typically, flow-through systems continually add filtered, algae-rich culture water at rates of up to 18 water replacements per day. Appropriately sized discharge screens (often referred to as 'banjo' screens) ensure that larvae are not flushed out of the system. Supan (2014) described setting up a flow-through larval system for oysters. Table 1.2 provides some useful parameters for the culture of *C. virginica*. The advantages of flow-through systems include a smaller footprint for tanks, elimination of oyster waste products, and oysters maintained on a continuous algal supply. The disadvantages include greater water flow, heat, and algal requirements. The fine points of larvae culture is beyond the scope of this chapter and the reader is encouraged to consult Helm et al. (2004) and Supan (2014) who provide a comprehensive discussion on approaches to the larval culture of oysters (see also Chapter 12).

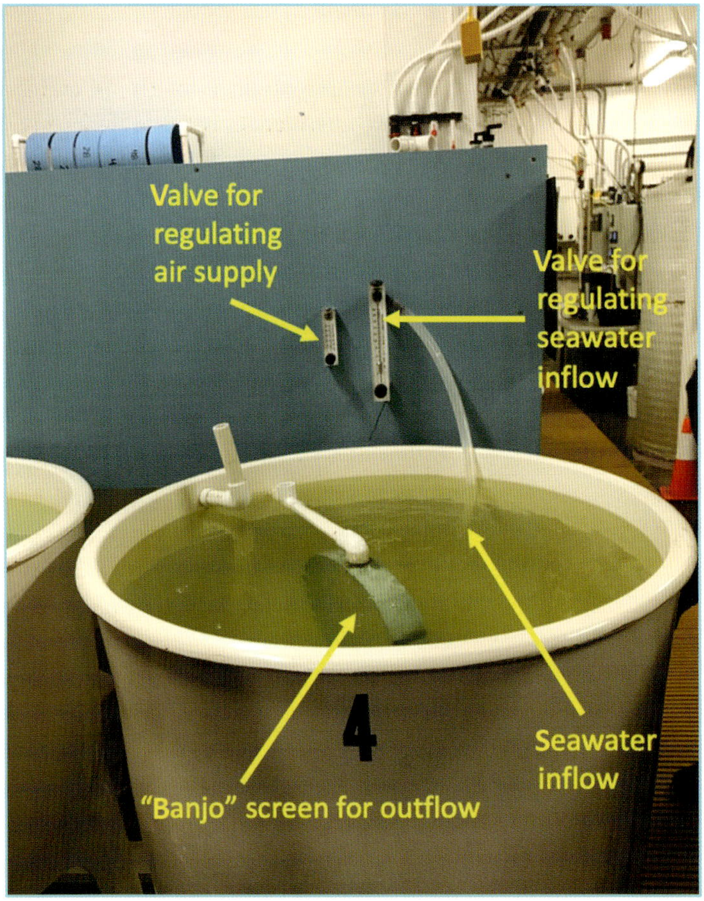

Figure 1.3 Larval continuous flow-through system. A 'banjo' screen combined with circulation via an air diffuser assist in preventing larvae from being impinged on the outflow screen. Courtesy of Pacific Hybreed, Inc.

Table 1.2 *Crassostrea virginica* larval culture parameters. Adapted from Creswell et al. (1990) and Margaret et al. (1993).

Approximate days/ development post spawn	Larvae size SL (μm)	Seive size (μm)	Larval density (static tank) (per mL)	Larval density (flow-through with 12–18 turnovers per 24 h) (per mL)	Phytoplankton density (cells.mL^{-1})
Eggs	45	22	10	0	-
D-hinge	75	45	5	200	20,000–25,000
2–3	120–130	63	5	150	20,000–25,000
4–5 (umbones developed)	135–170	90	5	130	25,000–30,000
6	180–200	125	5	130	30,000-40,000
day 8 (pediveliger)	240–260	150	4	100	50,000
10	260–280	165	4	100	60,000
day 11–12 (eyed larvae)	280–300	180	3	75	70,000–80,000
day 14 (setting)	330	212	1.5	50	100,000–150,000

Settlement and metamorphosis

The pediveliger stage in oysters commences late in the larval cycle, typically between 14 and 21 days from egg stage in modern facilities. For *Crassostrea* oysters, larvae are between 250–275 µm in length. Pediveligers are characterized by the presence of a pigmented 'eyespot,' a ciliated foot and upward spiral swimming behavior that is often associated with copious mucus production and clumping of larvae. Hatchery protocols are adjusted at this time to provide satisfactory conditions for successful settlement and metamorphosis. Accommodating for the cementation stage in the oyster life history requires that favorable substrate be made available to pediveligers with clean oyster shell providing the ideal substrate, facilitating a gregarious setting response (Fig. 1.4). Depending upon the end use of the oysters, whole shell or micro-cultch can be used as a setting surface, generating clusters of spat on shell or single seed, respectively.

To generate spat on shell clusters, pediveliger (Fig. 1.5) larvae are introduced into static setting tanks, stocked with clean, whole oyster shells (or other favorable substrate) with filtered seawater that is generously aerated. Hatchery technicians may monitor the rate of settlement and either remove the shell from the culture tanks when the desired number of oysters have set on individual shells or add additional larvae. Once the oysters have cemented themselves to the shell cultch and have been allowed a period of growth (typically at least 3 days), the shell bags, with the attached spat, can be transferred to the field nursery.

Oyster farmers targeting the 'half shell' market tend to pay particular attention to the shell shape and thus often obtain single oysters from the shellfish hatchery. Micro-cultch, composed of ground oyster shell chip is offered to pediveligers in lieu of whole shell. Resembling a fine powder, shell chip is typically ground from whole shell using a hammer mill followed by sieving on appropriately sized screens to produce micro-cultch in the 250–300 µm diameter size, essentially the same size as the pediveliger. Once cementation and metamorphosis have occurred, the oyster rapidly outgrows the micro-cultch, producing a "single" oyster. Single oyster seed are typically raised in various types of nurseries (for example, land-based upwellers, floating upweller

Figure 1.4 Pacific oyster spat set onto oyster shells contained in mesh cultch bags. Shells in bags are later spread onto beaches for grow-out. Courtesy of Taylor Shellfish Company.

Figure 1.5 An oyster pediveliger, competent to settle and metamorphose, is characterized by an eye spot and extended foot. Courtesy of Chris Davis.

systems, and so on) to a size more easily handled by oyster farmers (for example, sufficiently large to be retained on 2-mm mesh screens).

A variation of setting oysters on cultch or micro-cultch is 'remote setting' whereby eyed larvae are drained from the larval tanks and shipped to field setting nurseries oftentimes a great distance from the hatchery. Eyed larvae may be stored under refrigerated conditions (6–8°C) for at least 4 days before setting. The reduced cost of eyed larvae to the oyster farmer compared to 2–3 mm shell height post set can justify the cost of building a remote setting system.

Triploid oysters

The first development and application of triploidy in animals (organisms containing three sets of chromosomes in each cell nucleus) occurred in cupped oysters within the genus *Crassostrea* in the 1980s when researchers at the University of Maine modified triploid induction techniques developed for salmonids to produce triploid *C. virginica* (Stanley et al., 1984). Sexual reproduction in bivalves, including oysters, is unusual in that meiosis in eggs is not complete until the egg is fertilized and following two sequential meiotic chromosome divisions that release 2N and then 1N sets of chromosomes in the form of two polar bodies expelled from the fertilized egg. In diploids, the release of first and second polar bodies reduces the egg from an initial 4N state to 2N (diploid) and haploid (1N) state, respectively. The sperm adds another chromosome set and results in a diploid (2N) embryo. Induced triploid methods simply prevent the release of either the first or (more commonly) the second polar body. Historically, chemical induction of triploid oysters largely relied upon the application of either the fungal metabolite cytochalasin B or the purine, 6-dimethylaminopurine to inhibit the release/formation of a polar body. Unfortunately, 100% production of triploid larvae is rarely achieved using these methods due to asynchrony in developmental rates within individual embryos. The work initiated in Maine was further developed in the US Pacific northwest for Pacific oysters. Advantages to triploid oysters are due to their increased marketability due to the reduced capacity to produce gametes and tendency to become 'spawny' or 'watery,' as compared to diploids under similar environmental conditions. The Pacific oyster was particularly amenable to triploid development as marketability for triploids was enhanced during the summer months compared to diploids in the productive grow-out areas in Washington State (Allen and Downing, 1986) (Fig. 1.6). Triploid oysters also generally grow faster than diploids under conditions favorable for oyster growth as the energy normally contributed to reproduction is thought to be diverted toward enhanced tissue growth. To reduce the need for chemical treatments and to achieve

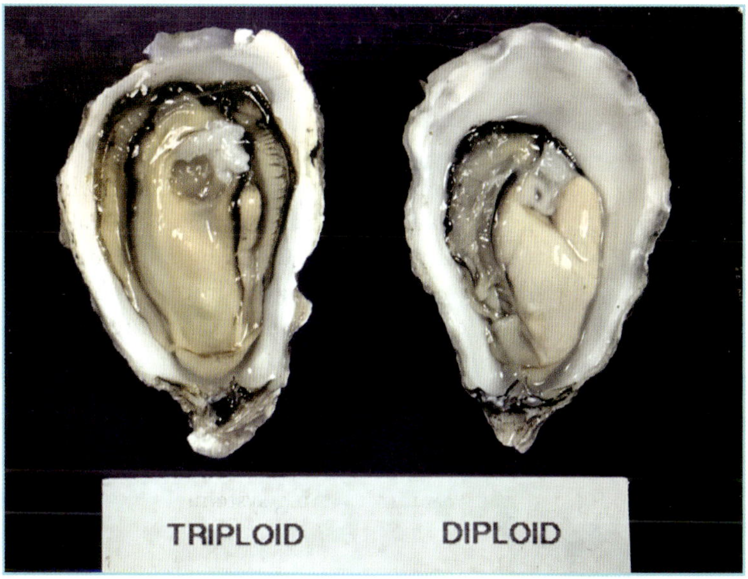

Figure 1.6 Diploid and triploid Pacific oysters. Note extensive development of germinal tissue in the diploid (female) compared to the triploid oyster. Courtesy of Joth Davis.

100% triploid induction, researchers at the University of Washington and Rutgers University undertook a new approach. Initially working with Pacific oysters and later eastern oysters, they were able to produce tetraploid (4N) oysters that when mated with diploid oysters, resulted in 100% "natural" triploid offspring (Guo et al., 1996). Generally, tetraploid males are crossed with diploid females. Today, all triploid production of oysters using tetraploid broodstock in matings with diploid females is commonplace and a mainstay of hatchery production for much of the industry in Europe, North America, and Australia. Triploid oysters are often observed to outperform diploid in many environments, including growing locations impacted by disease (Matthiessen and Davis, 1991; Dégremont et al., 2012). In recent years, however, high loss rates (up to 90%) in triploid oysters have been observed by growers in different parts of the world, related perhaps to increased susceptibility to multi-stress responses associated with excessive heat or cold, dissolved oxygen or disease pressure (Davis, J., unpub. obs.; Guévélou et al., 2019). The oyster farmer may desire to assess carefully their region and farm site for its suitability for triploid production and caution is recommended until triploid oyster mortality patterns are better understood and resolved.

Land- and field-based nursery culture of oysters

There are a number of options available for raising seed oysters from a smaller (more affordable) size to a more easily handled size for grow-out. Many growers, especially as they increase the size of their operation, may seek to add or expand their nursery operations. Typically, farmers may choose to operate a field nursery on the farm, employing fine mesh containers with an active program of grading and splitting seed, or opt to cultivate seed under more controlled conditions in some type of upwelling system.

Upwelling systems

The function of shellfish upwellers is to provide a secure growing environment for high density three-dimensional culture of juvenile (1–25 mm shell length [SL]) seed while providing optimal rates of water flow containing the natural seston that supports feeding, growth, and the removal of waste products. The benefits of three-dimensional upwellers compared to two-dimensional nursery systems, such as floating screens or raceways, include enhanced growth rates, reduced space requirements, exclusion of predators, and the ease and efficiency of handling large quantities of small oysters. These advantages may be offset by greater capital and operating costs as well as access and dependence on a reliable source of energy to run a system dependent on pumping seawater. Given the high densities of seed oysters held in upwellers, it is essential that the systems have alarms should a pump fail and the oysters are left in stagnant flow conditions. Under such conditions, levels of dissolved oxygen can rapidly become depleted resulting in mortality. When possible, upwelling systems should be designed to drain the water should a pump failure occur, thereby reducing the risk of loss given that oyster seed can survive out of water for considerably longer periods of time than under submerged, but stagnant conditions.

Upwellers consist of silos, bins, or 'bottles' constructed with an appropriate mesh size on the bottom to contain the seed while allow for adequate water flow up through a bed of shellfish seed (Fig. 1.7). Round silos are often constructed of thin walled (schedule 20) PVC pipe, plastic barrels and buckets, or custom fabricated fiberglass while rectangular bins usually consist of coated plywood, plastic sheet stock, or molded fiberglass. The silo bottom is covered by mesh fabric designed to contain the seed oysters while allowing for suitable water flow. It is critical that the mesh size (diagonal dimension) be no greater than half the shell height of the smallest seed oysters to assure full containment. The bottom mesh may consist of woven polyester monofilament fabric, extruded polyethylene plastic mesh netting, or stainless steel screening which is glued or mechanically fastened to the silo or bin.

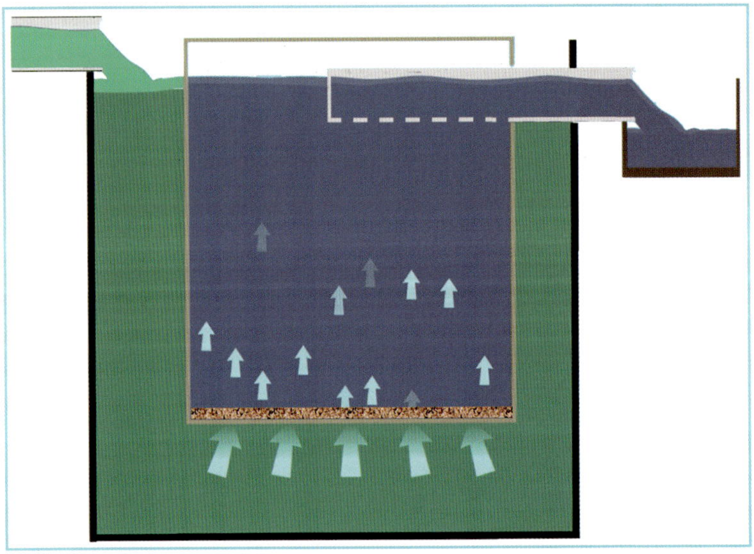

Figure 1.7 Illustration of a forced flow upweller system for oysters. Courtesy of Chris Davis.

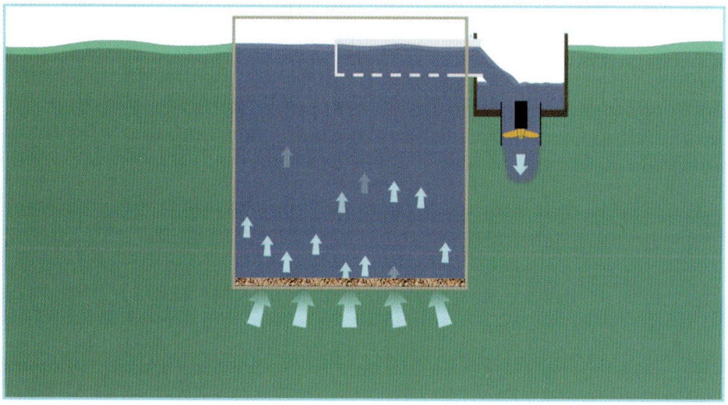

Figure 1.8 Illustration of a passive flow upweller system for oysters. Note the discharge pump lies downstream of the upweller. Courtesy of Chris Davis.

Land-based oyster upweller systems can provide a secure nursery culture environment both indoors and outdoors given suitable access to a seawater source. Most land-based systems employ a forced flow design whereby incoming water floods a tank containing the silos with water being forced through the silos in order to exit through a discharge trough (Fig. 1.7). Pumping requirements tend to be higher given the greater head (lift) needed onshore depending upon the elevation of the upwellers above sea level.

Floating **Up**weller **Sy**stems (FLUPSY) dramatically reduce energy costs given that water pumps require minimal lift to create the needed flow. Both forced and passive designs can accomplish this, although passive systems tend to be more efficient and require less maintenance and cleaning of biofouling organisms as they function by evacuating the water through the exit trough which contains the discharge pump (Fig. 1.8). Several US companies manufacture turnkey systems, although many oyster farmers construct their own to their specific design and needs (Fig. 1.9).

Water flowing through the upweller silos needs to be sufficient to deliver algae-rich water throughout each silo while minimizing downstream depletion effects such that oysters on the surface layer do not receive significantly less food than those closer to the water source.

Figure 1.9 A large multi-bin upwelling system in Washington State. Water flow in large systems are best driven by slow turning paddle wheels. Courtesy of Taylor Shellfish Company.

Figure 1.10 Small-scale tubular upweller system for use with juvenile oyster seed. Courtesy of Chris Davis.

Tubular upweller systems can be used for small oysters (2–10 mm SH) whereby the water flow can be adjusted to suspend and tumble the oysters in the water column (Fig. 1.10). Use of a hand-held fluorometer to monitor inflow and outflow levels can be useful in managing the seed size, density, and water flow conditions. Figure 1.11 illustrates the ease of direct measurement of outflowing water (5 µg/L) from an upweller silo. Note that incoming water was 9 µg/L indicating a 44% decrease in seston levels. Levels of seston depletion can easily exceed 80% if the flow conditions are too low for the size and quantity of oyster seed as illustrated by a 3-day time series of chlorophyll-*a* levels in an overstocked upweller (Fig. 1.12). Long experience with oyster nursery systems suggests that managing the upweller system to achieve seston reductions lower than 40–50% of ambient incoming water results in optimal oyster growth at a high density of seed. Channelizing of water flow through the oyster seed can occur under certain seed size, seed density, and flow rate conditions, depending upon the configuration of the silo or bin and can result in the uneven distribution of feed throughout the silo and thus uneven growth. Since biofouling can present a significant obstacle depending on the site, good farm husbandry requires regular cleaning and grading of seed to ensure uniform and optimal rates of growth. Access to a source of fresh water to regularly rinse the oyster seed will greatly reduce the establishment of fouling organisms such as tunicates. Maintaining the appropriate density of oyster seed in the upwelling system is important to monitor, whether in

Figure 1.11 Direct measurement (5 µg/L) of in situ chlorophyll-*a* florescence in the discharge of an oyster upweller. Note that the incoming water contained 9 µg/L chlorophyll-*a* based upon florescence. Courtesy of Chris Davis.

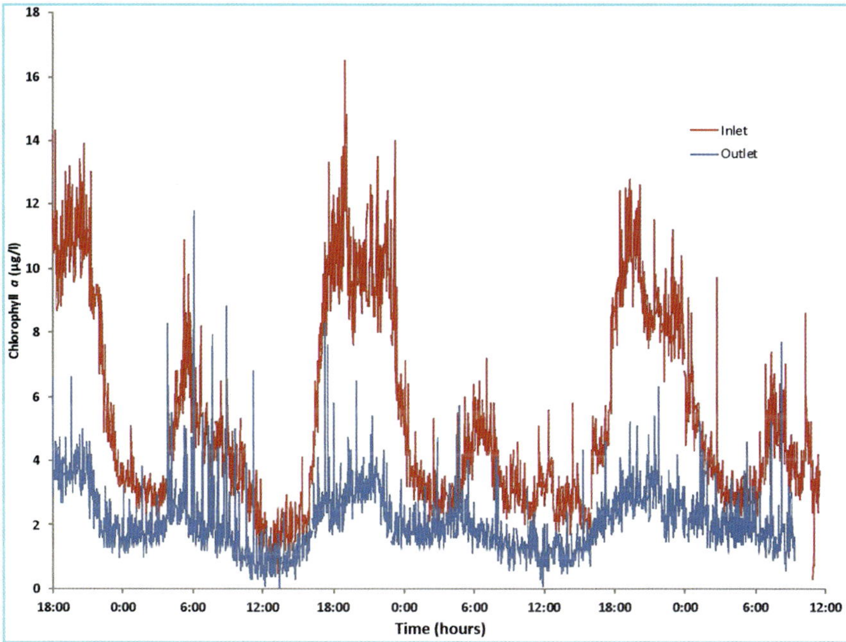

Figure 1.12 Time series (66 hours) of chlorophyll-a fluorescence for incoming and discharging water in an oyster upweller. Courtesy of Chris Davis.

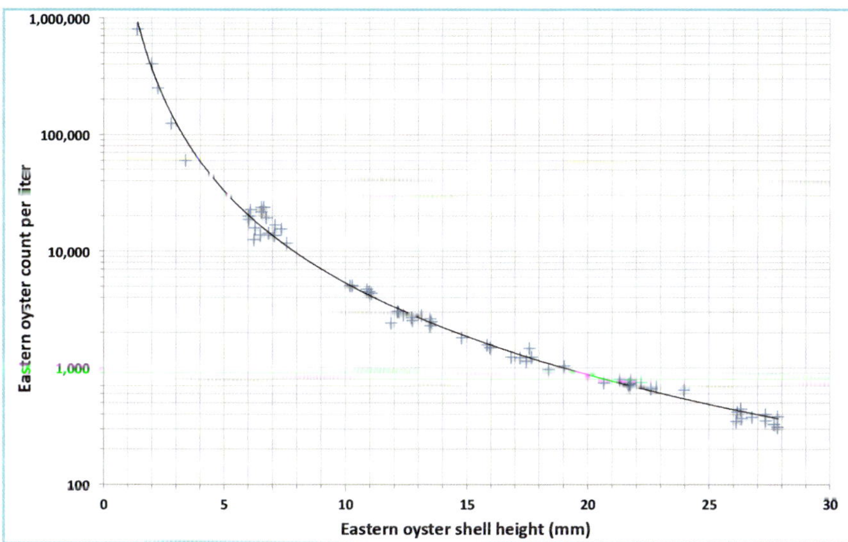

Figure 1.13 Relationship between shell height and count per liter for eastern oysters. Exponentially fitted curve $R^2 = 0.99$. Note that the y-axis is a log-scale. Courtesy of Chris Davis.

large FLUPSY or bottle-based upwellers. In all cases, the growth of seed is critically dependent on flow rate and seston quantity and quality supplying nutrition to oysters. The relationship between oyster size (shell height or biomass) and density (number of oysters per liter) of oysters in the nursery system should be calculated and density maintained within prescribed parameters for optimal system performance (Fig. 1.13).

Oyster grow-out

Site selection

Selecting a site for culturing oysters is the critical first step the new oyster farmer will consider. Siting of commercial grow-out locations depends on a variety of factors, which if not fully assessed will more than likely result in disappointing results. First, a practical note on permitting is necessary. Siting farms in the US and Canada, Australia, and New Zealand, and the European Union require working closely with local, regional, and federal agencies to obtain the permits necessary to establish a farm, cultivate, and sell oysters into commercial markets. Permitting requirements vary by country and region for all the different species being considered for cultivation and important considerations are covered elsewhere in this volume. From a production perspective, first and foremost, areas considered for oyster cultivation must be locations where seawater is reliably clean on a year-round basis with no significant sources of pollution from human habitation, large assemblages of birds and marine mammals, or which have a history of persistent biotoxins.

Oyster cultivation is often conducted in relatively protected bays and estuaries where a level of protection from prevailing winds and stormy weather is afforded. Siting oyster farms in natural assemblages of aquatic vegetation, including seagrasses is often discouraged by permitting agencies though oysters generally thrive in areas having sea grasses in the vicinity. Further, siting of oyster farms should be in areas that are not frequently closed due to biotoxins associated with harmful algal blooms, bacterial pathogens, viruses, or sources of industrial pollution. The volume of commercial and recreational boat traffic must be considered prior to siting an oyster farm, including assessing whether pollution from boating activities has the potential to impact oysters being cultured for human consumption. Proximity to marinas for growing out oysters to market size is often prohibited though nursery operations are often placed in protected anchorages where a power supply is available.

The expected annual range in seawater temperature, salinity, and availability of natural suspended food are very important considerations, and must be assessed in light of the biological requirements for the oyster species planned. As an example, Figure 1.2 illustrates the optimal annual temperature window for growing eastern oysters in productive estuaries along the coast of Maine, USA.

While a full consideration of all the criteria necessary to consider prior to siting an oyster farm is beyond the scope of this chapter, there are several excellent sources of information available (Boxes 1.2 and 1.3). Once the financial resources are in hand, the following considerations are necessary as a minimum when evaluating a site prior to conducting oyster farming. It should be confirmed that the:

- growing waters are safe for shellfish harvest
- substrate conditions are appropriate for the type of aquaculture anticipated
- seawater conditions (tidal currents, annual salinity, temperature, dissolved oxygen, carbonate chemistry conditions) are satisfactory for growing shellfish
- phytoplankton and other sources of shellfish nutrition are sufficient to grow oysters on an annual recurring basis

Box 1.2 Some additional practical resources for oyster farming.

- High density rearing of oyster larvae in flow through systems (Supan, 2014)
- Aquaculture management guide: manual for the identification and management of aquaculture production hazards. NRACAMG-2014 (Getchis, 2014)
- Contained culture cost analysis (Parker et al., 2014)
- Off-bottom culture of oysters in the Gulf of Mexico (Walton et al., 2013)
- Oyster hatchery techniques (Wallace et al., 2008)
- A practical manual for remote setting in virginia (Congrove et al., 2009)
- Small-scale oyster farming for pleasure and profit in Washington (Toba, 2002)
- *Shellfish Aquaculture and the Environment* (Shumway, 2011)
- *Oyster Culture* (Matthiessen, 2001)
- Methods for setting hatchery produced oyster larvae (Jones and Jones, 1983)

Box 1.3 A list of books/articles that address oyster culture.

Hector Bolitho, *The Glorious Oyster*. London, Alfred A. Knopf (1929)
Paul Brocchi, *Traité d'ostréiculture*. Paris, Librairie agricole de la maison rustique (1883)
William K. Brooks, *The Oyster*. Baltimore, MD, Johns Hopkins University Press (1891/1996)
Davide Carazzi, *Ostricultura e mitilicultura*. Milan, Ulrico Hoepli (1893)
Nicolas de La Casinière, *L'Huître*. Genève, Aubanel (2005)
J.J.M.C. Victor Coste, *Instructions pratiques sur la pisciculture*. Paris, Librairie de Victor Masson (1853)
J.J.M.C. Victor Coste, *Voyage d'exploration sur le littoral de la France et de l'Italie*. La Tremblade, le Musée Maritime (1855/1861/1993)
Eleanor Clark, *The Oysters of Locmariaquer*. Chicago, University of Chicago Press (1959)
Pierre Dalido, *L'huître du Morbihan*. Paris, Librairie Marcel Rivière (1948)
Birgit Damer, *Austern – Perlen des Meeres*. Weil der Stadt, Hädecke Verlag (2008)
David Erdal, *Local Heroes*. London, Viking (2008)
Mary J.K. Fisher, *Consider the Oyster*. New York, North Point Press (1941)
Roger Gachet, *Tout savoir sur les huîtres*. Nice, France Europe Éditions (2007)
David Gordon, Nancy Blanton, and Terry Nosho, *Heaven on the Half Shell*. Seattle, WA/Portland, OR, Washington Sea Grant Program/WestWinds Press (2001)
Michel Grelon, *Saintonge – pays des huîtres vertes*. La Rochelle, Éditions Rupella (1978)
Jacques et Ronan Guillet, *L'ostréiculture en Bretagne de 1850 à nos jours*. Spézet/Speied, Coop Breizh (2008)
Bertrand Hallier and Franck Vilboux, *Les huîtres, un océan de saveurs*. Strasbourg, Gyss & Clément (2006)
Robert Hedeen, *The Oyster*. Centreville, Tidewater Publishers (1986)
Rowan Jacobsen, *A Geography of Oysters*. New York, Bloomsbury (2007)
Christine Keiner, *The Oyster Question*. Athens, GA, University of Georgia Press (2010)

- proximity to pollution sources are not a factor for safe harvest of shellfish
- prevalence and occurrence of shellfish pathogens, fouling organisms, predators, disease and the presence of biotoxins are all within acceptable levels of risk
- site is protected from major storm damage and is reasonably secure and accessible.

Grow-out methods – extensive versus intensive?

There are a wide variety of grow-out methods used throughout the world to cultivate oysters and they are generally characterized according to the level of cultivation intensity, where intensity in this case refers to the costs associated with producing an oyster crop. *Extensive* oyster culture is a common approach to oyster farming worldwide and is characterized by a reliance on natural spat settlement, often supplemented by hatchery seed, little control over oyster density on the seafloor with little attention paid to predators and overall oyster density. Advantages to extensive approaches include low overhead and production costs, low labor costs, and overall reduced inputs. *Intensive* methods for rearing oysters are based on the containment of oysters in specialized gear that may be placed either on the seafloor, suspended in the water column, or floating at the water surface, collectively termed 'off-bottom'. A large variety of gear types are available to the oyster grower that serve to contain the oysters and enable the grower to largely control product quality such as growth rate, shell and meat qualities while providing protection from predators. Intensive methods are expensive, and often rely on specialized labor and the availability of hatchery-reared seed. Advantages to intensive oyster culture include increased productivity, better control over predators, use of genetically improved seed, and enhanced product quality. There are advantages and disadvantages to all grow-out methods that are dictated by considerations of relative cost, availability of space, access to markets and upland infrastructure (for example, sucking and processing facilities). These considerations are discussed in the next two subsections.

Extensive culture of oysters

The simplest and most seemingly cost-efficient extensive approach to oyster cultivation is to site farms in the vicinity of habitats that support reliable natural recruitment of oysters for both on-bottom, off-bottom, and longline and raft-based culture. These are farms of sufficient area that the economies of scale enable wild oysters that are recruited annually as pediveligers onto collectors and grown inexpensively to a harvestable size to provide mainly processed (for example, shucked and dried, frozen, or jarred) products. Oyster growing regions in eastern North America, China, South Korea, and Japan fit these criteria. Many operations blend the line between sea farming and wild harvest if spat initially wild caught on collectors is combined with hatchery-produced seed. Increasingly, hatchery seed is utilized on farms and in many cases this is due to the value of potential disease resistance as a result of breeding efforts. Farming methods depend on location, availability of local materials, the annual abundance of oyster larvae and cultural experience. In Asia, rafts constructed of bamboo stringers are used extensively to support strings of oyster, mussel, and scallop shells suspended in the water column. Intertidal sites in China utilize posts inserted into muddy ground that serve to attract oyster spat. In Korea, both rafts, and extensive longlines are used extensively. Spat collection techniques in open waters generally involve placing the cultch material on ropes into the water very close to the time that larval oysters are approaching the settlement phase of their life history. Once oysters settle and metamorphose to the juvenile stage they generally grow quickly and develop more robust valves. At this stage settlement substrates containing the oyster spat may be moved to more favorable locations for further grow-out. Grow-out to a harvestable size may be rapid and significantly less than a year in warm, productive environments, but 2 to 3 years are required to reach a harvestable size in cooler, temperate waters.

Raft and longline based culture using hanging cultch dominates production in Asian countries while intertidal production of oysters for the shucked markets mainly are common in North America. In the Pacific northwest of North America, natural settlement of oysters is accomplished by placing shell bags as bundles, each bag containing 250–300 Pacific oyster shells, on the intertidal beach at the mean low tide level just prior to forecasted settlement of oyster larvae (*C. gigas*) (Fig. 1.4). Once settlement occurs, the spat on shell are maintained in the cultch bag for a period of weeks to months in order that the juvenile oysters grow and "harden" (increase general hardiness and shell thickness) prior to planting. As the success of spat fall on natural collectors varies from year to year, growers are generally looking for 6–20 juvenile oysters per cultch piece, though this standard varies widely. Shell bags are often moved from the spat fall location to a more favorable site for grow-out prior to breaking the bags open and scattering the seed over suitable habitat for grow-out to harvest size.

Settlement substrates used for catching oyster spat are extremely varied depending on the species and where the oysters are grown. Collectors made of oyster shell, shells of other molluscs, wood, plastic, cement coated wood, PVC pipe, plastic plates, and coconut shells encompass the variety of substrates. Custom and ease of use in the specific farm setting required often dictates the method selected. Once settled, oyster spat is grown to a harvestable size using a wide variety of methods that mainly reflect local custom and the experience of the grower.

In parts of the world where natural spat fall may be locally sparse, sporadic, or unreliable on an annual basis, the use of hatchery produced seed added to natural production can significantly boost production though adding to production costs. Although hatchery production of oysters is covered elsewhere in this chapter, the use of hatchery-produced *C. gigas* pediveligers was pioneered on the North American west coast and is referred to as 'remote setting' and involves placing cultch bags packed with dry, clean oyster shells in bundles into tanks (for example, 5000 L) supplied with warm seawater and microalgae and adding pediveliger larvae competent to settle and metamorphose to the juvenile stage. The section on hatchery and nursery techniques further details the methods used for producing spat for grow out using remote setting techniques. A variation on growing cultched oysters to harvest size involves inserting the cultch shell through the strands of three-strand twisted line. These strings are bundled and placed into setting tanks and then suspending the line with cultch attached from either posts above the intertidal substrate or hanging from rafts. This approach is extensively used in settings when the ground for growing oysters is too soft to support oysters and has the added advantage of exposing oysters that set on both sides of the cultch to better growing conditions (Fig. 1.14). Oysters set on cultch material may be grown out on the bottom for later harvest and shucking for meat production. Harvesting involves either dredging for oysters at high tide or manually picking oysters at low tide and transporting in large tubs to shucking facilities for processing (Fig. 1.15).

Figure 1.14 Off-bottom intertidal longline culture of oysters in Washington State. Courtesy of Taylor Shellfish Company.

Predator control during the grow-out phase of extensive oyster culture often consumes the oyster farmer's time during this phase of production and must be considered by the grower. Predators of oysters include crabs, sea stars, gastropods, fishes, diving ducks, and marine mammals and vary by region as to their impact. The optimal approach to predator control for oysters grown unprotected on or above the seafloor is beyond the scope of this work, but the shellfish farmer has a choice to potentially build in a level of acceptable loss due to predators by planting more oysters than intended and/or handpicking predators as possible for removal with the appreciation that some loss of the crop will occur. This is often the approach used as control methods can be very expensive. On the US east coast where subtidal culture of *C. virginica* has been conducted for many decades, predator control for sea stars was a regular practice, accomplished by towing large fabric 'mops' over subtidal oyster beds and serve to collect sea stars feeding on the oysters. The non-native gastropod snail (*Ocinebrellus inornatus*), introduced from Japan in imported oyster seed, has plagued intertidal oyster growers on the North American west coast for

Figure 1.15 Pacific oyster culture in tray-based suspended culture in Tasmania, Australia. Courtesy of Joth Davis.

decades with control methods generally relying on handpicking and removing oyster drills at low tide to reduce impacts. Oyster farmers on the east coast of the US also contend with carnivorous snails, in this case, *Urosalpinx cinera*.

Methods for predator control vary around the world and growers finely tuned to the natural history and behavior of the suite of predators impacting their operations will contribute to their success. The most effective approaches are most often the most ecologically sustainable and utilize control methods that simply work to avoid predator interactions. The grower needs constantly to adapt their growing methods that simply work to diminish, discourage, and, hopefully, avoid interactions with predators. While increasing the cost of production, this may involve the use of temporary nets and barriers that are used only when predators are present to simply preclude interactions. Careful attention to the natural environment and the annual rhythm of life and what it brings in terms of both potential risk and reward to the oyster crop cannot be underestimated. In the US Gulf of Mexico, for example, oyster growers often cultivate oysters in areas where predation is reduced by periodic influxes of freshwater.

Intensive culture of oysters

Methods for on- and off-bottom and floating oyster cultivation have increasingly involved the use of hatchery-produced seed grown in specialized systems that has enabled large-scale production of oysters having excellent market characteristics (uniform size, shape, flavor, and high meat content). When appropriate intertidal or subtidal ground for premium half shell culture is not available, growers the world over have turned to using systems that enable oysters to be cultivated under largely controlled conditions. There are significant advantages to cage (bag)-based shellfish culture whether practiced on or off the seafloor. Protection from predators, ease of handling, control over the density of animals over the grow-out phase, potential for co-culture with other invertebrates or seaweeds, and control over inventory are all important attributes to consider. Chief disadvantages include significantly increased costs of the gear, often significantly greater requirements for controlling fouling of both the gear and the crop, and most importantly the necessity for increased handling of shellfish required to produce a premium product. The components associated with sea farming single oysters involve developing critical operating procedures over stocking, handling and grading of oysters from seed to market size.

Stocking, handling, and grading oysters for single oyster production

Single oyster seed is procured from a hatchery or nursery facility and culture usually requires the grower to have available a supply of seed oysters that are the appropriate size for stocking the grow-out gear. This requires the grower to either purchase seed oysters of a uniform size or produce them in a nursery operation for direct stocking. Most gear suppliers offer cages with different mesh sizes so the operator can outplant small seed oysters that are later split into grow-out cages with larger meshes. There is a wide variety of gear available for raising single oysters to a marketable size on intertidal beaches, off-bottom rails and racks, or in suspended culture. Most providers of gear offer cages with differing mesh sizes to accommodate oysters as they grow. The major task for the oyster farmer, therefore, is to either build or procure gear that enable he or she to manage size and density of oysters from the seed stage to harvest. The decision on what gear to build or purchase is dictated by cost, critical site considerations, and important extrinsic factors including expected wave action, durability, handleability, visual considerations, and fouling among others.

Single oyster production, unless practiced on the seafloor without protection from predators, requires a significant investment in handling and grading that if ignored will most often result in uniformly poor results. There are only a few 'commandments' associated with intensive oyster aquaculture. All are based on the biology and immobility of the animal and dictate how grow-out from seed to market size oyster should be approached.

> **Guideline 1** – Overstocking, or crowding, during the growth season will allow only a minority of oysters to grow well, with others having reduced growth and a substantial portion not growing at all or even dying. Overstocking will also result in increased variation in shell shape.

Nursery systems must be designed, therefore, to accommodate oyster seed at a size the farmer can direct stock into cage culture with a minimum of handling. Take note of the section above on nursery culture, whether practiced in bottle culture systems, upweller tanks, FLUPSYs, or intermediate cage culture. The latter approach will be addressed. The second rule of oyster culture, related to Guideline 1, addresses sizing and grading oysters.

> **Guideline 2** – Unless oysters are more or less the same size when stocked into cages, the larger oysters will grow significantly larger, intermediate sized oysters will grow a little and small oysters will not grow, all associated with the availability of suspended food to individual oysters in the cage.

The necessity to control density at all stages of production cannot be overstated. Stocking density of oysters in bags and trays is important to regulate. As a guide, trays and bags should be stocked at no more than 25% of the volume of the unit of culture. Commonly, oyster cages have a volume of 20–25 L and are designed to hold between 100–150 market size oysters at a harvestable size of about 75 mm shell height (for North American markets). This equates to a stocking density at harvestable size of no more than five oysters per liter. Specifications of the gear being considered should be assessed for this parameter and should assist in making fundamental decisions based on the number of cages deployed, farm layout and expected yields.

As oysters grow they must be restocked in bags or trays to allow for further growth. Throughout the cultivation process growers are faced with the periodic need to 'split' oysters as they grow, which is the simple process of dividing oysters (by size or weight) from one unit into two or more to reduce stocking density to avoid the effects of overcrowding. During the splitting process, growers may opt to replace damaged or heavily bio-fouled gear and/or stock oysters into units with a larger mesh size. Though splitting can be done manually simply, with regard to reducing stocking density, many growers use a size grader at this point to assist in the process. This allows a grower to stock like-sized oysters together, which assists with inventory management and allows larger oysters to be stocked into larger mesh bags or baskets. While some growers opt for home-made graders, many farmers utilize mechanical graders to efficiently process large numbers of oysters. These include shaker tables, rotating tubes, and computer-assisted 'vision' sorters. In all cases, graders produce multiple size grades of oysters for either harvest or restocking cages for further growth. Examples include rotary tumblers that utilize aluminum or plastic tubes filled with holes of slowly increasing diameter. Oysters are poured into one end of the grader and as the tube slowly rotates, oysters tumble through the tube (set at a downward angle), falling through the perforations based on their size. Other styles of mechanical graders used in the field utilize vibration to gently move oysters over horizontal screens of increasing diameter. Oysters are sorted by size as they fall through the screens into bins for further processing.

There are three main benefits to routinely sorting the oyster crop by size. First, oysters that are generally all the same size grow more uniformly. Second, oysters that are sorted using mechanical graders have reduced biofouling following grading resulting in cleaner oysters for either harvest or restocking for further growth. Third, grading oysters regularly can increase product quality significantly. Grading oysters regularly has the benefit of improving product quality through a process called 'self-trimming'. Rapidly growing oysters add layers of new shell at the valve margins that is often thin and frilly. Harvesting oysters in this condition (unless produced for the shucked market) can result in greatly reduced shelf life due to harvest induced damage to the fragile shells and loss of internal water (for example, shell liquor) upon exposure to air. Grading oysters via a controlled, tumbling action has the advantage of inducing gentle damage to the shells of live oysters, chipping away at the frill and valve margins without significantly impacting the oyster. Once returned to the sea, oysters will quickly repair the damaged shell by adding additional shell material resulting in thicker, stronger and more robust valves with deeper cups. Additionally, grading oysters regularly can produce oysters that increase the meat to shell volume. These induced changes to the oyster, all due to regular grading and tumbling, result in oysters having higher value for half shell markets.

Grow-out systems for single oysters

Over the last 20 to 30 years there has been a significant advance in the design, production and availability of gear specifically designed for raising oysters. Much of the equipment available today may also be adapted for a wide range of applications, including on-bottom, suspended, and floating operations. The oyster farmer has a wide variety of gear options to consider when setting up a new farm. Much of the available gear may be used in on bottom, off-bottom longline, in suspension, or floating on the sea surface. For example, cages designed for suspended culture of single oysters are also commonly used in floating culture or systems anchored to the intertidal, whether strictly on-bottom or longline based. Site and permitting considerations will usually dictate what equipment will be used and in what application.

Suspended and floating culture of oysters

For growing oysters suspended in the water column or floating on the surface, there are a variety of systems available. For suspended culture, Japanese pearl nets have long been in use all over the

Figure 1.16 Oysters contained in tubs, picked at low tide, are picked up by a crane and barge at high tide and transported to a shucking house in Washington State. Courtesy of Taylor Shellfish Company.

world. These consist of a wire framework supporting multiple layers of mesh-enclosed spaces for holding oysters from seed to market size. Lantern nets typically come in 5 or 10 tier configurations and are typically stocked with about 100 oysters per tier, or 1000 oysters in a 10-tier net. Nets are hung into the water column from rafts, longlines utilizing buoys for flotation or in mid-water configurations (subsurface longlines). Other suspended culture systems include trays that are similarly hung, often with attached weights, from buoys (Fig. 1.16).

Similarly, many systems are based on hanging trays of single oysters from buoys or rafts. Here, growers deploy rafts or longlines suspended by buoys to support subsurface gear, depending on local custom and regulations pertaining to floating aquaculture. A large variety of systems have been developed worldwide. All rely on simply holding oysters in dedicated gear for the months to years required for growth from seed to harvestable size or weight. Floating culture similarly includes gear deployed with flotation in order to maintain the systems on the sea surface. Most commonly, this gear is deployed such that the oysters are living just below the water surface (Fig. 1.17). Floating systems include series of floating bags/baskets, large buoyant cages, and cages that are designed to be flipped up on pontoons to allow air-drying to control biofouling (Fig. 1.18). These systems often have the capacity to flood the pontoons and sink the unit to the bottom to avoid unfavorable conditions (for example, winter ice, tropical storms) and then be emptied and refilled with air to re-float at the surface.

Growing oysters in the intertidal zone

Grow-out of oysters in the tidal zone is a common practice where environmental conditions are satisfactory or are required by permit. Environments that include freezing, excessive heat, or freshwater inputs during the low tide are not appropriate for tidal zone grow-out, but in many locations intertidal oyster farms may be very successful. Access to the farm during the time the tide is out enables labor and equipment to be deployed for maintenance of grow-out systems, predator control, grading, splitting and harvesting, and most other husbandry activities. Oysters

Figure 1.17 Longline culture of eastern oysters in floating cages in Maine. Courtesy of Chris Davis.

cultivated in the intertidal may be deployed in cages either directly on the tide flat (on-bottom cage culture), off-bottom on rails or racks or from longline systems (Fig. 1.19). Rail and rack systems consist of wooden, steel, or plastic frames inserted into the substrate that hold oyster cages a few centimeters to 0.5 m off the bottom (Fig. 1.20). Cages may be secured by straps or lines or other means to the frames. A large variety of approaches have been put to use by growers over the years that take into account site specific parameters that serve to simply increase efficiency during the scarce hours that the tide is low enough for workers to tend to systems.

Other approaches to intertidal grow-out include adjustable longlines where the culture gear is set at a specific tidal height by the grower, depending on the season and the conditions. In many longline systems, this height is adjustable, allowing the grower to adjust the immersion time experienced by the crop. This can be used to control biofouling, to regulate growth rates, and

Figure 1.18 An oyster culture (OysterGro) using a system of floating cages designed for regular rotation to expose cages and oysters to the air to control biofouling. Courtesy of Joth Davis.

Figure 1.19 Intertidal culture of Pacific oysters in an on-bottom longline system in Washington State. Courtesy of Joth Davis.

Figure 1.20 Intertidal trays used to culture Sydney rock oysters in eastern Australia. Courtesy of Joth Davis.

even determine the amount of wave action to which the oysters are subjected (either routinely or during storm events) (Fig. 1.21). In the Pacific northwest of North America where intertidal oyster culture is common, oyster cages, attached to intertidal longlines, may also be supplied with flotation. This enables the cage to rotate on the longline in concert with the tide (Fig. 1.22). When the longlines are covered, cages float in the up position and when the tide falls, the cages rotate into the down position. This daily activity, mediated by tidal height fluctuations, enables the oysters to tumble within the cage at regular intervals and results in higher product quality, similar to that attained by regular grading.

Figure 1.21 Intertidal culture of Pacific oysters using SEAPA trays and an adjustable longline system in South Australia. Courtesy of SEAPA.

Figure 1.22 Intertidal culture of Pacific oysters using a rotating-cage system in Washington State. Cages supplied with floats on one end of the cage float the cage up into the water column when the beach is covered. Cages fall back when next exposed, shuffling oysters on every low tide exchange. Courtesy of Joth Davis.

Diseases and pests

There are numerous diseases and pests that affect oysters at various life stages. In addition to predators and pests of oysters is the influence of a changing climate – warmer seas and air temperature will also likely threaten the capacity to grow oysters in the future. These impacts depend on geographic location, the host oyster species and its life stage, and environmental conditions. As mentioned earlier, many diseases have imposed substantial losses to oyster growers over the years. In fact, diseases and pests have contributed to the serial loss of production for specific species being cultivated in the world over the last 100 years. In western Europe, the native flat oyster (*O. edulis*) was depleted by disease and replaced in aquaculture production with the Portuguese oyster. This species was similarly depleted by disease and sea farming efforts switched to the Pacific oyster (*C. gigas*). As of 2019, Pacific oysters in Europe, Asia, Australia, and New Zealand are currently critically threatened by highly virulent microvariants of oyster herpes viruses, collectively referred to as POMS. The eastern oyster (*C. virginica*) has a long history of disease impacts including Dermo disease (the protozoan *Perkinsus marinus*), MSX disease (the protozoan *Haplosporidium nelsoni*) and ROD. Impacts of disease include a range of sub-lethal effects but often can lead to high levels of mortality in the crop. In most growing regions, there are substantial restrictions and checks on movement of oysters (seed or adults) to decrease the risk of spread of a disease to a new area. Methods to reduce disease prevalence once established in a crop have not generally been adopted due to the costs involved. Rather, most growers manage the effects of disease, either by planning harvest prior to expected losses to disease or by cultivating in areas less prone to disease. At least in the case with some off-bottom farms, some growers are raising oysters to market size before they succumb to disease. Despite these strategies, the predominant strategy to reduce losses to disease has been the development of disease-resistant strains through breeding. In North America, Europe, Australia, and New Zealand, selective breeding programs are in place that focus on the development of oyster broodstock lines resistant to diseases. These programs have been largely effective in reducing losses but require dedicated, long-term research and development efforts.

Growers also attempt to reduce the negative effects associated with competition for resources by other organisms. This typically manifests in off-bottom aquaculture as a biofouling community that en masse can reduce flow to the oysters by attaching to and clogging screens and meshes (adding significantly to flotation requirements for floating culture) as well as compete directly with the oyster crop for suspended plankton (see Chapter 16). Typical 'biofoulers' for oyster growers include barnacles, mussels, wild oysters (termed overset), macroalgae, ascidians, and encrusting bryozoans. In addition, there are organisms that invade the shells of oysters, affecting the shell appearance and possibly weakening the shell. These include mud blister worms and boring sponges. To manage these impacts on both the gear and the oysters, growers may utilize some mix of prevention and control techniques. To prevent establishment of these species, many growers use routine air-drying to impose lethal stress on the undesirable species. Growers may vary the frequency, duration, and timing of air-drying in response to the local site conditions. To control established species, growers may opt for a variety of solutions from highly labor-intensive physical removal (for example, by power washing) to large-scale immersion in dips that pose an osmotic challenge to the biofoulers (for example, brine dips, freshwater dips). Some growers have also used vinegar dips or very brief immersion in boiling water. Typically, growers find these solutions time-consuming and expensive.

Predators pose a fundamental challenge to on-bottom culture in many parts of the world and can also induce losses in off-bottom farms. Common predators include crabs, predatory gastropods, sea stars, some species of fish (for example, drums in the Gulf of Mexico, rays in the Chesapeake Bay, and predatory flatworms). For on-bottom culture, losses to predators can be catastrophic. Typically, predation levels tend to increase in higher salinity waters, so some growers opt to raise oysters at lower salinity sites if they have that option to reduce losses to predation (with relay to higher salinity sites as the crop nears market size). Growers have attempted a number of predator control methods for on-bottom farms, including exclusion, chemical applications, biological control methods, and mechanical removal (for example, sea star 'mops', crab traps). These measures are expensive solutions that yield mixed results. Off-bottom farmers often do not see high levels of loss to predation, as the oysters are protected within the bags or baskets. Despite this initial level of protection, many predators will simply recruit as larvae or enter as juveniles into the bags and grow quickly enough to become a threat to the oysters if not removed. Similarly, some predators, such as predatory gastropods, may inflict predation through the basket if not removed, while others such as predatory flatworms could easily enter a basket. In most off-bottom farms, regular tending of the crop allows growers to remove predators before substantial losses are realized.

Harvesting oysters

In terms of public health, oysters (and some clams) are a special case because they are often eaten live on the half shell and potentially harbor pathogenic bacteria and viruses. As a result, harvesting of oysters is managed with a suite of regulations, including proactive temporary harvest closures when public health criteria cannot be assured (for example, harmful algal blooms, heavy rainfall or high river flow, wastewater treatment plant failures). In some jurisdictions, a minimum size may be required for harvest, most typically 76 mm (3 inches) shell height, while many jurisdictions allow cultured oysters to be sold at whatever size the market demands. Universally, however, there are strict time/temperature limits in each jurisdiction, with specifics varying.

As an example, in the US these regulations are mandated by the Food and Drug Administration via the Interstate Shellfish Sanitation Conference and implemented at the state level through the National Shellfish Sanitation Plan (NSSP). Harvesters must properly tag all commercial harvest and transfer these to a licensed wholesaler. In the US wholesalers and shippers have strict requirements for refrigeration and proper storage to maintain product safety, with many regulatory agencies requiring specific forms of training to increase product safety (such as Hazard Analysis Critical Control Point [HACCP]), including time from post-harvest to refrigeration requirements and training. Some facilities use wet storage to hold stock or depuration tanks, but these are beyond the scope of this chapter, and readers are recommended to determine local regulations and guidelines.

Once harvested, oysters may be sold live in the shell, sold for possible raw consumption by humans, or shucked for the meat. All raw oysters pose a potential human health risk due to the presence of naturally occurring *Vibrio* bacteria, such as *Vibrio vulnificus* and *V. parahaemolyticus*, which can cause illness and even death (particularly in immune-compromised individuals). The species *V. vulnificus* causes 95% of all seafood-related deaths in the United States (Oliver, 2005). Infections from *V. vulnificus* can cause primary septicemia and are the result of consuming raw or undercooked seafood, or wound infections (Hlady, 1997; Oliver, 2013). The second species of concern, *V. parahaemolyticus*, is acquired primarily through ingestion of seafood, and causes acute gastroenteritis (Daniels et al., 2000; Iwamoto et al., 2010; Scallan et al., 2011). Since *Vibrio* spp. are endemic to marine waters and are concentrated in oysters during the filter-feeding process, there are no means to prevent the introduction of *Vibrio* spp. into oysters or other seafood products. This is in contrast to other human pathogens associated with seafood, particularly bivalve molluscs, such as norovirus, *Salmonella* spp., or *Campylobacter* spp. Those pathogens are associated with human fecal contamination, generally via sewage treatment plant discharge. The NSSP provides regulations and guidelines to ensure that shellfish in the US are not harvested from areas impacted by human sewage; however, these guidelines do not control for the naturally occurring *Vibrio* spp. pathogens, as they are autochthonous and have no association with human sewage. *Vibrio* bacteria are primarily managed through time/temperature requirements and/or *Vibrio* control plans as noted above. These programs are very effective, though do not eliminate risk. While many people prefer to eat oysters raw on the half shell, there are four US Food and Drug Administration approved post-harvest processing methods that eliminate *Vibrio* bacteria from oysters: high pressure, low temperature pasteurization, freezing, and irradiation. Each process has specific requirements that can be found in the NSSP. Oysters processed in this fashion are no longer alive, however, and are intended for markets other than for oysters harvested for live, raw consumption.

Emerging issues and opportunities

Despite the relatively straightforward process of cultivation, the oyster aquaculture industry is constantly assessing and adopting new types of gear and methods of production. Off-bottom gear manufacturers are constantly attempting to improve their equipment, with variations on proven gear types as well as innovative new gear designs. The industry is also always looking to increase efficiency through mechanization and reduction of labor, including interest in the development of automated technology and improved inventory management. In addition, there has been interest in the idea of using real-time data monitoring to inform decision-making on the farm.

The world over, hatchery operations are often viewed as proprietary components of a business model and as a result there is often very little exchange of information among operators, even though nearly all shellfish hatcheries suffer from production problems from time to time that significantly impact their operations. A movement toward standardization of fundamental protocols, equipment, and genetic practices, and greater exchange of information would be helpful in order to increase the overall efficiencies of production.

A changing climate including changes to the carbon cycle associated with anthropogenic carbon inputs are also impacting the capacity to grow oysters. As discussed earlier, the oyster industry in many parts of the world is grappling with emerging problems associated with changing oceanic and estuarine carbonate chemistry conditions associated with pH, alkalinity, and reduced aragonite saturation state to properly form shells in larval oysters. While impacts to hatchery operations and the production of larval oysters and seed have been mitigated by adopting mechanical systems that buffer incoming seawater, the long-term outlook for larval and especially seed operations that depend on natural supplies of seawater, is less clear. For oyster farming to move forward it is likely that genetic improvement for traits associated with both disease resistance and increased yield will remain important. For oysters to thrive in different environments in the future it may be necessary to direct breeding in response to simultaneous, multi-stressor laboratory and field challenges in genetic lines that include thermal stress, hypoxia and changing carbonate chemistry conditions.

Partly in response to these changing conditions, there has been interest in producing oysters in land-based systems, including recirculating systems. Nursery culture systems (covered earlier in this chapter) are in use in many locations and there has been interest in extending the nursery phase into a vertically integrated, land-based system for all components of grow-out of oysters to marketable size. Land-based systems offer potential and significant benefits pertaining to security, availability of labor and land (including the potential to grow oysters in inland locations), and control over water quality (pollution and biotoxins), predators, biofouling, and disease. While there is the potential to pump seawater (including processed, saline groundwater) into land-based tanks or troughs, economic considerations, mainly associated with the capacity to supply enough oysters for the operation to be profitable with the amount of algal nutrition required for growth to a marketable size, have tamped down enthusiasm for this approach.

References cited and further reading

Allen, Jr., S.K., and Downing, S.L. (1986) Performance of triploid Pacific oysters, *Crassostrea gigas* (Thunberg). I. Survival, growth, glycogen content and sexual maturation in yearlings. *Journal of Experimental Marine Biology and Ecology*, 102: 197–208.

Angell, C.L. (1986) *The Biology and Culture of Tropical Oysters.* ICLARM Studies and Reviews, 13. International Center for Living Aquatic Resources Management, Manila, Philippines.

Barton, A., Waldbusser, G.G., Feely, R.A., Weisberg, R.G., Newton, J.A., Hales, B., Cudd, S., Eudeline, B., Langdon, C.J., Jeffords, I., King, T., Suhrbier, A., and McLaughlin, K. (2015) Impacts of coastal acidification on the Pacific northwest shellfish industry and adaptation strategies implemented in response. *Oceanography*, 28: 146–159.

Bayne, B.L. (2017) *Biology of Oysters*. Developments in Aquaculture and Fisheries Science, 41. Elsevier, Amsterdam.

Beck, M.W., Brumbaugh, R.D., Airoldi, L., Carranza, A., Coen, L.D., Crawford, C., Defeo, O., Edgar, G.J., Hancock, B., Kay, M.C., Lenihan, H.S., Luckenbach, M.W., Toropova, C.L., Zhang, G., and Guo, X. (2011) Oyster reefs at risk and recommendations for conservation, restoration, and management. *Bioscience*, 61: 107–116.

Brooks, W.K. (1891) *The Oyster – A Popular Summary of a Scientific Study*. Johns Hopkins Press, Baltimore, MD.

Buestel, D., Robert, M., Prou, J., and P. Goulletquer. (2009) History, status and future of oyster culture in France. *Journal of Shellfish Research*, 28 (4): 813–820.

Carazzi, D. (1893) *Ostricultura e mitilcultura*. Ulrico Hoepli, Milan.

Congrove, M., Wesson, J.A., and Allen, Jr., S.K. (2009) A practical manual for remote setting in Virginia. VSG-09–01, VIMS Marine Resource Report No. 2009–1.

Cranford, P.J., Ward, J.E., and Shumway, S.E. (2009) Bivalve filter feeding: variability and limits of the aquaculture biofilter. In: Shumway, S.E. (ed.) *Shellfish Aquaculture and the Environment*. Wiley-Blackwell, Chichester, pp. 81–124.

Creswell, L., Vaughan, D., and Sturmer, L. (1990) Manual for the cultivation of the American oyster, *Crassostrea virginica*, in Florida. Harbor Branch Oceanographic Institution, Inc..

Daniels, N.A., MacKinnon, L., Bishop, R., Altekruse, S., Ray, B., Hammond, R.M., Thompson, S., Wilson, S., Bean, N.H., Griffin, P.M., and Slutsker, L. (2000) *Vibrio parahaemolyticus* infections in the United States, 1973–1998. *The Journal of Infectious Diseases*, 181: 1661–1666.

Dégremont, L., Frank-Lawale, C., and Allen, Jr. S.K. (2012) Triploid oysters in the Chesapeake Bay: comparison of diploid and triploid *Crassostrea virginica*. *Journal Shellfish Research*, 31(1): 21–31.

Dégremont, L., Garcia, C., and Allen, Jr., S.K. (2015) Genetic improvement for disease resistance in oysters: a review. *Journal of Invertebrate Pathology*, 131: 226–241.

de Melo, C.M.R., Durland, E., and Langdon, C. (2016) Improvements of desirable traits of the Pacific oyster, *Crassostrea gigas*, as a result of five generations of selection on the West coast, USA. *Aquaculture*, 460: 105–115.

Ferreira, J.G., Hawkins, A.J.S., and Bricker, S.B. (2007) Management of productivity, environmental effects and profitability of shellfish aquaculture the Farm Aquaculture Resource Management (FARM) model. *Aquaculture*, 264: 160–174.

Flimlin, G., Macfarlane, S., Rhodes, E., and Rhodes, K. (2010) Best management practices for the East Coast shellfish aquaculture industry. East Coast Shellfish Growers Association.

Food and Agriculture Organization of the United Nations (2017) Fisheries and Aquaculture Department, statistical collections: global aquaculture production. Available at: http://www.fao.org/fishery/statistics/global-aquaculture-production (accessed 25 March 2020).

Frieder, C.A., Applebaum, S.L., Pan, T.C.F., Hedgecock, D., and Manahan, D. (2017) Metabolic cost of calcification in bivalve larvae under experimental ocean acidification. *ICES Journal of Marine Science*, 74: 941–954.

Gallager, S., and Mann, R. (1986) Growth and survival of larvae of *Mercenaria mercenaria* (L.) and *Crassostrea virginica* (Gmelin) relative to broodstock conditioning and lipid content of eggs. *Aquaculture*, 56: 105–121.

Getchis, T. (ed.) (2014) Aquaculture management guide: manual for the identification and management of aquaculture production hazards. NRACAMG-2014, US Department of Agriculture.

Gosling, E. (2003) *Bivalve Molluscs: Biology, Ecology and Culture*. Fishing News Books. Blackwell, Oxford.

Goulletquer, P. (2004) Cultured aquatic species information programme. *Ostrea edulis* (2005 – 2019).

Cultured Aquatic Species Information Programme. In: FAO Fisheries and Aquaculture Department [online]. Rome.

Grant, J., and Pastres, R. (2019) Ecosystem models of bivalve aquaculture: implications for supporting goods and services. In: Smaal, A.C, Ferreira, J.G., Grant, J., Petersen, J.K., and Strand, Ø. (eds.) *Goods and Services of Marine Bivalves*. Springer Nature, Cham, pp. 507–526.

Grant, J., Cranford, P., Hargrave, B., Carreau, M., Schofield, B., Armsworthy, S., Burdett-Coutts, V. and Ibarra, D. (2005) A model of aquaculture biodeposition for multiple estuaries and field validation at blue mussel (*Mytilus edulis*) culture sites in eastern Canada. *Canadian Journal of Fisheries and Aquatic Sciences*, 62(6): 1271–1285.

Guévélou, E., Carnegie, R.B., Small, J.M., Hudson, K., Reece, K.S., and Rybovich, M.M. (2019) Tracking triploid mortalities of eastern oysters *Crassostrea virginica* in the Virginia portion of the Chesapeake Bay. *Journal of Shellfish Research*, 38(1):101–113.

Guo, X., DeBrosse, G.A., and Allen, Jr., S.K. (1996) All-triploid Pacific oysters (*Crassostrea gigas* Thunberg) produced by mating tetraploids and diploids. *Aquaculture*, 142: 149–161.

Guo, X., Ford, S.E., and Zhang, F. (1999) Molluscan aquaculture in China. *Journal of Shellfish Research*, 18(1): 19–31.

Guo, X., Wang, Y., Debrosse, G., Bushek, D., and Ford, S.E. (2008) Building a superior oyster for aquaculture. *The Jersey Shoreline*, 25: 7–9.

Hawkins, A.J.S., Duarte, P., Fang, J.G., Pascoe, P.L., Zhang, J.H., Zhang, X.L. and Zhu, M.Y. (2002) A functional model of responsive suspension-feeding and growth in bivalve shellfish, configured and validated for the scallop *Chlamys farreri* during culture in China. *Journal of Experimental Marine Biology and Ecology*, 281(1–2): 13–40.

Hawkins, A.J.S., Pascoe, P.L., Parry, H., Brinsley, M., Black, K.D., McGonigle, C., Moore, H., Newell, C.R., O'Boyle, N.T., O'Carroll, T., O'Loan, B., Service, M., Smaal, A.C., Zhang, X.L., and Zhu, M.Y. (2013) ShellSIM: a generic model of growth and environmental effects validated across contrasting habitats in bivalve shellfish. *Journal of Shellfish Research*, 32(2): 237–253.

Hedgecock, D., and Davis, J.P. (2007) Heterosis for yield and crossbreeding of the Pacific oyster *Crassostrea gigas*. *Aquaculture*, 272(Supplement 1): S17–S29.

Helm, M.M., Bourne, N., and Lovatelli, A. (2004) Hatchery culture of bivalves; a practical manual. FAO Fisheries Technical Paper. No. 471. FAO, Rome.

Hlady, W.G. (1997) *Vibrio* infections associated with raw oyster consumption in Florida, 1981–1994. *Journal of Food Protection*, 60: 353–357.

Iwamoto, M., Ayers, T., Mahon, B.E., and Swerdlow, D.L. (2010) Epidemiology of seafood-associated infections in the United States. *Clinical Microbiology Reviews*, 23: 399–411.

Jones, G., and Jones, B. (1983) Methods for setting hatchery produced oyster larvae. British Columbia Ministry of Environment, Marine Resources Branch, Info. Rpt. # ISSN 0229–8120; No. 4.

Kennedy, V.S. (2018) *Shifting Baselines in The Chesapeake Bay. An Environmental History*. Johns Hopkins University Press, Baltimore, MD.

Kennedy, V.S., Newell, R.I.E., and Eble, A.F. (1996) *The Eastern Oyster:* Crassostrea virginica. 2nd edition. University of Maryland Sea Grant Publications, College Park, MD.

Kesarcodi-Watson, A., Klumpp, D.W., and Lucas, J.S. (2001) Comparative feeding and physiological energetics in diploid and triploid Sydney rock oysters (*Saccostrea commercialis*): II. Influences of food concentration and tissue energy distribution. *Aquaculture*, 203: 195–216.

Langdon, C., Evans, F., Jacobson, D., and Blouin, M. (2003) Yields of cultured Pacific oysters *Crassostrea gigas* Thunberg improved after one generation of selection. *Aquaculture*, 220: 227–244.

Lannan, J.E. (1980a) Broodstock management of *Crassostrea gigas* I. Genetic and environmental variation in survival in the larval rearing system. *Aquaculture*, 21: 323–336.

Lannan, J.E. (1980b) Broodstock management of *Crassostrea gigas* III. Selective breeding for improved larval survival. *Aquaculture*, 21: 347–351.

Lannan, J.E., Robinson, A., and Breeze, W.P. (1980) Broodstock management of *Crassostrea gigas* II. Broodstock conditioning to maximize larval survival. *Aquaculture*, 21: 337–345.

Margaret, M., Hofmann, E.E., and Powell, E.N. (1993) Environmental effects on the growth and development of eastern oyster *Crassostrea virginica* (Gmelin, 1791), larvae: a modeling study. *Journal of Shellfish Research*, 12(2): 241–254.

Matthiessen, G.C. (2001) *Oyster Culture*. Fishing News Books. Blackwell Science, Oxford.

Matthiessen, G.C., and Davis, J.P. (1991) Growth, disease resistance and survivorship of diploid and triploid oysters in an MSX environment. *Journal of Shellfish Research*, 10: 379–388.

Matthiessen, G.C., and Davis, J.P. (1992) Observations on growth rate and resistance to MSX (*Haplosporidium nelsoni*) among diploid and triploid eastern oysters (*Crassostrea virginica* (Gmelin, 1791)) in New England. *Journal of Shellfish Research*, 11(2): 449–454.

Muranaka, M.S., and Lannan, J.E. (1984) Broodstock management of *Crassostrea gigas*: environmental influences on broodstock conditioning. *Aquaculture*, 39: 217–228.

Nell, J.A., and Perkins, B. (2005) Studies on triploid oysters in Australia: farming potential of all-triploid Pacific oysters, *Crassostrea gigas* (Thunberg), in Port Stephens, New South Wales, Australia. *Aquaculture Research*, 36(6): 530–536.

Newell, C.R., Brady, D.C., and Richardson, J. (2019) Farm-scale production models. In: Smaal, A.C, Ferreira J.G., Grant, J., Petersen, J.K., and Strand, Ø. (eds.) *Goods and Services of Marine Bivalves*. Springer Nature, Cham, pp. 485–506.

Newkirk, G.F. (1988) Response to selection for growth in *Ostrea edulis*: second generation. *Journal of Shellfish Research*, 7(1): 172.

Newton, K., Peters, R., and Raftos, D. (2004) Phenoloxidase and QX disease resistance in Sydney rock oysters (*Saccostrea glomerata*). *Developmental and Comparative Immunology*, 28: 565–569.

Oliver, J.D. (2005) *Vibrio vulnificus*. In: Belkin S., Colwell R.R. (eds.) *Oceans and Health: Pathogens in the Marine Environment*. Springer, Boston, MA, pp. 253–276.

Oliver, J.D. (2013) *Vibrio vulnificus*: death on the half shell. A personal journey with the pathogen and its ecology. *Microbial Ecology*, 65: 793–799.

Parker, M., Dill, S., Webster, D., and Meritt, D. (2014) Contained culture cost analysis. Oyster Aquaculture Technology Series AGNR-EM-14–07. University of Maryland Extension.

Proestou, D.A., Vinyard, B.T., Corbett, R.J., Piesz, J., Allen Jr., S.K., Small, J.M., Lie, C., Liu, M., DeBrosse, G., Guo, X., Rawson, P., and Gómez-Chiarri, M. (2016) Performance of selectively bred lines of eastern oyster, *Crassostrea virginica*, across eastern US estuaries. *Aquaculture*, 464: 17–27.

Rosa, M., Ward, J.E., and Shumway, S.E. (2018) Selective capture and ingestion of particles by suspension-feeding bivalve molluscs: A review. *Journal of Shellfish Research*, 37(4): 727–746.

Rose, J.M., Bricker, S.B. and Ferreira, J.C. (2019) Comparative analysis of modeled nitrogen removal by shellfish farms. *Marine Pollution Bulletin*, 91: 185–190.

Saurel, C., Ferreira, J.G., Cheney, D., Suhrbier, A., Dewey, B., Davis, J., and Cordell, J. (2012) Ecosystem goods and services from Manila clam culture in Puget Sound: a modelling exercise. *Aquaculture Environment Interactions*, 5: 255–270.

Scallan, E., Hoekstra, R.M., Angulo, F.J., Tauxe, R.V., Widdowson, M.A., Roy, S.L., Jones, J.L., and Griffin, P.M. (2011) Foodborne illness acquired in the United States – major pathogens. *Emerging Infectious Diseases*, 17: 7–15.

Schulte, D.M., Burke, R.P., and Lipcius, R.M. (2009) Unprecedented restoration of a native oyster metapopulation. *Science*, 325: 1124.

Shumway, S.E. (ed.) (2011) *Shellfish Aquaculture and the Environment*. Wiley-Blackwell, Chichester.

Silva C., Ferreira, J.G., Bricker, S.B., DelValls, T.A., Martin-Diaz, M.L., and Yanez, E. (2011) Site selection for shellfish aquaculture by means of GIS and farm-scale models with an emphasis on data-poor environments. *Aquaculture*, 318: 444–457.

Smaal, A.C., Ferriera, J.G., Grant, J., Petersen, J.K., and Strand, O. (eds.) (2019) *Goods and Services of Marine Bivalves*. Springer, Cham.

Stanley, J.G., Hidu, H., and Allen Jr., S.K. (1984) Growth of American oysters increased by polyploidy induced by blocking meiosis I but not meiosis II. *Aquaculture*, 37: 147–155.

Steele, E.N. (1964) *The Immigrant Oyster (*Ostrea gigas*) Now Known as The Pacific Oyster*. Warren's Quick Print, Olympia, WA.

Supan, J.E. (2014) High-density rearing of oyster larvae in flow-through systems. Southern Regional Aquaculture Center SRAC Publication 4311.

Toba, D. (2002) Small-scale oyster farming for pleasure and profit in Washington. Washington Sea Grant WSG-AS 0203.

Toro, J.E., and Newkirk, G.F. (1991) Response to artificial selection and realized heritability estimate for shell height in the Chilean oyster *Ostrea chilensis*. *Aquatic Living Resources*, 4: 101–108.

Waldbusser, G.G., and Salisbury, J.E. (2014) Ocean acidification in the coastal zone from an organism's perspective: multiple system parameters, frequency domains, and habitats. *Annual Review of Marine Science*, 6: 221–247.

Waldbusser, G.G., Voigt, E.P., Bergschneider, H., Green, M.A., and Newell, R.I.E. (2011) Biocalcification in the eastern oyster (*Crassostraea virginica*) in relation to long-term trends in Chesapeake Bay pH. *Estuaries and Coasts*, 34: 221–231.

Waldbusser, G.G., Brunner, E.L. Haley, B.A., Hales, B., Langdon, C.J., and Prahl, F.G. (2013) A developmental and energetic basis linking larval oyster shell formation to acidification sensitivity. *Geophysical Research Letters,* 40: 2171–2176.

Waldbusser, G.G., Hales, B., Langdon, C.J., Haley, B.A., Schrader, P., Brunner, E.L., Gray, M.W., Miller, C.A., and Gimenez, I. (2015) Saturation-state sensitivity of marine bivalve larvae to ocean acidification. *Nature Climate Change*, 5: 273–280.

Wallace, R.K., Waters, P., and Rikard, F.S. (2008) Oyster hatchery techniques. Southern Regional Aquaculture Center SRAC Publication 4302.

Walton, W.C., Davis, J.E., and Supan, J.E. (2013) Off-bottom culture of oysters in the Gulf of Mexico. Southern Regional Aquaculture Center SRAC Publication 4308.

Wang, H., Guo, X., Zhang, G., and Zhang, F. (2004) Classification of Jinjiang oysters *Crassostrea rivularis* (Gould, 1861) from China, based on morphology and phylogenetic analysis. *Aquaculture*, 242: 137–155.

Ward, J.E., and Shumway, S.E. (2004) Separating the grain from the chaff: particle selection in suspension and deposit feeding bivalves. *Journal Experimental Marine Biology and Ecology*, 300: 83–130.

White, J., Ruesink, J.L., and Trimble, A.C. (2009) The nearly forgotten oyster: *Ostrea lurida* Carpenter 1864 (Olympia oyster) history and management in Washington State. *Journal of Shellfish Research*, 28(1): 43–49.

Wijsman, J.W.M., Troost, K., Fang, J., and Roncarti, A. (2019) Global production of marine bivalves. Trends and challenges. In: Smaal A.C., Ferreira, J.G., Grant, J., Petersen J.K., Strand, Ø. (eds.) *Goods and Services of Marine Bivalves*. Springer Nature, Cham, pp. 7–26.

Yonge, C.M. (1960) *Oysters*. Collins, London.

You, W., and Hedgecock, D. (2019) Boom-and-bust production cycles in animal seafood aquaculture. *Reviews in Aquaculture*, 11: 1045–1060.

CHAPTER 2

Clam aquaculture

John Kraeuter and Brian Beal

Introduction

'Clams' are a diverse group of bivalve molluscs that inhabit marine, estuarine, and fresh water. They are typically infaunal and are found in sedimentary environments from the intertidal zone to the deep sea, but most cultured species are grown in shallow estuarine or marine waters. Some species, such as the giant clam (see Chapter 3), are atypical in that they inhabit coral reef systems and are more like other epifaunal organisms.

Clams have been utilized as foods since pre-historic times, but clam aquaculture has developed into a leading group of cultured bivalves and a significant source of food since the 1960s and 1970s when hatchery and field techniques were developed. The general culture techniques for clams are the basis of this chapter.

The first species to be cultured commercially were those that remain close to the sediment surface when they burrow (*Ruditapes* spp. and *Mercenaria* spp.). Culture of those that burrow deeper (for example, *Panopea* spp., *Mya* spp.) require special hatchery and/or field techniques. Much of the literature on clam aquaculture is in reports, manuals, or chapters of books and can be difficult to access. Manzi and Castagna (1989) summarized North American production and da Costa (2013) has recently reviewed information on many species. The general biology, including aquaculture of most species, has not been the subject of individual compendia. The biology of the northern quahog or hard clam could serve as a model for information on other high production species, such as *Ruditapes* spp.

Clam aquaculture includes many species, but worldwide production is dominated by the Manila clam (*Ruditapes philippinarum*) (in excess of 4,000,000 tonnes) which is cultured in Europe, North America, and Asia. Both *Sinonovacula constricta* (>800,000 tonnes), and *Meretrix lusoria* (>50,000 tonnes) are important species in Asia. The northern quahog or hard clam, *Mercenaria mercenaria* (>25,000 tonnes) has become the dominant species cultured on the US Atlantic coast. In the US, of all cultured invertebrates, hard clams (13%) and Manila clams (2%) comprise about 15% of production volume (total = 178,000 tonnes; FAO FishStat Plus (2009)), compared to eastern oysters (43%), Pacific oysters (16%), blue mussels (1%), and red swamp crayfish (25%).

Clam aquaculture

Most clam species being produced by aquaculture utilize hatchery-produced seed, but a few such as the Manila clam (*Ruditapes philippinarum*), grooved carpet shell (*R. decussatus*) and the Chinese razor clam (*Sinonovacula constricta*) may use seed harvested from the wild. Since the focus of this chapter is on aquaculture production, there is no further discussion of field-harvested seed, but the process of planting these seed would be similar to those of the same size produced in a hatchery.

One of the most important aspects of all aquaculture facilities is the siting of the facility. Siting for a hatchery/nursery system is generally different from the area selected for grow-out. Hatchery design and use are described in detail in Chapter 12. Typically, clam hatcheries are located on shore near a consistently high-salinity water source. Because it is expensive to pump water, the facility should be located as near the water level as possible without compromising the building to severe flooding. If the hatchery water system will not be utilized in a flow-through mode, a single system that is turned off and on would be sufficient, but if water flow is to be maintained on a continuous basis, it is desirable to have a redundant system that can be alternated allowing for a week between alternate system uses. By alternating the lines and intakes the unused lines become anaerobic thus killing newly settled fouling organisms. Flushing the system before use removes the anaerobic water. For small hatcheries it is desirable to have the dual systems including two intake lines and pumps. Nursery systems require constant flow and thus a dual system is essential unless an in-the-water nursery is utilized.

Utilizing suction pumps in colder climates with high tidal amplitude can result in freezing of the intake. When the seawater is close to freezing, the combination of very cold temperatures, spring tide lows, and a strong wind may cause ice to form in the intake system. The pump will continue to work, but because there is no water moving through the system, it can overheat sufficiently to melt PVC pipes. Eventually, the lack of water flow will also allow the discharge pipe to freeze. The easiest way to prevent freezing is to anticipate the extreme conditions, stop pumping, and drain the lines when the conditions dictate and tide is low.

These same systems can allow water to become supersaturated in dissolved gases, particularly nitrogen, to be incorporated into the clam tissue. This is because pumping from relatively deep water and introducing the water into shallow trays releases the pressure on the gases, but does not allow sufficient time for degassing. Gas supersaturation also can be associated with air leaks on the suction side of a pump. Degassing the water is particularly important during rapid warming that can occur in shallow bays in the spring. The water is being pumped from relatively deep sources and placed into shallow containers such as raceways. Simply allowing the water to be agitated before it is introduced into the tank through degassing columns, bag filters, heavy aeration, sluice boxes, or other means typically will prevent this problem.

Hatchery

Hatchery is a general term that may designate the building used for all bivalve rearing until the animals are placed in the field. Hatchery design and use are described in detail in Chapter 12. A restrictive definition is a building for processes such as: conditioning of the adults (broodstock), spawning, larval rearing and setting, and microalgal production. Cold, flowing seawater acts as a heat sump and that makes maintaining the hatchery temperature very costly, thus keeping the nursery system with its flowing seawater isolated from the hatchery is important. Hatcheries usually provide an area for growing unicellular algae as food, but algal production is not included in the description below (see Chapters 12 and 13). The hatchery building requires a source of seawater that can be modified to create the conditions needed for rearing microalgae, conditioning adults, and culturing larvae and post-set individuals. Hatchery water, depending on the volumes needed, may simply be filtered through a series of bag filters, but often a self-cleaning, rapid sand filter is the first stage in cleaning the water. After this initial filtration (sand filter and bag or a series of cartridge filters), the water may pass through an ultraviolet light system to eliminate living organisms, and then an activated charcoal filter to remove undesirable dissolved components. An alternative system for relatively small quantities of sterilized water is to use the rapid sand filter and then further filter the water with a tangential flow filter to remove fine and bacteria without altering seawater chemistry.

The most common loss of larvae in the hatchery is due to bacterial contamination. Treating the water as described above is important in controlling the bacterial levels, but one must also be sure that sieves, containers, and other materials are cleaned after each use. The simplest method is to use a biodegradable detergent and hot fresh water followed by air drying. It is important that brushes and other cleaning equipment receive similar treatment. A systematic method, such as having cleaned equipment inverted and assuming that containers in the upright position are contaminated, will prevent their inadvertent use.

Broodstock

The process of preparing adults to be ready to release gametes is called conditioning and is similar for most clam species. Broodstock animals to be conditioned are placed in a tank, supplied with filtered, heated seawater and plentiful food. The temperature is usually started at ambient and slowly raised (about 1 to 2°C/day) until a level is reached where the clams are adding gonadal material (for most temperate zone species this is in the range of 16–22°C) (see Table 2.1). Conditioning clams are then maintained at that temperature until the ripening process is complete. The entire process usually takes 4–6 weeks and near the end periodic ripeness checks can be made (see below). The culturist should be sure that the animals are not subject to rapid changes in temperature, salinity, or other stimuli that could cause them to release gametes prematurely. The tank used for conditioning of most species can be of any size or shape, and can be supplied with running seawater, but most conditioning tanks are batch cultures. Insulated tanks are often used to reduce the fluctuations in temperature within a narrow range. If a static system is used, then it is useful to be sure that some mechanism is in place to control water chemistry.

Table 2.1 Approximate conditioning temperature and salinity for various clam species. Blank spaces signify no information.

Species	Common name	Condition temperature (°C)	Condition salinity	Reference
Cyrtopleura costata	Angelwing	25–28	25–30	Gustafson et al., 1991
Ensis leei	Atlantic jackknife	16–18	30	Flanagan, 2013
Ensis magnus	Razor shell	18–20		da Costa, 2013
Ensis siliqua	Pod razor	28–20		da Costa, 2013
Leukoma staminea	Pacific littleneck			
Macrocallista nimbosa	Sunray venus	18–23	28–30	Sturmer et al., 2011
Mactromeris polynyma	Arctic surfclam	24		Davis, 1995
Mercenaria mercenaria	Northern quahog	22–23	25–30	Castagna, 2001
Meretrix spp.	Asian hard clam	23–26	30–32	Thiet and Kumar, 2008
Mya arenaria	Softshell	15–16	32	Beal, 2011
Panopea generosa	Pacific geoduck	11	29–32	Marshall, 2012
Paratapes undulatus	Undulate venus			
Ruditapes decussatus	Grooved carpet shell	18–20		Oeja et al., 2008; da Costa et al., 2013
Ruditapes philippinarum	Japanese littleneck	20	24–35	Goulletquer, 2005; Toba et al., 1992; da Costa et al., 2013
Sinonovacula constricta	Constricted tagelus			
Solen marginatus	Grooved razor shell	19	30–32	da Costa, 2009; 2013
Spisula solidissima	Atlantic surfclam	2–15	28–32	

The simplest systems use a biofilter supplied with aeration to enhance ammonia removal. This can be supplemented with periodic changes of water and siphoning off accumulated feces and pseudofeces. If water is exchanged, it must be of the same temperature and salinity as occurs in the conditioning tank or it may trigger spawning. This usually necessitates a second tank for preparing the exchange water.

Animals can be sacrificed periodically to test for ripeness, or a small hole can be drilled in the shell and gonadal tissue removed with a needle and syringe and placed on a slide to examine microscopically for gametes. This hole can be sealed with paraffin to maintain the shell integrity. Once clams have been conditioned, some species can be maintained at temperatures 3–5°C below the conditioning temperature to reduce the possibility of inadvertent spawning.

Conditioning special broodstock

Deep-burrowing genera (for example, *Cyrtopleura, Ensis, Mya, Panopea, Sinonovacula*) require some form of lateral pressure to maintain shell closure. Without external pressure on a geoduck or a softshell clam (normally provided by the natural substratum) the shell can gape. In the geoduck, this can cause tears in the periostracum (which covers the shell, siphon, and muscular mantle) making the clam susceptible to infection. In the softshell clam, the lack of pressure will

Figure 2.1 *Mya arenaria* held in a mesh sandwich for conditioning. The mesh provides lateral pressure to keep the clams from gaping. Courtesy of B. Beal.

cause the clam to gape and eventually die. The lateral pressure can be achieved with substrate, but the use of substrate such as sand creates significant maintenance difficulties because of the accumulation of organic materials (feces, pseudofeces) so that anaerobic bacteria may develop in the sediments. The use of substrate has also been recommended for some species of *Macrocallista*. As an alternative to substrate, rubber bands have been used on individuals. In other settings, a rigid mesh material (Fig. 2.1) that is placed on both sides of the animal and held tightly together with twist or cable ties, or densely packing the clams with the siphons upward in a container are alternative means of limiting shell gape.

Spawning

Once adults have been conditioned, they must be stimulated to release the gametes. This is accomplished by manipulating temperature, but in some cases sperm or chemical treatments such as serotonin may be used to effect spawning. The usual technique is to place the animals to be spawned in a shallow tray in seawater at the conditioning temperature. Once the conditioned clams open and begin to actively filter, the temperature of the water bath is then raised and lowered about 10°C, at intervals of about 30–45 minutes, often with the addition of additional microalgae, until animals release the gametes. Males often release gametes before the females. This sperm may also be used as a means to induce female clams to spawn by injecting a small quantity of sperm directly into the water entering the intake siphon of an actively pumping female clam. As soon as they can be identified, the males and females are placed in separate containers and the sperm and eggs are collected. The eggs can then be fertilized with sperm from selected males or the sperm mixed so that multiple males can fertilize the eggs of one or more females. Mixing of the sperm of several males is desirable to avoid genetic incompatibility between individuals. Care is needed to limit the amount of sperm introduced to the eggs to prevent multiple fertilizations of an egg (polyspermy). Sperm from a previous spawn may be kept in a refrigerator overnight and used the next day to induce spawning; however, the storage makes these sperm less viable for fertilizing the eggs.

Some, but not all, species can be strip spawned. This involves sacrificing the animal. The animal is opened, and the gonadal material is macerated with a scalpel, the gametes are rinsed into beakers, dispersed and the sperm added to the eggs. A third method is the use of serotonin as a spawning stimulus by injection into the adductor muscle with a hypodermic needle.

Once eggs have been fertilized, some hatcheries separate the eggs from sperm and other materials in the water by washing the eggs through a relatively large sieve (150–250 µm) and then onto a smaller sieve (10–25 µm) to retain the eggs. This must be done gently and the smaller sieve can be placed in a seawater bath that extends about halfway up the sieve walls to provide a cushion for the eggs. The collected eggs are then counted and dispersed directly into larval rearing tanks.

Counting the eggs is achieved by placing the egg/water mixture into a calibrated container, and carefully mixing with a plunger so that the eggs are evenly dispersed. A small calibrated sample (typically 1 mL) is then removed with a syringe or pipette, placed on a Sedgewick Rafter counting cell, placed under a microscope and the eggs counted. This process should be repeated several times and the counts averaged. The number of eggs in the initial container can then be obtained by multiplying the volume of the initial container by the average number of eggs per unit volume from the counting cell. It is possible to count eggs electronically, but the equipment is too expensive for all but the largest of hatcheries, and often more cumbersome than a simple microscopic count.

Fertilized eggs are then dispersed in a rearing container at a concentration of up to 20 eggs per milliliter of water.

Larval rearing

Rearing the larvae of most clam species is very similar, but optimum temperature and salinity vary with species (Table 2.2). The fertilized eggs are placed in a container (typically with a conical base) that has been filled with warm (22–25°C) filtered (1 µm and UV-sterilized) seawater to which microalgae has been added. The container may have some limited aeration to provide circulation, but until the animals reach the straight-hinge (D-) stage, this should be very gentle so as not to cause mortality. Stocking density varies, but can be high (60 or more per mL) for the first two days until the straight-hinge forms, but experience has shown that reducing the density of straight-hinge animals to about 10 per mL and reducing these further until there are about 1–2/mL at setting is best.

The denser the animals and the warmer the culture water the more likely it is to have bacterial infections. The typical control is to drain and discard the water on a regular basis. To refresh the water, the containers are drained slowly through an appropriate size sieve every other day (batch culture) until the animals are ready to settle. The tanks are then cleaned with hot fresh water and a biodegradable detergent, refilled with the filtered, ultraviolet sterilized

Table 2.2 Approximate rearing temperature and salinity of larvae for various clam species. Blank spaces signify no information.

Species	Rearing temperature (°C)	Rearing salinity	Rearing period (days)	Reference
Cyrtopleura costata	30	20 then 25	12	Gustafson et al., 1991
Ensis leei	18–20	30	10	Flanagan, 2013
Ensis magnus			20	da Costa, 2013
Ensis siliqua			14	da Costa, 2013
Leukoma staminea				
Macrocallista nimbosa	23–31	26–28		Sturmer et al., 2011
Mactromeris polynyma	16	29–32	26	Davis, 1995
Mercenaria mercenaria	20–24	30	10–14	Castagna, 2001
Meretrix spp.	23–25	27–32	7	Guo, personal communication
Mya arenaria	23–24	30–32	16–18	Beal, 2011
Panopea generosa	14–17	30–32		Beattie and Godwin, 1993
Paratapes undulatus				
Ruditapes decussatus	20		30	FAO, 2009, 2019
Ruditapes philippinarum	20–24	24–35	13–14	Goulletquer, 2005; Toba et al., 1992
Sinonovacula constricta				
Solen marginatus	19	30–32	8	da Costa, 2009
Spisula solidissima	18–20	28–32		

seawater that has been supplemented with algal food. The animals retained on the sieve are typically graded through a stacked series of sieves (smallest mesh on the bottom and largest on top), counted and then returned to the same tank or the various size larvae can be apportioned to separate tanks.

Increasingly, hatcheries have been experimenting with the use of a flow-through culture system for the larval stage. The larvae are maintained at much higher densities and water and food is added continuously. The effluent water exits through fine mesh filter cloth that is mounted on both sides of a disk forming a ‘drum filter’. The filter diameter must be large enough to allow water exchange without impinging the larvae on the filter. The smaller tank size allows relatively complete volume exchange every day or every other day. The advantages of this system are much higher densities of larvae, smaller tanks, and reduced effort due to lack of draining the tank to exchange water.

The most common problem encountered during larval culture is disease, usually due to high levels of *Vibrio* spp. bacteria. Careful husbandry is the best means of controlling outbreaks, and antibiotics are not recommended because of the unintended consequences of increasing the number of bacteria that are resistant to specific antibiotics. Recent experiments are beginning to yield recommended probiotic treatments, and this offers a potential means of reducing the effects of these bacterial infections. If an infection should occur, it is very difficult to control and usually it is best to abandon the culture, thoroughly clean all the equipment including air stones, and then start with a new batch of larvae.

Setting (settling)

Setting in many clams is recognizable because the late-stage larvae (pediveligers) begin to attach to the sides or bottom of the rearing container. During setting, the animals lose their swimming organ (velum) and develop a foot as well as the beginning of the adult shell. Most clam larvae attach to the substrate using a byssal thread that exists in the foot for a period of time during setting and post-set. This attachment causes them to clump together. Once clams reach this stage and take up a benthic existence, there can be a dichotomy between requirements for various species depending upon whether the species requires a sediment substrate for continued growth. The young animals are kept in a setting system for a week to three weeks. Some genera such as *Cyrtopleura*, *Ensis*, *Panopea*, and *Meretrix* appear to do better when a sand substrate is present, but this is an added complication when sorting seed, and the addition of sediment is often delayed until the nursery stage.

Typically, settling clams (usually 150–250 µm) are held in batch culture water similar to that used for larval culture, including the addition of cultured algae. The animals are held in this condition with daily or every other day complete water changes for periods from a week to three weeks. The process of setting is dependent upon the species and the facility. Setting may be in the larval rearing area, but larger operations often have a separate room or separate facility designed to keep the larvae and post-set activities isolated. In most cases, the animals are not supplied with running seawater because it would wash the larvae that have not yet reached setting size out of the system. While this stage is often similar to larval culture in that it is still a batch culture, the 'tank' may be a raceway, a deeper tank or upweller or downweller. If raceways are used, they are simply maintained as a tank supplemented with algae and no flow for several days. Following this period, a slow water flow is introduced and maintained, but a fine mesh screen is used to collect any individuals washed off in the effluent. In downweller and tray systems (Figure 2.2) the clams are held in a mesh-bottomed cylinder and the water is pumped from the tank housing the cylinders, often with an air lift, and the downweller effluent is discharged into the same tank and recycled through the unit.

Figure 2.2 Downweller nursery system (foreground), setting trays (background). Courtesy of B. Beal.

Post-set (nursery)

Following setting, the young clams may be moved to systems that are used to raise the small seed (>1 mm) to sizes that can then be planted in the field (usually >8–10 mm). This intermediate stage requires utilization of pumped seawater because it is no longer cost effective to use cultured algae as a food source. It is expensive to pump seawater uphill, so keeping the nursery system as close to sea-level as possible is important. There are several other considerations when directly using pumped water. The first is to determine what type of coarse filtration may be required, but more importantly, consideration must be given to being sure that the incoming water has been allowed to de-gas (see above).

How the seawater is supplied to the clams depends upon the type of containment system used. There are many different systems used for the nursery stage (Figures 2.3–2.9): downwellers,

Figure 2.3 Upweller system for newly-set clams. Courtesy of K. Pepperman.

Figure 2.4 Setting tank with wooden trays. Water flows across each tray by recycling water from the tank. Courtesy of B. Beal.

Figure 2.5 Post-set rearing tray in a protected salt pond. Courtesy of B. Beal.

Figure 2.6 Raceway system. Courtesy of Bill Dewey.

Figure 2.7 Land-based nursery upweller. Courtesy of M. Castagna.

Figure 2.8 Upweller/downweller system. Note pump inserted into the central raceway. Courtesy of Dan Ward.

Figure 2.9 Nursery trays. These may be mounted on the bottom or dug into the bottom. Courtesy of B. Beal.

upwellers, simple raceways, mesh trays held in tanks, or any of these suspended near the surface in well-protected areas outside the hatchery system. For any system in the open, some cover is usually appropriate as it helps to reduce predation by birds. Dark plastic film will retard the growth of algae.

Downwellers or upwellers that are powered either by the tides or pump systems may be on land or suspended in floating trays. At this stage, upwellers are generally favored because the water enters through the mesh and flows up through the seed clams. This provides three advantages: (1) the mesh keeps out larger particles and most potential predators; (2) most sediments and feces are carried out of the unit; and (3) all juveniles receive a nearly equal water flow. In these systems, water enters the upweller through the bottom and exits through a drain near the top. The drain may be fitted with an elbow that can be adjusted to regulate the water flow. Even though weekly to biweekly changes and cleaning are required, upwellers have the advantage of relative ease of maintenance and the assurance that individuals receive equal access to the flow. Fouling control for those species that can close tightly can be achieved by immersing the clams in freshwater for a period of time once a week during cleaning. In warm areas, the culturist needs to be sure that temperature shock from the typically colder fresh water is not an issue.

For those species that have been shown to do better with substrate such as *Macrocallista* spp., *Meretrix* spp., and *Mactromeris* spp., sediment can be added to raceways, or the organisms plus sediment can be placed in trays and added to the flowing water system. Appropriately designed raceways generally allow uniform sheet flow of the water to proceed horizontally down the length of the shallow container, but this means the individuals at the raceway head receive water before those at the discharge end. This effect can be mitigated during the sorting process when the clams are cleared of sediment, sorted by size, and reintroduced into the cleaned raceway. One advantage of raceways is that through time, sediments accumulate and cover the clams. This accumulation prevents most fouling organisms from attaching to the clam shells. Even with sediments, it is useful to control fouling by flooding the raceway with freshwater for a period of time once a week using the precautions indicated above.

Another method used by some hatcheries is to place seed (3 mm) in trays lined with window screening that are then floated on the surface of a protected cove or embayment. The small mesh can become fouled easily, but the addition of grazing snails (for example, periwinkles such as *Littorina littorea*) can help reduce fouling. In addition, covering the top of the tray with a piece of thin, black plastic sheeting can reduce algal fouling and potential predation by gulls and other shorebirds. Trays remain at the nursery site until clams reach plantable sizes of 12–15 mm shell length (SL).

Some growers who have their own hatcheries overwinter seed in very dense (hundreds per square foot) nursery plantings under fine mesh in the field. Because of the fine mesh, these plots require very close attention to avoid clogging of the mesh by sediments or fouling organisms. The plots may be harvested in the early spring and the seed are then placed back on raceways to grow to planting size. In other areas, usually in areas with colder water, growers will make a shallow water nursery area that is enclosed with lumber embedded in the bottom and covered with mesh. In some instances, these systems are maintained through the winter and most of the summer to assure that very large seed are available for planting in the mid to late summer of the second year. Other successful overwintering methods include the use of zippered, mesh bags (45 cm × 45 cm) to which high densities (2–3 kg) of juvenile clams such as *Mya arenaria* or *Mercanaria mercenaria* are added. Bags are then placed in wire (2.5 cm aperture) cages that either can be submerged in a flow-through tank in a hatchery or attached to a floating dock. Field use of these methods requires close attention to be sure that the mesh does not become occluded and thus reduce flow to the animals.

Both land-based and field nurseries have advantages and disadvantages. Field-based systems have limited equipment, and water exchange is free, but predators and vandals can be problems, the systems to prevent algal grown can be difficult to maintain, and sorting and cleaning are very difficult. In addition, if ice forms over the winter, it can cause losses of both equipment and animals. Land-based or in-the-water flow-through systems have substantial up-front equipment costs, and ongoing operational costs such as electricity. The advantages are predator control, ease of cleaning and grading the animals, and greater protection from human tampering.

Field grow-out

Nearly all grow-out of clams is in the intertidal or shallow subtidal areas that may or may not require leases. In the US, most states require leases. Acquiring a lease can be time consuming, so anyone interested in starting a clam farm should consult their local Cooperative Extension (US), state, or other appropriate regulatory agency for information on the particular area.

Eventually, all clams are placed in the natural environment to grow to market size. This is because it is cost prohibitive to raise algae or pump water to the larger organisms. While many techniques depend upon species and location, the primary purpose for all methods of the field grow-out is to retain the clams and protect them against predation. Meticulousness during installation and careful maintenance of these protection devices, especially during the first year, are essential to prevent unacceptably high losses. The primary consideration for planting is the size of the clams to be planted. Seed size also affects planting density (Table 2.3) and the size of mesh to be used. Site selection for the grow-out area is an important part of any successful clam aquaculture venture. The area should be relatively protected from strong wave action and in an area that is accessible. It is obvious that areas with high boat traffic, high human use, and/or are subject to ice scour are not suitable, and some species may be susceptible to high summer temperatures on intertidal flats. Bottom type is also an important consideration. Sandy bottoms with ripple marks are usually not useful grow-out sites because the ripple marks indicate the substrate is highly mobile. Seagrass beds are protected by regulation and are generally also off limits. Most clams are grown in the lowest part of the intertidal zone or in areas with water that is no deeper that mid chest high at low tide. Shallow water depth makes it relatively easy to install and maintain the protection devices. The deeper parts of this zone are used where soft mesh bag culture (Figure 2.10) is practiced, and shallower areas are utilized where the screen mesh is stretched over the bottom (Figure 2.11). The best bottom types for stretched screen mesh culture are typically sandy muds or muddy sands with a small amount of shell, but moderately firm mud bottoms may also be considered. Bag culture also takes place on sediments with a high proportion of fines. In some areas of the Pacific Northwest, intertidal sediments with sandy gravels have been used. Some species such as *Meretrix* spp. may require specific sediment types, but most clams can be grown in a wide range of sediment types.

There are five basic types of grow-out containment systems: screen systems, pens, trays, soft bags, and a specialized system of tubes used for geoduck culture. These containment or protection

Table 2.3 Planting density and size. Blank spaces signify no information.

Species	Planting density (clams m^{-2})	Planting size (mm)	Reference
Cyrtopleura costata		10–12	Gustafson et al., 1991
Ensis leei			
Leukoma staminea			
Macrocallista nimbosa	410–645		Sturmer et al., 2011
Mactromeris polynyma	236	23	Davis, 1995
Mercenaria mercenaria	430–645	10–15	Castagna, 2001
Meretrix spp.		10	
Mya arenaria	430–645	8–15	Beal, 2011
Panopea generosa	20–40	5–10	
Paphia undulata			
Paratapes undulatus			
Ruditapes philippinarum	325–645	10–15	Goulletquer, 2005; Toba et al., 1992
Sinonovacula constricta			
Solen marginatus	400	19	da Costa, 2009
Spisula solidissima	500–800	10–15	

devices rely upon some form of mesh to protect the clams, allow them to burrow, and provide access to the water for feeding. Mesh sizes are typically just slightly smaller than the clam seed. Some growers use a fine mesh (6.4 mm or ~¼ inch) that allows planting of smaller seed, but this requires extra attention because fouling organisms can easily clog the mesh. In most cases, mesh screens with a nominal size of about 10 mm (~ 0.5 inch) with 10 to 12 mm seed appears to be an optimal trade-off between seed size and fouling control.

The most common system is that of mesh screening stretched over the bottom. These may be any size, but are typically the manufactured width of the mesh is utilized and the length is determined by the growers needs. In new sites, the bottom may be prepared by removing any

Figure 2.10 Soft bag system. Note corner stakes to hold bags on the bottom, and corner for access. Courtesy of Leslie Sturmer.

Figure 2.11 Grow-out screen protective system. Note edges thoroughly embedded in the bottom. Courtesy of D. Murphy.

objects that might snag the mesh or cause it to be torn. Most installations require some form of edging along the mesh to help it remain in the bottom. This may be lead line, PVC pipe with rebar inside or tubular mesh bags filled with sand. In all cases the plot is marked out, and if new, raked to remove larger clams, and debris. The seed are then broadcast over the bottom before the mesh is installed. If the area is intertidal, this can be done at low tide when the area is bare of water, but in deeper areas the plot must be outlined with stakes and the seed carefully dispersed. Once the seed have been installed the mesh is stretched over the sediments (in deeper sites working with the tidal flow can be important). The edges of the stretched screen must be embedded in the bottom and sealed to prevent predators (mostly crabs) from entering. Methods to embed the mesh depend upon the site and bottom type. In soft muds the mesh can simply be forced into the bottom by walking along the periphery of the screening. With this method, it is important that all of the fringing screen is covered so it does not work its way out of the bottom. On harder bottoms it may be necessary to dig a trench around the plot where the edge of the screen is placed and then back-filled to secure the screen. If the sides are held in place with rebar or leadline that material must be embedded in the sediments as well or crab predators will burrow under any exposed edge. Some growers place floats under the mesh so it extends above the bottom when the tide is in, but this is usually not necessary if the screen is stretched tight across the substrate. In areas with high levels of barnacle or oyster recruitment the floats can become covered with these organisms and chafe through the mesh exposing the clams to predators. There are some sites where large intertidal flats of stable bottom are available that can support large equipment such as tractors. In these sites, specialized equipment can be utilized to plant the clams, spread and imbed the mesh in long rolls. In any system constant vigilance to the integrity of the mesh screening is essential.

Pens and trays are essentially screened systems covering the top and all sides but without (pen) and with (tray) a bottom. They are usually made of materials that are rigid, at least on the sides, and installed in the lower part of the intertidal. Tray systems are usually designed to completely encase the growing clams and often contain a system that allows them to be lifted onto a barge or other large vessel. Since pens lack a bottom, they must be harvested with conventional hand methods.

In deeper water the mesh may be in the form of a 1.2 m × 1.2 m (~4 ft × 4 ft) bag in which the clams are packed (Figure 2.10). The bag with the clams is placed on a prepared piece of bottom and held in place with some form of a pinning system at the corners. The mesh allows sediment to enter the bag as the clams grow. This system has been very successful in some areas and not useful in others.

The methods listed above are typical for many areas and species; however, some clam species require special handling. In some areas of China where *Meretrix* spp. are cultured, clams are placed in the center area (approximately one-third of the total area) of a large pond and the area for planting the clams is covered with sand and a fence of fine mesh that is embedded in the bottom extends well above the water surface and encircles the entire clam growing site. Another unique system is utilized for the high value geoduck clam in the Pacific Northwest (Figure 2.12). In this system, PVC pipes are installed in the intertidal zone and a small number of seed clams are placed in each tube. The tubes are then covered with a mesh, either individually or as a group to protect the seed from predators. Once these clams have grown and have burrowed deep in the sediments (1 to 2 years) both the mesh and pipes are removed (Figure 2.13). Lastly, at least one commercial operation prepares a deep-water site by dredging to remove debris, predators, and large clams. This prepared bed is seeded with large seed and left unprotected. While survival is lower with this method, the large scale and lack of other equipment offsets the losses.

All screen meshes must be inspected and maintained regularly, typically once per week, to ensure the edges are sealed into the bottom, there are no rips, the mesh has not been covered with sediment, and that fouling is not excessive. Fouling organisms (macroalgae, tunicates, mussels, gastropod eggs, barnacles, and other species; see Chapter 16) may attach directly to the mesh or may be tube dwelling organisms such as some amphipods that are able to take advantage of the stabilized sediments to construct their tubes. The tubes extend above the mesh and

Figure 2.12 Protecting geoduck clam seed in the intertidal zone of Washington State. Cages are kept in place for 1 to 2 years of a 4 to 6 year cycle. Courtesy of J. Davis.

Figure 2.13 Geoduck grow-out for years 2–6 after tubes protecting seed have been removed. Courtesy of Bill Dewey.

may reach a height that prevents clams from accessing the water. Many growers have adapted squeegees and brushes to remove fouling gently without damaging the mesh. In extreme cases, the entire screen can be removed and replaced. In areas subject to ice, protective netting should be removed for the winter over larger clams. For fall-planted small seed in areas that are less susceptible to winter ice, the mesh may have to be left on because ducks are significant predators of small clams.

Harvesting

Harvest is an essential part of any culture, and the methods utilized for harvest depend upon the size of the culture operation, methods of protecting the growing clams and local regulations. It is important for the culturist to know the local regulations on size at harvest and how they apply to the cultured organism. In some areas, size limits of wild harvest and aquaculture-produced clams for sale are the same, but in some jurisdictions these can be different.

For those species reared in mesh bags, harvest requires removing the clips holding the mesh on the bottom and retrieving the bag into the boat. Retrieval may be manual or assisted with a winch.

Clams reared under screen mesh, are typically harvested with some form of rake with an attached basket. The clams are extracted from the sediment with the rake teeth and then collected in the attached basket, brought to the surface and dumped into some form of container. For those species that burrow deeply into the sediments, such as *Mya arenaria* or *Ensis* spp. the 'rake' or 'hoe' may be used to turn over the top sediment layers followed by removing the clams by hand.

Figure 2.14 Mechanical harvesting of Manila clam plots. Courtesy of Bill Dewey.

In other areas, clams may be harvested with a suction device that consists of a pump, attached hose and a suction head. The water is pumped into the suction head through a narrow pipe, and directed toward the exit. Owing to the Venturi effect, the water flow entrains water, sediments, and clams which are collected in a bag or directed onto a conveyor system and into containers.

In areas with a stable sediment and large areas under mesh, the mesh is removed and the clams may be extracted with a device, derived from agriculture bulb harvesting, attached to a tractor. (Figure 2.14). The clams are moved onto a conveyor belt and into containers.

Deep-dwelling species such as geoducks require a method that avoids damage of the clam. A device is used consisting of a high-volume, low-pressure pump and a hose that attaches the pump to a section of metal pipe about 1 m long. The jet of water exiting the pipe is used to liquefy the sediment and the clam is removed by hand.

Most harvested clams are washed, graded by size, and smaller clams may be returned to the growing area for further growth. Grading is typically done by hand or, for clams with a thick shell, with a system in which the clams move down a rotating incline with a slot that gets progressively larger. The clams that drop through the slot, and are channeled directly into containers for sale.

Disease

Disease is covered in detail in Chapter 15, but a brief discussion of those diseases that have been associated with clam species is presented here. Exclusive of hatchery-related diseases that are typically associated with bacterial infections from species of the genus *Vibrio*, clams have not been subject to the variety of diseases that are typically associated with oysters and other species. Three important clam diseases have been reported: Quahog Parasite Unknown (QPX) in *Mercenaria mercenaria; Vibrio tapetis* and *Perkinsus olseni* in *Ruditapes philippinarum* and *R. decussatus*; and gonadal neoplasia in *Mya arenaria.*

QPX is associated with inappropriate stocks, or poor husbandry of the grow-out of hard clams in the northern part of the species's range. The disease has not been reported south of Virginia. In addition, hatchery-reared clam seed from southern stocks (those from Virginia and south) have been shown to be particularly susceptible when grown in northern areas. Other than proper stock selection and good husbandry, there is no known cure for this disease.

Brown ring disease in the Manila clam is caused by *V. tapetis*. The disease has caused significant losses of Manila clams that were imported into Western Europe from stocks that were introduced into the Pacific Northwest of the US with Pacific oysters. This disease does not appear to affect the native European littleneck clam *R. decussatus*, nor has it affected Manila clams in the Pacific Northwest or across most of its native range in Asia. As with QPX, this disease appears to have some genetic component that makes the clam susceptible when other stressors are present. Only recently has *P. olseni* been implicated in causing reduced populations of wild Manila clams in Japan (Waki et al., 2018).

Factors causing hematopoietic neoplasia in *M. arenaria* are unknown, but severe outbreaks have been reported in most areas where this species is endemic. In spite of these diseases, the impact of disease related mortality in clams has been less than that in oysters, and have been controlled by husbandry or use of appropriate stocks. In the future, development of disease resistant stocks will be important for clam aquaculture expansion.

Acknowledgments

Many individuals provided information for this article and in particular we would like to thank Chris Davis, Joth Davis, Bill Dewey, Leslie Sturmer, and Ximing Guo for substantial contributions. In addition, the content was substantially improved by the editorial efforts of Sandra Shumway.

References cited and further reading

Baldwin, R.B., Mook, W., Hadley, N.H., Rhodes, R.J., and DeVoe, M.R. (1995) *Construction and Operations Manual for a Tidal-Powered Upwelling Nursery System*. S.C. Sea Grant Consortium, Charleston, SC 29401.

Beal, B. (2011) Pictorial manual for the culture and grow-out soft-shell clams: a practical guide to farming clams in Maine's intertidal zone. University of Maine, Machias, ME, Downeast Institute, Beals, ME.

Beattie, J.H. and Goodwin, C.L. (1993) Geoduck culture in Washington state. Reproductive development and spawning. NOAA Tech Report, 106: 55–61.

Castagna, M. (2001) Aquaculture of the hard clam, *Mercenaria*. In: Kraeuter, J.N. and M. Castagna (eds) *Biology of the Hard Clam. Developments in Aquaculture and Fisheries Science*. Elsevier, New York, pp. 675–697.

da Costa, F. (2009) Culture potential of the razor clam *Solen marginatus* (Pennant, 1777). *Aquaculture,* 288: 57-64.

da Costa, F. (ed.) (2013) *Clam Fisheries and Aquaculture*. Nova Science Publishers, Hauppauge, NY.

da Costa, F. and Martinez-Patino, D. (2009) Culture potential of the razor clam *Solen marginatus* (Pennant, 1777). *Aquaculture*, 288: 57–64.

Davis, C. (1995) Suitability of the Stimpson's Surfclam (*Mactromeris polynyma*) for aquaculture development in northern New England. Final Report NSF-SBIR Award # 95606943. Phase I.

FAO (2009) FishStat Plus. Aquaculture production 1950–2007. Available at: www.fao.org/fishery/statistics/software/fishstat/en.

FAO (2018) *FAO Yearbook. Fishery and Aquaculture Statistics 2016/.* FAO, Rome.

Flanagan, M.P. (2013) Investigation of early development and importance of sediment choice in the hatchery production of razor clams, *Ensis directus*. Honors thesis, University of Maine. Orono. Honors College. 5–2013.

Goulletquer, P. (2005) Cultured Aquatic Species Information Programme. *Ruditapes philippinarum*. FAO Fisheries and Aquaculture Department [online]. Rome. Updated 1 January 2005.

Gustafson, R.G., Creswell, R.L., Jacobsen, T.R., and Vaughan, D.E. (1991) Larval biology and mariculture of the angelwing clam, *Cyrtopleura costata. Aquaculture*, 95: 257–279.

Hadley, N.H., Manzi, J.J., Eversole, A.G., Dillon, R.T., Battey C.E., and Peacock, N.M. (1997) A *manual for the culture of the hard clam* Mercenaria *spp. in South Carolina*. S.C Sea Grant Consortium, Charleston, SC 29401.

Helm, M.M., Bourne, N., and Lovatelli, A. (comp./ed.) (2004) Hatchery culture of bivalves – a practical manual. FAO Fisheries Technical Paper. No. 471. FAO, Rome.

Kaiser, M.J., Edwards, D.B., and Spencer, B.E. (1996) Infaunal community changes as a result of commercial clam cultivation and harvesting. *Aquatic Living Resources*, 9: 57–63.

Kaiser, M.J., Burnell, G., and Costello, M. (1998) The environmental impact of bivalve aquaculture: a review. *Journal of Shellfish Research*, 17: 59–66.

Leavitt, D.F. (2010) Biology of the Atlantic jackknife (razor) clam (*Ensis directus*, Conrad 1843). NRAC Publication 217–2010. University of Maryland, College Park, MD.

Luckenbach, M.W., Kraeuter, J.N., and Bushek, D. (2016) Effects of clam aquaculture on nektonic and benthic assemblages in two shallow water estuaries. *Journal of Shellfish Research*, 35(4): 757–775.

Manzi, J.J. and Castagna, M. (eds.) (1989) *Clam Mariculture in North America*. Developments in Aquaculture and Fisheries Science, 19. Elsevier Scientific, Amsterdam.

Marshall, R. (2012) Broodstock conditioning and larval rearing of the geoduck clam (*Panopea generosa* Gould, 1850). PhD Dissertation, University of British Columbia, Vancouver, BC.

McDonald, P.S., Galloway, A.W.E., McPeek, K.C., and VanBlaricomb, G.R. (2015) Effects of geoduck (*Panopea generosa* Gould, 1850). Aquaculture gear on resident and transient macrofauna communities of Puget Sound, Washington. *Journal of Shellfish Research*, 34: 189–202.

Munroe, D. and McKinley, R.S. (2007) Commercial Manila clam (*Tapes philippinarum*) culture in British Columbia, Canada: the effects of predator netting on intertidal sediment characteristics. *Estuarine, Coastal and Shelf Science*, 72: 319–328.

Ojea, J., Pazos, A.J., Martinez, D., Novoa, S., Garcia-Martinez, P., and Sanchez, J.L. (2008) Effects of temperature regime on broodstock conditioning of *Ruditapes decussatus. Journal of Shellfish Research*, 27: 1093–1100.

Sarkis, S. and Lovatelli, A. (2007) Installation and operation of a modular bivalve hatchery. FAO Fisheries Technical Paper. No. 492. FAO, Rome.

Shumway, S.E. (ed.) (2011) *Shellfish Aquaculture and the Environment*. Wiley, Ames, IA.

Spencer, B.E. (2002) *Molluscan Shellfish Farming*. Blackwell, Malden, MA.

Spencer, B.E., Edwards, D.B, and Millican, P.F. (1992) Protecting Manila clam (*Tapes philippinarum*) beds with plastic netting. *Aquaculture*, 105: 251–268.

Spencer, B.E., Kaiser, M.J., and Edwards, D.B. (1996) The effect of manila clam cultivation on an intertidal benthic community: the early cultivation phase. *Aquaculture Research*, 27: 261–276.

Spencer, B.E., Kaiser, M.J., and Edwards, D.B. (1997) Ecological effects of intertidal manila clam cultivation: observations at the end of the cultivation phase. *Journal of Applied Ecology*, 34: 444–452.

Sturmer, L.N., Scapa, J., Adams, C.M., and Creswell, R.L. (2011) Species diversification in Florida shellfish aquaculture: nursery and growout culture of the sunray venus clam. Final Report. Florida Sea Grant Program Project R/LR-A-45 Final Report 2008–10.

Thiet, C.C. and Kumar, M.S. (2008) Clam (*Meretrix lyrata*) hatchery manual. Aquaculture Research Sub-Inst. North Cent. (ARSINC) and South Australian Res. Dev. Inst (SARDI).

Toba, D.R. Thompson, D.S., Chew, K.K., Anderson, G.J., and Miller, M.B. (1992) Guide to Manila Clam culture in Washington. Washington Sea Grant Program, University of Washington, Seattle, WA.

Van Blaricom, G.R., Eccles, J.E., Olden, J.D., and McDonald, P.S. (2015) Ecological effects of the harvest phase of geoduck (*Panopea generosa* Bould, 1850) aquaculture on infaunal communities in southern Puget Sound, Washington. *Journal of Shellfish Research*, 34: 171–187.

Waki, T., Takahashi, M., Eki, T., Hiasa, M., Umeda, K., Karankawa, N., and Yoshinaga, T. (2018) Impact of *Perkinsus olseni* infection on a wild population of Manila clam *Ruditapes philippinarum* in Arisaka Bay, Japan. *Journal of Invertebrate Pathology*, 1513: 134–144.

CHAPTER 3

Culture of giant clams

Thane A. Militz and Paul C. Southgate

Introduction

Giant clams (Cardiidae: Tridacninae) are the largest living bivalve molluscs. They inhabit shallow water habitats throughout the tropical Indo-Pacific, where they are heavily fished for meat (adductor muscle and mantle tissue), as curios (shell valves), and to supply the marine aquarium trade (live juveniles). Heavy exploitation of giant clams has resulted in local extinctions in parts of their range and prompted the listing of all Tridacninae species by the Convention on International Trade in Endangered Species (CITES; Appendix II) since 1985, and on the International Union for Conservation of Nature (IUCN) Redlist since 1986. Depletion of wild populations of giant clams prompted research into the biology and culture of giant clams in the 1970s and 1980s, and culture methods were well established for the major species by the late 1980s. Initial culture efforts focused on the production of larger species targeted for meat and curios. But advancements in private aquarium technology, and increased demand for live juvenile giant clams in the marine aquarium trade, shifted production focus toward smaller species of giant clams with brightly colored mantles (Fig. 3.1). Yet

Figure 3.1 Common species traded for the international aquarium market include: (A) *Tridacna maxima*; (B) *T. noae*; and (C) *T. squamosa*. All have attractive mantle patterns and vibrant colors which are key determinants of demand and value.

despite more than 30 years of know-how for giant clam culture, and the relatively simple culture methods required, current global production of cultured giant clams is estimated to be only around 150,000 individuals annually, with approximately 70% of this supplying the marine aquarium trade.

Commercially cultivated species

There are 12 recognized species of giant clams within two genera (*Tridacna* and *Hippopus*), and seven species are routinely cultured for various markets (Table 3.1). With the majority of global giant clam production targeting aquarium markets, the more colorful species of *Tridacna maxima*, *T. derasa*, *T. squamosa*, and *T. crocea* are the most commonly produced and internationally traded species. The cumulative production of *Hippopus hippopus*, *H. porcellanus*, and *T. gigas* accounts for only around 20% of global giant clam production, with very limited international trade. These three species are primarily produced for restocking depleted wild populations as part of non-commercial government-led conservation programs. Despite *T. gigas* having the highest meat yield potential and having become locally extinct at many sites within its range, this species contributes less than 5% of global cultured giant clam production.

Table 3.1 Species of giant clam routinely cultured and their target markets.

Species	Aquarium	Meat and curios	Restocking
Hippopus hippopus	✓		✓
Hippopus porcellanus			✓
Tridacna crocea	✓		
Tridacna derasa	✓	✓	✓
Tridacna gigas	✓	✓	✓
Tridacna maxima	✓		
Tridacna squamosa	✓		✓

Unique biology requires unique culture methods

All species of giant clams are restricted to shallow tropical waters. They harbor endosymbiotic dinoflagellates, *Symbiodinium* spp. (zooxanthellae in the vernacular), within their tissues which contribute to the nutrition of the host via photosynthesis. Giant clams offer great potential for aquaculture since the symbiotic algae render giant clams almost self-sustaining, from a nutritional perspective, requiring only sunlight and natural seawater during the grow-out phase of culture.

Giant clams are cultured using land-based hatchery and nursery facilities with later transfer to ocean-based grow-out systems. Culture systems must account for husbandry requirements of both the giant clam and their zooxanthellae symbionts, by providing seawater with suitable chemistry and adequate photosynthetically active radiation (PAR) to support zooxanthellae (Table 3.2). The requirement for adequate PAR to reach the zooxanthellae residing in host tissues necessitates low giant clam culture densities along a horizontal plane, such as a tank bottom or seabed, to prevent self-shading and to maintain orientation toward a light source. For this reason, large bottom-surface area tanks are used exclusively for hatchery and nursery culture of giant clams (Fig. 3.2).

Table 3.2 Seawater chemistry and photosynthetically active radiation (PAR) parameter ranges suitable for culturing giant clams.

Parameter	Range
Temperature	26–32°C
Salinity	32–36 g/L
pH	7.9–8.2
Dissolved oxygen	>4 mg/L
Total ammonium nitrogen*	≤40 µM
Nitrite	<0.1 mg/L
Hydrogen sulfide	<1 µg/L
PAR*	500–1500 µE/m²/s

Note: **range varies based on size of giant clams*

Permits and permissions

Given the CITES status of giant clams, international shipments of giant clam products require an export permit from the relevant government authority to indicate that such export will not be detrimental to the survival of the species, that the specimens were not obtained in contravention of any state laws, and that they are shipped in a manner that minimizes the risk of injury. State laws on the harvest, culture, and trade of giant clams vary greatly among countries, and may even vary among jurisdictions within countries. Thus, prior to commencing giant clam farming, it is integral to understand local and international laws concerning harvest of broodstock, land- and ocean-based farming, and sale of giant clam products. Additionally, it is prudent to familiarize oneself with the respective permit application pathways and the feasibility of conducting commercial business within the estimated time frames of processing the required permits. It is common for multiple government agency approvals to be required for international shipments of live giant clams and their products.

Sources of stock

Spat collection

Spat collection is a commonly used method to obtain culture stock of bivalve molluscs such as edible oysters (Chapter 1) and pearl oysters (Chapter 9). It involves deploying a suitable material

Figure 3.2 Flat-bottomed, shallow tanks are a characteristic of giant clam culture facilities. They are commonly used for egg incubation, larval culture, and nursery culture in succession. The large surface area of the tanks is important for benthic juveniles once symbiosis with zooxanthellae is established. The masonry tanks shown here have retractable shade cloth covers. Courtesy of Martin Selch, Micronesia Management & Marketing Enterprises.

Figure 3.3 Spat collector after a one-year deployment at Reao Atoll, eastern Tuamotu archipelago, showing *Tridacna maxima* recruits with mean size of 1.5 cm (maximum size 4 cm) at a density of 348 clams per square meter. Courtesy of Vitore and Rosalie Teaka, and George Remoissenet.

or 'spat collector' to the ocean to provide a substrate for the settling larvae of the target species. Juveniles are later removed from the spat collectors to become farm stock. This method is rarely applied in giant clam culture but has been used successfully in areas of French Polynesia where very high-density aggregations (up to 500 clams/m^2) of *Tridacna maxima* naturally occur. Spat collectors are flat and may be several dozen meters in area. They are constructed from a wooden frame over which shade cloth and plastic netting are stretched. They are deployed horizontally, to present a large flat substrate, 1–2.5 m below the ocean surface, and are held in place with ropes and buoys. In this environment, spat densities of up to 400/m^2 can be achieved on spat collectors, with subsequent growth rates of around 3 cm/year reported (Fig. 3.3). Resulting giant clams are cultured to a size of 4 cm or more for export into the aquarium trade or cultured to larger sizes for human consumption.

Hatchery production

Broodstock selection

Giant clam culture relies on hatchery production of juveniles, which requires the use of sexually mature giant clams (that is, broodstock) to provide gametes. Broodstock can be selected from populations of wild or cultured giant clams. Factors influencing the selection of individual giant clams as broodstock depend on the farm's production aims. Farms producing giant clams for the aquarium trade must consider the patterning and color of the siphonal mantle of broodstock, which will influence demand and value of resulting progeny (Fig. 3.1). Farms producing giant clams for restocking wild populations will want to ensure that the genetic diversity of broodstock reflects the existing wild population into which the progeny will be stocked.

Giant clams are protandrous, simultaneous hermaphrodites and sufficient large broodstock are required to ensure successful spawning of eggs. Assessment of reproductive condition may be achieved using a gonad tissue biopsy, and the various stages of reproductive maturity have been described for most species; however, biopsy sampling can prove hazardous to giant clams because vital organs, including the pericardium overlaying the gonad, may be damaged during tissue extraction, and this means of assessing reproductive condition is only recommended in situations where surplus broodstock are available.

Because a single giant clam is capable of spawning sufficient quantities of eggs (up to 500 million) to stock an entire hatchery, limited numbers of broodstock are required for hatchery production. Despite this, a minimum of 10 broodstock is recommended to ensure adequate release of sperm and spawning of eggs occurs from different individuals to enable cross-fertilization, and to allow for selection of the best quality gametes. Greater quantities of broodstock (>20) are typically required when producing giant clams for restocking wild populations to ensure adequate genetic diversity in resulting progeny.

Collecting broodstock

In some situations, it may be possible to source broodstock from wild populations of giant clams close to the hatchery. Collection of adult *Tridacna gigas* and *T. derasa* individuals requires use of a rope harness or cargo net which, when connected to a winch or pull rope, can be used to haul clams aboard a boat for transfer to the culture facility (Fig. 3.4). Smaller individuals can be collected by hand while free diving. But for species adhering to solid substrata via byssus (for example, *T. crocea*, *T. maxima*, *T. squamosa*), it will be necessary to either cut the byssal threads with a knife or to chisel out the portion of substrate to which the clam is attached; pulling these species away from the substrate by force risks damaging the byssal apparatus. Once emerged from water, giant clams should be laid on their side to minimize mechanical stress on the visceral mass, which can be particularly damaging during a rough boat trip back to the hatchery. For return trips to the hatchery of less than 2 hours, the broodstock may be left on the boat deck, though longer transport times requires accommodating clams in a container of seawater that is regularly exchanged. Care should be taken to ensure that clams do not experience abrupt increases in temperature that could induce accidental spawning through heat stimulation (see text below). Sourcing broodstock from cultured giant clams held at ocean-based locations requires the same considerations.

Figure 3.4 Larger broodstock, such as this *Tridacna gigas*, generate logistic issues for collection and transportation to land-based culture facilities and transfer to holding tanks. In this case, the broodstock clam is placed in a rope harness for transfer to a broodstock tank and is overseen by one of the pioneers of giant clam culture Dr. Richard 'Rick' Braley. Courtesy of Rick Braley, Aquasearch, Australia.

Broodstock conditioning

Manipulation of environmental conditions, particularly water temperature and food availability, is commonly practiced in bivalve mollusc culture to improve broodstock gamete quality and yield over a relatively short period. Methods for conditioning giant clam broodstock have received relatively little research attention by comparison. This is partly due to the logistics and costs of adequately manipulating the volumes of water required to hold a sufficient number of these large bivalves. Additionally, all cultured species have tropical distributions and protracted spawning seasons that allow hatcheries to rely on the availability of mature broodstock under ambient conditions for the majority of the year.

Broodstock sourced from cultured or wild populations can be held year-round at ocean-based locations where ambient environmental conditions support growth and reproductive development. Locations that are 2–6 m deep, accessible by boat and where there is moderate coral growth, protection from breaking waves, and a stable substratum are generally suitable. Broodstock harvested from wild populations or retrieved from an ocean holding location during spawning season can immediately be induced to spawn without further conditioning. As spawning induction is non-destructive, spent broodstock may be returned to an ocean holding location to re-condition.

Land-based tanks receiving adequate sunlight and continuous exchange of filtered (ca. 50 μm) seawater can be used for interim holding of broodstock in the lead up to spawning induction (Fig. 3.5). It is important that water exchange is sufficient (one to two turnovers per day) and sunlight is partly reduced (PAR: 1000–1500 $\mu E/m^2/s$) through overhead structures (for example, shadecloth) to prevent accidental spawning through thermal stimulation or bleaching through loss of symbiotic zooxanthellae (Figs. 3.2, 3.4 and 3.6). Long-term holding of broodstock in land-based tanks is not recommended but may be necessary if the hatchery does not have access to suitable ocean-based locations for holding broodstock.

Spawning induction

After thorough cleaning of the outer shell surfaces of broodstock to remove fouling organisms and potential contaminants, broodstock clams are generally placed into separate spawning

Figure 3.5 Broodstock giant clams held in a flat-bottomed, shallow, fiberglass tank at Matangi Island, Vanua, Taveuni, Fiji. Courtesy of Claude Prevost, Civa (Fiji) Pearls Pte Ltd.

tanks containing static 1 µm filtered seawater. Plastic 60-L containers make suitable spawning tanks for most giant clam species, with at least two spawning tanks being utilized by each broodstock during spawning activity. Spawning induction of giant clams is normally stimulated by an injection of serotonin into the gonad (Fig. 3.7). A 2 mM concentration of serotonin is generally used and can be made by dissolving 0.17 g of crystalline serotonin (5-hydroxytryptamine creatine sulfate complex) in 250 mL of filtered seawater. Recommended dosages for broodstock of larger species (for example, *T. derasa*, *T. gigas*) range from 2–4 mL, while 0.5–1.0 mL is appropriate for smaller species (for example, *T. crocea*, *T. maxima*, *T. squamosa*, *H. hippopus*). Injection of serotonin into the gonad will typically elicit a spawning response within 10 minutes. If no response is obtained, a second injection of serotonin can be delivered.

Figure 3.6 Facilities used for giant clam culture must account for provision of adequate light, but also for the high light intensity and monsoonal rains in the tropical areas used for giant clam culture. Tanks are frequently covered by translucent roofing materials and shade cloth.

Like other bivalves, giant clams readily spawn in response to waterborne gametes released by other individuals. Introducing a suspension of ripe eggs or ova to the spawning tank near the incurrent siphon of the broodstock will stimulate spawning. Eggs can be sourced from an already spawning individual while ova can be obtained from a biopsy extraction or the gonadal material of a sacrificed broodstock. Exposure to ripe eggs or ova usually elicits a spawning response within 10 minutes.

Thermal stimulation has shown varying degrees of success in inducing giant clams to spawn,

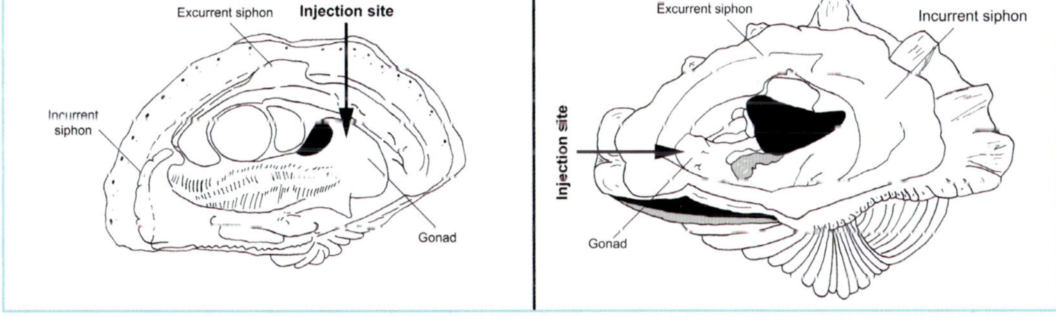

Figure 3.7 Diagrammatic representation of spawning induction of giant clams using direct injection of serotonin into the gonad. Care must be taken to minimize accidental damage to associated organs and the diagram shows the preferred injection sites for (A) *Tridacna* spp. and (B) *Hippopus* spp. Redrawn from Braley (1992).

Figure 3.8 Broodstock of *Tridacna noae* spawning eggs following induction with serotonin injection. Gametes are expelled following inflation of the mantle cavity with seawater and subsequent contractions of the adductor muscle to expel gametes from the excurrent siphon.

being most successful with broodstock sourced from populations that experience pronounced seasonal changes in seawater temperature. A rapid temperature change can be achieved by exposure to heated seawater (32–35°C) or by placing the giant clams on their side in full sunlight for 30 minutes to 1 hour exposure before returning them to the spawning tank. The release of sperm and spawning of eggs typically occurs 30 minutes to several hours after returning the giant clams to seawater of ambient conditions.

Regardless of the induction method chosen, contractions of the adductor muscle will prelude the start of spawning. Release of sperm occurs first which, after several minutes, will cease. Broodstock with ripe ovaries will then begin to spawn eggs (Fig. 3.8). Broodstock not yet sufficiently mature to spawn eggs can still be induced to release sperm, which may be necessary in situations with limited access to broodstock.

Survival and growth of progeny obtained from cross-fertilization is superior to progeny arising from self-fertilization. For this reason, once an individual clam has finished releasing sperm, it is generally transferred to a new spawning tank where the spawning of eggs will occur. Thorough cleaning of the outer shell surfaces and subsequent rinsing with filtered seawater occurs during transfer to the second spawning tank to prevent previously released sperm from contaminating the spawning tank used for egg collection.

It should also be noted that for larger species of giant clams, particularly *T. gigas*, it is not possible to easily transfer broodstock to a second spawning tank during spawning activity. In such circumstances broodstock may be held communally in a large spawning tank and their gametes collected in large plastic bags, as they are spawned, for removal to separate containers. Furthermore, to avoid the difficulties of transporting large broodstock of the larger species of giant clams (for example, *T. gigas* and *T. derasa*) to land-based hatchery facilities, they are sometimes induced to spawn in situ using serotonin injection. In this case, spawned gametes are also usually collected in large plastic bags, with use of SCUBA equipment, before transfer to the hatchery (Fig. 3.9).

Gamete fertilization

After spawning concludes, broodstock are removed from the spawning tanks and the isolated suspensions of sperm and eggs examined microscopically, to determine their quality and allow selection of the better-quality gametes for hatchery production. Preference should be given to sperm with high vitality and negatively buoyant spherical eggs that are relatively large in diameter (~100 μm; Fig. 3.10).

Fertilization should occur within 10 minutes following cessation of spawning. To fertilize

Figure 3.9 It is sometimes impractical to transport broodstock of the larger giant clam species to a land-based facility for spawning. Here, *T. gigas* at the Silaqui Giant Clam Ocean Nursery of the Marine Science Institute, University of the Philippines, are induced to spawn in situ using serotonin injection into the gonad. Resulting gametes are collected in large plastic bags for transportation back to the hatchery. Courtesy of Drs. Patrick Cabaitan and Cecilia Conaco, University of the Philippines.

the selected eggs, aliquots of the selected sperm suspensions are added, in small volumes, to the egg suspensions in the spawning tanks (for example, 1 mL of sperm suspensions per liter of egg suspension). In this manner, sperm suspensions of several different broodstock can be used to fertilize the eggs spawned from a single broodstock, taking care to avoid self-fertilization. After addition of sperm, the resulting egg/sperm suspension is gently agitated to ensure adequate dispersal of the sperm. Microscopic examination is used to determine the number of available sperm per egg and additional sperm can be added if necessary. It is important to minimize the number of sperm used to fertilize the eggs because excessive sperm may lead to polyspermy, affecting development of embryos and larvae with potentially negative water quality impacts. The success of fertilization is gauged 2 hours post-fertilization, with eggs commencing development as embryos (that is, two-cell stage or beyond) being considered successfully fertilized (Fig. 3.10). Good quality gametes should result in ≥85% of eggs being successfully fertilized, whereas batches of gametes with ≤50% fertilization success should be discarded.

Fertilized eggs are then left to settle for around 30 minutes, but if egg density is high (>100 eggs/mL), it will be necessary to divide the fertilized eggs among several containers to avoid oxygen depletion. Settled eggs are siphoned from the bottom of spawning tanks and through a 200 μm sieve mesh to remove debris; they are collected in buckets and transferred to egg incubation tanks.

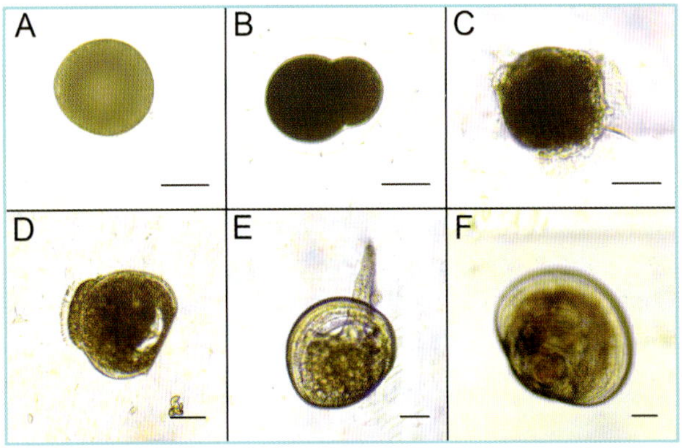

Figure 3.10 Development of *T. noae* embryos and larvae showing: (A) fertilized egg; (B) 2 hour old two-cell embryo; (C) 18 hour old trochophore; (D) 24 hour old D-stage veliger with extended velum; (E) 6 day old pediveliger with extended foot; and (F) 15 day old juvenile giant clam. Scale bars, 50 μm.

Egg incubation

Conical or cylindrical tanks (500–2000 L) with height and diameter of similar dimensions make ideal egg incubation tanks. Fertilized eggs are stocked into incubation tanks containing 1 μm filtered sea water to achieve a density of 20–25 eggs eggs/mL. Once stocked, water circulation within the incubation tank is provided by means of gentle aeration. Incubation tanks should be shaded and maintained in an insulated area to maintain a stable water temperature between 26 to 30°C. Owing to potentially high mortality of eggs and embryos during incubation, some farms administer a broad-spectrum antibiotic, such as streptomycin sulfate, at a concentration of 10 mg/L, during egg incubation.

Embryonic and early larval development of giant clams is similar to that of other bivalves but differs in terms of duration and aspects of larval morphology. The first motile stage is the trochophore, which is ciliated but lacks a shell and is unable to feed (Fig. 3.10). The shelled veliger stage is reached 24 hours after fertilization. They have a characteristic 'D'-shape shell and are known as D-stage or D-shaped veligers. They generally measure 130–170 μm in shell length and are relatively large compared to those of other cultured bivalves. Swimming veligers are 'selected' from incubation tanks 30–48 hours after fertilization. A partly submerged, large diameter 80–90 μm sieve mesh is used to concentrate swimming veligers that are siphoned from the water column of the incubation tank. Slower growing trochophores and larvae at the trophophore-veliger transition stage pass through the mesh, while moribund and dead larvae are left undisturbed at the bottom of the incubation tank. This 'selection' process ensures that only viable veligers are transferred to larval rearing tanks. Veligers retained by the mesh sieve are resuspended in a container of clean 1 μm filtered seawater, and their number estimated from microscopic counts of the larvae within multiple sub-samples removed from it. The volume required for addition to larval culture tanks to obtain the desired larval stocking density can then be calculated.

Farms lacking the necessary resources for egg incubation may stock fertilized eggs directly into larval rearing tanks at reduced densities (1–2 eggs/mL) without selection. This method is less hygienic because unfertilized eggs or dead embryos may accumulate on the bottom of the larval rearing tank, leading to increased bacterial load. Research with *T. gigas* has shown similar survival of D-stage veligers can be achieved regardless of stocking method chosen, but that there is a greater risk of mass mortality when selection does not occur.

Larval culture

Giant clams are atypical of bivalve molluscs in their larval rearing requirements. This is partly because the culture environment must consider the physiological requirements of both the giant clams and their zooxanthellae symbionts. Furthermore, larvae of giant clams have a relatively short planktonic larval period of four to seven days before settlement, followed by a protracted benthic larval period of another seven to ten days before larvae complete metamorphosis to become benthic juveniles. This development progression differs from most other cultured bivalves which generally have long planktonic larval phases before rapid metamorphosis into benthic juveniles after settlement.

Infrastructure

The larvae of giant clams are generally cultured in large (2,000–10,000 L), shallow (<1 m depth), flat-bottomed tanks of rectangular or circular shape, constructed from cement, fiberglass, plastics, or high density polyethylene (HDPE) liner laid within frames (Fig. 3.2). Giant clams take up their complement of symbiotic zooxanthellae during the larval phase and so the location of the larval rearing tanks must allow an adequate light intensity (PAR: 500–1000 $\mu E/m^2/s$) to reach the tank bottom. Because of the high electricity costs associated with illuminating tanks artificially, larval rearing tanks are commonly located outdoors or in greenhouse structures with transparent polycarbonate roofing. Greenhouse structures may be required at higher latitude locations or in areas receiving frequent heavy monsoonal rainfall to afford year-round production (Fig. 3.6). Larval rearing tanks should be filled with filtered (≤10 µm) seawater and provided with gentle aeration to facilitate water circulation within the tanks.

Husbandry

D-stage veliger larvae selected from incubation tanks are generally stocked into larval rearing tanks at a density of 1 larva/mL. If fertilized eggs were stocked directly into larval rearing tanks, no adjustments to the resulting larval density are made. Once stocked, larval rearing tanks must be carefully checked for bacterial strings that may develop in the water column and/or as patches of bacterial film (often white or pink in color) primarily on the tank bottom. If observed, the strings or patches should be removed through targeted siphoning. Prior to settlement, larval rearing tanks are most often maintained as static culture systems with no water exchange, using gentle aeration to provide water circulation (Table 3.3).

As early as 4 days post-fertilization, the number of larvae swimming in the water column will decrease as the planktonic veligers transition to a benthic existence (known as 'settlement' stage) (Fig. 3.10). Substantial mortalities typify the transition from veliger to pediveliger, and it is not uncommon for >90% of larvae to die if conditions on the bottom of the larval rearing tank are unhygienic or if larval energy reserves are inadequate. To estimate survival after settlement, a known area of the tank bottom may be siphoned through an 80–100 µm mesh sieve that retains pediveligers, and the contents examined using a microscope to count viable larvae.

Pediveligers are first apparent 5 days post-fertilization, alternately crawling on the bottom

Table 3.3 A generalized protocol for larval and early nursery culture of giant clams.

Days post-fertilization	Larval stage (behavior)	Rate of water exchange (per day)	Water filtration (µm)	Feeding	Zooxanthellae inoculation
1	Veliger (swimming)	None	–	No	No
2	Veliger (swimming)	None	–	Yes	No
3	Veliger (swimming)	None	–	Yes	Yes
4	Veliger (swimming/benthic)	None	–	Yes	No
5	Veliger + pediveliger (benthic)	1 turnover (optional)	≤10	Yes	No
6	Pediveliger (benthic)	1 turnover (optional)	≤10	Yes	No
7–9	Pediveliger (benthic)	1 turnover (optional)	≤10	No	Yes
10–15	Pediveliger + juveniles (benthic)	1 turnover (optional)	≤10	No	No
≥16	Juveniles (benthic)	1–2 turnovers	50	No	Optional

using their foot and swimming just off the bottom with a regressed velum (Fig. 3.10). Some farms commence water exchange of larval rearing tanks once the majority of larvae have settled (5 to 7 days post-fertilization) by gradual flow-through of filtered (≤10 µm) seawater at a rate of 100% of tank volume per day, using a screened (80–100 µm mesh) drain to prevent the loss of any remaining veligers in the water column. Beginning water exchange at settlement can help stabilize fluctuations in water temperature and salinity, and clear particulate food waste; however, if such factors are not of concern, farms may opt to continue static culture with gentle aeration for the entire larval rearing phase.

Pediveliger larvae may complete metamorphosis, indicated by the presence of ctenidia (Fig. 3.10), as early as 10 days post-fertilization. Completion of metamorphosis, however, is rarely uniform, and some pediveligers may persist in larval rearing tanks beyond 15 days post-fertilization. Completion of metamorphosis marks the end of larval rearing and the beginning of early nursery culture. As with the transition from veliger to pediveliger stages, the transition from pediveliger to juvenile clam may be accompanied by substantial mortalities. It is quite common for survival of fertilized eggs to juvenile clams to be less than 1%.

Feeding

Giant clam larvae are able to develop normally and metamorphose without an exogenous food source. Lipid is the principal energy reserve of larvae, and the major energy source for settlement and metamorphosis. Fertilized eggs of giant clams contain a high lipid content, with lipids comprising more than 60% of their total organic matter. The high lipid content of eggs is believed to facilitate development in the relatively nutrient-poor tropical waters where dissolved nutrient and particulate food availability is limited. These attributes support successful hatchery production of giant clams without nutrient or particulate food inputs. Despite this, it is widely accepted that provision of a particulate food source improves hatchery production. Feeding appears to be facultative in giant clam larvae and, although not obligatory in culture, is required for maximal larval growth and survival.

The veliger larvae of giant clams are demonstrable planktotrophs capable of digesting and assimilating ingested nutrients. Exogenous feeding begins between 24 and 48 hours post-fertilization and continues until settlement, when feeding activity noticeably declines. Bioenergetic studies and dietary trials strongly indicate that giant clam larvae cease feeding as pediveligers approach metamorphosis. The period of exogenous feeding in giant clam larvae is therefore restricted to the veliger and veliger–pediveliger transition periods, occurring 2 to 6 days post-fertilization. During this period of exogenous feeding, larvae will ingest a range of particulate foods, including unicellular micro-algae, yeasts, and formulated micro-diets.

The use of live micro-algae to feed veliger larvae of giant clams is a well-established practice. Veliger larvae readily ingest and digest golden-brown flagellates (Haptophyta) between 4 μm and 10 μm cell size, but selectively avoid ingestion of larger (7–20 μm) green flagellates (Chlorophyta) and diatoms (Bacillariophyta). For this reason, *Isochrysis* spp. are commonly used micro-algae in giant clam culture, fed to veligers either exclusively or together with *Pavlova* spp. Inoculation of larval rearing tanks with live micro-algae can occur immediately after stocking larval rearing tanks with selected veligers, or at 48 hours post-fertilization if fertilized eggs were stocked directly into larval rearing tanks. Optimal survival of veligers through metamorphosis has been achieved with densities of 1,000 to 100,000 cells/mL of *Isochrysis* spp.; however, densities toward the lower end of this range (for example, 1000–10,000 cells/mL) are recommended to limit fouling of the tank bottom with dead algal cells and to improve cost effectiveness.

Provision of live microalgae to giant clam larvae generally necessitates culturing large quantities of microalgae on-site at the hatchery. Considerable costs and technical resources are required to do this (see Chapter 13) and, for smaller farms or farms operating in remote locations, these requirements are major impediments. On this basis, a number of alternatives to live microalgae have been assessed as potential food sources for giant clam larvae. Commercially available highly concentrated microalgae has been shown to be suitable, eliminating the requirement for hatchery culture of live microalgae. Such products offer 'off-the-shelf' convenience and a nutritionally consistent larval food source. Successful production of *T. maxima, T. squamosa*, and *H. hippopus* were recently achieved when administering these products at similar cell densities as recommended for the feeding of live microalgae (Table 3.4). Because the microalgal cells composing these products are non-viable, sufficient aeration must be provided to larval rearing tanks to ensure that microalgal cells remain in suspension and available for larval ingestion.

Table 3.4 Suggested daily ration (cells/mL) of live microalgae for feeding giant clam larvae.

Days post-fertilization	*Isochrysis* sp. (cells/mL)	*Pavlova* sp. (cells/mL)
0	–	–
1	–	–
2	2000	2000
3	4000	4000
4	4000	4000
5	4000	4000
6	4000	4000
≥7	–	–

Zooxanthellae inoculation

Giant clams maintain ingested zooxanthellae cells within their zooxanthellal tube system that, once fully developed, extends from the digestive diverticular ducts of the stomach and terminates as a branching network in the siphonal mantle. The resident symbionts make substantial nutritional contributions to the host clam in the form of translocated photosynthates. The relative contribution of photosynthates to energy requirements of giant clams increases significantly with giant clam size. Even for juvenile giant clams of 0.1 g tissue weight, phototrophy has already become their most significant source of energy. For all species, except *H. hippopus,* the carbon obtained from photosynthates alone is sufficient to support the growth and metabolic requirements of these early juvenile clams.

As giant clams do not pass their symbionts directly to their progeny, the larvae or juveniles must acquire their compliment of symbionts from the environment. The easiest method of providing zooxanthellae to larval giant clams in culture is to extract zooxanthellae from the mantle tissue of adult giant clams and then inoculate larval rearing tanks with the resulting suspension of zooxanthellae. To extract zooxanthellae, pieces of mantle tissue cut from adult giant clams are washed in freshwater, then lacerated repeatedly to damage the zooxanthellal tube system of the mantle tissue, allowing the zooxanthellae to be washed out. By washing lacerated mantle tissue with filtered seawater and passing the resulting suspension through a 30–50 μm mesh sieve, larger pieces of dislodged mantle tissue and connective tissue can be removed from the zooxanthellae suspension (Fig. 3.11). Further isolation of zooxanthellae cells from mantle tissue debris can be achieved by centrifugation. This concentrates zooxanthellae into a pellet which can be resuspended in seawater before inoculating larval rearing tanks. Many farms, however, lack the necessary equipment for centrifugation, and instead inoculate larval rearing tanks with the sieve-filtered zooxanthellae suspension.

As an alternative to extracting zooxanthellae from mantle tissue, a monoculture of zooxanthellae can be maintained at the hatchery from which zooxanthellae are harvested as required to inoculate larval rearing tanks. Zooxanthellae cultures can be established by obtaining stock

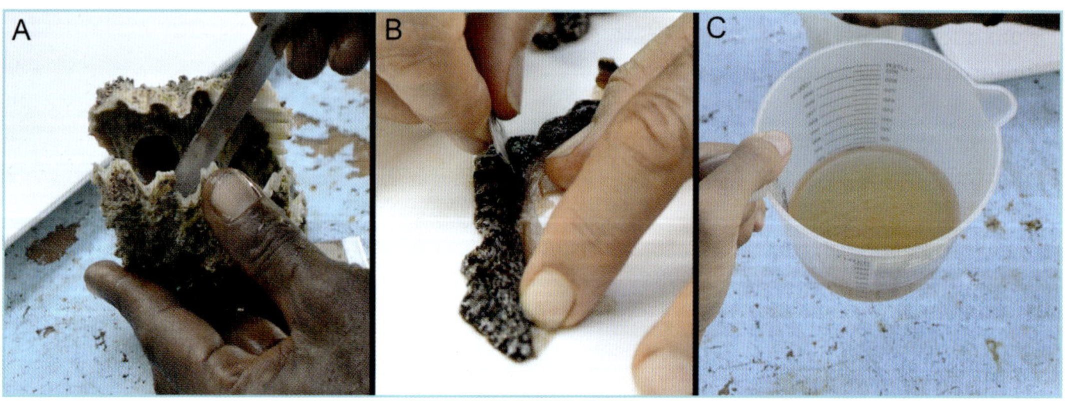

Figure 3.11 Zooxanthellae can be extracted from the mantle tissue of living giant clams through: (A) removal of a section of mantle tissue; (B) laceration of excised mantle tissue to release zooxanthellae; and (C) washing of exposed zooxanthellae into a collection container with filtered seawater that is used to inoculate larval culture tanks.

of previously cultured zooxanthellae or by culturing the zooxanthellae obtained from mantle tissue extraction. The culture of zooxanthellae requires similar technical expertise and resources as microalgae culture. A further alternative is to obtain zooxanthellae from fecal pellets sourced from the bottom of broodstock holding tanks (Fig. 3.5). Giant clams routinely expel surplus zooxanthellae from the zooxanthellal tube system through fecal pellets. These pellets containing viable zooxanthellae can be resuspended in filtered seawater to obtain zooxanthellae that can be used to inoculate larval rearing tanks, as detailed above. Although convenient, this method has obvious hygiene implications.

The optimal timing of zooxanthellae inoculations and appropriate zooxanthellae cell densities are not known. Achieving zooxanthellae densities of 10–50 cells/mL in larval rearing tanks appears sufficient for the majority of larval to acquire the symbionts, which can be observed microscopically as golden-brown spheres retained within the gut. It is general practice to first inoculate larval rearing tanks with zooxanthellae at three to four days post-fertilization. The zooxanthellae obtained by veliger larvae are not often retained through metamorphosis, and a second inoculation of larval rearing tanks with zooxanthellae should occur between 7 and 9 days post-fertilization, prior to metamorphosis. The rationale for providing cultured veliger larvae with zooxanthellae is based on recent molecular evidence suggesting that zooxanthellae begin translocating photosynthates 24 hours after symbiont acquisition by the giant clam larvae, even in the absence of the fully formed zooxanthellal tube system. Additionally, early inoculation of larval rearing tanks with zooxanthellae will establish a free-living population of zooxanthellae in the culture environment of the larval rearing tanks, providing larvae with continual access to the symbiont.

With completion of metamorphosis, zooxanthellae can be observed in rows extending from the region of the stomach toward the developing siphonal tissues of giant clam juveniles (Fig. 3.12). While larvae will metamorphose in the absence of suitable symbionts, juvenile giant clams lacking resident symbionts will soon perish. If a large portion of newly metamorphosed juveniles are found lacking symbionts, a third inoculation of the larval rearing tank with zooxanthellae will be necessary.

Figure 3.12 Juvenile giant clam with transparent shell valves showing zooxanthellae (golden-brown spheres) accumulated within the gut and migration and proliferation of zooxanthellae throughout the zooxanthellal tube system within the developing mantle tissue. Courtesy of Claude Prevost, Civa (Fiji) Pearls Pte Ltd.

Nursery culture

Early nursery culture

Early nursery culture of juvenile giant clams encompasses the period of culture immediately following metamorphosis until juveniles attain sizes of around 1 cm, typically at 5 or 6 months of age under ideal conditions. The first half of early nursery culture is a period of apprehension for most giant clam farmers as the juveniles only become visible to the naked eye at around 2 months of age. Prior to this, accurately gauging survival and growth by visual observation proves challenging. Most farms leave juveniles undisturbed in the larval rearing tanks during early nursery culture as handling risks damaging the newly formed byssus of the juveniles.

Filtered (50 μm) seawater exchange begins on completion of metamorphosis, if not already started during larval rearing, at a rate of one to two turnovers of tank volume per day. Removal of particles and plankton >50 μm from the seawater supply is necessary to avoid introducing the larvae of predatory gastropods (for example, cymatiids and pyramidellids) into the culture environment; however, continuing to restrict the passage of smaller particles <50 μm is disadvantageous as this prevents juvenile giant clams from feeding on particulate organic matter that may be present in the source water. Care also needs to be given to water circulation within the tank, ensuring adequate water exchange along the tank bottom where the giant clams are located. Water exchange along the tank bottom can be improved by reducing the water depth of the tank to 30–50 cm, placing a larger pipe sleeve (with notches cut out of the bottom edge) over the drainage pipe to ensure draining from the tank bottom, and increasing aeration along the tank bottom.

A reduced level of light intensity (PAR: 500–1000 $\mu E/m^2/s$) is maintained during the first two months of early nursery culture to restrict fouling of the tank bottom by filamentous algae. Small juvenile giant clams have very thin, transparent valves and mantle tissue (Fig. 3.12) so that reduced light intensity does not proportionately reduce the benefits derived from symbiont photosynthesis. After two to three months of age, however, juvenile giant clams have noticeably calcified valves and require higher light intensities (PAR: 1000–1500 $\mu E/m^2/s$). Continuing to shade juveniles beyond this point, allowing only 10–30% light transmittance, is detrimental to growth and survival.

Demonstrable production improvements and cost-benefits have not yet been proven for the continued feeding of particulate foods to juvenile giant clams. Feeding of juvenile giant clams after metamorphosis does not appear to be necessary for successful production when allowing particulate organic matter in the source water to enter tanks through water exchange. Giant clams are efficient filter feeders, capable of retaining most particles between 2 μm and 50 μm. Increasing the pore size of water filtration equipment to 50 μm during the early nursery phase will allow passage of potential food particles into the culture tanks, while still excluding the larvae of potential predators and competitors.

Later nursery culture

Once juvenile giant clams obtain a shell length of around 1 cm it becomes necessary to distribute them among nursery tanks to avoid overcrowding in the larval rearing tank (Fig. 3.13). Nursery tanks of the same or similar design to the tanks used for larval rearing are ideal, requiring a large bottom-surface area that receives adequate light (PAR: 1000–1500 $\mu E/m^2/s$), and water exchange (100–200% per day) with filtered seawater (50 μm) is recommended.

As juveniles, all giant clams attach to the substrate using byssus, and removing them from larval rearing tanks involves carefully cutting the byssus with a sharp knife or razor. Pulling juvenile clams away from the substrate by force will damage the byssal apparatus and may result in impaired growth or mortality. Stocking densities of nursery tanks range from 200–500

Figure 3.13 Juvenile *Tridacna maxima* (0.5–1.5 cm) ready to be harvested from larval rearing tanks and distributed among nursery tanks to avoid overcrowding. Courtesy of Claude Prevost, Civa (Fiji) Pearls Pte Ltd.

juveniles/m^2, being dependent on available hatchery space and how long juveniles will be maintained in nursery tanks before transfer to ocean grow-out systems or sale. Maintaining a relatively high density is advantageous as there will be less available substrate for fouling algae; however, if the mantles of juvenile giant clams begin to extend and overlap each other, the density needs to be further decreased.

If later grow-out of juveniles is to occur in the ocean, it is ideal to seed juvenile clams onto a portable substrate when stocking nursery tanks. Bathroom tiles (30 cm × 30 cm), coated with a cement mixture made from coarse beach sand are a suitable substrate enabling subsequent transfer to ocean grow-out culture systems. Commonly used alternatives include gravel (20–30 mm in size) filled trays (Fig. 3.14) and concrete slabs (Fig. 3.15). Substrates without crevices for predators to hide in (for example, tiles and concrete slabs) are ideal, while concrete slabs have a further advantage in that they can be fabricated on-site at the hatchery to required dimensions.

Figure 3.14 Culture of giant clam (*Tridacna maxima*) juveniles in shallow raceways at a government hatchery facility at Toloa, Samoa. The clams are held in gravel-filled trays within raceways.

Control of biofouling

Development of filamentous algae and cyanobacteria mats on the bottom of nursery culture tanks is the principal cause of mortality for juvenile giant clams. Fouling algae compete

Figure 3.15 A farm technician harvests *Tridacna squamosa* cultured at high density on concrete slabs at Arno Atoll, Marshall Islands. The clams are cultured to supply the international aquarium trade. Courtesy of Antoine Teitelbaum, Aquarium Fish Nouvelle-Calédonie, New Caledonia.

with the symbiotic zooxanthellae for waterborne nutrients, reduces light availability through shading, and impairs water circulation on the tank bottom (Fig. 3.16). Biofouling becomes most problematic when the light intensity is increased after the first 2 months of early nursery culture. The primary methods of managing fouling are through manual removal or stocking herbivores into nursery tanks.

Biofouling is manually removed from nursery tanks by wafting the tank bottom with a hand or paddle. This should detach most cyanobacteria and filamentous algae from the tank bottom which can subsequently be removed through targeted siphoning. When siphoning algal debris, it is necessary to siphon into a container with high sides to allow retention of any juvenile clams that were accidentally dislodged and siphoned from the tank. Juvenile giant clams which collect on the bottom of the container are returned to the nursery tank after cleaning. As the juvenile giant clams age (≥1 cm shell length), this technique can be modified to include brooming the tank bottom with a soft bristle brush to detach fouling algae resistant to gentle wafting.

Manual removal of biofouling from nursery tanks should not be undertaken until juvenile giant clams are at least 2 months of age. The disturbance and probable tearing of byssus of early juveniles associated with manual removal methods can have a detrimental effect on juvenile survival. After 2 months of age, nursery tanks can be cleaned at regular intervals (every 1 to 2 weeks) or as required to control biofouling.

The ideal means of managing biofouling is through the introduction of herbivores that purposely feed on algae while avoiding the juvenile giant clams. Employing the use of herbivores to manage algal fouling reduces the burden on hatchery staff who would otherwise be undertaking effort-intensive manual cleaning. Additionally, if appropriate herbivores are selected (Table 3.5), herbivory will be less stressful on juvenile giant clams than manual removal. Juvenile trochid or cerithiid gastropods are suitable herbivores which can be introduced to nursery tanks when juvenile giant clams are two months of age. Co-culture of the topshell or trochus, *Rochia nilotica*, with juvenile giant clams has been very successful, with the benefit of producing two commercial

Figure 3.16 Fouling algae can rapidly develop in nursery culture tanks where it competes with the symbiotic zooxanthellae for waterborne nutrients, reduces light availability through shading, and impairs water circulation on the tank bottom. Control of algal fouling is an important part of giant clam nursery husbandry and can be carried out manually or with the assistance of herbivores. Courtesy of Claude Prevost, Civa (Fiji) Pearls Pte Ltd.

products (see Chapters 7 and 8). Only smaller herbivorous gastropods (<2 cm shell length) should initially be introduced to nursery tanks; the radular teeth of larger individuals can easily dislodge and damage juvenile clams less than 0.5 cm shell length. Once juvenile clams obtain a shell length of 1 cm, fish can also be stocked in nursery tanks to help manage algal fouling. Rabbitfishes (Siganidae) and surgeonfishes (Acanthuridae) have been used successfully in this regard, as well as salinity adapted tropical (freshwater) aquarium species. Before stocking herbivorous fish into nursery tanks, it is advisable to trial the fish in a separate tank with a small number of similar sized giant clam juveniles to ensure they are not opportunistic omnivores.

Table 3.5 Organisms successfully used to manage algal fouling during the nursery culture phase of giant clam culture.

Family	Species
GASTROPODS	
Aplysiidae	*Aplysia* spp., *Dolabella* spp.
Cerithiidae	*Cerithium* spp.
Tegulidae	*Rochia nilotica*
Turbinidae	*Lithopoma* spp., *Turbo* spp.
CRUSTACEANS	
Amphipoda	*Ampithoidae* spp., *Maeridae* spp.
Isopoda	*Sphaeromatidae* spp.
URCHINS	
Toxopneustidae	*Tripneustes* spp.
FISH	
Siganidae	*Siganus* spp.
Acanthuridae	*Acanthurus* spp., *Ctenochaetus* spp.
Blenniidae	*Atrosalarias* spp., *Ecsenius* spp., *Salarias* spp.

Nutrient additions

Given the important nutritional role played by the zooxanthellae symbionts of cultured giant clams, as well as the giant clams themselves, nursery culture of giant clams must also consider appropriate husbandry of their symbionts. This includes consideration of the nutritional needs of the zooxanthellae which, as microalgae, are ammonia and phosphate limited. Increasing the availability of inorganic nitrogen significantly increases the density of zooxanthellae within giant clam tissues and results in enhanced tissue growth of juvenile giant clams. It is evident, however, that the response to increased nitrogen is dependent on the size of juveniles, with such benefits only being realized once juveniles exceed 1 cm in shell length. The benefits of nutrient enrichment of giant clam nursery culture tanks must also be balanced with the risk of encouraging proliferation of fouling algae. The current recommendation is for twice weekly additions of ammonium sulfate or ammonium chloride to elevate nitrogen concentrations in nursery tanks to 40 μM. This regime is sufficient to stimulate improved growth rates but will not stimulate algal fouling to the degree of more frequent enrichment regimes. Further amendment of dosage and frequency may be required to manage biofouling and suit local conditions.

Grow-out culture

Nursery culture concludes when juvenile giant clams reach a large enough shell size for transfer to ocean-based grow-out systems or for sale. In aquarium markets, juvenile giant clams are commonly traded in the size range of 2.5–6.0 cm, with such sizes being reached after 1–1.5 years of land-based nursery culture. Thus, production for the aquarium trade typically concludes with nursery culture, with no requirement for ocean-based grow-out. In contrast, production of giant clams for meat, curio products, and conservation-oriented programs that restock reproductively mature giant clams requires much larger giant clams are required, and this necessitates a grow-out phase of culture. To obtain larger giant clams in an economically sensible manner, a period of ocean-based grow-out culture follows the land-based nursery culture phase. This allows the hatchery to repurpose nursery tanks for production of successive cohorts and takes advantage of favorable growth rates typically achieved with natural, unfiltered seawater.

Figure 3.17 Some species of giant clams, such as these *Tridacna gigas* at James Cook University's Orpheus Island Research Station, can be successfully grown in intertidal habitats. The advantages of this include minimized husbandry inputs because of reduced fouling and predation resulting from regular emersion, and ease of inspection and access to culture stock. Courtesy of Dr. Rick Braley, Aquasearch, Australia.

Ocean-based culture

An appropriate site for deployment of ocean-based grow-out culture systems must be selected. Intertidal sites are ideal for *Hippopus hippopus* and *Tridacna gigas* (Fig. 3.17), while subtidal sites should be selected for *T. crocea*, *T. derasa*, *T. maxima*, and *T. squamosa*. Subtidal sites should be shallow (<10 m), easily accessed by boat, and have good water visibility (>5 m). Additionally, both intertidal and subtidal sites must be located away from areas receiving freshwater runoff or strong wave action. While coral reefs represent the natural habitat for giant clams, such sites tend to contain high densities of giant clam predators

(Table 3.6) and should be avoided as grow-out sites. Sandflats, seagrass beds, or other locations with firm substrata tend to yield the best results. Consideration also needs to be given to site security, as theft of giant clams and/or culture equipment are commonly reported challenges of ocean-based grow-out.

A range of ocean-based culture systems have been developed for the grow-out of giant clams. These include benthic and elevated cages of various designs and construction for deployment at either intertidal or subtidal sites (Fig. 3.18). The use of materials in cage construction depends on local availability, economics, scale of culture and local conditions (Table 3.7). The use of a cage-like structure with 10–20 mm mesh to protect juvenile giant clams is necessary to exclude predators. Stocking juvenile giant clams into ocean-based grow-out systems without predator exclusion measures can result in 60% mortality during the first week of grow-out. In addition to predator exclusion, cages help retard strong water movement that may displace juveniles. Giant clams must remain caged until suitably sized (>20 cm) to overcome risks of predation.

Table 3.6 Giant clam predators known to be problematic during ocean-based grow-out culture. Adapted from Calumpong (1992).

Family	Species
FLATWORMS	
Stylochidae	*Stylochus* spp.
GASTROPODS	
Costellariidae	*Vexillum plicarium*
Cymatiidae	*Cymatium* spp.
Fasciolariidae	*Pleuroploca trapezium*
Muricidae	*Chicoreus* spp., *Tenguella granulata*, *Thalessa aculeata*
Pisaniidae	*Pollia fumosa*
Pyramidellidae	*Tathrella iredalei*, *Turbonilla* sp.
Volutidae	*Melo* sp.
OCTOPUS	
Octopodidae	*Octopus* spp.
CRUSTACEANS	
Callianassidae	*Gonodactylus* sp.
Carpiliidae	*Carpilius convexus*, *C. maxulatus*
Diogenidae	*Dardanus* spp.
Portunidae	*Thalamita* spp.
Xanthidae	*Atergatis* spp., *Demania* spp., *Leptodius sanguineus*, *Lophozozymus pictor*, *Zosimus aeneus*
FISHES	
Balistidae	*Balistapus undulatus*, *Balistoides* spp., *Pseudobalistes* spp., *Rhinecanthus* spp.
Lethrinidae	*Monotaxis grandoculis*
Labridae	*Choerodon* spp., *Cheiliunus* spp., *Halichoeres* spp.
Tetradontidae	*Arothron* spp.
RAYS	
Myliobatidae	*Aetobatus narinari*

Juvenile giant clams are generally transferred to ocean-based grow-out systems after 1–1.5 years of nursery culture, once they have obtained a size of around 3–6 cm. Earlier transfer of juveniles is generally accompanied by substantial mortalities, relative to juveniles retained in land-based nursery tanks. Juveniles pre-seeded onto portable substrates (such as gravel-filled trays, cement-coated tiles, and cement slabs) during nursery transfers can immediately be transferred to ocean-based grow-out systems when desired. Juveniles attached to the bottom of nursery tanks must first be detached and seeded onto a portable substrate 1 month prior to field transfer to enable secretion of an appropriate quantity of byssus to adequately adhere the juveniles to the new substrate. Transferring unattached juvenile giant clams to ocean-based grow-out systems

Figure 3.18 Raised platforms or cages have been used widely for juvenile culture of giant clams. Raising culture units from the substrate reduces risks from predatory gastropods and crustaceans. Such culture units may also attract fish aggregations that may help reduce algal fouling but may also result in dislodgement or predation of culture stock.

can result in 80% mortality because they are considerably more vulnerable to wave action and predation during the first week of grow-out.

Stocking densities during grow-out vary with giant clam size. For juvenile giant clams (3–6 cm) a stocking density of 200 giant clams/m^2 is appropriate, while larger giant clams (7–8 cm) will require a lower stocking density of 60–100 giant clams/m^2. Growth should be assessed regularly and juveniles redistributed as required to avoid overcrowding.

Monitoring and husbandry of ocean-based grow-out systems should be performed weekly (Fig. 3.19). Cages are cleaned of algal fouling and accumulated sediment. Both the cages and surrounding area are inspected for predators which are removed. The cages are also checked for damage and repaired as required. Survival of giant clams during ocean-based grow-out was shown to be positively correlated with such husbandry activities, particularly the frequency of removing predators (for example, cymatiid gastropods) and fouling algae from cages. Larger clams that no longer require protection from a culture unit, should be similarly inspected and their density thinned as required (Fig. 3.20).

Under stable environmental conditions, annual growth rates of the adductor muscle and other soft tissues of giant clams increase exponentially until the onset of reproductive maturity (5 to

Table 3.7 Materials used to construct cage-like structures for grow-out culture of giant clams. Their relative availability, cost, and durability are noted from the perspective of a farm based in the Indo-Pacific region.

Cage material	Availability	Cost	Durability
Frame			
Bamboo/wood	Locally available	Low	1–2 years
Iron rod	Imported	Moderate	3–5 years
PVC pipe	Imported	High	>5 years
Mesh			
Fishing net	Locally available	Low	1–2 years
Galvanized wire mesh	Imported	High	1–2 years
PVC-coated wire mesh	Imported	Very high	3–5 years
Polyethylene mesh	Imported	High	>5 years

6 years of age). Harvesting at the age of maximal biological production is of obvious importance for farms producing giant clams for meat yield; however, longer culture periods may be advantageous if a lucrative enough market can be found for giant clam shells, which continue to grow past the point of reproductive maturity.

Use of cultured giant clams

Aquarium market

The majority of current giant clam production targets aquarium markets in the US and Europe. These markets demand juvenile giant clams in a 2.5–6.0 cm shell length size range. By supplying small juveniles, the cost of transporting live giant clams, which must be packaged with seawater (Fig. 3.21), to international markets is greatly reduced. Additionally, small juveniles can be more easily accommodated in private aquaria, which are generally less than 400 L.

Patterning and color of the siphonal mantle have the largest influence on demand and price. Bright blue, green, and orange mantle colors are highly desired, particularly when combined with intricate mantle patterns. In terms of species, *Tridacna crocea*, *T. maxima*, and *T. squamosa* are the most heavily traded in aquarium markets because they generally have attractive mantles and attain relatively small maximum sizes. These species are more suited for long-term residence in private aquaria than *T. gigas* or *T. derasa*, which can quickly outgrow the confines of typical private aquaria. Additionally, the most heavily traded species are those amendable to culture. Other species with desirable physical characteristics (for example, *T. noae*) are not traded in aquarium markets simply because challenges in culturing these species limit production.

Typical prices obtained for live juvenile giant clams on export to aquarium markets are shown in Table 3.8, although these may vary depending on

Figure 3.19 Regular monitoring of giant clams during ocean-based grow-out is required to address predation, theft, algae fouling, and juvenile performance. Weekly inspection is recommended. The photo shows *Tridacna gigas* juveniles cultured intertidally in shallow, plastic mesh cages at Orpheus Island, Australia. Courtesy of Dr. Rick Braley, Aquasearch, Australia.

Figure 3.20 Regular maintenance of cultured giant clams, like these *Tridacna gigas* at the University of the Philippines, Silaqui Giant Clam Ocean Nursery, includes reduction in density to minimize overcrowding and ensure appropriate illumination. Courtesy of Drs. Patrick Cabaitan and Cecilia Conaco, University of the Philippines.

Figure 3.21 International transportation of live giant clams for the aquarium market requires specialized packaging. This encompasses individual packaging in a minimum volume of seawater, oxygenation of the packaging, and transportation within an insulated container to maintain water temperature. Under these conditions, clams can be successfully transported from the south Pacific to northern Europe.

freight costs. Giant clams with exceptionally patterned and well-colored mantles can be sold for a price premium. Recently, price competition between mariculture production of giant clams and those sourced from wild fisheries (for example, Vietnam) has impacted profitability, with some farmers indicating that the aquarium market is currently oversaturated.

Table 3.8 Prices typically obtained for live juvenile giant clams of varying sizes upon export to aquarium markets.

Species (size)	Price (US$/clam)
Tridacna crocea	
2.5–4.0 cm	3.00–5.00
5.0–7.5 cm	7.50–15.00
Tridacna derasa	
2.5–4.0 cm	1.25–2.00
5.0–7.5 cm	2.50–3.50
Tridacna gigas	
2.5–4.0 cm	3.50–5.00
5.0–7.5 cm	6.00–7.00
Tridacna maxima	
2.5–4.0 cm	2.50–3.50
5.0–7.5 cm	4.50–6.25
Tridacna squamosa	
2.5–4.0 cm	2.50–3.50
5.0–7.5 cm	4.50–6.25

Restocking programs

A number of countries have developed culture programs to replenish locally extinct and diminished giant clam stocks. In the Philippines, for example, more than 70,000 giant clams have been restocked at more than 40 sites, in an effort to replace locally extinct populations. This effort has focused primarily on *T. gigas*, the giant clam species most vulnerable to extinction (Figs. 3.9 and 3.20). Restocking has resulted in juvenile recruitment at some sites indicating that restocked clams can potentially replenish local populations. In many cases, however, poor survival, limited production, hatchery expenses, poor local governance and reliance on government funding have resulted in cost-ineffective production and eventual termination of activities.

Meat and curios

The commercial market for giant clam meat is currently limited to high-end restaurant establishments that utilize the adductor muscle. Market preference is for adductor muscles originating from older clams (≥3 years), because the adductor muscles of younger clams are too small, soft, and watery for most preparations using giant clam meat. Demand for the other soft tissues of giant clams (for example, mantle) is limited, presumably due to the distinctive 'seaweed' taste afforded by the presence of zooxanthellae. Operations producing giant clam meat also obtain sizeable shell valves (20–25 cm for *T. derasa* and 40–55 cm for *T. gigas*) as a by-product of farming. The shell valves of giant clams have commercial value in the curios trade. Despite the existence of commercial markets for both the meat and shell, an analysis of markets for these products across the Indo-Pacific concluded that demand and prices were insufficient to support profitable giant clam aquaculture operations. The economic prospects of farming giant clam meat and shell require careful examination before allocating resources to such an endeavor. Hatchery production of giant clams to stock small community-run farms for non-commercial local consumption is increasingly popular in developing countries. This activity may be associated with locally managed marine areas and, as well as improved food security, may stimulate natural recruitment depending on the harvest age of the giant clams.

References cited and further reading

Braley, R.D. (ed.) (1992) *The Giant Clam: A Hatchery and Nursery Culture Manual.* ACIAR Monograph No. 15. Australian Centre for International Agricultural Research, Canberra.

Calumpong, H.P. (ed.) (1992) *The Giant Clam: An Ocean Culture Manual.* ACIAR Monograph No. 16. Australian Centre for International Agricultural Research, Canberra.

Lucas, J.S. (1994) The biology, exploitation, and mariculture of giant clams (Tridacnidae). *Reviews in Fisheries Science*, 2: 181–223.

Militz, T.A. (2019) Aquaculture in the aquarium industry. In: Lucas, J.S., Southgate, P.C., Tucker, C.S. (eds.) *Aquaculture: Farming Aquatic Animals and Plants.* 3rd edition. Wiley-Blackwell, Oxford, pp. 587 616.

CHAPTER 4

Scallop aquaculture

César Lodeiros, Leslie-Anne Davidson, Michael Dadswell, Guilherme Sabino Rupp, and José Manuel Mazón-Suástegui

Introduction

There are some 400 pectinid species, commonly referred to as scallops, that are important members of benthic communities from tropical to boreal regions. Only 10% of these species, however, have been commercially exploited. Compared with oysters, mussels, and clams, the culture of scallops is a relatively new activity, beginning in Japan in the 1930s, and developing in the 1960s on a large scale, when the natural beds were already overexploited. The culture methods were developed in Mutsu Bay (Aomori Prefecture) given the ingenuity of the fishers, who observed high settlement of the scallop *Mizuhopecten yessoensis* (= *Patinopecten yessoensis*) on their fishing nets. Scallops were then confined in growing baskets until they reached about 30 mm and subsequently deposited on the bottom (hanging and sowing), to stabilize the production of the Japanese pectinid beds. The Japanese scallop culture techniques were disseminated (Ventilla, 1982) and are generally used in other parts of the world, but the specific procedures vary significantly according to the species and the environmental conditions of other culture areas. Some of the factors to consider in defining the specific techniques are the source of juveniles to begin culture, the environment, the growth rate, the sources of mortality, and the market size of the species. Some fast-growing species (for example, *Argopecten* spp., *Nodipecten* spp.) are harvested after one to two years of culture, other species from temperate areas (for example, *M. yessoensis, Placopecten magellanicus, Pecten maximus*) usually require 3 to 5 years to reach a commercial size.

Data provided by FAO (2019) indicate that the production of cultivated scallops during 2017 was 2,185,243 metric tons (MT), 4.5 times more than wild capture. This production is equivalent to 12.6% of the total worldwide production of molluscs, which has yielded economic benefits of about US$576.5 million, almost 20% of the global total of cultured molluscs.

Recent worldwide scallop production by aquaculture is shown in Table 4.1. China is the largest producer growing more than 90% of the total harvest. Although referred to as Scallops nei (pectinids without sufficient information), the highest production in China includes an introduced species, the North American bay scallop *Argopecten irradians* (58%), the native Chinese or zhikong scallop *Azumapecten farreri* (previously *Chlamys farreri*, 20%), followed by *Mizuhopecten*

Table 4.1 Recent aquaculture production of scallops (in metric tons) (FAO, 2019).

Species	Scientific name	2011	2012	2013	2014	2015	2016	2017
America								
Atlantic lion's paw	*Nodipecten nodosus*	0	0	0	0	0	10	10
Peruvian calico scallop	*Argopecten purpuratus*	63,231	30,580	72,695	59,242	25,987	24,522	166,33
Scallops nei	Pectinidae	300	215	116	160	31	38	163
Asia								
Scallops nei	Pectinidae	1,244,983	1,360,689	1,539,226	1,576,378	1,710,080	1,849,887	2,007,529
Yesso scallop	*Mizuhopecten yessoensis*	119,028	185,006	168,578	185,794	250,016	217,816	138,893
King scallop	*Pecten maximus*	75.28	75.9	68.81	46.622	77.41	48.9	36.11
Europe								
Mediterranean scallop	*Pecten jacobaeus*	0	0	0	0.04	0.02	0.07	0.04
Queen scallop	*Aequipecten opercularis*	1.5	0.7	7.4	6.8	4.3	7.9	12.3
Variegated scallop	*Mimachlamys varia*	0.6	0.7	0.6	0.6	0.4	0	0
Yesso scallop	*Mizuhopecten yessoensis*	725	504	1311	2400	2000	3745	5463

yessoensis (= *Patinopecten yessoensis*, 19%) and a smaller fraction (3%) of the noble or huagui scallop *Mimachlamys nobilis* (Guo and Luo, 2016).

The most important species produced in Japan is *M. yessoensis*. In the subtropical Pacific region of South America, the Chilean or Peruvian calico scallop, *A. purpuratus*, is the cultivated species. In Canada and the US, the species mainly produced is the sea scallop, *P. magellanicus*, while in Europe the king scallop, *P. maximus*, and the queen scallop, *Aequipecten opercularis*, are cultured. Recently, other species started to be grown in small production such as the Mediterranean scallop, *P. jacobaeus* and two species of *Nodipecten* spp. are currently emerging as candidates for aquaculture. In Brazil, production of the lion's paw, *N. nodosus*, has begun, and in Mexico there has been some production of *N. subnodosus*. These species are also being considered for aquaculture in Ecuador, Venezuela, and Colombia. Other species cultured, but at low production levels include the Catarina scallop, *A. ventricosus*, in Mexico, and the southern Australian scallop, *Pecten fumatus*, in Australia during the 1990s.

A general description of the techniques commonly used for scallop culture and the main factors to take into consideration while establishing a scallop culture operation are provided in this chapter.

Scallop culture – general techniques

The first step in scallop culture is obtaining spat, seed or juveniles, which can come from wild collection or be produced in a hatchery. The subsequent culture can be divided into three phases: an initial nursery management for growth of the spat (post-larvae that do not have a manageable size, normally ≤ 2 mm), to seed size (normally ≥ 5 mm), then a field-based intermediate cultivation phase where the juvenile scallops are cultured in high densities, and a third phase of final grow-out, where the scallops are cultured at lower densities until they reach commercial size. Seed are normally cultivated in the sea with suspended techniques, using, in most cases, the long line system (Fig 4.1), or rafts in well-protected areas (Fig 4.2). Another option is bottom culture, where young scallops are confined in the bottom with fences, or near the bottom in cages or baskets.

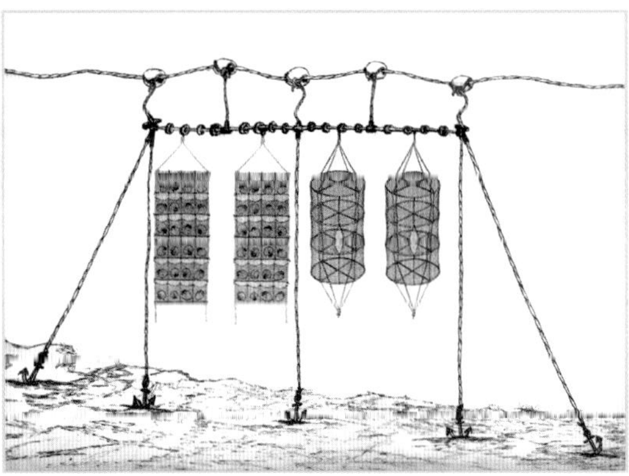

Figure 4.1 Simple longline culture system for scallops culture.

Spat production

Natural spat collection

Since the mid-1950s scallop culture has been particularly successful based on wild seed collection. This method uses the deployment of artificial collectors, upon which the competent larvae (pediveliger or larvae with eye spot) attach and undergo the metamorphosis. These collectors consist of synthetic filaments contained inside a plastic-mesh bag, which will

Figure 4.2 Commercial mussel culture raft used in scallop culture (Spain).

attract the pediveligers for settlement and metamorphosis. The collector bags can vary in size and filament type, but the mesh opening should not be greater than 3–5 mm. In this way the competent larvae that are attached by byssus will grow larger than the opening and, when the attached phase ends (behavior of most of the scallops), the scallop will detach from the filament, but be retained inside the bag (Fig. 4.3).

The materials used for spat collection are in general monofilament nets or Netlon™. Monofilament gill net can be purchased or acquired from commercial gill netters when they replace netting. Old gill net is best since the roughened surface collects scallops better. The typical commercial artificial collector developed to collect spat consists of a mesh bag filled with gillnet or two pieces of Netlon™ (Fig. 4.3). Each piece of Netlon™ can measure 40 cm × 100 cm and is made of highly resistant and reusable polyethylene net. The bag is made of polyethylene with a mesh size of 3 mm and, if maintained properly, can last up to 10 years; however, some growers replace them every year. Commercially available spat collectors range in several colors, however, green or blue are the most efficient in the spat collection.

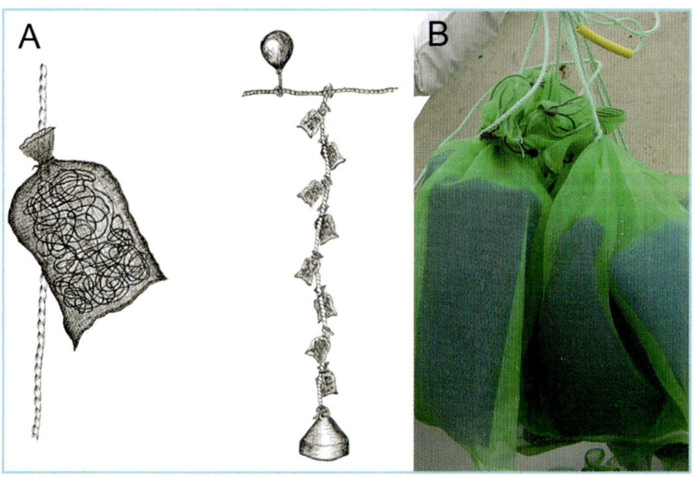

Figure 4.3 Plastic collectors for juvenile scallop collection in the wild: (A) monofilament normally contained in plastic bags; (B) commercial collector with Netlon™. Courtesy of Gaspar Soria.

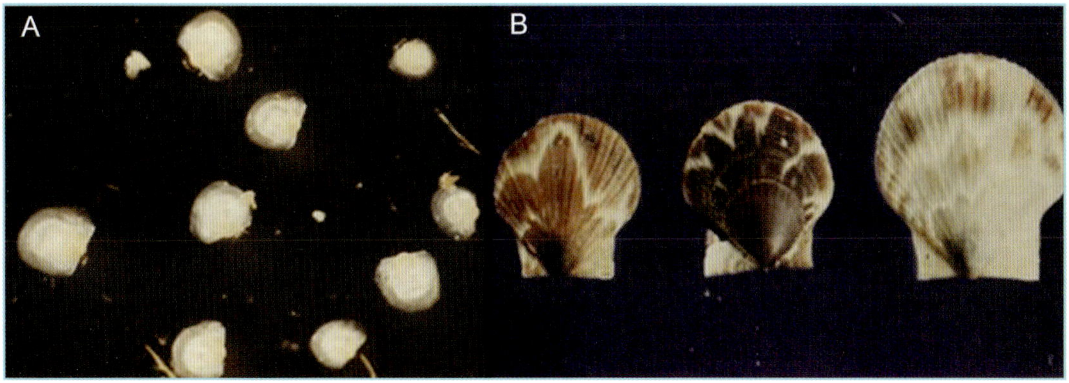

Figure 4.4 Scallop spat (*Placopecten magellanicus*): (A) newly-settled; (B) 3 months old.

The collector bags are suspended on long lines below the wave action and preferably 6 m from the bottom at the culture site during a period determined by the reproductive cycle of the species. Studies to demonstrate the key settlement periods are necessary for successful spat collection. Monitoring the gonad index and the availability of larvae in the plankton are the methods normally used to determine when the collectors should be deployed. Normally, collectors are deployed at spat collection sites about two to three weeks after the adult scallops begin to spawn. Newly-settled scallop spat will only be visible under a microscope and should look like white or transparent scallops (Fig. 4.4A). Within a few months for cold water species, and around one month for tropical species, the scallop spat will become pigmented and acquire the distinctive scallop appearance (Fig. 4.4B). The aim is to collect >500 scallop/bag. Settlement of predators is one of the concerns during spat collection, especially in temperate waters. Culturists should experiment with different deployment times in their particular area to determine periods of predator settlement and time of collection to obtain the largest number of spat.

The harvest of early juveniles to start the intermediate culture stage must be done in time to avoid overpopulation in the collection bag. If thousands of juveniles/bag are obtained, they can leave the byssus, detach from the collector mesh, and look for another site to re-settle, but they can also fall to the bottom of the bag, provoking high mortality rates. After intermediate culturing of juveniles, they must be harvested to start (in another suspended or bottom art, a third grow-out stage and several months later, finally to harvest market sized scallops. This implies manipulation of seeds, juveniles and adults, which increases costs and mortality, so some improvements must be developed, especially in sites where several thousands of juveniles per bag are collected. As an example, a 'lantern net type' collector has been used for collecting and the intermediate culture of Catarina scallop *A. ventricosus*. This lantern is manufactured by hand with galvanized wire rings and different types of mesh in walls and bottom (Fig. 4.5A), but the best results have been attained with square monofilament mesh 5 mm × 5 mm (Fig. 4.5B). This mesh type and opening size reduces the area available for settlement of larvae and reduces the number of spat/lantern. As a result, two months after settlement of wild larvae, the juvenile scallops are suitable for the subsequent grow-out period on bottom or in suspended nets. The greatest advantage of 'lantern net type' collector is that the handling of spat, seed, and juveniles is highly reduced, decreasing the mortality associated with early culture.

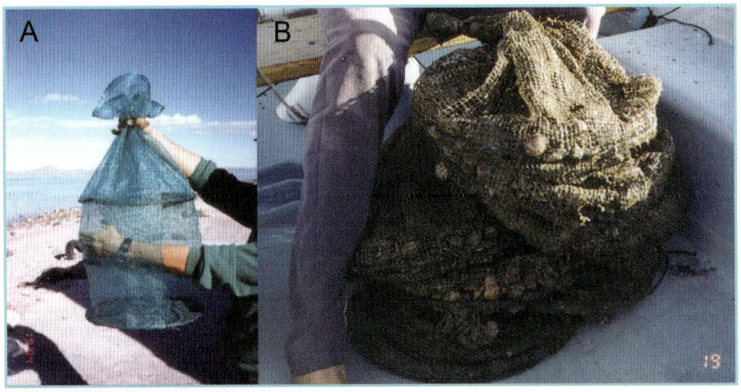

Figure 4.5 Collecting and nursery device. Onion bag mesh (A) or monofilament-type mesh (B) and galvanized wire have been used to hand-build 'lantern type' units to collect spat and to harvest pre-adult scallop (*Argopecten ventricosus*). Courtesy of J.M. Mazón-Suástegui.

The collection of wild seed from the natural environment allows, for certain species, the availability of juveniles in quantities large enough to permit production at the industrial level; however, the success of spat collection is dependent upon the reproduction period of the species and larval abundance, which are frequently affected by population fluctuations. In other instances, obtaining seed or juveniles depends on the behavior of the species. For example, the tropical scallop *Euvola ziczac* detaches from the substrate at a small size (~1 mm), so bags with 5 mm mesh are not effective in retaining the juveniles, and to use a smaller mesh (1 mm) is impractical due to rapid clogging. Thus, species without a significantly strong or prolonged byssal-attachment phase have proven difficult to collect from the wild. Conversely, *Nodipecten nodosus* maintains the ability to attach byssally until the adult phase, but the small populations and low abundance of larvae results in unsuccessful wild spat collection. In these cases, where natural collection is impractical, the alternative is hatchery production and the technology has already been developed.

Hatchery spat production

Hatchery production consists of several stages, such as broodstock conditioning, induction to spawning, embryonic and larval development, settlement and metamorphosis, each of them requiring specific diets, physical-chemical conditions, densities, prophylaxis, and settlement techniques, which are dependent on the species under culture. Many bivalves are amenable to culture using general hatchery techniques; however, for scallops there are some specific differences in relation to other bivalves (for example, oysters, clams, mussels) especially in the settlement

Figure 4.6 Plastic collector bag: (A) on-cultch scallop spat; (B) and cultch-less seed; (C) of Catarina scallop *Argopecten ventricosus*. Courtesy of J.M. Mazón-Suástegui.

procedures. Unlike oysters which permanently attach to the substrate, post-metamorphic scallops attach by production of the byssal threads. Therefore, when the larvae become competent showing a conspicuous eyespot and pedal activity (pediveliger), usually the artificial spat collectors (Fig. 4.6) are placed inside the culture tanks for larval settlement and metamorphosis where they are maintained for a certain period before transference to nursery culture systems. The methods of producing bivalve seed in the hatchery are described in detail in Chapter 12.

Nursery culture

The intermediate step between the controlled hatchery environment and the grow-out in the sea is known as nursery culture and is a critical phase in scallop culture. The first phase of nursery culture includes the period from metamorphosis to approximately 0.5–1 mm shell height, when post-larvae are cultured in the hatchery facilities, usually attached to monofilament collectors or other substrates. Unlike other bivalves that strongly attach to the substrate, scallop post-larvae are extremely fragile and easily detachable from the collectors, being difficult to manipulate. Therefore, culturing them requires nursery strategies different from those used for some other molluscs. The second phase of nursery culture, until juveniles reach about 5–10 mm in shell height, can be either land-based downwellers, recirculating waters, flow-through systems, but principally upwellers (Figs 4.7 and 4.8), or carried out in the sea. The land-based approach requires spat to be maintained in semi-controlled conditions, and demands feeding with high quantities of microalgae, which results in high costs. In addition, high mortality rates may occur due to nutritional deficiencies of the cultured phytoplankton. Therefore, transferring spat to the sea as early as possible is a desirable strategy to reduce costs of production and minimize mortalities. The sea-based nursery consists of transferring spat collectors inside nylon mesh bags (similar to those used for natural spat collection) or in mesh-lined plastic trays directly to the sea, when spat are generally larger than 1mm, or about 8–10 days after larval setting in plastic mesh. In the sea-based nursery, spat may be subject to loss during transport and mortalities due to environmental factors such as salinity and temperature exceeding the limits of tolerance, smothering, predation or handling. Therefore, the specific conditions of nursery culture vary with species and location and studies are necessary to determine the best nursery culture strategy.

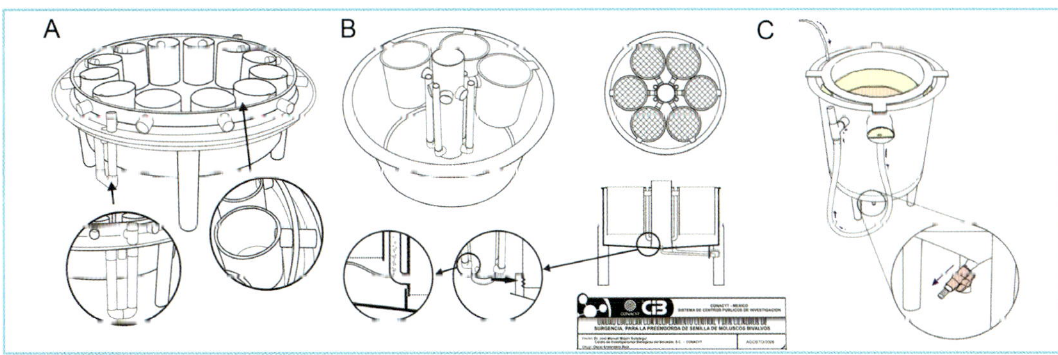

Figure 4.7 Nursery upwelling units in recirculating pattern. Seawater flows by means of airlifts and accessories, such as (A) perimetral channel, (B) central cylinder, or (C) external tubing. Water and microalgae added continuously to produce scallop seeds of a marketable size (3–5; 5–7 mm). Units were designed by J.M. Mazón-Suástegui.

Figure 4.8 Hatchery upwelling units for seed production of Catarina scallop *Argopecten ventricosus*, and lion's paw *Nodipecten subnodosus*. Nursery upwelling units operated by means of airlifts: (A) upwelling units with perimetral chanel at left and with central chanel at righth; (B) fiberglass upwelling cylinder; (C) cultch-less seeds. Units were designed by J.M. Mazón-Suástegui.

Intermediate culture

The intermediate culture starts when the juveniles, obtained either by hatchery production or natural collection, reach a size at which they can be handled (usually >10 mm) and can be transferred to pearl nets or small-mesh lantern nets, which are normally suspended from long lines or rafts (Fig. 4.9).

One important aspect to consider during the intermediary culture phase is the stocking density. There are two approaches pertaining to the stocking density for intermediate culture; stocking at a lower density to achieve rapid growth, or at a higher density to obtain maximum harvest numbers with less space and labor. The individual grower can decide on their best strategy based on their aquaculture space, available nets, and personnel; however, the first strategy is costlier (need of personnel), the second requires net cleaning and a knowledge of the fouling conditions at the aquaculture site (that is, when nets can be deployed without becoming heavily fouled; see Chapter 16). The intermediate culture is normally undertaken in high density, usually covering above 50% of the base area of the net.

There should be changes of nets to reduce the scallop density, remove predators such as starfish, crabs and gastropods, and avoid accumulation of fouling in the meshes, which decreases the circulation of water within the net and consequently, the availability of food. There is no specific size that determines the completion of the intermediate culture, and it is dependent on the species. Usually a size between 40 and 50 mm in shell height, respectively for small and large species, can be considered adequate to establish the subsequent phase of final grow-out.

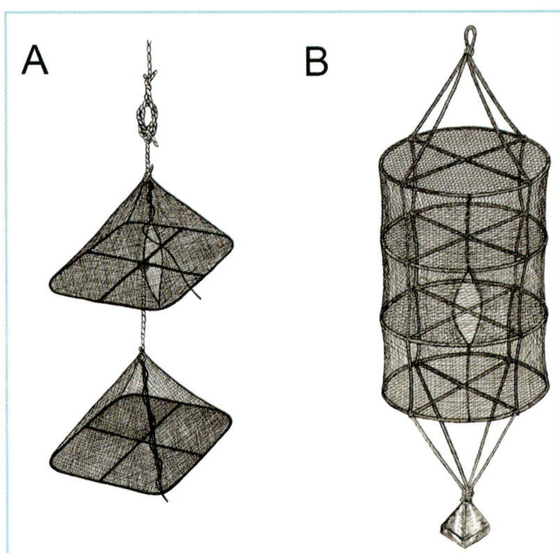

Figure 4.9 Classic enclosures for the suspension culture of scallops: (A) pearl net; (B) lantern net.

Final grow-out

Suspension

Final grow-out of scallops depends on the market size and consumer preference, and is the phase with the most variations in terms of the type of enclosures used. In general, lantern nets are a preferred strategy for some growers because they are available in larger mesh sizes (20–50 mm square) which allows for greater water flow (and therefore faster growth of scallops); they are relatively rapid to fill and close and they have given better growth performances. Lantern nets normally come in two types, five floor or ten floor, and the grower can use the type that suits their site best. In deeper areas, ten-floor nets are an excellent option for final grow-out because these nets take advantage of optimizing the use of the water column. The density of scallops in the enclosures at the final grow-out is normally less than 50% of the enclosure base coverage, and the net mesh used is frequently larger than 15 mm. When scallops are removed from nets after intermediate growth, there is usually a wide range in size. An optimum strategy can be to sort the scallops by shell height and put the smaller ones back into small-mesh lantern nets and the larger ones into larger mesh lanterns. Using this strategy, the grower can obtain market size scallops more quickly and improve cash flow.

In some cases, lantern nets are not well-suited for grow-out since the adults do not produce byssal threads and can move freely, so they are easily injured or killed from cutting into each other. Single-compartment cages have been developed and greatly improve survival and growth. Other culture devices can be used for maintaining scallops individually, diminishing the competition for space among the organisms. The pocket net is one of them and consists of structures made of plastic mesh in the form of pockets, where one or two organisms are placed (Fig. 4.10A). Another method is ear-hanging, which consists of suspending scallops from rope with a nylon pin that holds the scallops individually, through an orifice (1–1.5 mm) drilled in the projection of the anterior shell (Fig. 4.10B). Ear-hanging can only be used for final grow-out and uses less

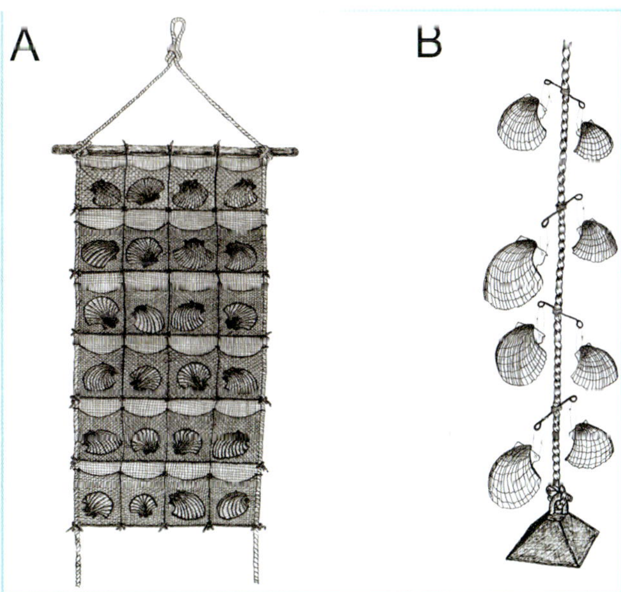

Figure 4.10 (A) Pocket net and (B) ear-hanging culture methods.

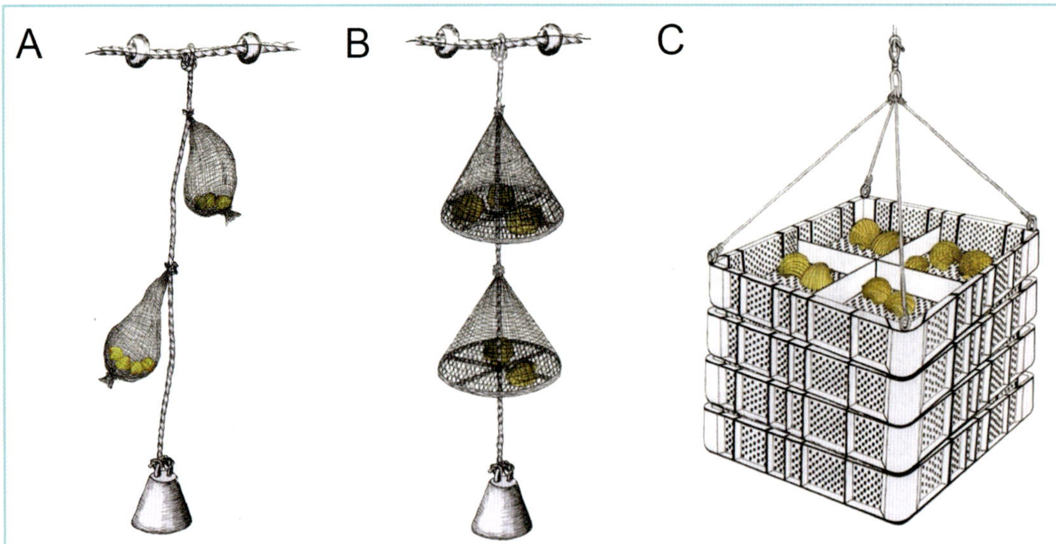

Figure 4.11 Different scallop culture enclosures: (A) bags; (B) pyramidal baskets; (C) nestier trays.

space than other methods. Ear-hanging lines, however, require much more labor to deploy than net gear. Ear-hanging lines tend to sway in the current and should have a 1 kg weight at the bottom of the line and be attached to the suspension line at least 0.5 m apart from each other to prevent entanglement. Although ear-hanging methods have been comprehensively tested and used, their use has now declined. Cultivation bags or variants of pyramidal baskets or simply other baskets for bivalve cultivation, such as the classic nestier trays for oysters (Fig. 4.11) have also been used, but their use has been limited compared with the use of lantern nets.

A recent revision of the cultivation methods for final grow-out of scallops in temperate and tropical zones showed that lantern nets are the most used enclosure, and suggested a stocking density of 33% coverage of the lantern net floor area. The growth of cultured scallops in the tropical region was significantly faster than the temperate regions; however, survival was significantly lower compared to the scallops cultured in the cold-temperate regions. The lower growth rate in the temperate regions was attributed to the decrease in growth rate during winter, while the observed predation in the tropical region may have caused low survival of the tropical species.

Bottom culture

An alternative to the suspended method is bottom culture. In this system, the scallops obtained by natural spat collection or after the intermediate culture are placed on, or close to, the bottom, either enclosed in baskets or laterally limited with meshes. Also, the scallops can be released at low densities in a specific area for final growth and be harvested by capture with fishing techniques. This last approach can be referred to as 'sea ranching'. Another method developed for bottom culture involves the establishment of structures with cages that are placed on the bottom (Fig. 4.12); however, in waters with strong currents, the cages can become unstable if their height is greater than their width. Also, the cages should have a structural gap of at least

Figure 4.12 Cages used in bottom culture (Canada).

60 cm between the bottom of the cage holding the scallops and the seafloor to avoid the organic and inorganic matter under the cage that accumulates and potentially smothers the scallops.

In cages placed near the bottom, the growth performance of some species, for example *Placopecten magellanicus*, is lower than in suspended culture; however, this is species dependent. In another example, the tropical scallop, *Euvola ziczac*, showed more rapid growth in bottom than suspended culture, mainly due to stress caused by wave action in suspended culture.

Scallop culture – factors to consider

Site selection

Apart from the selection of appropriate culture techniques, there are several factors to consider for successful scallop production. The simple fact of the existence of the species in the area is not enough to determine an adequate cultivation site. Ideally, the site should include the integral components to permit success in the sea-based production phases, as well as in processing and commercialization. In practice, it is often common that the site is suitable for one phase of cultivation, but not others. For example, the hatchery or wild spat collection is carried out in one site, and other phases (intermediate culture and grow-out) in another location, or the processing is carried out in an area different from the sea-based production. Also, certain logistics aspects, such as availability of electricity, transportation access, easy and safe access to the sea, legal issues for concession of marine areas for aquaculture, among others are some important factors to take into consideration.

Logistics should never, however, overpass biological and environmental criteria in selecting the culture site. The site selection criteria should include the biological feasibility of cultivation to permit the best growth and survival, which are influenced by the environmental factors that precisely modulate these physiological conditions. Among the factors that must be taken into account are temperature, salinity, dissolved oxygen, predation, fouling, and availability of food, all of which modulate growth of scallops in culture conditions, as well as other factors associated with the culture (use of different meshes), and all of them are dependent on the cultured species. There are several detailed studies that address the environmental influence in scallops and modelling studies that assess carrying capacity for aquaculture siting (see Chapter 10). Apart from these, other factors inherent to the quality of the product (for example, contamination and toxins) must be considered, as well as the feasibility of harvesting, transporting, and processing the product.

Scallop transportation

Transportation is one of the processes that can cause mortality in seed, juvenile, and adult scallops, either with the purpose of sowing the culture in the field, or as the harvested product at a marketable size. Good practices are necessary for this and there are protocols established for the transport of these organisms. The response to stress associated with transport is different for each species given its condition and metabolic adaptation. For example, molluscs whose habitat is intertidal, such as oysters and mussels, generally have greater resistance to desiccation. These species are capable of hermetically sealing their valves and can survive several days out of the water, depending upon temperature and other factors. They have developed an ad hoc anaerobic metabolism that allows them to survive out of water, unlike the molluscan species that inhabit subtidal niches such as scallops, which do not have that capacity.

Regardless of the behaviors described above, which are species-specific, the transport method and handling of the organisms change according to their life phase. It is different for eyed setting larva, crawling or metamorphosed larva, newly-settled larva (spat), juveniles (seed), or broodstock. The transport of seed in water with aeration and/or bubbling of oxygen is appropriate only for short periods (a few hours), since, in the presence of water, the organisms continue to respire aerobically, filter water, and feed. Owing to these physiological processes, the enclosed environment rapidly becomes inappropriate with the presence and decomposition of organic material (feces) and the accumulation of nitrogenous metabolites produced by excretion, such as ammonium and nitrites, which have high toxicity. As a result, the concentration of dissolved oxygen decreases, and favorable conditions for bacterial growth develop. The transport of scallops out of water, but in a wet-cold environment, is effective but also has limitations. The method is known as 'sandwich' type because it includes several superimposed foam rubber plates moistened with seawater, damp seaweed, or wet paper towels. Transport in water is also an alternative and it is feasible if a healthy closed recirculation system is available.

Wild collected scallop seed can be infested with parasites and local diseases, but laboratory seed can also present pathogens. In both cases, prior to transport, it is recommended to perform pathological analyses to minimize the risk of dispersion and introduction of diseases to the destination sites. If a declared disease or the presence of a pathogen is suspected, the organisms must undergo a controlled quarantine, and be subject to subsequent pathological evaluation. Depending upon the results, the decision will be made to eliminate or release that specific batch of larvae, juveniles, or adults for transport and planting.

Juveniles

Generally, juvenile transport has been accomplished in water or in a cool and humid environment. Low temperatures are necessary to reduce metabolic activity, stress associated with handling, and energy consumption. Depending on the species the temperature will range from 5°C to 20°C. For example, when transporting *Placopecten magellanicus*, greater survival is obtained if the temperature in the transportation containers hovers around 5°C and the ambient air temperature is between 2°C and 10°C when transferring the spat from the container to the culture gear. When scallops are transported out of water, a humid environment can be attained with wet natural or artificial materials such as seaweeds or foam plates. As an example, the 'sandwich-type' method for scallop transport applies to adults and juveniles and consists of placing them between superimposed foam rubber plates and tied together forming a package, inside an

Figure 4.13 Sandwich-type method for transporting scallop seeds in a humid and cold environment, between superimposed foam rubber plates. Sandwich method designed by J.M. Mazón-Suástegui.

insulated isothermal container, with gel ice packs to maintain a low temperature (Fig. 4.13). This method has proven successful in transporting juveniles of *Argopecten ventricosus* for up to 16 hours and *Nodipecten subnodosus* between 8 and 10 hours. In the tropical zone it is suggested to carry out studies to define the duration and thermal conditions of transport, as well as to introduce the transport device in a plastic bag and inject oxygen, since this may prolong the transport time up to 15 to 24 hours.

A typical transport unit for juvenile scallops in water may consist of an insulated isothermal container, equipped with a 12-volt electric pump and plastic tubing lines that allows a continuous filtering and recirculation of seawater. If necessary, water temperature can be lowered by introducing on demand, ice bottles made with seawater to avoid risks of reducing salinity. An advantage of this method is that hatchery seed can be pre-counted in the laboratory and placed in small plastic-mesh bags, just the amount required for the sowing of each module of culture in the field. Also, plastic-mesh bags gently press seeds, avoiding valves opening and self-trapping of two or more scallops, that provokes injuries in soft tissues and mortality. Juveniles of *N. subnodosus* have been transported with practically zero mortality (99% survival), since the unit keeps operating in the field, maintaining recirculation, filtration, and water quality during the time required for planting (Mazón-Suástegui et al., 2011).

Adults

The transport of adult scallops is generally broodstock adults for use in the hatchery and live scallops to the market, for fresh consumption without freezing or industrial processing. Depending on the species, locality, and environmental conditions, isothermal boxes equipped with ice packs are used, placing the adult scallops in the same way as the seed, between seaweeds or foam rubber plates moistened with seawater (Fig. 4.14). Also, for species that are not capable of effective hermetical sealing, it is recommended to place them with the cup-shaped shell down to help prevent desiccation.

The transport in water of broodstock adult scallop can be done as described above for juveniles. The total biomass of the organisms should be considered, and depending on the duration of the trip, make the necessary adjustments. The previously described recirculating, filtering and aeration system is basically applicable for broodstock adults, since it is very important to avoid changes in temperature that can produce unwanted spawning during the trip and deteriorate the quality of water.

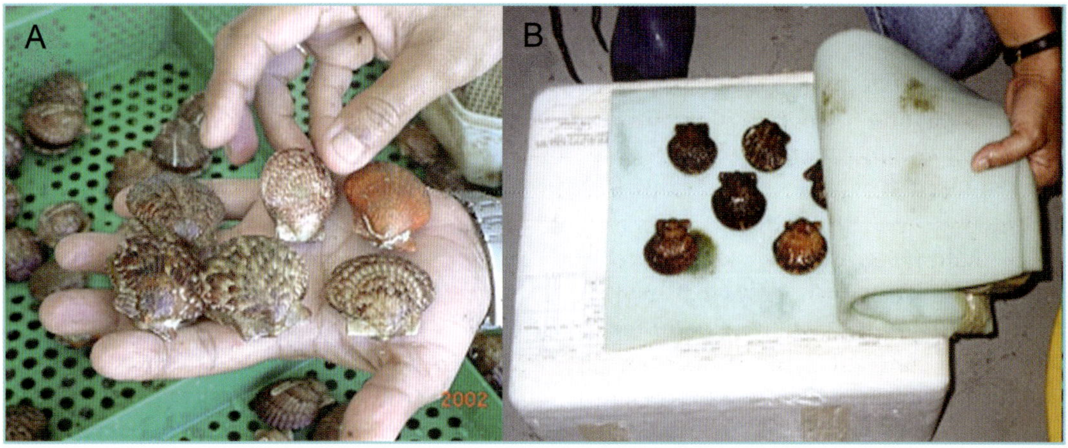

Figure 4.14 (A) Transport of broodstock adults (*Argopecten ventricosus*) applying the sandwich-type method, with (B) moistened foam rubber plates, plastic mesh, and adhesive or elastic bands. Sandwich method designed by J.M. Mazón-Suástegui.

Environmental and associated factors

Temperature

Temperature is the most influential physical factor on growth, reproduction, and survival of bivalves. Scallops are highly sensitive to extreme temperatures, which produces physiological stress, resulting in slow growth, mortality, and reduced production. High temperatures cause greater sensitivity to diseases, and even escape responses to predators.

Scallops are adapted to the specific temperature range of their natural environment. The species from temperate zones are adapted to mild temperatures (<18°C), and tropical and subtropical species are adapted to warmer temperatures (>18°C); however, each species has an optimal temperature range to metabolism development, and temperatures usually over 5°C of that range, can affect the growth and survival of scallops under cultivation (for more information on the optimal temperature range for the scallop metabolism refer to MacDonald et al. [2016]).

Salinity

Salinity is a factor that clearly limits the distribution of scallops and it is an important consideration in aquaculture. Most species live in high salinity seawater and do not survive at low salinities. Scallops, unlike most other bivalves, are incapable of closing their valves due to the byssal notch opening, and the osmotic concentration of the hemolymph comes into equilibrium with the concentration of the external medium after a certain period, thus causing physiological stress and later mortality. Salinity tolerances vary among scallop species; however, most are considered stenohaline.

Oxygen concentration

Unlike other bivalves, such as oysters, most scallops are poor oxygen regulators even though there are certain species that are capable of limited oxyregulation, resisting low oxygen concentrations. Oxygen levels should be maintained at saturation to ensure healthy scallops.

Seston and associated factors

Scallops are filter-feeding organisms and rely on phytoplankton as the main source of nutrition. Several types of particles have been identified in the stomach contents of scallops, including those of terrestrial origin. Seston is a mixture of organic particles with high nutritional value (phytoplankton) and inorganic matter with low nutrition value. Several studies have established that pectinids are capable of active phytoplankton particle selection. This system is generally activated in the presence of high inorganic seston (>20 mg/L) causing excessive production of pseudofeces resulting in reduced growth. Therefore, it is recommended that the cultivation site does not contain high inorganic fraction of seston, and that the organic portion contains high nutrition particles (phytoplankton). There are no specific data for the amount of phytoplankton suitable for scallop cultivation, but it has been suggested based on a comparison of sites in temperate and tropical regions that the phytoplankton content, estimated in chlorophyll-*a* concentration, should exceed 1 $\mu g.L^{-1}$ (Saxby, 2002). In any case, the culture site must have adequate water flow that provides enough phytoplankton to allow the entry and residence of the particles in the culture enclosures without disturbing the scallops. In general, a water flow velocity of <10 $cm.s^{-1}$ is considered adequate (see Chapter 10).

Wind and wave action

Seasonal prevailing winds are a factor that must be considered in siting of aquaculture sites. Scallops do not grow well under strong wave action, or in areas with high concentrations of suspended sediments (particulate inorganic matter) (Robinson et al., 2016). Depending on how the containment system is suspended in the water column, wind-induced waves can transfer the energy down into the containment structures causing a jerking motion, which disturbs the scallops and results in an extra expenditure of energy to counteract the disturbance. Submerging the head ropes of the long lines to 2 to 4 m below the surface will generally decrease disturbance. The effect of wave action is also dependent on the species, since those that maintain the byssus throughout their life cycle, for example, *Nodipecten nodosus*, can counteract disturbance by attaching themselves to the basket and to conspecifics minimizing the motion effect in suspended culture. On very exposed sites, even bottom-cultured scallops can suffer abrasion due to the movement of gravel and sand. Pectinid culture sites should be should be placed in areas protected from large waves (bays, behind islands); however, if a culture site is not protected from wave action, using rigid gear may reduce mortality.

Sediments

In bottom culture, the sediment type is especially important. Scallops live mainly on gravelly or sandy substrates and do not do well on muddy bottoms. Selection of a poor bottom will result in increased predation and emigration of scallops. Suitable bottom type is necessary for successful bottom culture.

Ice

In temperate zones, ice cover and ice flow are also a factor to consider when selecting a site. Generally, it is a good practice to submerge culture gear during the winter to avoid entanglement of gear with the ice.

Depth

Depth is another important factor to be considered. Several studies indicate the adequate depths for culture of various pectinids, however, depth is not a single factor, and several factors interact, such as temperature, oxygen, salinity, quantity and quality of phytoplankton, and other environmental factors such as fouling organisms. Thus, the establishment of the culture depth is dependent on the species and on the site characteristics and is critical to successful culture.

Fouling

Fouling is a biotic factor that can have major negative effects in scallop culture. The proliferation of epibiontic organisms, such as sea squirts, bryozoans, barnacles, and also algae, interferes with food availability, either by competition or by clogging the mesh of the culture gear, causing a decrease in water flow, accompanied by the decrease of oxygen, and the possible accumulation of excretion of the confined organisms. The accumulation of fouling on the shells of the cultured scallops can produce a mechanical effect that reduces the ability to open and close shell valves normally, which in turn interferes with filtration rates, and can cause rupture of the hinge. In some areas, fouling can be managed by cleaning and deploying the culture gear in a timely manner as to avoid fouling settlement (see Chapter 16).

Predators

The presence of predators is somewhat dependent on the system used and its level of protection. In open bottom culture or sea ranching, the main predators are starfish, fish, crustaceans, carnivorous snails, and octopuses. There are several techniques that have been used in the past to control predators including mopping, trapping, brine, lime, fencing, and hand picking.

In suspended culture, the main predators are fish, especially of the Balistidae family, and particularly in the tropics, where they can break the mesh of the baskets with their teeth and access the bivalves. Resistant meshes should be used or some kind of protection of the baskets.

Regardless of the culture region, scallops are maintained for several months in the culture devices and after a few months, they will likely be exposed to multiple predators. Starfish, crabs, and gastropods enter the nets as larvae, settle and later metamorphosis, and then grow and reach a size at which they may devastate the growing population. For example, in *Euvola ziczac*, predatory gastropods of the family Ranellidae (previously known as Cymatiidae), recruited within the pearl nets can produce a total mortality of the juvenile scallops in just 1 month, with a predation rate of nearly one scallop per day.

The number of predators may be reduced or avoided completely in suspension culture through good husbandry practices, such as handpicking and changing of nets with an adequate frequency. Since this is part of the overall husbandry process of managing crop densities, this approach has generally been successful if the labor costs are manageable. In any case, the culture lines should be kept off the bottom.

Diseases

Scallop larvae are particularly subject to infections by various microorganisms (viruses, bacteria, heterokont microalgae, and dinoflagellates). Special attention must be paid to the site selection and installation of a hatchery, since larval pectinids are usually much more sensitive to diseases than other species of bivalves (oysters for example), and to *Vibrio*-type bacteria. Although *Vibrio* are important agents found in the at-sea cultivation site, parasites are considered a major source of mortality. The presence of recognized pathogenic organisms should be criteria for rejection of a culture site. Advances in the field of scallop health research have followed the broader improvements in the ability to identify disease agents in other aquatic animals. Significant progress has been made in diagnostic capability, particularly the ability to distinguish between infectious pathogens and cryptic histopathological lesions in scallops using molecular techniques. A separate chapter in this book focuses on diseases of bivalve molluscs (Chapter 15). Avoiding these pathogens is a task that should always be pursued.

Contamination and toxic accumulations

The marine environment, especially the coastal waters, receives a wide range of industrial, agricultural, and domestic wastes containing contaminants. Also shelf waters receive direct contaminant inputs from the atmosphere and from offshore industries including drilling and mining, producing a wide range of biological effects in organisms including biochemical, immunological, physiological, and bioenergetic responses to stress, altered growth and reproductive output, and histopathological effects. These scenarios may have an immediate impact on the scallops, so it is essential to study the presence of contaminants in the growing area.

As other suspension-feeding bivalves, scallops accumulate toxins associated with harmful algal species, and this may cause serious threats to public health, especially since many species bind the toxins in various tissues for extended periods and become unsuitable for human consumption even when the blooms are over. The accumulation of toxin depends on the scallop species and the toxin source but in general the toxins do not accumulate in the adductor muscle, but mainly in the digestive gland, gill, and gonad. A study of the presence of microorganisms and harmful algae blooms is advisable before choosing the culture site.

In conclusion, with regard to the environmental factors, suitable areas for scallop cultivation should have levels of chlorophyll-*a* greater than 1 μg/L, low inorganic fraction of seston, high levels of dissolved oxygen, little variation in salinity, low currents but enough to allow an adequate flow inside the baskets, and temperatures within the optimum range of the species. The waters must be free from organic and inorganic contaminants and with absence of historic records of parasites and pathogens for bivalve molluscs. Additionally, areas protected from strong winds and high waves are preferable to avoid negative effects on cultured organisms. For bottom culture, the gravel/sandy sediment type is preferable. The cultivation site should be away from estuarine areas and subject to a narrow range of fluctuation in environmental factors.

Practical husbandry information

The aquaculturist must start by selecting a culture site that has the environmental conditions that meet the scallop species requirement. If the grower plans to use SCUBA equipment (to place anchor systems and to check for fouling and weakening of anchor lines), it is wise to select a site considering the type of substrate, current velocity, and turbidity.

Once the culture site is selected and secured, the gear can be purchased or made. If the grower plans to employ suspension culture, the long lines must then be constructed. Usually, suspension lines should be placed approximately 20 m apart to avoid tangling of the suspension gear when the line is lifted, and wind or current carries the vessel sideways. Tangled gear causes the scallops to fall into one side of the net or cage resulting in smothering or cutting.

The grower will need a crane to lift the suspension line, either hand operated or hydraulic. Suspension lines are then worked with a star-wheel roller. Two rollers, one on the bow and one on the stern of the vessel can be used or a single roller will work if on a power crane. Two rollers are usually best, so the vessel remains parallel to the suspension line as the grower runs along it. Long lines or cages should be submerged about 3–5 m below the surface to minimize the effect of wave action on scallops and potentially avoid warm surface water. Plastic buoys are placed along the suspension line tied close to the line with a few indicator buoys with 3–5 m lines to the surface to indicate if the line has not sunk to the bottom when loaded with gear. Excessive movement in the nets or cages tends to retard growth and the scallops may cut into each other with their shells if moved around too much by wave or current action; however, rigid cages can reduce the effect of the wave or current action.

Most aquaculturists will grow the scallops in three phases: initial culture, intermediate culture, and final grow-out to market size. Square pearl nets are the recommended gear for early grow-out of scallops in suspension and lantern nets for intermediate and grow-out to market size. The rigid cages placed in suspension or on the bottom can also be used, however, their cost effectiveness has not yet been confirmed.

When using pearl or lantern nets it is imperative to keep them in suspension. Even an overnight contact with the bottom can lead to extensive mortality. Preparing pearl nets on land will save time on the aquaculture site. Nets can be tied together in strings of 5 to 15, depending on the depth of the site. Lengths over 15 nets can lead to net damage when the pearl nets become heavy with scallops and fouling. When dealing with small scallops (<30–50 mm) square pearl nets can be laced up ahead of use. Round pearl nets must be tied as filled and are more time consuming to use on the aquaculture site. The square net is laced from the top down but a section in the middle of the lacing (~50 cm long) can be open. Scallops can be introduced through this opening which closes after weight comes on the net. In some species, if nets are not laced properly scallops will swim out. Growers should examine all nets after each use for holes in the mesh from wear and tear or predator damage. Holes must be patched to prevent scallops from escaping. When handling any grow-out gear used in suspension the grower should be careful not to quickly turn the gear onto its side when removing it from the water causing all the scallops to fall together on one side. Scallops grown in suspension have thinner and sharper edged valves than scallops in the wild. When they drop to one side of the gear the sharp shells cut into scallops that have not closed, which results in injury and delayed mortality. Gear should be lifted from the water so that it remains as level as possible giving all scallops time to close. *Placopecten* spp. usually clap vigorously three to four times when taken from the water and then close tightly.

These management practices are generalized and should be adapted to the species to be cultivated. If the cultivation of the species is not known, experimental studies should first be done at a small scale and then extended to trials at a large enough scale to mimic a commercial operation. It is recommended that the studies are carried out together with economic analysis to evaluate the factors affecting profitability of different culture methods and to identify the research directions most likely to aid in establishing culture operations, including an economic analysis to choose the optimal size for harvesting the scallops. All this information and the projections of long-term market demands can forecast the profitability and feasibility of scallop farming on a large scale.

Concluding remarks

The techniques and factors to consider summarized in this chapter are those mainly used for scallop culture in many parts of the world. Japan was the pioneer country to develop scallop culture techniques, for the species *Mizuhopecten yessoensis*. These techniques have been successfully adapted in other countries and for other species and currently, China is the world's main producer and exporter, largely supplanting Japan's production. Some countries in Europe, North America, and South America also developed small-scale scallop aquaculture operations, supplying mainly local markets, and becoming socially and economically important at the local level. Considering the increasing demand and the appreciation for the flavor of scallops, the market continues to expand; aquaculture is the only way to meet this growing consumption. Through research and development, aquaculture production is expected to continue to increase and participation among countries considered to be emerging in scallop culture will widen.

Acknowledgments

We appreciate the illustrations by Mr. Leandro José Ramírez Díaz. The manuscript was substantially improved by the editorial efforts of Sandra Shumway.

References cited and further reading

FAO (2019) Global Aquaculture Production 1950–2017. Fishery Statistical Collections, Fisheries and Aquaculture Department. 19 April 2019. Available at: http://www.fao.org/fishery/statistics/global-aquaculture-production/es.

Guo, X. and Luo, Y. (2016) Scallops and scallop aquaculture in China. In: Shumway, S.E., and Parsons, G.J. (eds) *Scallops: Biology, Ecology, Aquaculture, and Fisheries.* 3rd edition. Elsevier Science, Amsterdam, pp. 937–952.

MacDonald, B., Bricelj, M., and Shumway, S. (2016) Physiology: energy acquisition and utilization. In: Shumway, S.E., and Parsons, G.J. (eds) *Scallops: Biology, Ecology, Aquaculture, and Fisheries.* 3rd edition. Elsevier Science, Amsterdam, pp. 301–353.

Mazón-Suástegui, J.M., Maeda-Martínez, A., Robles-Mungaray, M., De La Roche, J.P., Rupp, C., Mendes-De-Bem M., Velasco, L.A., and Freites-Valbuena, L. (2011). Advances in the production of juvenile *Nodipecten* spp. In: Maeda Martínez, Alfonso N., and Lodeiros-Seijo, C. (eds.) *Biology and Cultivation of Scalloped Mollusks of the Genus Nodipecten.* Editorial Limusa S.A de C.V., Cuauhtémoc, pp. 275–311.

Robinson, S.M.C., Parsons, G.J., Davidson, L.A., Shumway, S., and Blake, N.J. (2016) Scallop aquaculture and fisheries in eastern North America. In: Shumway, S.E., and Parsons, G.J. (eds) *Scallops: Biology, Ecology, Aquaculture, and Fisheries.* 3rd edition. Elsevier Science, Amsterdam, pp. 737–779.

Saxby, S.A. (2002) A review of food availability, seawater characteristics and bivalve growth performance at coastal culture sites in temperate and warm temperate regions of the world. Fisheries Research Report No. 132, Department of Fisheries, Western Australia 42.

Ventilla, R.F. (1982) The scallop industry in Japan. *Advances in Marine Biology*, 20: 309–382.

CHAPTER 5

Mussel aquaculture

Carter Newell, Kevin Heasman, Aad Smaal, and Zengjie Jiang

Introduction

Annual mussel production from China, Europe, New Zealand, and North America currently exceeds over 1.5 million metric tons (MT), and a variety of culture methods from bottom cultures, poles, longlines and rafts are used. Other important areas of production include Chile (over 90,000 MT) and Thailand, which are not included in this chapter. While the landings, species, and methods differ among countries, it is useful to compare some of the major world mussel producers in order to get an understanding of how to grow mussels profitably and sustainably. Most successful operations around the world have adapted, through technology transfer or technology development, methods that produce the best results in specific sites. For this reason, comparative case studies of the important topics of annual production, site selection, seed collection, hatcheries, seeding, cultivation methods and stocking densities, harvesting, processing, predators, biofouling, and environmental and social interactions are presented. The mussel seed is sourced from bottom beds, suspended ropes or nets, or hatcheries, and the mussel species include *Mytilus edulis*, *M. galloprovincialis*, *M. coruscus*, *Perna canaliculus*, *M. trossulus* and *P. viridis*. But in all cases, the important steps of proper site selection, husbandry practices, mechanization, and scale of operations provide a common theme to success in mussel aquaculture around the world.

History and landings

North America

There has been a long history of wild mussel (*M. edulis*) harvesting in the US due to the abundant natural resources in New England and Maine, peaking in landings over 35,000 MT a year in the 1980s and 1990s, and declining to under 5000 MT recently. Massachusetts also had some landings over 5000 MT in the early 1990s from deep water beds in over 20 m of water near Nantucket Shoals.

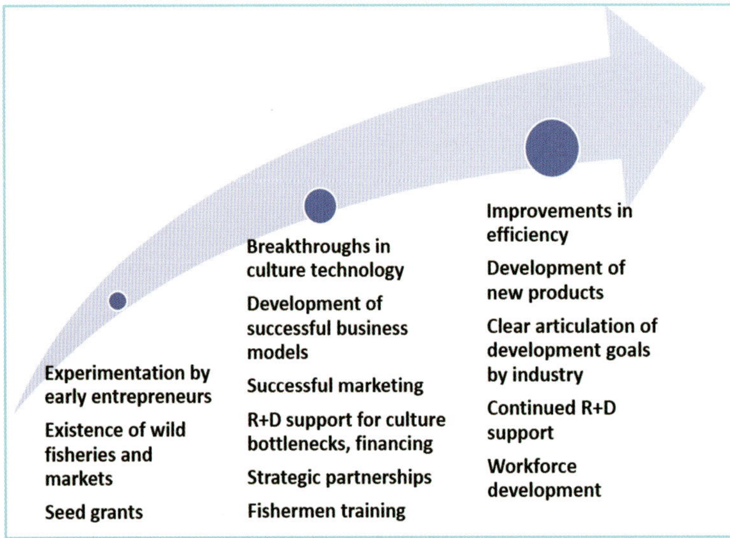

Figure 5.1 Essential elements of successful shellfish aquaculture development strategies. Courtesy of Northern Economics Inc. (2015).

The mussel aquaculture industry began in the US with the first mussel farm lease site by Ed Myers and research being conducted into the feasibility of rope and bottom culture in the 1980s in Maine and Washington State. Pioneering companies using wild harvest and bottom culture (Great Eastern Mussel Farms, Blue Gold) and rope culture (Penn Cove) helped introduce higher quality mussels to the US market in the 1980s. Within a decade, however, the market was dominated by mussels from Prince Edward Island (PEI), which received development support from the federal and provincial governments of Canada. A review of case studies that show the essential elements of successful shellfish aquaculture development strategies was recently carried out for the state of Alaska (Fig. 5.1). Mussel culture on the U.S. west coast also involves another species, *M. trossolus*, and sometimes growers use hatchery-reared *M. galloprovincialis*.

Landings of wild, bottom cultivated, and rope cultivated mussels in the US are currently 3000 MT (~6 million pounds), with about 75% of the farmed mussels coming from rafts in Washington State and on the east coast and Maine with a mix of bottom culture and raft culture. Rafts are the primary form of suspension culture in the US due to the presence of predatory ducks; rafts can support predator nets to keep the ducks away. The raft technique was developed in Spain, where they grow 225,000 MT/year. In contrast, mussel farming in nearby Canada has grown to over 25,000 MT, 80% of which are grown in PEI using a longline technique and individual droppers. For the fresh farmed mussel market in the US, Canadian imports, primarily from PEI and Newfoundland, account for the over US$40 million or 80% of the live mussels consumed.

Mussel consumption in the US in 2015 was estimated at ~50,000 MT with a value of US$150 million, with approximately one-third of the mussels coming from Maine (fresh wild and farmed), 25% from Canada (fresh farmed), and about 40% coming from New Zealand and Chile (frozen). A projection of mussel demand for the future shows a promising potential for Maine raft grown mussels to increase to 5000 MT over the next decade, and also coincident with a decline in the harvest of wild mussels in that state.

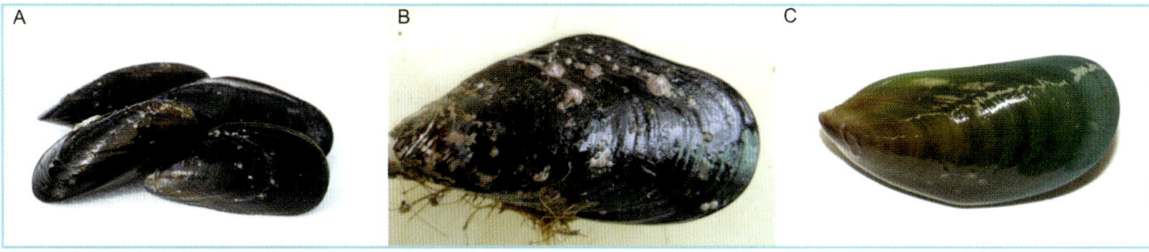

Figure 5.2 The three important mussel species cultured in China: (A) *Mytilus galloprovincialis*; (B) *M. coruscus*; and (C) *Perna viridis*.

China

China is the largest mussel producer with more than 878,000 MT in 2016. According to FAO statistics, the total mussel production from aquaculture in China accounts for over 45% of global mussel production. Approximately 50 species of mussels are found along the coast of China. The most important species being cultured are the blue mussel, *M. galloprovincialis*, the thick shell mussel, *M. coruscus*, and the green mussel, *P. viridis* (Fig. 5.2).

Shandong Province is the largest producer of cultured mussels, followed by Zhejiang, Fujian, Guangdong, Liaoning, and Jiangsu provinces. Mussel aquaculture was initiated in the early 1970s and rapidly developed in the early 1980s when farmers improved the technologies of hatchery, seed collection, and on-growing. Mussel production has been declining since the early 1990s because of the low economic value compared with high-value species such as scallops and oysters, among others.

New Zealand

Mussel farms are distributed around New Zealand at a number of sites (Fig. 5.3). Marlborough Sounds and Coromandel produce the majority of the green-lipped mussels (*Perna canaliculus*) in New Zealand (Table 5.1). These two areas also have the highest production per hectare. The current production has fluctuated between 86,000 MT and 101,000 MT per year between 2011 and 2016. This equates to a free on board (FOB) value of NZ$191 million and NZ$268 million (Table 5.2).

Table 5.1 Farm numbers, sizes, and mean production for the current inshore farming areas as of 2016. Source: Aquaculture New Zealand.

	Farms	Total area (ha)	Average production (MT) from each region	MT/ha/year
Region				
Northland	5	20	200	10
Auckland	40	100	4250	42.5
Coromandel	100	1450	24,750	17.07
Nelson Bays	15	300	4750	15.83
Marlborough	565	3000	55,000	18.33
Canterbury	15	150	1750	11.67
Southland	20	200	2750	13.75
Total	**760**	**5220**	**93,450**	**17.9**

Europe

In Europe, two species of mussels are produced, the blue mussel,

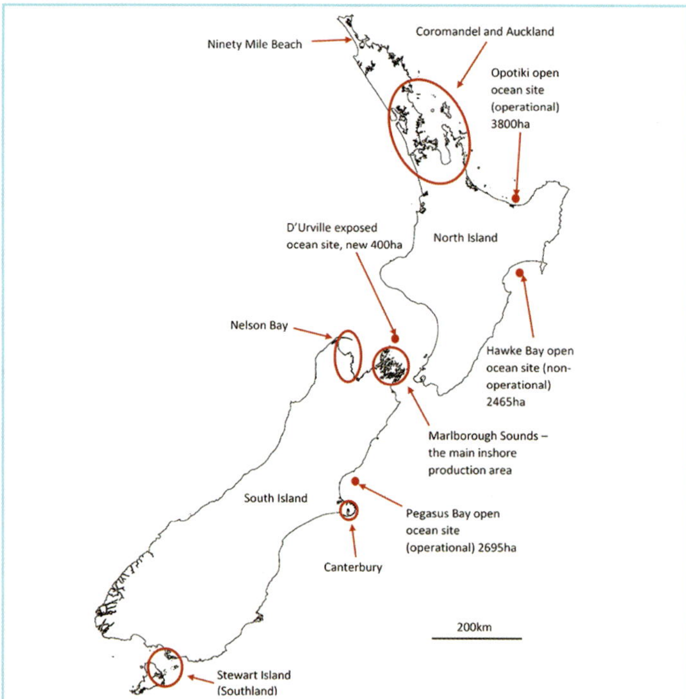

Figure 5.3 The mussel production sites of New Zealand. Courtesy of Kevin Heasman.

Mytilus edulis, and the Mediterranean mussel, *M. galloprovincialis*. The species *M. trossulus* is also found in Europe, predominantly in the Baltic, and is not appreciated for aquaculture. Harvesting and cultivation of mussels in Europe dates back to historic times. The ancient Romans stored shellfish in marine ponds, in Greece, shellfish were cultured already by 500 BC. Descriptions of pole (*bouchot*) culture in France date back to 1235. Mussels were on the menu in Italy and Spain more than 2000 years ago, although production in the Spanish rias in Galicia emerged only after 1950. In the Netherlands, wild mussel fisheries were described in the 17th century; lease sites were first granted in 1823 in the Western Scheldt estuary. In Norway, mussels were harvested in the past as duck feed rather than for human consumption. After a period of trial and error, partly due to harmful algal blooms, mussel production in Norway has been limited to 2000 MT; new applications have been developed as part of integrated multi-trophic aquaculture with salmon farming. In the UK and Ireland, mussels have been fished and cultivated since historic times as small-scale bottom culture in the many estuaries and loughs. Mussels were popular in England

Table 5.2 Quantity, unit price, and free on board value of Greenshell™ mussel (GSM) production from 2012 to 2016. Source: Statistics NZ.

	2012	2013	2014	2015	2016
Quantity (MT)	33,600	28,700	33,100	27,900	33,300
Revenue FOB (NZ$ million 2016)	191	180	252	223	267
Unit price (NZ$/kg)	5.68	6.29	7.65	8.04	8.03

as food, while in Scotland they were used for bait, and in Ireland they were not popular for consumption either, maybe due to hygiene problems.

Mussel production gradually evolved from wild fishery to culture by the end of the 19th century. Improved infrastructure such as train connections facilitated shellfish transportation. Traditional mussel culture predominantly consisted of bottom culture in north-western Europe; along the French Atlantic coast, characterized by a large tidal range and high current velocities, intertidal poles were used: the bouchot culture. Suspended culture systems have been in use in the Mediterranean since historic times, such as the Italian pergolare (mussels seeded into mesh socks) system: stakes with ropes in between. This type of suspended culture is still in use in the French Thau Lagoon and various bays in Italy. Raft systems as used in the Galician Rias in Spain were first developed along the Mediterranean coast in Spain in the 1930s. Longline suspended culture is presently the dominant mussel culture technique, and is nowadays also used in offshore conditions.

Total mussel production figures, based upon FAO data, show a steady increase since the start of data collection in 1950. After a peak in the 1990s (800,000 MT) there was a general decline in aquaculture and wild fisheries yields down to about 600,000 MT. Production mainly comes from aquaculture. Wild fisheries reached a peak in the 1990s with a maximum harvest of 183,000 MT in 1993. Production later decreased to ca. 50,000 MT. Also, aquaculture production has decreased since 2000 and amounted to 476,000 MT in 2015.

According to FAO data since 1995, landing values have increased with the yields and amounted at maximum to US$600 million in the years 2008 and 2011. In 2015 landing values for *M. edulis* were US$233 million and *M. galloprovincialis* US$216 million.

The price per kilogram of all landed mussels, that is, both species from culture and capture, shows an increase over the period 1995–2008 and a stabilization since at around 1–1.2 $ kg^{-1}. The price increase and the decrease of local supply implies a market pull. Indeed, nowadays 20% of demand is satisfied by imports. Yet the market demand provides an opportunity to increase the local production.

Countries and regions

In Europe, mussels are cultured in 23 countries. The main production of *Mytilus edulis* is from aquaculture in the Netherlands and in France (at about 50,000–60,000 MT), while in Denmark, production (at about 40,000–50,000 MT) is based upon wild catch. Ireland, the UK, and Germany produced around 10,000–15,000 MT in 2015. In Norway and Sweden, production was 2700 and 1500 MT, respectively. The production *M. galloprovincialis* includes the output of Spain at 225,000 MT (raft culture), Italy at 64,000 MT, Greece at 18,000 MT, France at 10,000 MT in 2015, and limited amounts from Portugal and Bulgaria.

The other European countries that produce mussels have annual yields of less than 1000 MT. Small amounts of *M. edulis* are produced in Iceland and the Channel Islands, and *M. galloprovincialis* is cultured in areas along the Adriatic Sea, in Croatia, Slovenia, Montenegro, Bosnia, and Albania as well as along the Black Sea in Romania, Ukraine, and the Russian Federation. Given the extent of the coastline and the vicinity of consumer markets, there might be quite a potential for expansion in these areas.

Site selection

A full discussion of site selection is available in Chapter 10.

North America

Owing to the effects of waves on aquaculture structures and vessels, mussel aquaculture sites for both suspension and bottom culture have, until recently, been limited to areas which are not exposed to long period ocean waves. An approach to mapping large coastal areas in relation to wave energy was performed in a modeling study in Penobscot and Machias Bays in Maine, where large areas of the coastline were identified where the significant wave height was less than 2 m (Panchang et al., 2008). The use of hydrodynamic flow models is also useful for finding coastal areas where currents are adequate to provide enough flow through the farms to minimize the depletion of algae and detritus particles that the mussels feed on. Optimal tidal velocities for mussel rafts (25–35 cm s^{-1}) were identified in Maine where the currents are strong enough to prevent food depletion in the rafts but also provide optimal velocities for mussel feeding (Newell et al., 2001). Optimum velocities are related to the effects of aquaculture structures such as mussel rafts on water velocities through the rafts (Newell and Richardson, 2014).

Water velocity, on a bay scale, can be modeled as a primary screening tool, where, for example, the lime green areas in the map of Blue Hill Bay, Maine have mean depth averaged water velocities of 25 cm s^{-1} (Fig. 5.4).

Water velocity is important, not only in maximizing the production capacity of the farm, but also for the environmental effects it has on benthic organisms due to organic enrichment from feces, pseudofeces, and drop-off. Since food supply to large populations of mussels is a direct function of the volume flux of water through the farm, it is important to choose sites that have

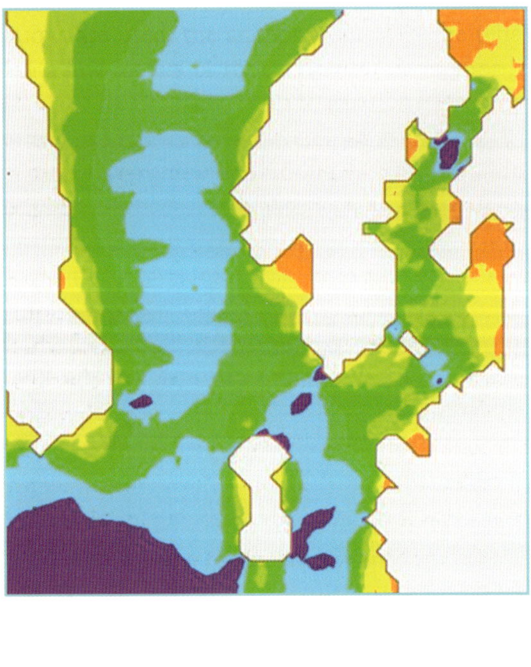

Figure 5.4 Modeled water velocities in Blue Hill Bay, Maine, with decreasing values from purple, blue, green, yellow and brown. The green represent good areas for mussel rafts. Courtesy of Carter Newell.

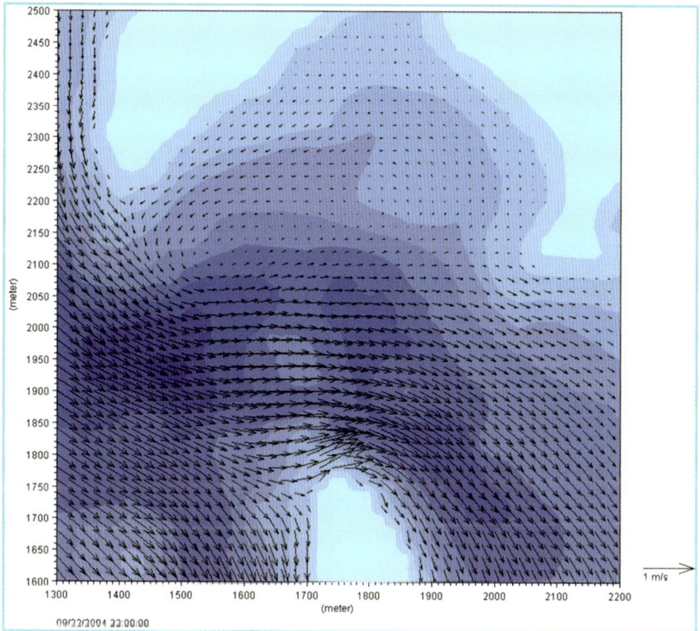

Figure 5.5 Water velocity vectors in Clark's Cove, Bristol, Maine.

adequate currents to support the nutritional needs of the bivalves. Resuspension and dispersal of organic matter is also facilitated by tidal water velocities of over 25 cm s^{-1} and waves, as well as by bioturbators such as crabs, lobsters, and starfish. In addition, for aerobic decomposition of organic matter, oxygen flux to the benthos is a function of water velocity. For bottom culture, water velocity dictates the optimal seeding density on bottom plots (assuming the mussels are spread evenly and not planted in dense, multi-layer patches on the bottom). Thus, for example, in Mud Cove, Maine, mussels could be spread at twice the bottom density in areas of the lease site that had twice the mean water velocity. An example of a convenient output of depth averaged tidal velocities in the Damariscotta River, Maine area, is shown in Fig. 5.5, where the length of the arrows and their direction indicate the magnitude and path of the water flow through the farm. As a rule of thumb, the blind-ended shallow coves have much lower flows than the deeper channels and banks going into coastal embayments. The water depth or bathymetry of the growing areas are useful indicators of the flow in those areas, and coarser sediments also reflect higher flow rates or wave resuspension.

Using an approved water quality classification is a key aspect of site selection for mussel farms, where most mussels are grown in FDA-approved shellfish growing areas, where no depuration is required. The Maine Department of Marine Resources (DMR) has developed a GIS tool which shows existing and pending aquaculture lease sites in relation to a layer of water quality classification, which is updated monthly. Other layers included in this GIS tool are seabird nesting sites, eelgrass beds, and other features that would limit the ability of the applicant to obtain a private lease site at any given area. Knowing the history of harmful algal blooms (HAB) in an area is also useful for site selection. Certain regions where upwelling occurs may stimulate blooms of HAB species such as those in the genus *Alexandrium.* Other areas may be prone to fall blooms of *Pseudonitzchia* spp.

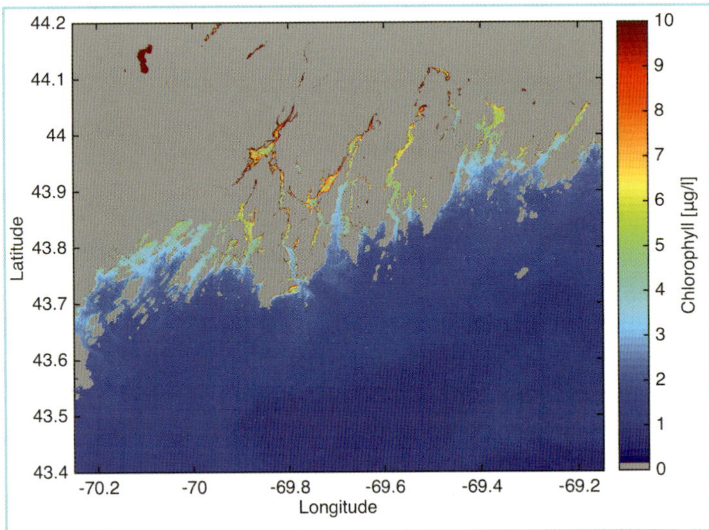

Figure 5.6 Landsat image (corrected) showing chlorophyll-*a* along mid-coast Maine on August 23, 2016. Source: University of Maine.

Mussel growth is a function of environmental growth drivers, or the conditions under which animals can feed and grow. For blue mussels, while they survive over a wide range of conditions, the most rapid growth and best survival rates are in temperatures below 20°C, in salinities over 25, with chlorophyll-*a* levels over 2 µg L^{-1}, and without a large amount of resuspended sediment that dilutes the phytoplankton and organic detritus particles with inorganic matter (clay and silt), which is indigestible. Mussel growth and meat yields are highest in areas characterized by a large percentage of phytoplankton particles in water samples analyzed using a flow cytometer. In general, river mouth areas that have low salinity and lots of sediment are avoided for mussel farming; however, the low salinities result in low abundances of starfish and therefore they might be good areas for seed collection. Recently, autonomous oceanographic buoys have been able to collect seasonal long-term data records of chlorophyll and high-resolution satellites, such as Landsat and Sentinel, have been able to capture large coastal areas in sufficient resolution in the nearshore area to be useful. Images (Fig. 5.6) demonstrate that chlorophyll-*a* levels are higher in estuaries than in the open ocean, due to the geomorphology that provides better light availability (shallower) and a longer residence time for the phytoplankton, allowing them to bloom. They also reflect daily and fortnightly variations in food availability in macrotidal estuaries.

Recent work is allowing the prediction of bivalve growth using the model Shellsim (Hawkins et al., 2013) for mussels, oysters, and other species in a GIS system based on corrected satellite images of temperature, chlorophyll-*a*, and turbidity (and derived particulate organic matter, or POM).

Europe

The selection of sites for mussel culture in Europe has, for the most part, been established historically, as there is a long history in mussel culture. This is the case for bottom and pole culture in northwestern Europe. Sites have been in use for many years and, in general currently, no new space is available for expansion due to regulations as well as other users. For example, bottom culture has been developed in the Netherlands since 1870 when the government issued licenses for the use of culture plots in the south-western Delta area. Through trial and error, the farmers determined which sites could be used for culture. Boundary conditions are factors such as salinity and turbidity, while local factors such as current velocities are critical: the flushing away of the mussels by high currents or wave action sets the maximum and supply of food and oxygen and removal of silt sets the minimum required currents. Farmers use the characteristics of the sediment as a proxy for the local hydrodynamics. For bottom culture, local adaptations are sometimes needed, due to coastal engineering works or due to changes in the local geomorphology. This is, however, a complex process. Selection of sites to compensate for decrease in quality is a tailor-made process in which government and other users, for example, shrimp fishers, are involved. New sites are selected by the farmers, and after a test phase successful new plots are registered and old plots are withdrawn. Most of the traditional pole, or bouchot culture sites date back to the 13th century. Owing to siltation in some areas, farmers shift to longlines rather than bouchots, as in Aiguillon Bay. In general, the extensive use of coastal areas by other users limits the opportunities for expansion of the existing traditional bottom and pole mussel cultures.

Suspended cultures, for example, raft and longline rope culture, depend on water depth and need some protection from hydrodynamic forces. Fjords, loughs, and rias are used for large-scale suspended cultures and abandoned ferry harbors or other protected sites are used for small-scale longline cultures. The use of longlines dates from more recent times in comparison with bottom, bouchot, and raft culture systems. Site selection is based on boundary conditions, that is, protection from waves and storm damage, but also on space availability, that is, competing claims of other users. In addition, water quality (salinity, pathogens), food availability, and risk of harmful blooms can be limiting as well. For example, in some Scottish lochs, low salinity in the top layers, due to freshwater discharges, has promoted the development of a population of *M. trossulus* on the culture ropes that has less market value. As a consequence, farms had to be abandoned.

The concept of carrying capacity may provide a framework for optimizing site selection (Smaal and van Duren, 2019).

- Physical capacity defines the boundary conditions of a given site for shellfish culture.
- Production capacity concerns the production potential of a site, given the food production and transport potential, and the stock size of food competitors.
- Ecological and social carrying capacity determine the extent to which shellfish farming has no unacceptable impacts on the ecosystem, and to what extent the local society accepts a certain size of shellfish culture in the area.

Seed collection

North America

A number of studies have been carried out on the timing of mussel larval settlement on rope collectors, and more recently related to mussel raft culture. The timing and distribution of mussel larvae, pediveligers, and drifting juvenile mussels were studied in Maine where it was found that, while settlement of larvae (about 250 μm) peaked in Maine coastal waters in June and early July, drifting juvenile mussels up to 2 mm long could be found in high current areas during other times of the year. Ropes are often coiled to increase the surface area of ropes near the surface (Fig. 5.7) where spatfall tends to be heavier, and often the settlement of starfish is lower than in deeper waters. Growth of mussel seed on collectors is a function of the density on the ropes (that is, mussels grow to a larger size if the density is lower (1000–5000 per meter) than if the density is higher (>5000 per meter) and the water temperature is higher (seed grows faster in warmer water).

Growers have used different types of ropes, Christmas tree rope, and fuzzy rope, as well as frayed rope, strips of mesh, or even nets suspended from high density polyethylene (HDPE). Seed is also collected on single droppers off longlines, continuous longlines, or on 16 m × 16 m predator nets put around mussel rafts, where the seed can be power washed off the nets and attached to ropes using biodegradable cotton. On Maine mussel rafts, the use of mussel pegs prevents the seed from falling off at high densities, but it has been found that smaller diameter (1/2 or 5/8 inch) rope limits the number of mussels collected and therefore results in better growth rates of the seed than 'fuzzy' or 'X-mas tree' rope. When ropes are put out for seed collection, they are usually weighted to prevent tangling – with the case of the 13 m long dropper ropes, plastic weight cases filled with cement and with holes for the ropes to go through are used. Some growers use hollow seed collector droppers with spikes in them to weigh them down, others use a New Zealand-style continuous longline system with weighted ropes inside old circular salmon pens (to prevent duck predation), and in some cases, mussel settlement on the grow-out ropes can be recovered during harvesting and grading, and then put right back out on the ropes after harvesting.

If collectors are put out too early, they may get covered with fouling organisms (diatoms, tunicates, and others) which compete for the surface on the ropes with the mussel spat, and the yield may be poor. The timing of spat settlement can be estimated through water sampling and looking for pediveligers, the numbers of which increase prior to spatfall. A good practical guide to monitoring spat through water sampling is given in Brown et al. (2000).

Figure 5.7 Mussel seed on a coiled pegged rope in September. Courtesy of C. Newell.

It is important to harvest and grade the mussel seed before it grows too large and falls off the ropes when a dense settlement occurs. If a good grow-out density of mussels on ropes is 500 m^{-1} and seed density is 5000 m^{-1} or more, it must be thinned and graded, and put back on grow-out ropes in a seeding process called 'socking' or 'retubing.'

Seeding the grow-out ropes

Mussel seed ropes can be harvested using a continuous longline system where the seed is stripped off the fuzzy ropes, and reattached to a continuous rope at the right density (Fig. 5.8), harvested using a conveyor, or using a Spanish-style basket (Fig. 5.9), which prevents the mussels falling off the ropes when they leave the water and their weight is three to four times greater than in the water (preventing 'drop-off'). The mussel seed is removed from the ropes, declumped, and graded into smaller and larger sizes so that the seeded mussel rope, or tube, is a smaller diameter for smaller seed and a larger diameter for larger seed. In the New Zealand-style system (Fig. 5.8), the density of the seed on the rope is controlled mechanically. In the case of the Spanish-style system (Fig. 5.10), the Aguin socking machine can be adjusted to control the rate at which the rope is fed and also the volume of mussels feeding the rope with a screw auger. In the Italian-style system (Fig. 5.11) or similarly, in PEI, Canada, seed is fed into individual socks which have mesh large enough for the seed to crawl out.

Predation of mussel seed may occur from predatory fish, diving ducks, or starfish, which often settle 1 to 2 months after the primary mussel spatfall. In Scotland, large tanks have been built that allow the longlines to be dragged through the tanks

Figure 5.8 Mussel seeded onto ropes from bulk bags using a New Zealand longline system in Scotland. Mussels were previously harvested, stripped, and graded and are attached to the double longline system as they are fed through the machine wrapping the seed on with cotton. Courtesy of C. Newell.

Figure 5.9 Seed lines ready to be thinned. Courtesy of C. Newell.

Figure 5.10 Retubing mussel seed onto pegged grow-out ropes using an Aguin A2 machine and biodegradable cotton in Galicia, Spain. Courtesy of C. Newell.

which contain quicklime, which kills the starfish. In Maine, the coiled seed collector ropes are taken out of the water and air dried overnight on the raft timbers to kill the starfish.

If left untreated, starfish can decimate the mussel crop and significantly reduce yields per rope. It is also noteworthy that seed collected in areas that have some freshwater input and lower salinity often do not have starfish on them. Mussel seed is protected from diving ducks and fish by placing predator nets around the rafts. In Maine, predator nets with 75 mm squares of 3 mm twine are effective, and in Washington State, smaller sized mesh is needed for the fish predators.

China

Depending upon the abundance of mussel spat in nature and the cost of hatchery seed, mussel seed are either collected from the wild or produced in hatcheries. In terms of the blue mussel, hatchery-produced seed

Figure 5.11 Italian sock system. Courtesy of C. Newell.

were initially used during the 1970s to supplement wild-set spat, but now there is a reliable and abundant supply, so the blue mussel seed were generally obtained from the wild; however, much of the thick shell mussel (*Mytilus coroscus*) farming relies upon hatcheries for the source of seed due to the unpredictability of wild seed supply.

Natural spatfall

Mussels have two spawning seasons per year in China, in spring and in autumn, consequently producing spring spat and autumn spat. The spring spawning season lasts about 3.5 months (from late March to early July) and the autumn spawning season about 2.5 months (from early October to mid-December). Spring spat constitute the main stock and there are usually very few autumn spat. Mussel spat fall is largely dependent upon spring spat.

Spat collection activities require suspending spat collectors in the water just prior to the predicted maximum spat fall. Timing is important, as deploying the collectors too late will miss the set, and deploying too early will allow fouling by other marine organisms that would compete for space with the setting mussels. The exact timing of settlement varies between sites, but the main spat fall typically occurs as an annual maximum between late April and early July.

A variety of spat collectors are employed, including polyethylene and palm-coconut fiber ropes. Recently, sliced tires have been widely used. The length of spat collectors is normally 1.5–2 m. Normally, 250–300 spat collectors were hung on each headline 0.4–1.0 m apart. In the large-scale kelp aquaculture sites, mussel larvae occur in May and June to early July, just about the same time that kelp is harvested. Kelp longline rafts made of straw are very good for collecting spat. So, the farmers there developed a creative spat fall approach by keeping the kelp longline rafts in the sea after harvesting the kelp. This approach has proven to be an effective means of collecting a large number of spat. In August and September, plenty of seed mussels 5–30 mm in length can be found on these longlines. Once mussels set, they typically remain on the spat ropes until they grow to become 0.7–1.0 cm seed mussels within about 2 months at which time they can be stripped.

Mussel hatcheries

Most of the seed of *Mytilus coruscus* were collected directly from naturally occurring stocks until the successful artificial breeding implementation in 2008. Shellfish hatcheries are described in other chapters in this book, and involve choosing and conditioning mussel broodstock to spawn, rearing single celled phytoplankton to feed the broodstock, the larvae, and the mussel spat, and providing substrates for the mature larvae to settle.

The microscopic phytoplankton should be cultured indoors. The whole culture unit normally takes up one-third of the area of the hatchery. The light, temperature, salinity, and filtration of the water must be strictly controlled during the cultivation process. The algal production should be carried out about 2 weeks before broodstock is transported into the hatchery tanks. Several algal species, including *Platymonas* spp., *Phaeodactylum tricornutum*, *Nitzschia closterium f. minutissima*, and *Dicateria zhanjianggensisi* are suitable feed for broodstock.

Broodstock enhancement

At the beginning of spring (in the middle of February), the broodstock of thick-shelled mussels (above 7 cm) are moved to indoor tanks with natural and filtered sea water. The density of broodstock is about 3.0–4.5 kg m^{-3}. The water temperature is increased by 0.5°C per day until it reaches 20°C. The mussels are fed on microscopic phytoplankton at a concentration of 30,000–50,000 cell mL^{-1}, four to six times daily. The water is exchanged (80%) daily.

Spawning

Spawning is usually induced by thermal shock, whereby the mussels are removed from the water for 3 hours and then put back in the tanks with fresh sea water at 23°C. About 2 hours later, the release of gametes will start. One animal's release will stimulate others to follow. The mussels should be removed from one tank to another when the eggs in the water reach a concentration of 2–30 mL^{-1}. Aeration is maintained after spawning and surface foam is removed using 150 μm silk at the same time. The aeration is stopped 1 hour after spawning and manual mixing of the sea water every 1–2 hours implemented.

D-stage larval rearing

About 40 hours after fertilization, the fertilized eggs develop into the characteristic D-shaped larvae with two shells. Once mussel larvae develop to the D-larvae stage, they can be reared to metamorphosis in other tanks. If conically based containers are employed, then aeration should be provided because it has been shown to improve larval survival in the early stages. Cages made of 45 μm silk are used to collect D-larvae to transfer to another tank.

The D-larvae are reared at 21–23°C. They are fed with microscopic phytoplankton, *Nitzschia closterium f. minutissima, Pavlova virdis, Isochrysis* spp., with 30,000–50,000 cell mL^{-1} five to six times daily. The water is changed 50% daily. Initial stocking densities of approximately 10 larvae mL^{-1} appear to be optimal, allowing good survival rates, although larvae can be cultured at higher concentrations. The growth rate of larvae is temperature sensitive, with higher temperatures resulting in higher growth rates.

Metamorphosis and spat settlement

Toward the end of the planktonic larval phase, mussel larvae develop a characteristic pigmented 'eye spot', usually at a shell length of 200–260 μm. They then also develop a foot, which rapidly becomes functional in crawling. The velum starts to degenerate and is replaced by buds that will later form the gills (gill buds). Larvae at this stage are known as pediveligers.

The pediveliger larvae explore potential places to settle by crawling around on the substrate using their foot. If the pediveliger does not come into contact with a suitable substrate, it withdraws its foot and may swim away while the velum is still intact. Larvae are able to delay metamorphosis for several weeks enabling them to seek more suitable locations. When 50% of larvae have developed eye spots, the larvae are ready for settlement.

The substrates (fishing net, coconut palm rope) onto which the mussels will settle can be disinfected and then introduced directly into the larvae tank. It should be about 10 kg substrates per cubic meter. Settling larvae need to be fed a daily ration of 50,000–60,000 cell mL^{-1} of a mix of microalgae. Water exchange should be about 70–80% daily. Deployed in setting tanks, mussels are transferred at a 1 mm size to a nursery, where they will remain until they reach 6–10 mm; then the spat is moved outdoors into grow-out systems.

Seeding

After two months of growing on the nursery rope, the juveniles reach about 0.7–1.0 cm, the spat collectors are retrieved, and the mussel seed is stripped from them. Socking is usually done manually on shore by wrapping the seed mussels around the grow-out rope with a piece of mesh (Fig. 5.12A and 5.12B). The grow-out rope is 2.0–2.5 m long and made of polyethylene with 3.0–3.2 cm diameter or the rope hinged with four palm ropes of 1.2 cm diameter. The density is controlled at between 750 and 950 mussels per meter of rope. Generally, the socking mesh is made of polyethylene with 0.5–1.0 cm diameter (Fig. 5.12C). When the seed have firmly attached to the grown rope after 1–2 days, the mesh can be taken away. Nowadays, the farms in Zhoushan Island of Zhejiang Province are starting to use the biodegradable socking material (such as cotton) to protect the environment and to save on the cost of labor. The grow-out ropes are then suspended from the surface headropes where they will remain until harvest time.

With regard to the transplanting process of seed mussels to growing ropes, some farmers prefer to insert masses of seed between the strands of growing ropes and then fasten them to the

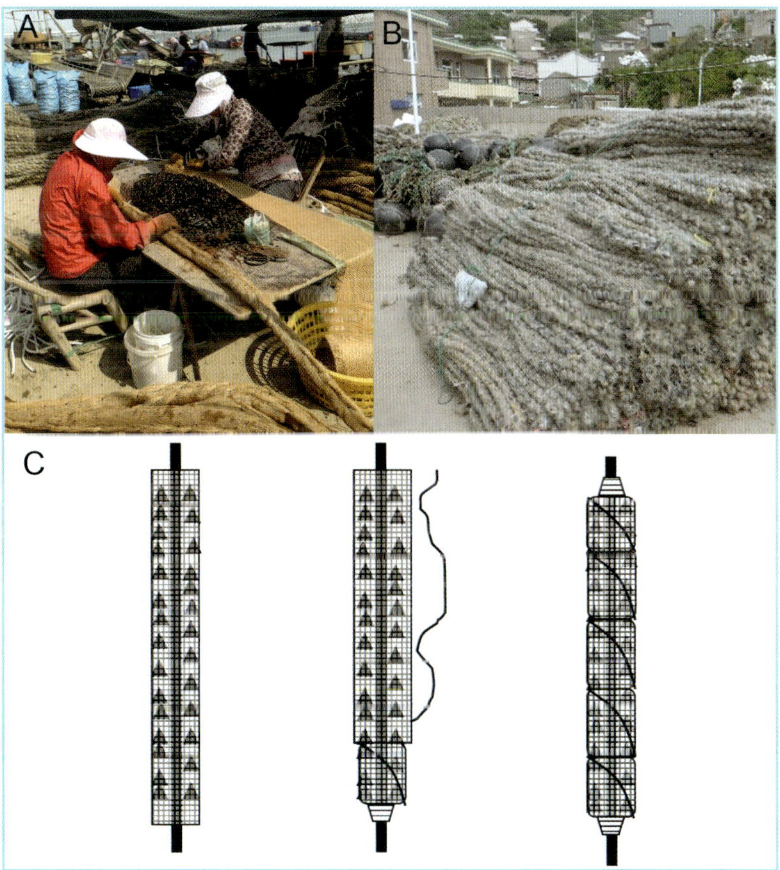

Figure 5.12 (A) Socking the seed mussels and wrapping the juvenile mussels. (B) The grow-out ropes. (C) Schematic of the bags. Courtesy of Zengjie Jiang.

head ropes. Sometimes the spat collectors on which spat have already been collected are twisted around growing ropes and the spat then attach to the growing ropes later. At other times, the transplanting is done by putting growing ropes in a container, a shallow tank or a boat, scattering some seed mussels on them, and pouring seawater into the container or keeping water running over them. Seed mussels settle on the growing ropes in an hour or so.

Most farmers in China consider that the time and cost expended in 're-socking' is uneconomical and normally they prefer to allow 'natural thinning' to occur as stock falls from the droppers under its own weight.

New Zealand

At the very start of mussel culture in New Zealand, the mussel seed was collected from surfaces on which they had naturally settled. It soon became obvious that it would not suffice in the long term and alternative sources were sought. A spat source was identified from an expansive beach area on the west coast of the far north peninsula of North Island of New Zealand. This beach (called Ninety Mile Beach, even though it is not ninety miles long) has periodic strandings of organic beach cast material (hereafter called Kaitaia weed) consisting mainly of fragments of seaweed, sponges, hydroids, and other small animals including mussel spat. This occurs mainly from July to December (mid-winter to early summer). Mussel spat recovered from the Kaitaia weed, ranges in size from 300–1500 μm, and their numbers can be up to 4 million per kilogram of Kaitaia weed. The Kaitaia weed is deposited ashore in large quantities, that is, hundreds of tons. This Kaitaia weed is managed within a quota system to avoid environmental impact of depriving the ecosystem of this organic resource. The Kaitaia weed is collected, bagged (damp but not wet) and transported by refrigerated vehicle to the main growing sites (Coromandel and Marlborough Sounds) where it is placed on seeding vessels and taken out to the farm for seeding. This process may take up to 24 hours. About 80% to 85% of the industry spat has traditionally been derived from this source.

On arrival at the site where the spat will be on-grown, the Kaitaia weed is transferred to a spat rope (Fig. 5.13). This process requires a mussel stocking (sock). This mussel stocking is made of cotton and is in the form of a continuous tube ~10 cm in diameter unstretched. The stocking is fed onto the outside of an 80 mm pipe. The spat growing rope is pulled through the 80 mm pipe, which draws the stocking over it. At the point where the pipe and the stocking come together, the Kaitaia weed fragments are uniformly dropped. The result is that the stocking holds the Kaitaia weed onto the rope. Once in the water the mussel spat will move around attempting to find good water flow and a good settlement position and a proportion of the spat will attach onto the rope. The stocking will rot away after 4 to 6 weeks and the Kaitaia weed floats free. The spat that attached to the rope (estimated at between 1–4% of the spat available in the Kaitaia weed) remain as the growing product. This is only successful due to the large starting numbers of spat found in the Kaitaia weed.

The available physical space on the rope allows for large numbers of small spat to settle initially but as they grow they run out of attachment space and thin out rapidly. By the time they are at a size where they can be easily handled without mortality and for ease of distribution (15 mm to 20 mm in length), they will be graded and thinned to a stocking density of 600 to 1200 per meter of rope. As they increase in size (35 mm to 60 mm) they will be graded and thinned again, to approximately 160 to 200 per meter of rope (depending upon site) and on-grown for production. The seeding procedure is completed using an adjustable conveyor.

Figure 5.13 A good seed rope with consistent seed coverage and size. The macroalgae (*Undaria* spp.) is seasonal and in low quantity is not an issue. Courtesy of Kevin Heasman.

It takes a specific set of circumstances for good spat to be cast onto Ninety Mile Beach. First the spat must have been produced and settled in the diverse substrate that will end up being part of the Kaitaia weed. Then the weather conditions have to be right to bring it ashore where it can be collected (collecting Kaitaia weed at sea, away from the beach is illegal). Although Kaitaia weed comes ashore in most years, there have been years where the Kaitaia weed does not come ashore and/or the spat load is very poor causing challenges for the industry.

There have been rare occurrences where the collection of the spat has been influenced by other challenges. For example, there was an instance in the early 2000s where a bloom of the very toxic alga, *Gymnodinium catenatum*, occurred at Ninety Mile Beach. Since this toxic alga was not found in the growing areas further south, the transport of Kaitaia weed from Ninety Mile Beach to the growing areas was banned. Even after the bloom had disappeared, it had to be proven that the cysts of the bloom would not be transported via the Kaitaia weed to the growing areas. This took some time and influenced production cycles in the years following.

An alternative spat source, which has traditionally provided up to 20% of spat, is what is referred to as 'local catch.' This spat is caught from January to June (early summer to late autumn) in specific areas identified to have natural spat settlement. Small spat-catching ropes are placed in the water during the spat season. These are collected every week and inspected for spat. If spat settlement numbers are increasing, the monitoring frequency is increased. If a good spat fall is developing, all industry members are notified and spat-catching apparatus is deployed. The apparatus may range from core string on frames to lead cored 'spat-catching rope.' The structure of spat-catching rope may vary from a rope with a short stiff brush like surface to a rope with multitudes of frilly loops over 25 cm in cross section, which provide massive surface area for the spat to settle. The ropes are lead cored to get them to sink. Leaving ropes for extended periods in the hope of catching spat has the advantage of knowing the catching apparatus is in the water when the spat do appear; however, a rope that is fouled will not catch spat and the rope has to be recovered, put through a cleaner and stored for the next event. With hundreds of kilometers of spat-catching rope deployed at each event, if the spat are not caught there are significant volumes of rope to recover and clean, thus there is an art to the procedure.

Settled spat need to be managed. Poor conditions, for example, poor food availability, low currents, or over-settlement by fouling will induce the spat to 'sail' away. They exude a mucus strand and use the water current to 'sail' away to greener pastures. Spat ropes with over 100,000 spat per meter can be naked of spat in days if they are unhappy.

Hatchery spat

Mussel hatcheries are a new development in New Zealand. This followed an extensive research period closing the Greenshell™ mussel (GSM) lifecycle, developing a breeding program, and methods of mass production with dependable survival. Industry members worked side by side with researchers from the Cawthron Institute in the laboratory for a number of years before building a hatchery. The hatchery began operating in 2015 and aims to produce sufficient spat to harvest 30,000 MT of live mussels. The spat is produced in batches and the larvae offered spat-catching ropes as a substrate to settle in settlement tanks. Once settlement has taken place and the spat conditioned, they are taken to the grow-out area in the Marlborough Sounds. Hatchery-produced spat is selectively bred and has been shown to grow 80% faster than wild Kaitaia spat. There is still ongoing research on the cryopreservation of gametes to improve management options, mussel physiology and feeding efficiency, influence of ocean acidification, and spat requirements. Spat retention post hatchery is also a research focus as this obviously has significant implications on the size of the hatchery, cost of production, and management of production.

Europe

Seed supply for mussel culture is traditionally based on fishery on wild beds for bottom culture or collection of seed from wild stocks, such as those on rocky shores in the Spanish rias. The production efficiency of a culture type determines the required amount of seed. For bottom culture, the relatively high losses during the culture cycle require more seed than for other culture types per unit end product. Natural seed beds on the bottom are vulnerable for predation by bottom dwelling shrimp, crabs, and starfish, hence wild seed can be scarce in some years. In the water column, mussel recruitment is more stable over the years as predation is much lower. In the Netherlands and Germany this has led to the development of seed mussel collectors (SMC) for bottom culture by using ropes and nets in the water column (Fig. 5.14). After a series of experiments running since 1998, the SMC application was boosted by pressure from environmentalists. There was concern over the negative impacts of bottom seed fishery in the Wadden Sea, an area protected for nature conservation. After a court case in 2008, the mussel industry made an agreement with the government and an NGO to stepwise reduce bottom seed fishery and scale up SMC. In the Netherlands, SMC techniques have been applied at increasing scale since 2008 and provide, since 2015, about 20,000 MT of seed annually, that is 50% of the annual need. Yet the cost of the collector techniques is about six times higher than wild seed fishery. It has been demonstrated that seed fisheries from wild beds, particularly in areas vulnerable to storm damage and starfish predation was not as detrimental as originally thought. In a large-scale experimental BACI study, it was shown that on plots closed to seed fishery, mussels were lost due to natural forces. There was no difference in stock size on closed and open plots, while the seed that was fished was transplanted to culture plots in the same area, thereby contributing to the total mussel stock. The same study showed relatively high biodiversity on the mussel culture plots, no less than on wild beds (Smaal et al., 2020).

The collector technique is now practiced in the Oosterschelde (NL) and in the Dutch and German Wadden Sea. The ropes or nets are connected to pole anchors and are deployed in the period March–November. Main spatfall occurs normally in between May and June, but may extend through the entire summer season. Seed harvesting occurs one to three times over the season. For harvesting the seed, specific equipment has been developed. To reduce losses, harvesting occurs submerged. By applying the venture principle, seed is sucked from the nets or ropes and transported on-board the mussel vessel. Subsequently seed is brought to the bottom culture plots either in the Wadden Sea or the Oosterschelde. The seeding technique consists of flushing the seed from submerged openings in the ship, while sailing in circles on the culture plot. One shipload can be up to 80,000 kg. Recent studies have shown that the seeding method may give rise to high initial mortality as seed is not equally distributed. So ongoing research aims at improving seed survival during and after seeding.

Seed collection with ropes is also practiced in the bouchot culture. In this case ropes are used in between the poles in horizontal orientation. The ropes are wound around the poles after spatfall, or the spat is first socked and then wound around the poles. For suspended culture, seed is normally collected on seed ropes as part of the rope culture system. If additional spat is required, this is collected from wild stocks in the area. There have been experiments with hatchery-produced seed. It was demonstrated that triploid seed could be produced, but commercial exploitation is not (yet) feasible. Also, the use of recirculation systems for seed production has been tested, and this could be a promising technique for mussel seed nurseries (Kamermans et al., 2013).

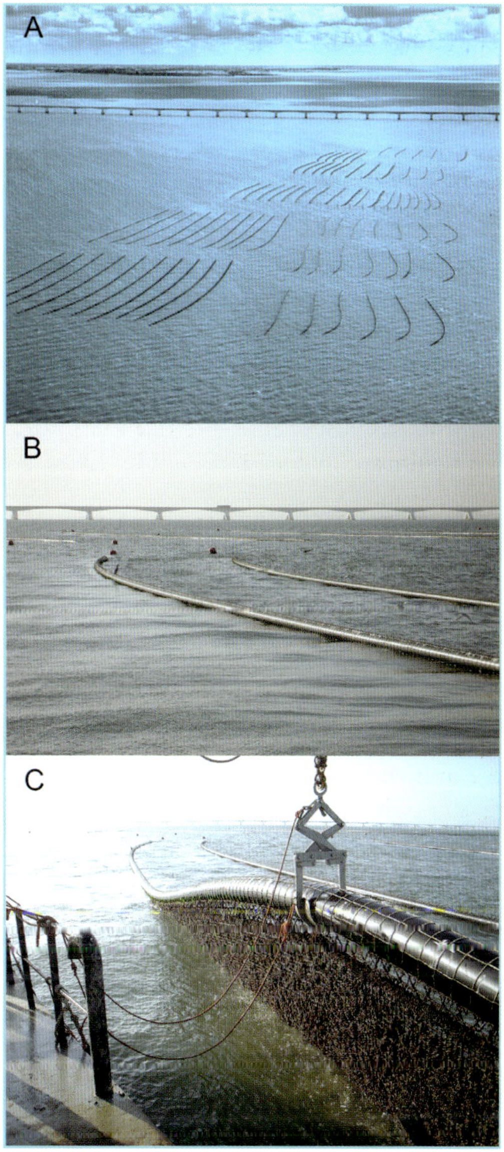

Figure 5.14 Seed mussel collectors in the Netherlands. Courtesy of Aad Smaal.

Farming systems

North America – structures and stocking densities

In North America, longline systems are used in eastern Canada (Fig. 5.15) such as PEI where the mussels are grown in shallow bays and short droppers are used; however, in Maine, Washington

State, and Alaska, rafts are used because they allow for predator nets to be hung around the mussels to prevent predation from diving ducks, fish, and sea otters. The first commercial scale mussel rafts were imported from Kames Fish Farm, Scotland, by Great Eastern Mussel Farms in the 1990s with a dimension of 12 x 12 m (40 x 40 ft) and a production capacity of 30 MT of mussels. The Maine mussel working group involved a number of growers along the Maine coast and early development of this type of raft was summarized in a guide to Maine mussel raft culture. A company called Maine Aquaculture Equipment built a modification of these rafts (about 50) called the Maine Mussel Raft (Fig. 5.16). Since then this modular raft, about 2.5 times smaller than the typical Spanish raft, has become the method most widely used in Maine. Mussel droppers on pegged ropes are attached at approximately 0.6 m on center to 75 x 100 mm timbers (Fig. 5.17), allowing for about 350, 14 m long ropes to be used on each raft. The rafts are often moored with two to four rafts in a row, with chain or rope in between each raft allowing for access on each end as well as the sides of the rafts. A novel submersible mussel raft has recently been patented (Figs. 5.18 and 5.19), and is being commercialized by Pemaquid Mussel Farms. The design is modular, based on the Maine mussel raft, but using pipe pontoons which can be flooded with water or air to adjust the raft buoyancy. It is 12 x 15 m (40 x 50 ft), with 450 droppers and can grow 60 MT of mussels in a crop. The advantages of these rafts are that they eliminate damage from drift ice, they minimize mussel drop-off from storms since the rafts are submerged under the waves, the mussels can feed without interruption, and the seed can crawl evenly along the ropes. This results in two to three times higher yields on the mussel ropes, where SCUBA observations have shown constant

Figure 5.15 Longline system in PEI with individual droppers. Courtesy of Department of Agriculture and Fisheries, Prince Edward Island, Canada.

Figure 5.16 Floating Maine mussel raft with 350, 14 m (45 foot) ropes and predator nets. Courtesy of C. Newell.

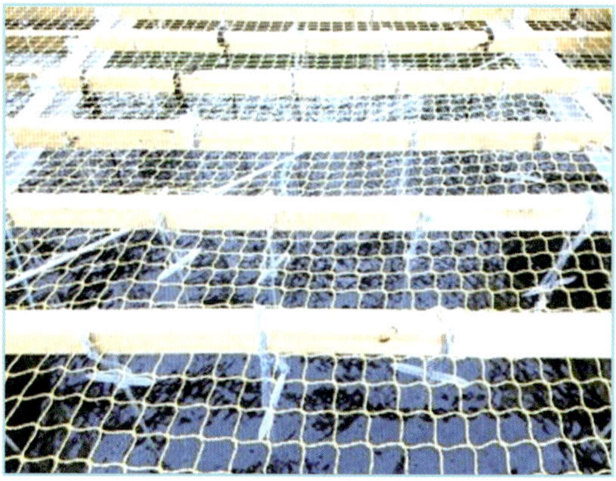

Figure 5.17 Ropes attached to timbers. Courtesy of C. Newell.

Figure 5.18 Submersible rafts (floating). Courtesy of C. Newell.

feeding during storms (Fig. 5.20). Yields vary from 3 kg to 15 kg m^{-1}, with the highest numbers found on submersible rafts.

The hydrodynamics of mussel rafts have been investigated in detail (Newell and Richardson, 2014) and have shown that adequate flow through the rafts for rapid and even growth of the mussels would occur if the rafts were angled slightly to tidal current directions, the predator nets were maintained to reduce fouling, and water velocities at the sites were over 25 cm s^{-1} (resulting in flow inside the rafts at about 5 cm s^{-1}). The modeling of mussel rafts and depletion of food particles within the rafts has resulted in the development of new tools for site selection in Maine and elsewhere.

Figure 5.19 Submersible mussel raft (sunk). Courtesy of C. Newell.

Figure 5.20 Submersible mussel raft rope with mussels feeding. Courtesy of Heather Perry.

Mooring systems

Mussel longlines and rafts use granite blocks, cement blocks, steel anchors, or augers. In general, the moorings are placed at a 3/1 scope related to water depth (that is, 90 m of scope from the anchor to the raft or end buoy for a water depth of 30 m), and heavy chain (that is, 25–40 mm) is either shackled to the anchor or shackled to the mooring staple. Growers use 37 or 31 mm superpro or polysteel rope (it is chafe resistant) with thimbles attached to the chain, and then the ropes are shackled to the rafts or tied into the buoys. A number of cement molds are used to pour anchors, often 3 MT or more, but the problem with blocks (cement or granite) is that they can slide on the seabed. In Maine, Pemaquid Mussel Farms uses 1 MT steel plow anchors which have a holding power of 20 MT (Fig. 5.21) in a system that uses the anchors, chain, rope, tension buoys, and chain on the rafts.

It is important that the mooring system is tensioned. With the steel anchors, ropes are attached to the back of the anchor so it can be pulled after setting. In addition, if tension buoys are used, the anchor system can be tensioned by picking up a tension buoy and removing a few meters of rope going to the anchor line. More advanced raft systems are using mooring grids, similar to salmon farms, where, for example, nine rafts can be placed in a 3 × 3 grid using just 16 anchors. Anchoring rafts together, two to four in a row, may result in snap loads from storm waves, where the rafts jerk against each other and mussels stop feeding or may dislodge from the ropes.

Figure 5.21 One ton steel plow anchor. Courtesy of C. Newell.

The mussel industry is moving toward steel auger anchors which are installed by a submerged hydraulic drill. The mooring ropes are attached to rings which rotate around the auger shafts when the anchor is installed. Mooring installations are carried out by contractors from New Zealand or

Iceland, and auger length and diameter is determined from the sediment type and the resistance of the auger as it is installed.

China: grow-out system and stocking densities

Off-bottom culture with the longline system is the most popular method used in China. The typical longline systems consist of a single horizontal headrope supported by red or grey plastic floats or white polyfoam buoys at regular intervals along the center segment (Fig. 5.22). Rope of 2.4 cm diameter is commonly used for the longline ropes. The longlines are moored at both ends on the seafloor with concrete or stake anchors. The length of the longline can be varied to suit the location, but it is generally 150 m with 100 m head ropes. Two polyethylene anchor ropes of 2.4 cm diameter are attached to the edges of the head rope. Floats with a 30 cm diameter are spaced appropriately to support the mass of growing mussels on the head rope. As the mussels grow and get significantly heavier, more floats are added to keep the mussel ropes suspended in the water column. Generally, they are spaced at 2–3 m apart. The separation between longlines is site specific and largely dependent upon the size of the servicing vessel. Sufficient clearance should be allowed to enable the vessel to work. The gap between adjacent longlines is generally 15–20 m. In some locations, the longlines may be kept slightly submerged in order to avoid disturbance from wave action. The rope on which the mussels are grown is called 'droppers', is usually 0.4 cm in diameter and made of palm fiber or sliced waste tires. Dropper lines are then suspended from the main line into the water column. 'Droppers' are generally between 2.0–3.0 m in length depending on the depth of the water and tied to the head ropes at points between 1 m and 2 m apart, depending upon local tidal conditions.

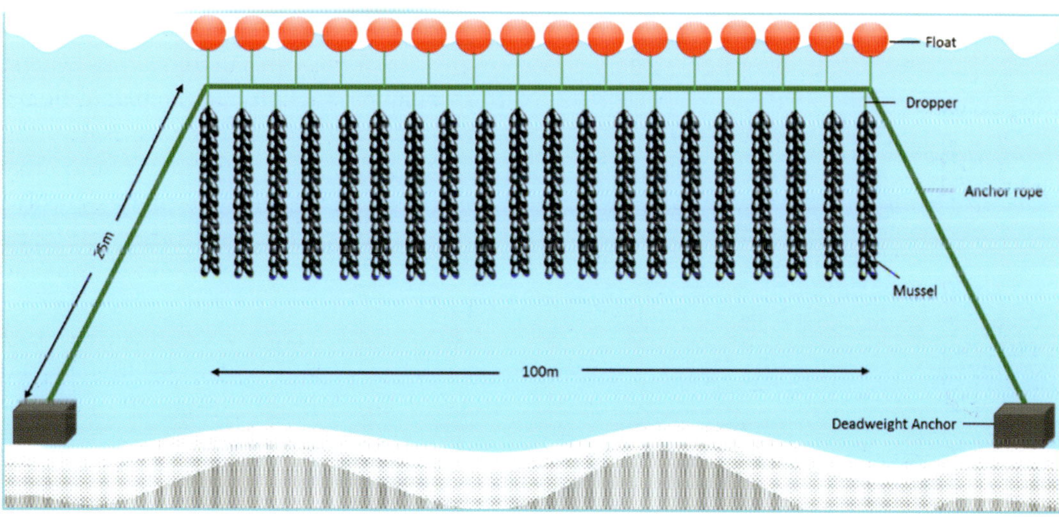

Figure 5.22 Schematic of a basic mussel longline system in China. Courtesy of Zengjie Jiang.

New Zealand

Current farming systems and open ocean expansion

Mussels are farmed using the New Zealand longline system, the general structure of which is discussed below. There are variations to this theme depending on location and farmer preference. For greater structural detail please refer to Goseberg et al. (2017). A traditional inshore New Zealand mussel longline consists of: a mooring; chain; warp; bridal and backbone (Fig. 5.23). The mussel longline holding the buoys or floats is called the backbone. Since this backbone has a header rope on each side of the float to support the production line it is referred to as a 'double backbone'. Typically, the backbone ranges from 100 m (328 ft) to 200 m (656 ft) long. A mussel longline (production rope) is hung from the backbone. It is attached on one side, descends down to 10 m (32 ft 10 in) or 15 m (49 ft 3 in), loops up to the opposite side of the backbone and crosses the gap between the buoys and then descends down again. In this way, between 3000 m (3,281 yds) and 4000 m (4,374 yds) of mussel long line is hung from a 100 m backbone. The longline hanging from the backbone can produce between 6.5 kg (14 lb 5 oz) to 13 kg (28 lb 10 oz) per meter (or 4 lb 12 oz per ft to 9 lb 8 oz per ft) depending on site and situation. The time period to produce this yield ranges from 12 to 20 months, again depending on the site and location. The target size of these mussels is 80 mm to 100 mm for the half-shell market. In 2003, consents for the offshore aquaculture space were lodged and the process of obtaining this space started. Open ocean sites in this instance are sites that experience significant wave heights of over 4.5 m at least once a year with a maximum potential of 7.6 m. There were numerous reasons for the departure from the sheltered waters including user conflict and a desire to increase farm size. The proposed farms were in waters ranging from 6 km to 20 km off the coast in water depths ranging from 25 m to 70 m.

In 2005, the first experimental ropes were installed into the open ocean waters of Hawkes Bay. The first open ocean structure was based on the traditional mussel backbone, but it was influenced by the systems used in the Coromandel in the north-eastern corner of New Zealand North Island. The Coromandel generally has higher energy than that of other culture sites such as the Marlborough Sounds. As the industry changed from inshore to more exposed sites the backbones

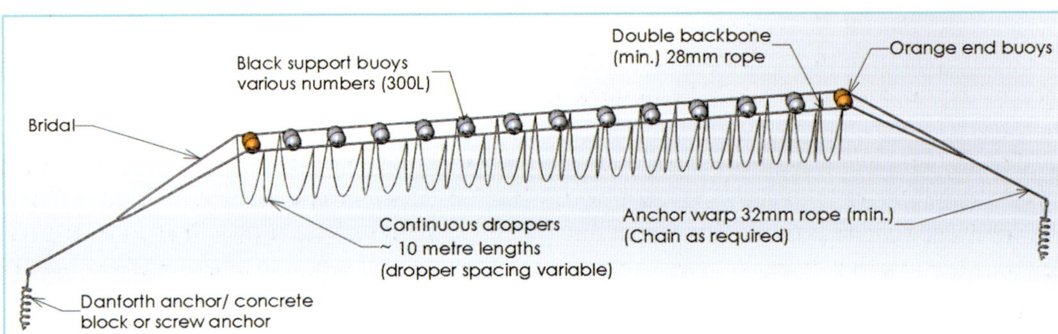

Figure 5.23 Diagrammatic cross section of a New Zealand surface inshore longline – not to scale. Courtesy of N. Scott.

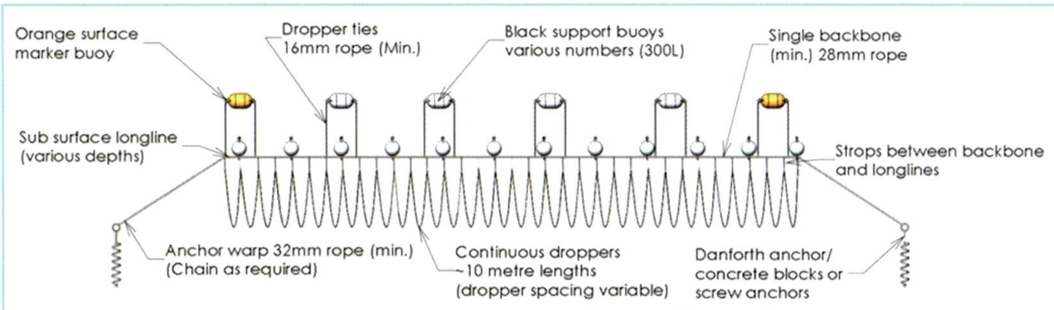

Figure 5.24 Diagrammatic cross section of a New Zealand subsurface longline. Note: more recent systems have more floats on the submerged part of the longline. Courtesy of N. Scott.

were cut down to a single header rope (single backbone, Fig. 5.24) and the production longline is draped in loops along the single header rope or backbone. The backbone is submerged and has approximately 66% of the flotation attached directly to it. Additional flotation is positioned on the surface with strops extending from the surface floats to the backbone. The length of the strop dictates the depth of the backbone.

The surface floats will provide 30–40% of the flotation and also give some indication of the load the backbone is bearing and when additional flotation will be required. The backbone is installed loosely enough so that a hook can be lowered from a vessel to snag the backbone and the resultant apex of the snagged backbone brought to the surface. The spacing between the droppers may be increased when compared to the inshore systems and may be as much as 1 m (3 ft 3 in) apart. By submerging the ropes, this system provides some protection from the wave energy experienced in open ocean situations; however, the surface floats still transfer energy to the backbone, which can result in production losses and increased maintenance. In addition, there is an issue arising from the additional requirement in flotation as a result of mussel growth: once the farm has a large number of backbones on it, the management of flotation will increase significantly.

Research in New Zealand is currently being carried out to develop new systems and structures for mussel and other shellfish farming in these areas of high energy. To date there are three prototypes being considered. It is also intended to include elements of technology to increase efficiency, for example, self-regulating buoyancy and the ability to report the growing systems status to a central control. A system that is already being tested is the 'Set and Forget' (S&F) system (Fig. 5.25). This system, developed by the Cawthron Institute in conjunction with an open ocean farming operation, is a fully submersible double backbone system which will be deployed and recovered from the surface.

The S&F system has a similar configuration to the open ocean double backbone but where the bridal meets the backbone, there is a mooring directly below it (screw anchor) (Fig. 5.25). There are additional intermediate anchors spaced every 35 m along the backbone. These intermediate moorings are threaded through a mechanism in the S&F buoy. There are also single S&F attachment mechanisms on each warp with a surface float. The idea is to fully seed the longline on the backbone. The backbone is then floated with sufficient buoys to support the intended harvest mass. Once seeded the backbone is pushed below the surface to the desired depth and the S&F mechanism engaged. The mechanisms on the warps are tightened to ensure the backbone does

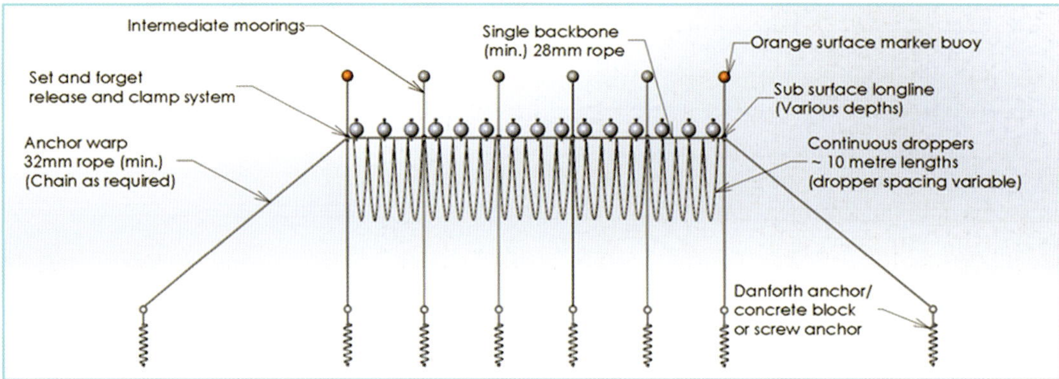

Figure 5.25 Diagrammatic Set and Forget (S&F) system – next generation submersible backbone under testing. Courtesy of N. Scott.

Table 5.3 The open ocean farm sites and the sizes of each site.

Open ocean farm site location	Farm size
Opotiki, Bay of Plenty, North Island	3800 hectares (9390 acres)
Hawke's Bay, North Island	2465 hectares (6091 acres)
Pegasus Bay (Christchurch, South Island)	2695 hectares (6659 acres)
D'Urville Island, South Island	400 hectares (988 acres)
Clifford Bay, Blenheim, South Island	400 hectares (988 acres)

not collapse toward the center. The mussels can then be left until they are due to be harvested. No intermediate flotation is required. At harvest the mechanisms are released using a surface driven unit (physically not electronically) and the backbone rises to the surface to be harvested.

As of 2017 there was just one, 10,000 ha (24,700 acres) of permitted open ocean or exposed water space. Development into this space is beginning to gain momentum. At the time of writing, there were 170 longlines in standard configuration. The majority of these ropes are in one farm that has shown good potential as a spat-catching site as well as a grow-out site. The local spat are caught between August and December and reach harvest size (80 mm) in 12 to 14 months. A list of the current open ocean sites in New Zealand is presented in Table 5.3.

Europe

Bottom culture

The culture cycle of bottom culture of mussels starts with fishery or collection of seed. This is harvested and spread out on culture plots in intertidal of subtidal areas. Not all culture plots have the same characteristics and quality, so the farmer needs a strategy to transplant the mussels according to growth and survival rates, the risk of predation, and the availability of culture plots. The latter depends on the time of harvesting of the previous cohort. On average, the bottom

Figure 5.26 Bouchot culture in France. Courtesy of Aad Smaal.

culture cycle takes 2 to 3 years and mussels are usually transplanted at least two to three times before harvesting. Seeding density varies from 5–10 kg/m, depending on size. Predator control regards predominantly starfish and crabs. Bird predation is also an important factor, but in most areas birds are protected and cannot actively be removed. Starfish predation is handled by collecting the starfish with mops or purse-seines, or by harvesting the mussels and leaving them in vessel hold while the hold is filled with fresh water from a nearby source overnight. The next day the mussels are seeded on another culture, while the starfish have died. Crab predation is handled by adding crab pots to the culture plots and harvesting the crabs. Capelle (2017) presents an extensive overview of the characteristics of mussel bottom culture.

Bouchot culture

Bouchot mussel culture in intertidal areas consists of poles that are substrate for socked mussels (Fig. 5.26). Seed is collected by horizontal ropes in between the poles, and wound around the poles. Grading and restocking are site-specific and occurs one to three times per culture cycle that takes 2 to 3 years. Stocking density is 12–20 kg/m. Predation is prevented by specific devices that hamper bottom fauna, such as crabs and starfish, from attaching to attach to poles. Fouling is not a large problem, as the poles are exposed to air at low tide that limits densities of predators, such as tunicates.

Suspended culture

Suspended mussel culture makes use of ropes submerged in the water column with mussels attached to the ropes or in socks. The mussels make use of the water column in three dimensions for food supply and discard of feces, pseudofeces, and dissolved nutrients. Mussel seed is normally socked with the holding rope in the center, and they attach to the rope while the sock breaks down. Grading and restocking are usual to avoid mussels dropping from the ropes when size and biomass increases. The ropes can be connected to a raft, to buoys, or to stakes that stick into the sediment. Stocking density can be 5–10 kg/m. Compared with the other techniques, growth rate is higher, hence the culture cycle is relatively short, in most culture areas less than one year.

Harvesting and processing

North America

Harvesting rope-grown mussels involves specialized vessels that utilize hydraulic machinery to hoist the ropes from longlines or rafts via harvest conveyors, line haulers, or baskets (Fig. 5.27). The vessels then strip the mussels off the ropes, separate them from clumps in a process called declumping, wash and grade the mussels by size, and either do further processing on-board or bring them to shore for purging (to remove grit in the meats), depuration (if the growing areas are not approved), debyssing (removing the byssal threads), inspecting (to remove broken mussels or imperfections such as barnacles or limpets on the shells), weighing, bagging, chilling, and icing for shipment. In some countries (that is, PEI, Canada) plastic mussel socks are just cut from the ropes and shipped to the plants where all the processing occurs. In Maine and Washington State, there is more of a tendency to process (up to bagging) at sea, where the small mussel seed recovered from harvesting can be reattached to the grow-out ropes as described above. In Maine, one company uses a stripping harvesting basket (Fig. 5.28) into bulk bags, and a predeclumper and brush machine from Spain (Fig. 5.29) followed by a debysser, grader, and inspection belt from the Netherlands on the harvest vessel. Then a bagging and tagging machinery from Italy is used in a plant onshore (Fig. 5.30).

Processing plants ship live mussels in packages ranging from 1–25 kg, in mesh bags chilled in ice, or in modified atmosphere packages (MAP) approved in Europe and Canada but not in the US, despite extensive research showing they are safe. Live mussel MAP has the advantages of a drip-free consumer pack, improved shelf life, and a tamper free package. A typical bagging line

Figure 5.27 Specialized mussel harvest and processing vessel. Courtesy of Penn Cove Shellfish.

Figure 5.28 Harvesting basket with stripper. Courtesy of C. Newell.

Figure 5.29 Processing line on vessel. Courtesy of C. Newell.

involves a bin tipper, a weighing conveyor, and a bag closer/tagger provided by supplier (Fig. 5.30). Further processing involves removing the meats from the shell, or shucking, and freezing the mussel meats or frozen mussel meats on the half-shell. Most frozen mussel products sold in the US are imported from New Zealand and Chile, where advanced technologies are used to process the mussels (including high pressure processing [HPP]) to separate the meats from the shells.

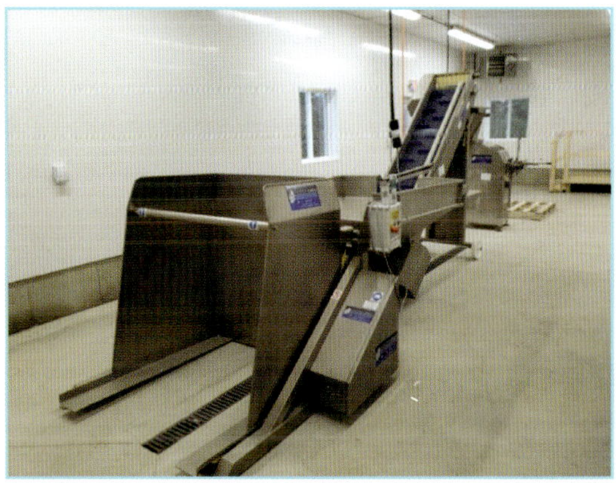

Figure 5.30 Mussel bagging line from Italy. Courtesy of Pemaquid Mussel Farms.

China

During March and April some of the mussels are harvested and the remainder are allowed to grow to 60–70 mm for harvest by September and October. Depending upon the size of the business, harvesting may be by hand or fully automated (Fig. 5.31). Normally, the mussels are harvested by a set of star wheels, each dropper is raised from the water and the mussels removed either by hand or by machine. The function of the star wheel is to allow the vessel to pull alongside the head rope and remain in proper position. By running the star wheel forward or backward, the vessel moves along the head rope, without having to engage the propeller. Each dropper will produce between 25–50 kg of mussels over a production cycle, which is usually between 12–18 months. Mussels with full gonads are considered most desirable on the market. Harvested mussels are then transferred to a shore-based facility. Washing, declumping, debyssing, and size-grading of the mussels virtually always takes place prior to packaging. Each stage requires specialized equipment, these are often placed in series with conveyors in between and are usually hydraulically operated (Fig. 5.31B). Mussels are then transferred to a wholesale packing and dispatch center in bulk or they may be packed in bags or other display containers by the

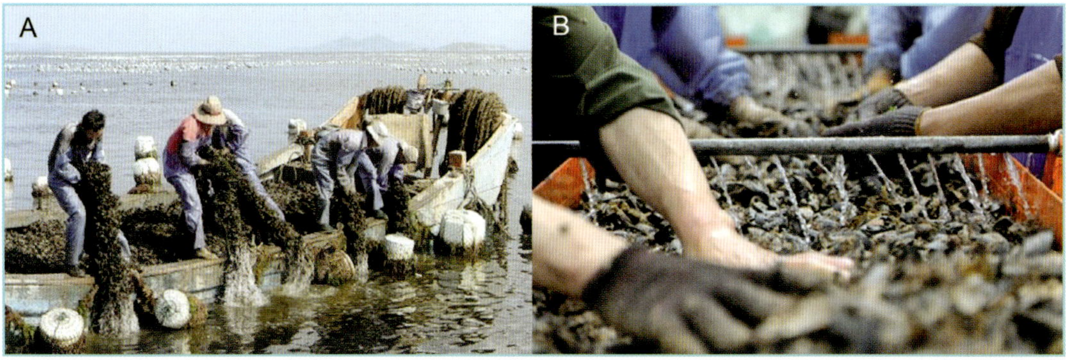

Figure 5.31 (A) Harvesting on the sea and (B) declumping on conveyor. Courtesy of Zengjie Jiang.

Figure 5.32 (A) Drying process and (B) finished 'Dan-cai'. Courtesy of Zengjie Jiang.

Figure 5.33 Processing the 'butterfly meat'. Courtesy of Zengjie Jiang.

grower in chilled transport conditions. Some mussels are sold fresh on local markets, but most are steamed and dried into a traditional product called 'Dan-cai' (Fig. 5.32). Mussel meat that is dried without cooking is called 'butterfly meat' (Fig. 5.33). Some mussels are cooked for the frozen half-shell mussel (Fig. 5.34) and exported globally.

New Zealand

Figure 5.34 (A) The processing of half-shell mussels and (B) the frozen product. Courtesy of Zengjie Jiang.

Harvesting of a backbone is a coordinated effort that is initiated by notification from the sourcing boat. The sourcing boat is a reasonably fast vessel which can lift a backbone and pull part of the continuous line to the surface. Mussels are then collected from the shallowest and deepest mussels on the dropper and assessed for condition. If the condition is suitable, the sourcing vessels will notify the production manager and a harvesting vessel sent to harvest the mussels. The harvesting vessel (Fig. 5.35) may range from 17 m (small companies) to 35 m long and up to 6 m wide. They are generally a mono-hull, but may be in catamaran configuration. A mono-hull will have the cockpit, galley, and sleeping quarters (if configured that way) at the front of the vessel. Immediately behind the cockpit, at deck level, the working area begins which extends to the transom of the vessel. The main motors of the vessel generally also drive the hydraulic system. The hydraulic system drives all the machinery required for harvesting, seeding, and handling of the mussels.

On arrival at the backbone to be harvested, the harvesting vessel will lift the backbone to eye level and place the backbone on hydraulic wheels which will move the vessel slowly along the backbone. The start of the continuous production line is threaded over a ramp which feeds the rope through a stripper. The stripped rope is automatically cleaned, washed, and coiled into bags. The stripped mussels go through a horizontal rotating declumper, which also acts as a washer. All detritus is discarded overboard. The declumped mussels are fed from the declumper onto a conveyor that passes the mussels through a sorting mechanism. The sorting mechanism separates the blue mussels from the Greenshell™ mussels (GSM) (using video technology and paddles). The GSM are then graded and collected in 1 MT bags. The bags are distributed around the deck with an on-board hydraulic crane. The large vessels can harvest up to 110 MT a day with a crew of five. The bags of whole, live mussels are then taken to the processing facility where they are

Figure 5.35 A typical harvesting boat. Courtesy of Kevin Heasman.

Table 5.4 The quantity of the main mussel products produced in New Zealand in 2016.

Product	Mass (MT)	Value (NZ$)
Half-shell frozen	27,295	209,291,769
Mussel meat frozen	3999	32,763,372
Powder from broth	215	11,995,141
In brine or salted	64	1,990,921
Smoked	37	809,430
Chilled (not frozen)	8	203,625
Capsules	3	199,802

taken off the vessel and sent to a large chiller for later processing. In some instances, the chiller may be at the wharf site or the mussels may require transport to the processing facility.

The mussels have three main size ranges for market: small, less than 50 g live weight; medium, between 50–65 g live weight; and large, greater than 65 g. Mussels are generally harvested at a live mass of 30 to 35 mussels per kg. The target market is the half-shell market. Half-shell mussels are produced through automated shuckers or manual shucking. New Zealand-made automatic shucking equipment can shuck 1 MT of mussels an hour per machine, producing a half-shell product that is the main NZ seafood product. There is a variety of value-added products that can be produced from mussels (Table 5.4). Most processing plants will have hand shucking of half-shell mussels. In these cases, the mussels are blanched and one shell removed by hand. Mussels not suitable for half-shell sale are generally sold as frozen mussel meat.

Europe

For harvesting mussels from the bottom, a mussel dredge is used (Fig. 5.36), consisting of a steel rectangular frame with a net and a bottom rod to lift the mussels from the sediment. The dredge is pulled over the culture plot and when filled is lifted and unloaded in the vessel. Bottom culture vessels are ca. 30–40 m long and can carry ca. 100 MT. The vessels (Fig. 5.37) are designed to sail over long distances as culture areas can be spread over a large spatial scale. For example, in the Netherlands, culture plots of a company can be located in the Wadden Sea in the north as well as in the Oosterschelde in the southwest, more than 100 km apart.

Mussels are separated on-board from other components, such as starfish and crabs. After harvesting, mussels are landed for processing on land. In the Netherlands, harvested mussels are brought to the auction in Yerseke (where there is a concentration of all the mussel processors and the major distribution center for mussels in the Netherlands), and sold to a mussel trading company. The mussels

Figure 5.36 Mussel dredge. Courtesy of Aad Smaal.

are subsequently brought to specific plots for rewatering. They remain there for a few days to eliminate silt and to recover before further processing takes place. When the mussels are brought on land, they undergo another rewatering treatment in controlled flow-through systems with cooled UV treated seawater. Processing in the plant occurs under low temperature conditions. Mussels are declumped, byssus threads are removed, visual quality control occurs, and mussels are graded into different size categories. Mussels are packed fresh for restaurants and retail. Most companies apply modified atmospheric packaging (MAP) to increase shelf life. When mussels are packed live, MAP helps to keep the shells closed and provides a high humidity drip-free package both of which have increased sales significantly. While millions of kilograms are sold this way in Europe and Canada, the US FDA prohibits sales of MAP mussels, despite considerable research which shows that live mussel MAP is safe.

Figure 5.37 Mussel bottom culture vessel. Courtesy of Aad Smaal.

Bouchot mussels are harvested with a special device that pulls the mussels from the poles. Mussels are collected in large bags and processed on land. As they grow off-bottom, no rewatering is required to get rid of silt.

Mussels from suspended culture are harvested by stripping off the mussels from the rope system. The mussels are put in big bags on-board the vessel, no rewatering is required and further processing occurs on land.

In the case of raft culture, as is practiced in Spain, mussels are harvested by collecting the ropes and the mussels with a basket that prevents the mussels from dropping off the rope. On the raft or on-board, the mussels are separated from the rope. Mussels cultured in waters of quality B for shellfish culture, such as in Spain, require depuration for 24 hours, through which chlorine treatment guarantees sanitary quality for the consumer.

In addition to processing mussels for the fresh market, there are various types of production of preserved mussels. Traditionally mussels are cooked and the meat is preserved in cans or bottles. Bottled mussels are preserved with citric acid and vinegar. Novel developments are focused on convenience food, for example microwave-ready mussels with a shelf life of up to 4 weeks.

Predators and biofouling

There is more detail on biofouling in Chapter 16.

North America

The main predators of farmed mussels are starfish and sea ducks (suspension culture) and starfish, ducks, and crabs (bottom culture). On the US west coast, there are fish predators as well as sea otters. Growers using mussel rafts have successfully excluded sea ducks and otters using predator nets hung around the raft perimeter, using a similar approach to salmon farms. Recent work (Varennes et al., 2013) has shown that a maximum net mesh size (stretched) of 15 cm (6 inches) can eliminate eider duck entanglements in the nets. Starfish often settle a few weeks after primary settlement of mussel spat, and can easily eliminate crops if not managed. Successful management includes air drying, freshwater treatments, or using quick lime. Starfish mops and roller dredges have also been effective in reducing starfish densities in bottom culture.

Biofouling by barnacles and sea squirts on mussel ropes have caused significant mussel losses because the barnacles make the shells unsightly, and the sea squirts crowd out the mussels on the ropes, causing them to fall off, and they compete for food. Previous (unpublished) work has shown that mussel ropes that are established with small seed mussels completely covered (no empty spaces) will resist sea squirt settlement whereas ropes with any clean surfaces (that is, unseeded ropes, cotton, pegs, and so on) will get colonized. Therefore, a strategy can be to deploy seeded lines either before or after the peak in sea squirt settlement (usually June to August) so the ropes are established and fully covered with mussels. In PEI, Canada, power washing systems have been developed to remove sea squirts from their short mussel dropper lines.

China

The on-growing mussels should be checked regularly to remove any predators settled onto the droppers (for example, starfish and crabs). Other marine organisms that have planktonic larval stages, such as barnacles, tubeworms, and sea squirts, may also settle onto the mussel ropes or onto the mussel shells themselves. They can compete with the mussels for food, and if settlement is particularly dense, they can increase mortality and lower the quality. 'Soft fouling', for example, sea squirts and macroalgae or predators like starfish, normally are controlled by raising the droppers from the water and allowing them to dry in shady conditions before they are resubmerged. 'Hard fouling', such as barnacles and tubeworms, is more difficult to control. The current method farmers use is called settlement avoidance, which is practiced by raising the droppers from the water when settlement is predicted to occur.

New Zealand

Detailed studies of fouling have occurred in New Zealand and the range of species and mitigation strategies are listed here. For more information on biofouling, Chapter 16 goes into detail on this topic for all the bivalve species.

Predation of spat/seed is limited to seasonal events of snapper feeding on newly seeded stock. Snapper increase in numbers to spawn in the spring and have learned that tearing the stocking on newly seeded ropes provides bite-sized mussels to feed upon. In some areas, the seeding of ropes during this 2–3 month period is avoided to reduce losses of this kind. Crabs can colonize mussel spat ropes and feed on spat. They are thought to have an impact but the level of impact has not been quantified.

Spat ropes that come into contact with the seabed can be colonized by high numbers of starfish which influence survival. This is a rare occurrence as ropes seldom come into contact with the seabed. Sea squirts (both solitary and colonial) are commonly found on the mussel ropes.

Methods of control are more mitigation than a physical remedy. Site selection and correct positioning of spat ropes in the water column are used to reduce the infestation and therefore impact of these fouling species.

The thecate hydroid (mussel beard, *Amphisbetia bispinosa*) is a periodic problem. It can grow in dense 'thickets' on the mussel ropes smothering small mussels and reducing growth in larger mussels. It is also a problem as it clogs machinery, which is severely restrictive in the highly automated New Zealand mussel industry.

The jeweled anemone, Corynactis australis, can settle and multiply in large numbers. These animals do not appear to have a significant influence on mussel growth (indeed they may assist by intercepting the larvae of fouling species), but at harvest the nematocysts (stinging cells) are liberated and can cause inflammation and discomfort to the eyes and lips of workers. This is not a common issue, however, and relatively local in occurrence.

The Asian seaweed (Wakame), *Undaria pinnatifida*, was thought to be introduced via shipping and has gone from being classified as a pest (with extensive effort to eradicate it) to an accepted part of the system. It is generally not a significant issue, but can reduce water flow and hinder rope handling. Farmers may harvest it or offer the *Undaria* to interested parties. It is currently being harvested and made into fertilizer. It is not farmed at present, but farming is a possibility in the future.

The New Zealand pea crab, *Pinnotheres novaezelandiae*, is not a common issue in Greenshell™ mussels (GSM). It can be found in local patches within populations of GSM, but in terms of local species it is more prevalent in local surfclams. The Mediterranean blue mussel, *Mytilus galloprovincialis* is apparently far more desirable to pea crabs as blue mussels are generally highly infected.

Other issues which are probably specific to New Zealand are the fouling by barnacles and tube worms. The majority of New Zealand mussels are sold in a half-shell (that is, one valve with the tissue in it) and the aesthetics of the shell is part of the attraction and market appeal. Organisms that corrupt this visual appeal are a marketing issue. In addition, there are now mechanized automatic shell openers that have issues orienting the shell if it is fouled reducing efficiency and factory output.

A major issue for New Zealand mussel farmers is the increasing incidence of over-settlement by the accidentally introduced Mediterranean blue mussel. The spat from the blue mussel over settles the GSM spat and results in mixed seed production. The seed has to be graded before reseeding or thinning. The industry has developed video technology which selects out the blue mussel based on color and discards it resulting in a nearly 100% GSM reseed. Since the blue mussel is not a targeted species, at harvest it is not normally in synchronous condition with that of the GSM and so normally in poor condition. It is also a low value species which is preferred fresh as against frozen in the countries where they are traditionally eaten. Since New Zealand is a long distance from most countries, it is not possible to get a fresh product or a frozen product to market within economically viable means. As a result, it is estimated that between 7,000–10,000 MT of blue mussels are discarded annually. The fouling of floats and growing structures by the blue mussel is also a major time-consuming issue, which adds to the cost of farming. The blue mussel is not going away, and it is acknowledged that it will increase its presence in years to come. Therefore, increased effort is being directed toward reducing its impacts and finding potential economic solutions. The blue mussel has not been seen at the operating offshore sites as yet. Obviously, as the amount of equipment going out onto the site increases and the number of vessels, both shellfish related and recreational, increases the potential for a blue mussel incursion on to the site will increase. One advantage of the open ocean sites, however, is that spawning blue mussel larvae will be far from the site reducing inoculation pressure. In addition, the water carrying fouling species entering the farm generally has only one pass through the farm, unlike inshore bays where tidal currents may pass through the farms several times increasing the opportunity for fouling organisms to settle. Mechanisms to reduce introduction of the blue mussel to the open ocean sites (for example, effective boat antifouling, introducing only clean or new gear), and the careful selection of spat and hatchery spat will reduce the incidence of blue mussels on these sites reducing their impacts. Offshore farms have not yet recorded blue mussels on their sites.

Open ocean site fouling

Fouling on the open ocean sites is generally less than that of the inshore sites. There are a number of reasons for this. Most of the sites in New Zealand are in areas where the substrate is mud/sand/cobble that does not provide a substrate on which many of the normal fouling species can exist. Therefore, the parent stocks that would normally produce the larvae and juveniles are some distance away (near the shore). Consequently, organisms that have short larval period durations (for example, some colonial sea squirts) cannot reach the farm site before the larvae run out of resources and perish. The typical fouling on the open ocean sites are organisms with long-lived larvae stages such as barnacles, hydroids, and anemones.

Another reason fouling is not as much of an issue in New Zealand in general is the fact that the mussels are thinned twice before grow-out and are graded and cleaned at each event. The grow-out takes 12 to 24 months from spat to harvest so the fouling organisms seldom have more than one season to establish and dominate.

The open ocean sites also have the advantage that the water carrying fouling larvae generally has a single pass at the farms before continuing their journey out the other side of the farm. In inshore farms, the water movement has a stronger tidal influence and may be driven through a farm several times providing multiple opportunities for the larvae to settle. The close proximity of the inshore structures (±15 m apart) as against the open ocean structures (±150 m apart) also increases fouling settlement opportunity in the inshore waters.

Europe

In Europe, the main mussel predators are benthic animals like starfish, crabs, and shrimps, as well as diving birds such as eider ducks. Fish predation by wrasses is also observed on Mediterranean longlines; other loss factors are fouling by tunicates and loss of stock due to hydrodynamics.

Environmental and social interactions

North America

Commercial scale mussel aquaculture results in both environmental and social interactions, the scale of which depends, in part, on the scale of the operations. Any negative water column and benthic impacts are mostly ameliorated by locating farms in areas with moderate water flow (that is, depth averaged current speed is 20–40 cm s^{-1}) because the depletion of phytoplankton is limited in spatial extent, organic deposition is dispersed, and there is a sufficient flux of oxygen to the benthos to prevent anoxia (Findlay and Watling, 1997). The ecosystem benefits of mussel farms on water quality, such as controlling eutrophication, reducing turbidity, and removing nitrogen, as well as the benefits of the farms on biodiversity are due not only to the grow-out lines but also the invertebrates, fish, and birds attracted to the buoys, floats, rafts, and other structures, all of which provide a strong case for utilizing bivalve filter feeders such as mussels to help restore ecosystem health in coastal zones. The use of blue mussels or ribbed mussels has been considered as a strategy to improve water quality in various parts of the world including the US, Sweden and elsewhere. In areas with more expansive mussel aquaculture, such as New Zealand, progress has been made in understanding the factors that influence social carrying capacity, and a careful consideration of existing fisheries, viewscapes, and other uses through local knowledge is an important prerequisite to a successful mussel farm development at new sites.

Canada produced the east coast shellfish sector strategic plan, and a National Aquaculture Strategic Action Plan Initiative (NASAPI), which utilizes the principles of sustainable development.

- Environmental protection: mandating that maintenance of healthy and productive ecosystems is prerequisite for all aquaculture ventures.
- Social well-being: a social license to operate.
- Economic prosperity: responsible, market-driven sectoral growth and development.

In addition, several mussel farms in Canada are organically certified based on the 2012 standard recently upgraded in 2018 (National Standard of Canada, 2018).

Mussels can provide local communities with a variety of important ecosystem services, such as water filtration, nitrogen removal, and carbon storage. Mussels are filter feeders and have the capacity to filter large volumes of the surrounding seawater. It has been reported that a single blue mussel can filter ~36 L of seawater per day. In doing so, they improve water clarity by filtering phytoplankton, microorganisms, detritus, and reduce turbidity. Mussels also play an important role in the cycling of nutrients. They incorporate nitrogen from the phytoplankton which use dissolved inorganic nitrogen into their tissues and removed excess nutrients from inshore coastal waters by harvest. In recent years, interest has increased in exploring the possible role of shellfish culture as a method for enhancing the removal of carbon from the ocean. The mussel culture infrastructure also provides new habitat and food chains, on a seasonal basis, for many forms of marine life. Polychaetes and crustaceans, in particular, are known to gather in these habitats, which in turn provide a valuable food source for marine fish (Smaal et al., 2019). Because mussel culture is closely connected to ecosystem processes with various types of interactions, this knowledge can be used to combine mussel culture with ecosystem management goals.

China

In China, in order to ensure the sustainable development of the mussel aquaculture industry, farmers have organized spontaneously various kinds of specialized cooperative associations for mussel aquaculture and processing. Normally, the cooperative model they use is a family contract system. The association helps the farmer solve the difficult problems facing the single family through technical promotion, standardization, brand creation, market development, and improvement of the standards of living.

New Zealand

Given the large scale of mussel farming in New Zealand, there is increased awareness and more programs in place to ensure sustainable development of the industry on many levels. Mussel farming started in the Marlborough Sounds in the 1970s. The Marlborough Sounds are fjords, which provided good protection and good growing sites. As expansion continued, mussel farming became established in Coromandel, Stewart Island, Canterbury, and Nelson Bays (which include Tasman Bay and Golden Bay) (Fig. 5.3). Both Coromandel and Golden Bay are high energy areas and a bit more exposed than other sites. There are some spat-catching areas for local spat in Tasman Bay which are only occupied for part of the year. At present, it is estimated that 2% of the surface area of Marlborough Sounds are covered by mussel farms, less for other areas. At the start of the enterprise, water space was relatively easy to secure with tenure on sites set at 25–30 years. There are environmental requirements that have to be met to keep tenure. As time has progressed, there is increasing recreational usage and holiday home development, which has put considerable pressure on allocation of water space. Social license objections include: interference with recreational fishing grounds; navigation; visual impacts; vessel noise; general environmental implications (actual and perceived); and pollution derived from mussel farming (rope offcuts, damaged floats).

In response, companies associated with the Marine Farming Association (Blenheim) [www.marinefarming.co.nz], Marine Farming Association (Coromandel) [www.coromandelaquaculture.co.nz], and Aquaculture New Zealand [www.aquaculture.org.nz], have education and public programs to facilitate information transfer and points of public contact. Companies within the marine industry (not just mussel producers) have instigated their own programs as part of their social commitments. For example, they have a program where each company 'takes ownership' of a stretch of beach or bay and have field (and family) days where they clean up all (not just aquaculture) rubbish. In addition, there are company-driven stringent protocols on the vessels strictly controlling the handling and discarding of aquaculture-associated waste. Some companies have taken the research further and are developing biodegradable equipment, such as lashings (small ropes that are used as float ties and production line bindings) which, if inadvertently lost, will biodegrade.

It is estimated that there are over 3000 people directly employed by the mussel industry and many more indirectly associated. The mussels in the Marlborough Sounds are landed at the port of Havelock. The survivability of this town is reliant on mussels and calls itself 'The Greenshell™ capital of the world'. Tourism vessels that provide tours of the Sounds and mussel farms have developed, and in the case of the Bay of Plenty site at Opotiki, there is a significant direct link between the marine farm and the community with which it is associated. This is a commercial Maori enterprise, which extends past the standard commercial benefits by having community development as one of its cornerstones. Maori have a very close association with the sea, which is integrated into their culture, ethos, and folk law, and this extends to mussel farming activities. Some mussel farming sites appear to attract many species of fish. Over 700 recreational fishing boats have been recorded on a single day over a holiday period catching fish within mussel farms in the Coromandel. Similarly, fishing is also undertaken near farms in other localities. Fishing is also becoming popular at the open ocean sites, as the open ocean sites have proven to increase the density of fish and other species. This is not entirely associated with the fish attracting device (FAD) effect, as juvenile fish have been seen to use the structures as a nursery. The farm structures essentially act as midwater reefs.

Research targeting the sustainable growth of the mussel industry in New Zealand

Mussel aquaculture research and development in New Zealand is a model for success since it is outcome oriented and comprehensive. It is described in more detail here because of the comprehensive nature of the programs. Almost all research has at least one mussel farmer involved. Indeed, general government funding requires significant interaction and input from the mussel industry for funding to be allocated. Significant research has taken place since the late 1990s. There are a number of papers completed by various universities around New Zealand, but by far the most research on the Greenshell™ mussel (GSM) has been completed by the Cawthron Institute (www.cawthron.org.nz) and associated investigators.

Completed or ongoing projects include research into:

- closing the life cycle and developing methods and technology for research and commercial larvae production
- cryopreservation of gametes and larvae
- probiotic research in support of larval rearing
- selective breeding for qualities requested by the mussel industry (for example, fast growth, disease resilience)
- gynogenesis in support of the selective breeding program and for commercial all female stock
- triploidy for the production of fast-growing animals that could potentially be harvested all year around
- food conversion by shellfish (physiology and genetic associations)
- extending the shelf life of live mussels through physiological cues
- developing health markers
- pathogen identification and detection to reduce transmission risks
- developing of smart farming and tools to get real time information from the farms
- forecasting tools for seafood safety (environmental, harmful algal blooms and toxins)
- ocean acidification effects on GSM at all stages including genetic effects
- spat retention to reduce current losses at the early farming stages (that is, sea-based nurseries)
- seafood safety program: pre- and post-harvest research to ensure safety (which includes development of real time monitoring of toxic algae and pathogens, emerging threats and contaminants, such as anthropogenic toxins)
- developing of high-value products (bioactives including nutraceutical, pharmaceutical and components with other levels of action)
- understanding biofouling threats, developing biofouling control methods and risk management
- developing sustainable farming options to reduce environmental footprints (for example, integrated multi-trophic aquaculture [IMTA]).

Nutraceuticals

The GSM, as do other mussel species, has a unique assemblage of lipids which have health benefits to both human and domesticated animals. The market for mussel extract oils (lipids) and powders is growing. These include Seatone™, Biolane™, Lyprinol®, and Biolex®, and others. The value of mussel oil has now risen to more than NZ$3000/kg and mussel powder is presently worth around NZ$50–100/kg. The health benefits of GSM and its extracts were originally thought to be due to the long-chain polyunsaturated fatty acids (LC-PUFA) such as eicosapentaenoic acid (EPA) and docosahexaenoic acid (DHA) that they contain; however, other components such as other lipids present at low levels as well as bioactive carbohydrates may also be important and may differ following different extraction processes.

Europe

Many studies over the past 30 years in Europe were focused on the possible environmental impacts of mussel culture. This is a consequence of culture sites being located in areas that are now protected for nature conservation. EU-Natura 2000 rules require that a permit is needed for seed fishery and for aquaculture, and the permit will only be issued if the assessment makes clear that the activities have no negative impact on the maintenance goals of the area. Hence, knowledge of the industry's impacts on the environment is needed to secure licenses to produce. Interestingly, this has promoted research on the functions of bivalves in the ecosystem, that addressed not only possible negative but also positive impacts (Smaal et al., 2019).

It was demonstrated that mussels and mussel culture can have a positive impact on the quality of the ecosystem. For example, the filtration capacity of mussels can be used to reduce turbidity and the production of feces and pseudofeces results in nutrient regeneration that stimulates phytoplankton turnover and the productivity of the ecosystem. The uptake and accumulation of nutrients by the mussels is used to mitigate eutrophication and the same function is used in integrated multi-trophic aquaculture (IMTA) to compensate for nutrient input through fish farming. Mussel beds provide feed for birds as well as often being biodiversity hotspots; it was demonstrated that the same holds for mussel culture plots. These and other ecosystem services have been reviewed in the book *Goods and Service of Marine Bivalves* (Smaal et al., 2019). It shows indeed that mussel culture is closely connected to ecosystem processes with various types of interactions, and this knowledge can be used to combine mussel culture with ecosystem management goals.

As a framework for managing mussel culture in the natural and social environment, the concept of carrying capacity is often practiced. While carrying capacity considerations are important in other areas around the world, leadership in this area has occurred primarily in Europe, Canada, and New Zealand. Although the concept has constraints, it is a useful tool to specify what is meant with the ecosystem approach to aquaculture (EAA), which identifies four categories of carrying capacities (see 'Site selection' section or Smaal and van Duren, 2019). Various authors have pointed out that ecological and social carrying capacity have a common basis, as the category of unacceptable ecological impacts is a social construct. Social carrying capacity deals with stakeholder involvement, why mussel farming is considered acceptable or unacceptable and by whom. It is evident that this is context dependent and quite different in different societies.

Given the large variation in mussel culture communities in Europe, no overall social pattern can be identified. Yet, the level of organization of the mussel culture in Europe is well developed. In most countries there are national organizations of farmers, supported by locally organized farmer groups. There are several national and international journals, platforms and conferences, and there is a European Mollusc Producer Association (EMPA). The EMPA is a member of the Aquaculture Advisory Committee of the EU.

Social interactions

For the practice of mussel farming, tools like models and indices are available to evaluate environmental impacts as well as ecosystem services. Successful application, however, also depends on the societal context of mussel aquaculture in a certain area. This depends on the local situation, the attitude of the farmers, the government, the local economy, the level of acceptance by the community and the infrastructure for stakeholder communication. An analysis of performance of aquaculture in public media may give useful indications of the license to produce.

Conclusions

The mussel industry is currently positive and progressing. It will be limited in the short term by constraints on the availability of inshore space. The use of open ocean space is developing and needs some time to gain momentum. The industry-led science is driving improvements in spat production, product enhancement, management efficiencies, open ocean farming methods, nutraceuticals, and sustainability.

References cited and further reading

Bardach, J.E., Ryther, J.H., and McLarney, W.O. (1972) *Aquaculture: The Farming and Husbandry of Freshwater and Marine Organisms*. John Willy & Sons, New York.

Bayne, B.L. (1976) *Marine Mussels*. Cambridge University Press, Cambridge.

Brown, C., Courtier, C., Parsons, J., Nichols, J., Struthers, A., McNeil, S., Pryor, M., Moret K., and Zovik, T. (2000) A practical guideline for mussel aquaculture in Newfoundland. Marine Institute, Memorial University, Newfoundland.

Capelle, J.J. (2017) Production efficiency of mussel bottom culture. PhD thesis, Wageningen University, the Netherlands.

Findlay, R.H. and Watling, L. (1997) Prediction of benthic impact for salmon net-pens based on the balance of benthic oxygen supply and demand. *Marine Ecology Progress Series*, 155: 147–157.

Goseberg, N., Chambers, M.D., Heasman, K., Fredriksson, D., Fredheim, A., and Schlurmann, T. (2017) Technological approaches to longline- and cage-based aquaculture in open ocean environments. In: Buck, B.H. and Langan, R. (eds.) *Aquaculture Perspective of Multi-Use Sites in the Open Ocean: The Untapped Potential for Marine Resources in the Anthropocene*. Springer International Publishing, Cham, pp. 71–95.

Gosling, E. *The Mussel* Mytilus*: Ecology, Physiology, Genetics and Culture*. Elsevier, Amsterdam.

Hawkins, A.J.S., Pascoe, P.L., Parry, H., Brinsley, M., Black, K.D., McGonigle, C., Moore, H., Newell, C.R., O'Boyle, N., O'Carroll, T., and O'Loan, B. (2013) Shellsim: a generic model of growth and environmental effects validated across contrasting habitats in bivalve shellfish. *Journal of Shellfish Research*, 32(2): 237–253.

Island Institute (1999) *The Maine Guide to Mussel Raft Culture*. Eastern Maine Development Corporation. Rockland, ME.

Kamermans, P., Galley, T., Boudry, P., Fuentes, J., McCombie, H., Batista, F., Blanco, A., Dominguez, L., Cornette, F., and Pincot, L. (2013) Blue mussel hatchery technology in Europe. In: Allan, G. and Burnell, G. (eds.) *Advances in Aquaculture Hatchery Technology*. Woodhead Publishing, Cambridge, pp. 339–373.

Lutz, R.A. (ed.) (1980) *Mussel Culture and Harvest: A North American Perspective*. Elsevier Scientific, Amsterdam.

National Standard of Canada (2018) Aquaculture – general principles, management standards and permitted substances lists. Canadian General Standards Board. CAN/CGSB-32.312–2018.

Newell, C.R., Wildish, D.J. and MacDonald, B.A. (2001) The effects of velocity and seston concentration on the exhalant siphon area, valve gape and filtration rate of the mussel Mytilus edulis. *Journal of Experimental Marine Biology and Ecology*, 262(1): 91–111.

Newell, C.R. and Richardson, J. (2014) The effects of ambient and aquaculture structure hydrodynamics on the food supply and demand of mussel rafts. *Journal of Shellfish Research*, 33(1): 257–272.

Newell, C.R., Brady, D.C., and Richardson, J. (2019) Farm-scale production models. In: Smaal A., Ferreira J., Grant J., Petersen J., and Strand Ø. (eds.) *Goods and Services of Marine Bivalves*. Springer, Cham: pp. 485–506.

Northern Economics, Inc. (2015) Economic analysis to inform the Alaska mariculture initiative: phase 1 case studies'. Prepared for the Alaska Fisheries Development Foundation.

Panchang, V.G., Jeong, C., and Li, D. (2008) Wave climatology in coastal Maine for aquaculture and other applications. *Estuaries and Coasts*, 31(2): 289–299.

Smaal, A.C. and Van Duren, L.A. (2019) Bivalve aquaculture carrying capacity: concepts and assessment tools. In: Smaal A., Ferreira J., Grant J., Petersen J., and Strand Ø. (eds.) *Goods and Services of Marine Bivalves*. Springer, Cham, pp. 451–483.

Smaal, A., Ferreira, J.G., Grant, J., Petersen, J.K. and Strand, Ø. (eds.) (2019) *Goods and Services of Marine Bivalves*. Springer, Cham.

Smaal, A.C., van Stralen, M.R., and Craeymeerssch, J. (2020) The impact of mussel seed fishery on dynamics of wild subtidal mussel beds in the western Wadden Sea. *Journal of Sea Research* (in press).

Varennes, É., Hanssen, S.A., Bonardelli, J., and Guillemette, M. (2013) Sea duck predation in mussel farms: the best nets for excluding common eiders safely and efficiently. *Aquaculture Environment Interactions*, 4(1): 31–39.

CHAPTER 6

Abalone aquaculture

Peter Cook

Introduction

Abalone are large gastropod molluscs, belonging to the exclusively marine family, Haliotidae. They are found on shallow, subtidal rocky shores in most tropical and temperate waters, notable exceptions being both polar regions, the east coast of North America, and the west coast of the Indian subcontinent. Although, in the past, some authors suggested that there were about 75 species of abalone worldwide, in the most recent and authoritative account of abalone taxonomy, Geiger and Owen (2012), recognized only 56 valid species, with 18 additional sub-species.

Some historical literature suggests that abalone may have been exploited by coastal communities since pre-historic times, but large-scale commercial exploitation has only been recorded since the early 1970s. Of all of the species worldwide, less than one-third are of commercial importance, the others being either too small or too rare for commercial exploitation. A typical, commercially exploited species, *Haliotis midae*, from South Africa, is illustrated in Fig. 6.1.

Biology

On temperate and tropical shores, abalone occur from the low intertidal to a depth of about 30 m, and tend to favor areas supporting seaweed beds, particularly kelps. They are found attached to rocks or boulders, or in crevices and cracks in rocks. They have a strong, muscular foot that allows them to adhere to the substratum, even in conditions of moderate to strong wave action. Some very large abalone have been estimated to live up to 50 years.

Figure 6.1 The South African abalone, *Haliotis midae*, showing inside and outside of shell.

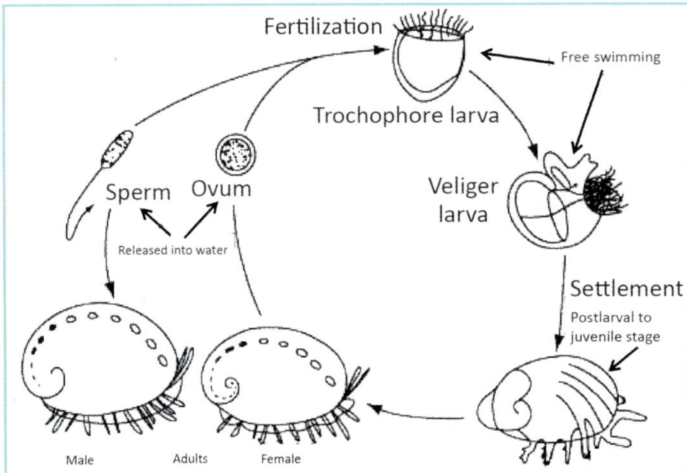

Figure 6.2 The life cycle of abalone.

Abalone adults are macroalgal herbivores, feeding by trapping drift algae under the muscular foot. A file-like radula is used to scrape algae from rocks or to rasp the surface layers from macroalgae. Most species show preferences for particular types of algae, with various types of kelp being the most commonly consumed. Some abalone will also eat small invertebrates attached to the algae. Post-larvae and very small juveniles graze on benthic diatom films and encrusting algae that coat rock surfaces.

The sexes are separate and, in most species, a 1 : 1 sex ratio is maintained. The gonad tissue becomes visible when the shell reaches about 30 mm shell length (SL), and the gonads can be seen under the edge of the mantle membrane or under the foot. When immature, the gonad is generally a brownish color, but it turns green in females and cream-white in males as they mature. Most species become mature at between three to five years.

Spawning is usually in response to external cues, such as water temperature or photoperiod, and this helps to ensure that males and females spawn synchronously. Being broadcast spawners, the eggs and sperm are released into the water where fertilization occurs externally (Fig. 6.2). It has been suggested that to ensure successful fertilization, a minimum population density is required in the natural environment. Although this varies between species and between different locations, an example quoted for California suggests that a minimum population density of 2000/ha is required (California Department of Fish and Wildlife, 2014).

Abalone fisheries and farms

In most areas where abalone are common, they have formed the basis of important fisheries, generally being exploited for their meat or for their shell. In many cases, such fisheries have been over-exploited. In the statistics that follow, figures are adjusted to represent 'in shell' weight.

Legal landings from abalone fisheries have gradually decreased from more than 27,000 metric tonnes (MT) in the late 1960s to only about 6500 MT in 2014, with production in 2016/2017 remaining at about the 6,500 MT level (Fig. 6.3). Over-exploitation, illegal harvesting, disease, increased predation, and habitat degradation have all contributed to this decline. Following such

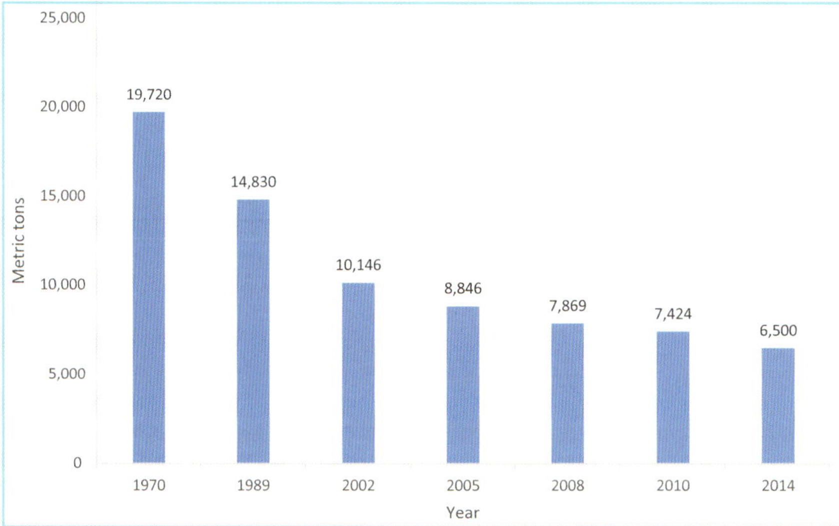

Figure 6.3 Abalone production from fisheries worldwide, 1970–2014.

declines, several fisheries have suffered severe quota restrictions, or have been completely de-commercialized. In California, for example, the commercial and recreational exploitation of all species of abalone has recently been prohibited. The abalone fishery in Australia has also declined over the past few years but, in this case, the decline followed an outbreak of a serious abalone disease in both farmed and wild stocks. Even in the Philippines, where there is only a very small abalone fishery, recent evidence has suggested a decline in catches following over-exploitation.

The abalone fishery in South Africa has suffered from many years of illegal over-exploitation by poachers. As a result of this, the fishery was de-commercialized in 2008. Although the fishery was re-opened to a small (150 MT) experimental quota in 2011, its future remains uncertain for as long as the illegal exploitation continues. According to Markus Bergener and co-authors (2018) of TRAFFIC, a nonprofit organization that monitors illicit wildlife trades around the world, the total South African legal abalone harvest in 2015 was 105 MT. while, in the same year, it was estimated that the illegal take was 3477 MT. Since 2001, an estimated 75 million abalone, about 40,000 MT, have been illegally harvested. In support of this suggestion, a recent survey of South African dried abalone being sold in Hong Kong suggested that 65% of dried abalone was illegally sourced, 33% was derived from legal aquaculture, and only 2% from the legal fishery (Britz, pers. comm. 2018).

South Africa is not the only country in which abalone poaching occurs. Cook and Gordon (2010) suggested that in 2008, the worldwide illegal catch, totaled about 5300 MT, representing over 60% of the total legal catch from fisheries. Although a significant proportion of the illegal catch comes from South Africa, other countries, such as Australia and New Zealand, also contribute large quantities of abalone to the illegal market. Although it is difficult to obtain accurate figures for illegal harvesting, anecdotal evidence suggests that up to 40 MT of abalone, worth more than US$3 million, is poached annually from South Australia alone. Last year, Asian countries imported twice as much Australian abalone as Australians are legally allowed to catch or farm, while in New Zealand, it is estimated that only two-thirds of abalone shipments are legally caught.

According to some recent estimates, the black market trade in abalone in California is estimated at US$25 million annually, with San Francisco being suggested as the hub of the trade. A spokesman for the Special Operations Unit of California Fish and Game, has suggested that the majority of abalone poaching is now operated by organized gangs. In Mexico, there is empirical evidence of a sizable abalone black market and of substantial sectors of fishing communities dedicated to poaching.

Based on these anecdotal reports, it is estimated that the current annual level of illegal abalone exploitation worldwide is in the region of 7000 MT.

On a more positive note, several countries, such as Canada, the USA, and South Africa, have been using hatchery-produced abalone to reseed depleted populations in an attempt to rebuild abalone stocks. In Washington State, for example, efforts to rebuild stocks of the pinto abalone (*H. kamtschatkana*) appear to be yielding positive results, while in California, projects under the banner of 'conservation aquaculture' are aimed at restoring populations of white (*H. sorenseni*), black (*H. cracherodii*), and green (*H. fulgens*) abalone. In South Africa, over US$2 million has already been spent on attempts to reseed depleted abalone populations (*H. midae*) in both the Western and Eastern Cape.

While worldwide legal fisheries have been declining, farm production has been rapidly expanding in several countries. In the 1970s, farm production was almost negligible and it is only since about 2000 that the really big increases have taken place. Table 6.1 shows changes in abalone production in various countries between 2010 and 2016/2017. While farm production in most countries, over this period, has either been stable or has grown very slowly, production in China and Korea has increased very rapidly. Smaller, but significant, increases have also occurred in South Africa, while in both Chile and Taiwan there was a disease setback in 2015 but, since then, the industry in both countries appears to be recovering.

Table 6.1 Estimated farm production in various regions (MT) in 2010, 2015, and 2016/2017.

Region	2010 production	2015 production	2016/17 production
China	42,373	115,397	139,697
Korea	5000	9400	16,042
South Africa	1023	1400	1685
Chile	1500	700	1468
Australia	500	900	969
Taiwan	300	171	300
Japan (seeds only)	200	200	200
USA (including Hawaii)	200	362	175
New Zealand	90	100 (estimated)	60
Mexico	33	30	23.5
Europe	10	15	30
Thailand	10	8	8
Philippines	4	4	5
Total	51,043	129,287	160,662.5

Abalone farming technologies

There are several different phases in the farming process.

1 Broodstock males and females are collected and conditioned.
2 Males and females are induced to release gametes synchronously.
3 Eggs and sperm are brought together to induce fertilization.
4 Larvae are cultured through a free-swimming stage for several days.
5 Larvae are settled onto suitable substrate where they undergo metamorphosis.
6 Post-larvae are grown through to juvenile stage.
7 Juveniles are grown to market size in various types of grow-out systems.

Broodstock

Farms have a choice to either hold a captive, breeding population of abalone adults, or to collect suitable mature adults from the wild. Farms that maintain and condition broodstock, keep them under controlled hatchery conditions until their gonads are fully developed, while those that rely on collection of wild broodstock bring ripe adults into a hatchery during their natural spawning season.

It has been shown that very specific diets are required during the conditioning process and that formulated feeds that maximize growth rates are not necessarily appropriate for producing viable, high quality eggs, and larvae. Specific dietary lipids play an important role in gametogenesis and must be included in conditioning diets, sometimes in the form of red seaweeds, which have been shown to be beneficial. Red seaweeds contain arachidonic acid, which is a precursor of the prostaglandins that play a vital role in molluscan reproduction.

The time that it takes for gonads to develop is dependent on water temperature and access to adequate and appropriate feeds, and there is a positive linear relationship between the visual gonad index (VGI) and conditioning time. The biological zero point (BZP), is the temperature below which no gonad development occurs and by subtracting the BZP from the daily water temperature and summing this figure over the conditioning time (in days), it is possible to describe the effective accumulative temperature (EAT, expressed as °C × days) for completion of gametogenesis and spawning. As an example, the EAT for *H. discus hannai* is about 1000°C-days for males and about 1500°C-days for females. By keeping adults at warmer than ambient temperatures, similar to those present during the spawning season, and with access to adequate food, gonad development can be accelerated.

Spawning induction

To obtain successful fertilization, it is essential that males and females are induced to spawn synchronously. A number of different methods can be used to induce spawning, including the manipulation of water temperature, short periods of air exposure, the addition of hydrogen peroxide, and the application of UV light.

In the past, short periods of desiccation, or thermal shock, were commonly used to induce spawning, but both methods are unreliable. More effective and reliable methods are to irradiate seawater with UV light or treat it with hydrogen peroxide at slightly elevated pH levels. In both cases, spawning induction appears to be in response to the presence of free oxygen radicals in the seawater.

If UV light is used to induce spawning, ripe males and females are placed separately into filtered seawater in relatively small plastic containers, generally with three to four individuals per container. Filtered, UV-irradiated seawater is passed through the container at about 0.5 L/min. The intensity of UV light required depends on the species but, for example, for *H. discus hannai*, an intensity of 2.42 Watts/L is generally used. Spawning usually occurs after about 2 hours of exposure. If the hydrogen peroxide method is used, the container is emptied and then re-filled with 3% (v/v) hydrogen peroxide, buffered with 2 mol/L Tris-buffer. The animals are left in this solution for up to 2 hours, and then the buffered seawater is replaced with clean, filtered seawater as soon as there is any sign of gamete release.

Fertilization

Fertilization is achieved by adding a small volume of water containing about 25,000 sperm/mL, to a bucket containing eggs. The ratio is between 10 to 100 sperm per egg. After gentle mixing, the fertilized eggs are allowed to settle and then, after 5 to 10 minutes the water is siphoned from the bucket and replaced with clean, filtered seawater at the same temperature. The fertilized eggs are then washed several times with clean, filtered seawater before being transferred to hatching tanks.

Development of larvae

Fertilized eggs remain at the bottom of the hatching tank during pre-hatch development and no aeration is applied at this stage. Egg density on the bottom of the tank of 200/cm^2 is recommended. Water is exchanged at least once per day. The time between fertilization and hatching varies between species and depends on water temperature, but it is generally between 10 to 24 hours. After hatching, the free-swimming veliger larvae generally accumulate near the water surface. Gentle water exchange and aeration can be applied at this stage. Banjo sieves, with a very fine mesh (not more than 100 μm), are used to prevent the veligers from being washed out of the hatching tanks.

A few days after hatching, larvae are siphoned from the hatching tank, care being taken not to suck up unhatched eggs or dead larvae from the bottom of the tanks. Free-swimming larvae are transferred to rearing tanks at a density of between 5 to 20 larvae/mL. Larval tanks receive continuous gentle aeration and water exchange. The larvae are lecithotrophic and, therefore, do not need to be fed. Time spent in larval tanks depends on water temperature but, generally, is between 5 to 7 days. High mortalities occur at this stage, often reaching up to 90%. Once the larvae have reached metamorphic competence (3 to 8 days), they are ready to settle onto a suitable substratum. Settlement of abalone larvae involves larval attachment (a reversible behavior) followed by metamorphosis (irreversible physical changes). The length of time that larvae can remain free-swimming before settlement depends on the endogenous supply of yolk. Generally, if they do not find a suitable settlement substratum within approximately a week post-hatching, survival is very low.

Figure 6.4 Plastic settlement plates held vertically in a frame, and a large corrugated plastic feeding plate, with juvenile abalone attached to the underside.

Settlement

Hatchery-produced veliger larvae are transferred into settlement tanks at a density of 1000–2000/m^2 of available settlement surface. In these tanks, larvae are usually settled onto clear, plastic plates, normally held in batches by being fixed vertically into racks (Fig. 6.4).

A range of different techniques can be used to induce settlement and metamorphosis, the most common of which are:

- a film of diatoms and/or bacteria
- a coat of the green, encrusting algae, *Ulvella lens*
- allowing abalone juveniles to graze on diatom-coated plates before placing the plates into settlement tanks (the juveniles leave a mucous trail on the plates)
- addition of 10^{-6} molar gamma-amino butyric acid (GABA) to the tank water.

When diatom films are used to induce settlement, it is important that the species and size of the diatoms are appropriate. Diatoms such as *Navicula*, *Cocconeis*, *Amphora*, and *Nitzschia* genera have been found to be suitable for first-feeding post-larvae as these are small pennate diatoms with a diameter of about 10 μm or less. In some farms, particularly in Australia, *Ulvella lens* has become a commonly used replacement for diatoms to induce settlement and as a food source for first-feeding post-larvae.

If pre-grazed settlement plates are used, conspecific abalone leave a trail of mucus which becomes populated with bacteria. The mucus probably contains chemicals that trigger attachment or metamorphosis, but such chemicals have not been positively identified. When GABA is used to induce attachment and/or metamorphosis it may bind to larval receptors or may act as a neurotransmitter inhibitor, but its precise mode of action remains unclear. Some hatcheries use antibiotics together with GABA but, for environmental reasons, this practice is generally discouraged.

Post-larvae

After settlement onto plates, post-larvae begin to feed initially on the biofilm of diatoms and bacteria. At first settlement, farmers generally attempt to achieve a post-larval density of about $10/cm^2$. Following this, many farms achieve high post-larval densities by adding a daily supply of diatoms. By the end of the post-larval feeding period, density has often been reduced to about $1/cm^2$. At about 3 to 5 mm SL in the wild, the post-larvae gradually transition onto a diet of macroalgae. In a hatchery, however, unless the operation has access to supplies of seaweeds, this transition is often onto small pellets of a formulated diet.

There are many different methods used to culture juveniles. A common method (for example, in China), is to place horizontal feeding plates, usually plastic sheets with access holes drilled into them (Fig. 6.4), above the vertical settlement plates. Juveniles still feed on the diatoms adhering to the vertical settlement plates but will gradually transition onto the food provided on the horizontal feeding plates. An alternative method, used particularly in South Africa, is to transfer juveniles, at about 3–5 mm SL, to an intermediate weaning stage where juveniles have access to both diatoms and formulated pellets. Small inverted plastic cones are placed on the bottom of a shallow tank and aeration is applied beneath the cones. Diatoms, algae, and formulated pellets are added to the tank. Animals remain under the cones during the day and emerge to feed on the algae and diatoms at night.

In the wild, when juveniles reach 5–10 mm SL, they start to feed on large seaweeds, eating an average of about 20% of their body weight per day. Some farms still use seaweeds to feed juveniles, but worldwide the use of formulated feeds is becoming more popular. Where seaweeds are used at this stage, the most commonly used genera are *Ecklonia*, *Laminaria*, and *Macrocystis,* often supplemented with *Ulva* or *Gracilaria*. Whenever possible, red seaweeds (for example, *Porphyra* spp.) are included in the diet of juveniles as they have been found to promote good growth rates.

Grow-out systems

The type of grow-out system used at any particular location is usually determined by local water conditions and the specific requirements of the abalone species being farmed. Important factors include:

- sea conditions at the farm location
- water quality of the local seawater
- availability and cost of land
- the cost of labor
- local and national environmental regulations.

The most commonly used grow-out systems are:

- sea cages
- land-based tanks with pump-ashore water
- intertidal ponds
- sea ranching.

Figure 6.5 An off-shore abalone farm in China using net cages.

In China, where most of the world's farmed abalone are now grown, the majority of production is undertaken in off-shore cages (Fig. 6.5), this being one of the most cost effective systems available.

Off-shore cages are also used in Chile to grow abalone (Fig. 6.6) although, in this case, the cages are much larger and usually require a barge with an on-board crane to lift them for feeding, cleaning and harvesting.

In countries such as South Africa and Australia, where sea conditions generally preclude in-sea culture, land-based tanks are commonly used. The precise design of such tanks is, however, very variable. The designs most commonly used in Australia (slab tanks) and in South Africa (deep tanks) are illustrated in Fig. 6.7. The South African deep tank is particularly suitable where the abalone are fed mainly on seaweeds, whereas the Australian slab tank was designed primarily for use with artificial, pelleted feeds.

Figure 6.6 Large cage used in an off-shore abalone farm in Chile.

Figure 6.7 (A) A slab tank as used in Australia. (B) Deep grow-out tank as used in South Africa (a rack of white settlement plates can also be seen near the tank).

In the South African deep tank system, abalone are usually held in nets or plastic baskets, suspended into the water. Surface area is maximized by use of a series of vertical plates, usually held in place by a rack. Seaweed, as food, is inserted between the vertical plates. There is often also a horizontal plate located over the top of the vertical plates, and this can also be used as a feeding platform.

The Australian slab tank system is based on very large concrete tanks in which the abalone are kept in very shallow water, perhaps only about 20 cm deep. Overall tank dimensions can be as much as 15 m long by 5 m wide, water movement being maintained by a water-dumping bucket system, located at one end of the slab. Some recent variations of this system (used, for example, in New Zealand) use plastic tanks of similar dimensions that can be stacked several layers high.

In some countries, abalone are grown on near-shore reefs, in various types of ranching operations. The juveniles are produced in hatcheries and then moved into the sea when they are about 50 mm SL. This method has been used in Japan for many years, natural reefs being used to deploy the juveniles. In other countries, such as Australia and South Africa, ranching is a more recent innovation and utilizes either natural reef, or artificial substrate designed specifically for this application. The advantage of this method of abalone production is that, once deployed into the sea, relatively little labor is required to maintain them until they are ready to be harvested. A disadvantage, particularly in South Africa, is that it is difficult to secure the animals from illegal exploitation during their time on the reef.

To ensure the economic viability of a farming operation, it is important that the animals achieve an acceptable growth rate. In land-based tanks, a growth rate of up to 30 mm per year can be achieved during the first grow-out year, but the rate generally reduces after this. It can take animals up to 4 or 5 years in grow-out to reach marketable size. A notable exception is the tropical abalone, *H. asinina*, which grows very quickly in warm water, and can achieve 60 mm SL after a year in grow-out.

Abalone farmers have a choice of feeding their stock either on natural seaweeds, on formulated feeds, or on a combination of both. In countries such as South Africa, where huge quantities of harvestable kelp are available, this is often the feed of choice. In other countries, such as Australia, less seaweed is available and, therefore, it is more common to utilize formulated feeds. Both systems have their advantages and disadvantages and, as an example, the characteristics of a seaweed diet and formulated feed used in South Africa, are listed in Table 6.2.

Table 6.2 Characteristics of a seaweed and formulated diet (including % proximal composition) used in South Africa. Courtesy of Peter Britz, Ichthyology and Fisheries Science, Rhodes University, South Africa.

Component	Seaweed (e.g. kelp)	Formulated feed
Protein (%)	10–17	26–34
Fat (%)	Negligible	<5
Ash (%)	25	10
Carbohydrate (%)	62	50–60
Feed conversion rate	15 (wet); 2.4 (dry)	1
Cost (US$)	103/wet MT	140/MT

The use of formulated abalone diets was first developed in Japan, but now most farms throughout the world use artificial diets at some stage of their production cycle. Much research is still being carried out on the development of artificial diets, particularly aimed at reducing manufacturing costs, while also improving performance. Manufactured abalone diets generally contain high levels of proteins and carbohydrates, and low levels of fat and fiber. Some formulated diets are manufactured to include feeding attractants. Additives to reduce leaching of nutrients are also important.

The formulation of abalone diets is still the subject of intense research, particularly aimed at reducing costs while, at the same time, maintaining dietary efficiency. Of particular importance, in recent years, has been the reduction of fish meal and fish oils in diets, and their replacement by plant-based, or synthetic components. In addition, it has been recognized that different stages of the abalone life history require different dietary formulations, a particularly important change occurring as the animals become sexually mature, usually at about 30 mm SL.

Diseases and parasites

Over the past two decades, diseases in wild abalone have become better understood but, as aquaculture has increased, a wider range and more frequent incidences of diseases, in both wild and farmed abalone, have been noted. Factors such as over-stocking of farms and deterioration of water quality near farms, have probably contributed to the spread of parasites and diseases.

Withering syndrome is a well-known abalone disease, that was originally found in California in 1986 in black (*H. cracherodii*) and red (*H. rufescens*) abalone, where it caused up to a 99% decrease in black abalone populations. Although originally described as a viral disease, it is known that it is closely associated with a Rickettsiales-like prokaryote (RLP); possibly suggesting that it is actually the RLP that is responsible for the symptoms of the disease. The disease attacks the lining of the gut, inhibiting the production of digestive enzymes. Infected animals consume their own body mass, causing the muscular foot to wither, which reduces the ability of the animal to cling to rocks and makes it far more vulnerable to predation. Withering syndrome has had a devastating effect on Californian abalone populations and came close to completely wiping out all populations of some of the local species. Intensive recovery efforts have recently improved the situation and it may be that, in the future, with continued conservation actions, populations in the wild may be restored.

Another important abalone disease is caused by a boring sabellid worm (*Terebrasabella heterouncinata*), which is endemic to South Africa. In the mid-1990s it was also discovered in

California, possibly resulting from the illegal importation of South African abalone into the USA. The worms bore small holes into the growing edges of abalone shells, resulting in the shells becoming deformed. Infected individuals show reduced growth rates and poor meat yield, probably because of the extra energy that is channeled into shell repair as opposed to meat growth.

Abalone viral ganglioneuritis (AVG), was discovered in abalone farms in Australia in 2005. Confined mainly to Victoria and Tasmania, this had devastating effects on both wild and farmed abalone populations. AVG can cause high mortality rates of up to 90%, in all age classes. AVG was declared a notifiable disease in Australia under the *Diseases of Livestock Act* 1994. A similar disease has occurred in several other countries, including Taiwan. AVG affects the nervous system of abalone and results in curling of the foot, swelling of the mouth, and eventually weakness and death. In Australia, AVG affects both blacklip (*H. rubra*) and greenlip (*H. laevigata*) abalone, as well as the hybrid between the two. Since the discovery of AVG, very strict biosecurity protocols have been imposed on both the abalone farming and fishing industries and, so far, the improved biosecurity appears to have helped to prevent further spread of the disease.

The mudworm (*Polydora* spp.) infests abalone shells. It bores into the shell and causes internal mud blisters. Blisters affect the adhesion of the body to the shell and this can cause significant mortalities on farms, sometimes up 60% of the stock. Mudworms can be controlled by good animal husbandry, including keeping the animals off the bottom and preventing the build-up of excessive bottom sediments. Infected animals are usually culled to prevent the spread of the parasites.

Worldwide abalone market

Over the past few years, the most significant change in abalone production has been in China where the vast majority of the world's abalone are now produced. In 2010 there were over 300 abalone farms in China, with the largest producing over 1000 MT/year. Although falling prices and Chinese austerity measures have resulted in the closure of some farms, the total production for 2016/2017 was estimated at 139,697 MT/year. Over the past few years, many Chinese abalone farms have become more efficient by changing from land-based to sea cage operations, and innovations in low-cost seed production have helped to establish a viable industry. Disadvantages of sea-based farms include that the nets are subject to biofouling that requires regular cleaning of cages, and that farms are subject to the vagaries of weather and sea conditions, but the advantage is that abalone in-sea cages can be grown to market size at much lower costs than in land-based farms.

Over the past decade or so, Korea has become an important producer of farmed abalone. Prior to about 2000, only small quantities of abalone were farmed, and the production method mainly used suspended baskets in land-based farms. Production has recently evolved to a more efficient methodology, utilizing off-shore cage farms. This change has resulted in rapid increases in volume of production, going from about 1000 MT in 2003, to 9400 MT in 2014, and to 16,042 MT in 2016/2017. The majority of Korean production is *H. discus hannai*, the species that is most popular, and commands the highest price, in the Japanese market.

Farmed abalone production in the USA (including Hawaii) was about 362 MT in 2016/2017. It is unlikely, however, that total production along the Californian coast will increase much

Figure 6.8 Aerial view of an abalone farm on the coast of California.

in the future because of very high land values and high compliance costs. An aerial view of a Californian abalone farm is shown in Fig. 6.8, illustrating the quantity of coastal land required. In Europe, abalone farming is a very small, but growing, industry. Farms are located in the UK, the Channel Islands, Ireland, France and Spain. In Ireland farms are mainly experimental, producing mostly *H. discus hanni,* but it seems unlikely that production will exceed about 3 MT in the near future, while in the Channel Islands there are two farming companies, currently producing less than 1 MT/year. French farms are located in Brittany and Vendee, and these have the advantage of being located in regions where small, natural populations of abalone occur, and have been fished for some years. Farm production from the French farms is likely to increase over the next few years.

Other countries that produce significant quantities of farmed abalone are Chile, Japan, South Africa, Australia, Taiwan, and New Zealand, with smaller quantities being produced in Mexico, Thailand, the Philippines, and Oman.

Even though there have been huge recent increases in abalone farming, global demand still exceeds supply. The total supply of abalone available to the world market in 2015 was more than five times that which was available in the 1970s. What effect this has on the international abalone market is unknown. China is both a major producer and consumer of abalone, and, while China has often been regarded as a huge marketing opportunity, over-production in China could also represent a risk to other producers.

The future of abalone farming, both in China and the rest of the world, and the predicted level of production, are still the subject of much speculation. What is certain, however, is that to be competitive in the future, farms both inside and outside of China will need to reduce the cost of production by implementing more efficient production methods, in particular by reducing the costs of labor, and by the use of less expensive and more efficient feeds. Recently, some farms have adopted international environmental certification standards and, besides the possibility that this will increase market access, it may also help them to improve efficiency and reduce production costs.

References cited and further reading

California Department of Fish and Wildlife (2014) www.wildlife.ca.gov/Conservation/Marine/Red-Abalone-FMP (accessed 15 April 2016).

Cook, P.A. and Gordon, G.H. (2010) World abalone supply, markets and pricing. *Journal of Shellfish Research*, 29: 569–571.

Geiger, D.L. and Owen B. (2012) *Abalone: World-Wide Haliotidae*. Conch Books, Dürkheim.

Hahn, K.O. (1989) *Handbook of Culture of Abalone and other Gastropods*. CRC Press, Boca Raton, FL.

Leighton, D.L. (2000) *The Biology and Culture of Californian Abalone*. Dorrance Publishing Company, Pittsburgh, PA.

Okes, N., Burgener, M., Moneron, S., and Rademeye, J. (2018) Empty Shells: An assessment of abalone poaching and trade from southern Africa. TRAFFIC Report, September 2018.

Shepherd, S.A., Tegner, M.J., and Guzman del Proo, S.A. (1992) *Abalone of the World: Biology, Fisheries and Culture*. Fishing News Press, Oxford.

CHAPTER 7

Conch aquaculture

Queen conch, Strombus gigas, *and fighting conch,* Strombus alatus *and* Strombus pugilis

Megan Davis

Introduction

The queen conch, *Strombus gigas*, is the largest molluscan gastropod (18–23 cm shell length [SL], 7–9 in; 2.3 kg, 5 lbs) of the six *Strombus* species. The queen conch has spines on each whorl of the spire and a glossy deep pink flared aperture (Fig. 7.1). They are found in the shallow seagrass beds and sandy flats of the territorial waters of at least 36 countries and dependent territories including Florida, The Bahamas, Bermuda, the Caribbean Islands, and the northern coasts of Central and South America (Fig. 7.2). The queen conch was originally described and named by Linnaeus 1758 as *Strombus gigas*, which was the accepted name for over 200 years up until it was recently reclassified to *Aliger gigas*. It is known by various common names throughout the region such as caracol rosado (Mexico, Honduras, Columbia), carrucho (Puerto Rico), cobo (Cuba), lambi (Hispaniola, French Antilles), karko (Netherlands Antilles) and caracol de pala (Venezuela).

Figure 7.1 Shell of adult queen conch, *Strombus gigas*. Courtesy of Tom Smoyer.

The queen conch hatch from an egg mass as planktonic veligers (larvae) and feed on phytoplankton for about 3 weeks (Fig. 7.3). After metamorphosis the juveniles remain buried for the majority of their first year. As herbivorous gastropods, the juvenile and adult conch feed on a variety of epiphytic algae and diatoms associated with macroalgae, such as *Batophora oerstedi*, seagrass *Thalassia testudinum*, and the sandy substrate. During the first couple of years, the juvenile conch add length to their shell and remain in the seagrass beds. Conch juveniles that do not have a flared lip

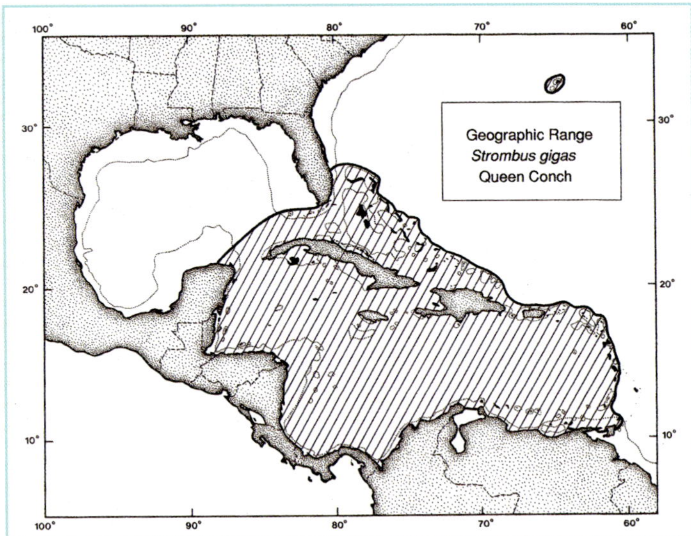

Figure 7.2 Geographic region where queen conch are found. Courtesy of Bonnie Bower-Dennis.

are sometimes referred to as rollers. As the flaring lip begins to form the sub-adult conch begin to migrate to deeper water near the reef tract and to form spawning aggregations. They complete their lip flare at about 3 years old and they are sexually mature at 4–5 years old when their lip has a thickness of ≥15 mm (0.59 in). Queen conch are estimated to have a life-span of 25–40 years; at this stage their lips are very thick (24–28 mm (0.94–1.10 in) or thicker) and they are often called samba conch.

The queen conch is deeply rooted in the island culture due to its great social heritage and strong economic importance as a fishery species in The Bahamas and the wider Caribbean. The queen conch is a prized delicacy and has long been harvested both for its meat and for its beautiful shell. It is one of the most important benthic fisheries in the Caribbean region, second only to

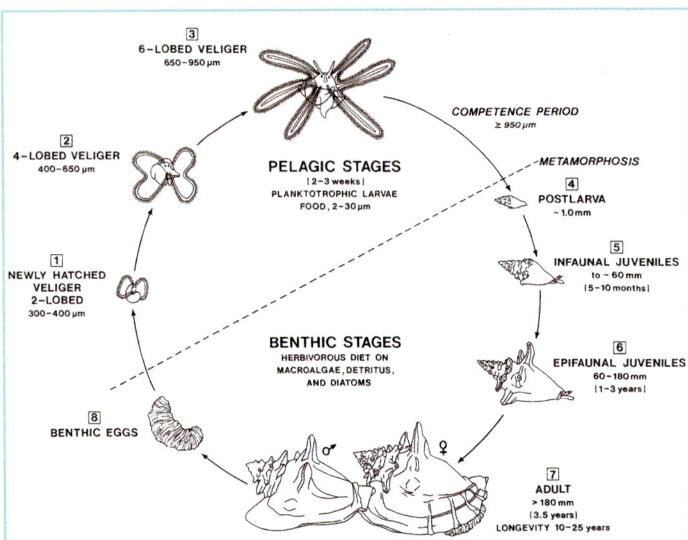

Figure 7.3 Queen conch life cycle stages from eggs to adult conch. Courtesy of Bonnie Bower-Dennis.

spiny lobster. Although a commercial fishery for queen conch has existed for many years, a combination of factors led to the dramatic increase in the size of the fishery since 1970s. Traditionally conch were harvested by fishermen working in small dinghies (Fig. 7.4). Using glass-bottom buckets to gain a clearer view, the fishermen would scan the seafloor for conch. When a conch was located it was retrieved using a long, hooked pole. With the advent of snorkel gear and in some cases hooka and scuba along with the ability to store large numbers of conch in freezers the fishery grew rapidly. It is the steady rise in tourism in The Bahamas and Caribbean region over the last 40 years that has fueled a tremendous increase in the demand for conch. Owing to heavy fishing pressure, the conch populations have been significantly depleted throughout much of their range and in 1992 the Convention for the International Trade of Endangered Species (CITES) placed queen conch on its Appendix II species list. As a result, countries that export conch now require certification from CITES that the annual harvest does not exceed sustainable limits and that management plans are in place. Conch were also placed on the 1994 International Union for Conservation of Nature (IUCN) Red List as a commercially threatened species. Fishing regulations that include harvesting conch with a minimum lip thickness of ≥15 mm (0.59 in), closed season during their breeding months, quotas, moratoriums, and/or no-take marine protected areas are in place throughout the geographic area where conch are found. It is not clear whether fishing restrictions alone will allow queen conch populations to recover or keep pace with the fishing demand. Therefore, as a means to help offset fishing pressure, methodologies to culture queen conch larvae to the juvenile stage for stock enhancement and for grow-out markets began to be developed in the mid-1970s and are now well-established. These culture techniques have also been used to study larval ecology, population genetics, and fisheries oceanography, which aid in regional fisheries management plans for this pan-Caribbean species.

Figure 7.4 Traditional queen conch fishing methods. Courtesy of Caicos Conch Farm.

There are five other *Strombus* conch species in the region, which include milk conch, *S. costatus*; Florida fighting conch, *S. alatus*; West Indian fighting conch, *S. pugilis*; hawkwing conch, *S. raninus*; and roostertail conch, *S. gallus*. The techniques to culture queen conch have been applied to culturing several of these other *Strombus* spp. through their larval stage to early juvenile or adult stage.

One of the other promising conch species for the seafood and aquarium markets are the

Figure 7.5 Shell of adult Florida fighting conch, *Strombus alatus*. Courtesy of Tom Smoyer.

Florida and West Indian fighting conch. The Florida fighting conch are found from the coasts of North Carolina to south Florida and throughout the Gulf of Mexico. The West Indian fighting conch inhabit south Florida through the Caribbean basin. Both species are smaller than the queen conch (7–10 cm SL; 2.8–3.9 in SL) and are not a regulated species under CITES. They are mature at one year of age, which is much faster than the queen conch (Fig. 7.5).

This chapter focuses on the aquaculture of the queen conch, *S. gigas*, and the two fighting conch species, *S. pugilis* and *S. alatus*.

Culture techniques

The development of queen conch aquaculture began with the early hatchery work in Los Roques, Venezuela in the 1970s. In the 1980s other small hatcheries at University of Miami in Florida and throughout the Caribbean, including Puerto Rico and Turks and Caicos, made significant contributions to queen conch culture. Based on techniques to grow conch on a small-scale, large-scale, or commercial techniques for growing queen conch commercially were developed at the Caicos Conch Farm in Providenciales, Turks and Caicos Islands throughout the 1980s to 2000s. In 2016 and 2017, an experimental conch hatchery for restoration was operated by Hummingbird Cay Foundation and Florida Atlantic University Harbor Branch Oceanographic Institute (HBOI) in the Exuma Cays, Bahamas (Fig. 7.6). In Mexico there is an experimental conch research hatchery in operation at Cinvestav Unidad Merida and conch hatchery research was conducted at Instituto Nacional de la Pesca in Quintana Roo. A production scale conch hatchery was started at the Curacao Sea Aquarium in 2020. Starting in 2019, a pilot-scale queen conch hatchery was initiated and will be fisher-operated at the Naguabo, Puerto Rico Fishing Association with assistance from HBOI and Conservación ConCiencia. The commercial Caicos Conch Farm in the Turks and Caicos Islands closed in 2013 following a change in management

Figure 7.6 Experimental queen conch restoration hatchery at Hummingbird Cay, Exuma Cays, Bahamas. Courtesy of H. Forrest Thomas.

Figure 7.7 Ariel view of the Caicos Conch Farm at Providenciales, Turks and Caicos Islands. Courtesy of Caicos Conch Farm.

and degrading facilities due to several tropical storms and hurricanes. Based on the techniques developed at the Caicos Conch Farm it was shown that over one million queen conch can be raised annually in the hatchery and nursery of a commercial conch farm (Fig. 7.7). Although the technical problems associated with conch culture have been resolved, conch culture has not expanded in the Bahamas and Caribbean region due to the initial high capital cost of establishing a large-scale conch production facility.

The methodology developed for growing the fighting conch in captivity was based on the techniques to culture the queen conch (Tables 7.1 and 7.2). The only large-scale hatchery and nursery facility for culturing fighting conch was established in 2000 at HBOI in Florida and operated up until the hurricanes of 2004 and 2005. This facility produced up to 100,000 fighting conch juveniles per year for the marine aquarium market. In 2019, a fighting conch breeding system was established in the HBOI Integrated Multi-Trophic Aquaculture (IMTA) system. This species may prove to be a good species to supplement the queen conch food market since their meat has a similar taste, texture and appearance to the queen conch.

Table 7.1 Culture criteria for queen conch.

Stage	Time	Size	Survival (%)	Stocking density	System
Egg farm	April–November	18–23 cm	100	1 conch/10 m^2	Enclosed field site
Egg mass	May–November		100	1/container	Incubation tank
Larval culture	2–3 weeks	0.3–1.0 mm	10	Begin 100/L; end with 10/L	Conical tanks
Metamorphosis	2 weeks	1.0–3.5 mm	50–60	3500/m^2	Shallow tank with screen trays
Nursery, indoors	5–6 months	3.5–40 mm	75	1700/m^2	Shallow tank w/sand substrate
Nursery, outdoors	8–9 months	4–9 cm	90	75–150/m^2	Shallow pond or raceway w/sand substrate
Grow-out	1.5–2 years	9–20 cm	90	1–2/m^2	Sub-sea pasture

Table 7.2 Culture criteria for fighting conch.

Stage	Time	Size	Survival (%)	Stocking density	System
Broodstock	Year round	7–10 cm	100	18 conch/m^2	Shallow tanks w/sand substrate
Egg mass	May–November		100	1/container	Incubation tank
Larval culture	2–3 weeks	0.3–1.0 mm	10	Begin 100/L; end with 10/L	Conical tanks
Metamorphosis	2 weeks	1.0–3.5 mm	50–60	3500/m^2	Shallow tank with screen trays
Nursery, indoors	3 months	3.5–20 mm	75	1700/m^2	Shallow tank w/sand substrate
Nursery, indoors or outdoors	2–7 months	2–6 cm	90	75–150/m^2	Pond, raceway or tank w/sand substrate

Broodstock and egg mass procurement

Queen conch

Adult conch have separate sexes (Fig. 7.8), fertilization is internal, and they are sexually mature at approximately 4 years old or possibly older. They begin flaring their lip at 3 years old and after the lip has fully flared and has a thickness of ≥15 mm (0.59 in) they are sexually mature. It is difficult to differentiate male and female conch from outside appearance; however, female conch are typically larger than male conch. This can be used as a general indication of sex, unless fishermen have altered this relationship by preferentially harvesting the larger females. A more accurate, but time-consuming technique to differentiate sexes is to place conch on their sides with their lip pointing up. As the conch rights itself, their sex can be noted: a verge for male and egg groove for female (Fig. 7.8). The conch can also be held with their aperture pointing down and the conch will come partially out of their shell and their sex can be noted (Fig. 7.9). Another method to determine sex is to observe conch in their mating or copulating position (female in front, male behind) (Fig. 7.10) or view females laying an egg mass (Fig. 7.11). The crescent-shaped, sand-covered egg masses can be found under the anterior lip of the female conch. The female will lay an average of nine egg masses/season and each egg mass contains approximately 500,000 eggs (Fig. 7.12). As the conch is laying the egg mass, the thin sticky strand is covered with sand grains to camouflage the egg mass from predators during the three to five-day incubation period.

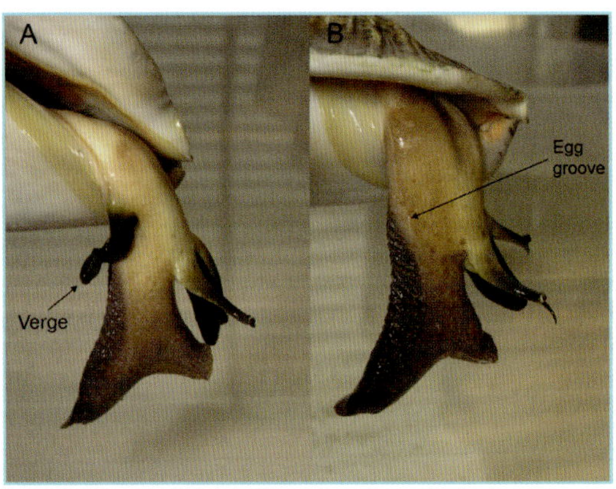

Figure 7.8 External sex organs of the milk conch, *Strombus costatus*: (A) male; and (B) female; these are similar to those of the queen and fighting conch. Courtesy of Tom Smoyer.

Figure 7.9 Determining the sexes of the conch by holding the shell to allow the conch to come partially out of their shell (A) queen conch and (B) fighting conch. Courtesy of Megan Davis and Tom Smoyer, respectively.

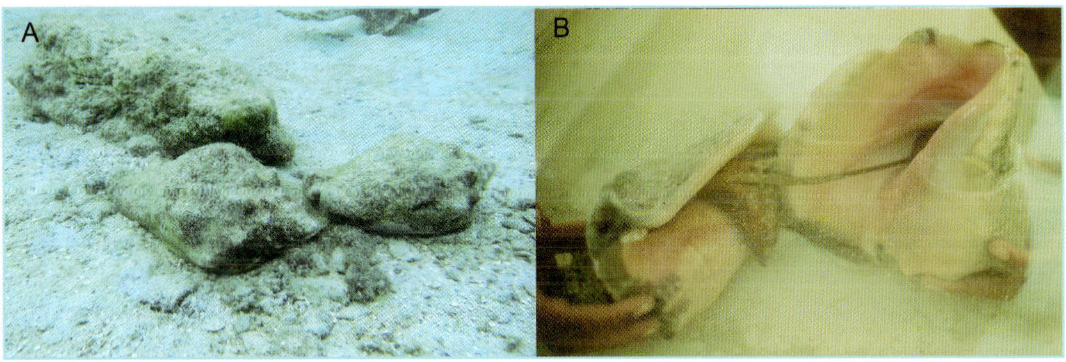

Figure 7.10 (A) Queen conch copulating in the field with the female on the left or in front with an egg mass being laid, and male on the right or behind; (B) copulating in the breeding tank showing the verge of the male with the female on the left and the male on the right. Courtesy of Michiel van Nierop and Foundation for PRIDE, respectively.

A critical requirement for successful hatchery production of any species is a reliable source of eggs. Captive breeding and egg laying of hatchery-reared queen conch, raised at Florida Fish and Wildlife Conservation Commission, was accomplished at HBOI in 2001. This was the first reported laying of egg masses by queen conch in captivity. A total of four *S. gigas* conch (1 conch/10 m^2, 1 conch/100 ft^2) were stocked at 1 : 3 female : male ratio and held in the circular breeding tank with an elevated crushed coral aragonite sand (1–3 mm dia.) substrate (Fig. 7.13). The breeding tank was on a recirculating water system and was kept at 27–29°C (81–84°F), salinity of 32, pH 8.0, and

Figure 7.11 Queen conch laying egg mass under the shell lip in the front area of the conch shell. Courtesy of Caicos Conch Farm.

Figure 7.12 Queen conch egg mass contains approximately 500,000 eggs. Courtesy of H. Forrest Thomas.

ammonia 0.06–0.10 ppm. The breeding tank was inside a shaded greenhouse and artificial lighting was supported above the tank on a 12 hour cycle (08:00–20:00). The queen conch broodstock were fed approximately 50–70 g/day/conch of prepared diet. The feed was a mixture of koi chow (36%), dried *Ulva* seaweed (16%), freshwater (42%), and a gelatin binder (5%). A total of four egg masses were laid by the single female conch in captivity and the hatched veligers were cultured successfully to the juvenile stage. Further investigations on temperature, photoperiod, density, and sex ratios are needed to achieve a reliable method to induce egg laying for queen conch in captivity.

Since production of eggs from a captive queen conch breeding program is not yet routine, queen conch hatcheries need to rely on wild adult conch populations as a source of egg masses. The availability and number of egg masses can be uncertain in some locations due to environmental factors and fishing pressure that can reduce breeding aggregations below sustained levels

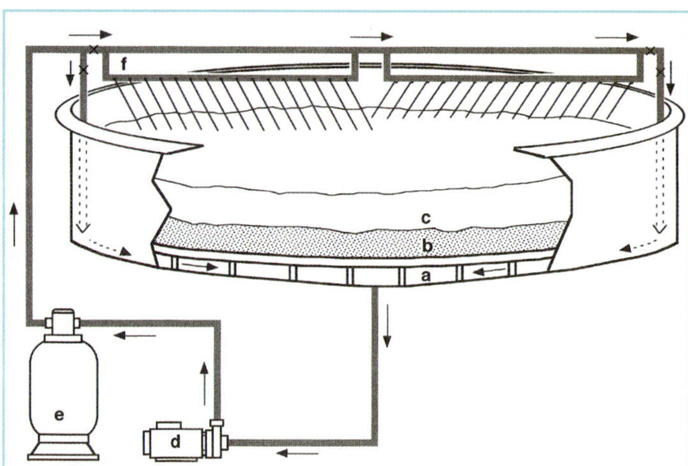

Figure 7.13 Conch breeding tank used at Harbor Branch Oceanographic Institute: (a) undergravel support; (b) sand substrate; (c) water column; (d) pump; (e) sand filter; and (f) spray bars. Courtesy of Jackie Aronson.

(<56 adult conch/ha; 23 adult conch/acre) for breeding and spawning. Populations of >100 adult conch/ha (40 adult conch/acre) is a desirable density to ensure breeding.

In the early 1980s through to 2006 the Caicos Conch Farm in the Turks and Caicos Islands developed and used a reliable method for obtaining conch egg masses. The Conch Farm established and used an underwater enclosure for breeding adults that was called an 'egg farm'. The farm was stocked with 120 adult conch and approximately 400 egg masses were laid during each breeding season. Ten egg masses per week would be adequate to support a commercial hatchery. Egg farms may also prove to be a method to increase egg and larvae productivity in the wild to assist with restoration of overfished populations. In 2019, an egg farm for restoration purposes was established and tested in a Marine Protected Area (MPA) in the Exuma Cays, Bahamas. Approximately 250 adult conch were translocated from two nearby fishing areas and reestablished in a historical breeding ground in the MPA. This partnership project included HBOI, The Exuma Foundation, the Bahamas National Trust, Bahamas Department of Marine Resources, and local Bahamians.

The criteria to select an egg farm site are the following: (1) area is or was historically inhabited by adult conch; (2) the location is typically in a back reef area; (3) substrate has calcareous sand and coral rubble with low-organic content; (4) grazing areas such as seagrass habitat; (5) water movement is less than 1 knot; (6) depth is typically 4–8 m (13–26 ft), and (7) location is within 3 km (2 mi) of the hatchery facility, for ease of access and security. In addition, for restoration purposes the egg farm should be located in an area where the oceanographic currents are known to assist with dispersal of the larvae to their juvenile habitats.

There are many fencing materials that can be used to enclose an egg farm to retain the adult conch. One recommendation that was used in 2019 for the egg farm in The Bahamas was a vinyl coated, rigid 11-gauge wire mesh (openings 7.6 × 7.6 cm; 3 × 3 in) with a height of 0.6–0.8 m (24–30 in). The mesh was held perpendicular to the substrate with reinforcing rods spaced 1.2–2.4 m (4–8 ft) depending on the current (Fig. 7.14). Anchors were also added to assist with securing the fence in place. Thick cable ties were used to secure the rods to the fencing material. The egg farm fencing can be configured into a large circular enclosure and the size will depend on the number of egg masses needed for the hatchery or restoration purposes. The size of the egg farm used in the Turks and Caicos and The Bahamas was 1400 m^2 or 0.14 ha (15,000 ft^2). The Bahamas egg farm was 42 m in diameter (138 ft in diameter) circular enclosure. The egg farm in the Turks and Caicos had coral reef and boulders that formed the majority of the enclosure boundaries and the two large openings were closed off with fencing. The fencing used for the Caicos egg farm was orange polypropylene mesh with a lead line on the bottom and 15 cm in diameter (6 inches in diameter) floats on the top. Sandbags can be used if there are gaps where the fencing meets the substrate. It is not necessary to remove the

Figure 7.14 Fencing material for a queen conch egg farm. Courtesy of Megan Davis.

fence during the non-breeding season, unless it is located in an area where hurricanes or winter storms would cause damage.

The egg farm should be stocked annually with sexually mature conch (lip thickness ≥15 mm; 0.59 in) 1 month prior to the breeding season. It is preferable if the conch stocked in the egg farm are from nearby populations. It might be possible to overwinter the adult conch in the egg farm if the enclosure has adequate food and is protected from the weather; however, if the conch are overgrazing the egg farm and are not removed in the non-breeding season, there might be low egg production the following breeding season. The 6–8 month breeding season is typically from March to October with the highest activity between June to September when the water temperatures are the warmest (28–30°C; 82–86°F). Based on wild populations, the sex ratio should be 1 : 1 with a density of 1 conch/10 m^2 (1 conch/100 ft^2). This is equivalent of 1000 conch/ha (405 conch/acre). If the egg farm is 1400 m^2 or 0.14 ha (15,000 ft^2) then 140 conch (70 males and 70 females) should be stocked in the egg farm. This would be a sufficient size egg farm for a commercial hatchery. Conch are transported to the egg farm by boat underneath wet burlap or towels. The conch can be held out of the water for 5 hours and sometimes longer if they are well shaded and bathed with seawater. Conch should be bathed with seawater every one to two hours during the transport. This can be accomplished by keeping the conch with aperture up and using a bucket to pour seawater over the conch (Fig. 7.15). By using this technique, the conch retain small amounts of water in their aperture opening. In a previous study, conducted by HBOI, a large number of adult conch were shipped from the Turks and Caicos Islands to Florida inside Styrofoam boxes, which contained seawater moistened straw-like packing material. The conch were placed aperture down and an ice pack was included in the shipping box (Fig. 7.16). The conch were in transit for approximately 20 hours and there was no mortality.

Figure 7.15 Bathing queen conch with seawater during transport. Courtesy of Megan Davis.

Excursions to the egg farm one to two times/week are necessary to collect egg masses, observe the conch, check the fence, and for security purposes. An underwater sign can be used to alert unexpected visitors not to disturb the egg farm conch and site (Fig. 7.17). Egg masses are usually collected from underneath a female, because they are hardier for transport, the development of the embryo is minimal and hatch date can be estimated. It takes approximately 24 to 36 hours for a female to lay an egg mass and the eggs hatch in 3 to 5 days. Since each egg mass can have approximately 500,000 eggs it is not necessary to collect a full egg mass; a 1/4 to 1/2 of an egg mass will yield a substantial hatch. For restoration purposes, multiple egg masses should be used in the hatchery to ensure genetic

Figure 7.16 Queen conch in shipping box. Courtesy of Megan Davis.

diversity. For transport to the hatchery, each egg mass is placed in an individual plastic bag filled with seawater and placed inside a shaded bucket of seawater. If the egg masses are being transported for several hours on the boat, the water in the bucket should be changed every 1–2 hours to keep the egg masses at a steady temperature. A freshly laid egg mass can be shipped for 24–48 hours to other laboratories or hatcheries. Each portion of an egg mass (1/4–1/2 egg mass) is placed in seawater inside a beverage-sized thermos (1 L; 0.26 gal). Once the eggs have arrived at the laboratory they should be acclimated to the temperature of the new seawater and put into a bucket or egg incubator to continue their development (see description below).

Figure 7.17 Underwater sign to alert unexpected visitors not to disturb the adult queen conch. Courtesy of Laura Issac Norton.

Fighting conch

The Florida and West Indian fighting conch have similar characteristics to the queen conch in regards to their sexual morphology, internal fertilization, and egg mass laying; however, they demonstrate a mating behavior unique to their species. The males have a tendency to guard an egg laying female or copulating female and will challenge any male who attempts to mate with her. The challenge occurs in the form of a jousting tournament with their proboscis or they will use their shell to push another male away, hence the name 'fighting' conch.

The fighting conch have shown to copulate and lay eggs on a regular basis in captivity at

HBOI. The females tend to be slightly larger than the males and their shell lengths range from 7.7 to 9.6 cm SL (3–3.8 in SL) for females and from 6.3 to 9 cm SL (2.5–3.5 in SL) for males. Since fighting conch are considerably smaller than the queen conch, a breeding program can be set up efficiently in a condensed amount of space for the fighting conch. The broodstock system can be configured in fiberglass tanks that are 0.61 m W × 3.1 m L × 0.61 m D (2 ft W × 10 ft L × 2 ft D) (Fig. 7.18). Each tank can have 0.5–1.0 cm (0.20–0.39 in) of crushed coral aragonite sand (1–3 mm in diameter) placed on the bottom of the tank. The flow rate should be 3–5 L/min (0.8–1.3 gal/min) and the water can be plumbed with 2.5–3.8 cm in diameter (1.0–1.5 inches in diameter) PVC that allows the water to flow in one end of the tank and exit through a 5 cm in diameter (2 inches in diameter) standpipe at the other end of the tank. The water depth can be 15–20 cm (6–8 in). There should be up to five airstones in each tank. Another arrangement is to have the sand placed in individual polypropylene baskets (0.57 m × 0.57 m, 0.32 m^2; 1.9 ft × 1.9 ft, 3.6 ft^2 each) that have a mesh diameter of 0.6 cm (1.5 in) (Fig. 7.19). These baskets need to have window screen fastened to the bottom to contain the sand. It is best to elevate the baskets above the tank bottom using 0.2 cm (0.5 in) PVC pipe. Each of the baskets should have their own water flow (0.5 L/min), which allows for downwelling through the sand. An airstone should be placed in each basket to ensure additional circulation. The breeding system can be on a recirculating system, which includes a small sump with biofiltration media and an inline or submersible pump. The sand also provides additional biofiltration. Every one to two weeks the sand should be cleaned to remove feces and leftover feed. An aquarium gravel vacuum siphon

Figure 7.18 Tank system for broodstock fighting conch. Courtesy of Megan Davis.

Figure 7.19 Tank system with baskets for broodstock fighting conch. Courtesy of Megan Davis.

hose can be used for tanks that have the sand directly on the bottom or a spray bar can be used to clean the sand in the elevated baskets. The conch do not need to be removed during the cleaning of the sand substrate.

The stocking density for the adult fighting conch in a previous HBOI captive breeding study was 0.2 conch/m^2 where there were four males and four female fighting conch. The females laid 58 egg masses in a 26-week breeding season or 0.56 egg masses/week/female. In 2019, a captive breeding experiment conducted at HBOI in an integrated multi-trophic aquaculture (IMTA) system showed that the stocking density could be as high as 18 fighting conch/m^2. Therefore, approximately 30 conch can be placed in one of the fiberglass tanks or six conch per basket. In this study the average egg production was 0.28–0.42 eggs masses/week/female. Sex ratio can be 1 : 1 or the ratio of females can be higher than males (2 : 1). The female lays approximately 76,000 to 182,000 eggs/egg mass. The success of a fighting conch captive breeding program is to have ideal temperature and photoperiod ranges (same as described above for queen conch), steady supply of feed, and sandy substrate that is cleaned on a regular basis. Fighting conch require the same general water quality parameters as the queen conch, however, they have tolerance for murkier water, higher temperatures, and lower salinities. The adult fighting conch can be fed 2 g/day/conch of the same prepared diet: koi or catfish chow (36%), dried *Ulva* seaweed (16%), fresh water (42%), and a gelatin binder (5%). In the HBOI IMTA system, the fighting conch also received bio-flocculated feeds that were naturally in the system.

Egg mass handling and hatching

In the hatchery the strands of the sand-covered egg masses of the queen or fighting conch are placed in seawater and slightly teased apart with gentle finger pressure. This will ensure that the developing eggs have a good flow of water passing by the strands in the incubator (Fig. 7.20). As mentioned above it is not necessary to incubate a whole egg mass if less eggs and larvae are needed in the hatchery or experimental system. A full queen conch egg mass has approximately 500,000 eggs and a full fighting conch egg mass has approximately 76,000–182,000 eggs. Each egg mass incubates for three to five days in an individual upwelling screen (60–75 µm) PVC container (6 inches in diameter × 6 in high; 15 cm × 15 cm) (Fig. 7.20). The flow rate is 0.1 L/min/egg mass (0.03 gal/min/egg mass). The incubator can be set up on a simple recirculating system, where a sump with biological filtration material is positioned below the tank and a submersible or inline pump returns the

Figure 7.20 Incubator with containers for the conch egg masses. Each container holds one egg mass and the water upwells through the screen on the bottom of the container. Courtesy of Megan Davis.

Figure 7.21 Conch egg masses incubating in buckets and containers with gentle aeration. Courtesy of Megan Davis.

water back to the egg masses. Egg masses can also be incubated in a bucket (4–19 L; 1–5 gal) with gentle aeration and water exchanges once per day (Fig. 7.21).

Embryo development time is dependent on temperature; with optimal temperature 27–29°C (81–84°F). It is recommended that each day a small piece of egg mass is collected from each egg mass in the incubator and observed using a dissecting microscope to determine the stage of development. On the day of hatching the capsulated veligers have velar lobes that are well-defined with cilia, red pigment on the foot, and the developing embryos steadily rotate (Fig. 7.22). Based on many years of observation, the veligers have been observed to hatch from the egg masses in the evening at approximately 21:00 and close to 90–100% will hatch the first night. Some egg masses require 2 nights of hatching. In the late afternoon it is optimal to place the egg mass inside a clean egg mass incubator container and hang the egg mass container inside the larval tank about 10 cm (4 in) below the water surface. As the veligers hatch they swim out of the container into the larval tank. The following morning the egg mass container with the remaining sand and egg mass debris should be gently removed from the larval tank to avoid the egg mass debris from falling in the tank.

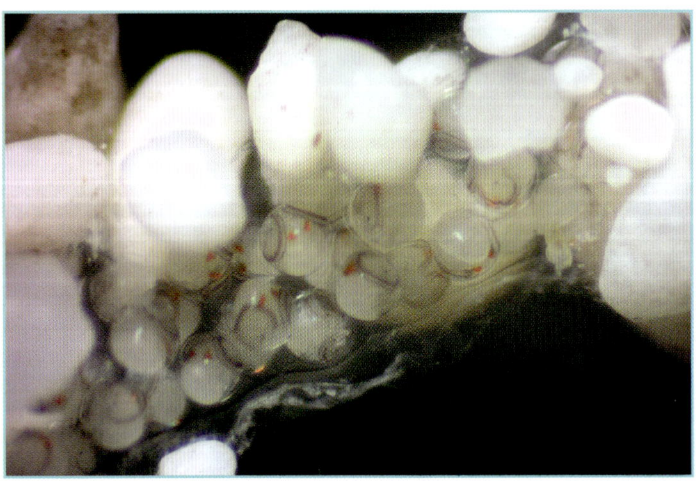

Figure 7.22 Sand-covered egg mass strand showing capsulated veligers ready to hatch. The velar lobes are well-defined, the foot has reddish-orange pigment, and the eye spots are visible. Courtesy of Megan Davis.

Hatchery and larval rearing

Queen and fighting conch larvae are also known as veligers and the two terms are used interchangeably. Both of these conch have the same larval development and will be described together. After hatching the conch are planktonic veligers. These larvae bear little resemblance to adult conch; however, they do have a shell that is the same shell that develops with them their entire life. The veligers have velar lobes covered with short bristles, called cilia, that serve three functions. The rapidly beating cilia help the veliger swim, exchange oxygen for respiration, and also create feeding currents that carry food toward the mouth of the veliger.

With optimal feeding regime, temperature, stocking density, and water quality the larval cycle is approximately 14–21 days long for queen and fighting conch. The emerged veliger is 300 µm SL and has two velar lobes, which divides into four by the third to fourth day and begins six lobes by the sixth to seventh day (Fig. 7.23). These lobes continue to elongate until the 1 mm SL (0.04 in) veliger is competent for metamorphosis. The queen and fighting conch look similar, however, the fighting conch veligers have dots on the tips of their six lobes (Fig. 7.23).

Veligers can be cultured in a variety of containers and tanks. For experimental studies conch can be raised in 1 L (0.25 gal) polypropylene tri-corner beakers or small buckets (4–19 L; 1–5 gal) (Fig. 7.24). The smaller container does not require aeration, however, gentle aeration using an air stone or Pasteur pipet at the end of an airline in the bucket is necessary to keep the veligers in suspension. For larger production the larvae are cultured in 500–2000 L (130–530

Figure 7.23 Life stages of *S. gigas* conch larval and post-larval development: (A) newly hatched two-lobed veliger (350 µm); (B) 4 day old four-lobed veliger (450 µm); (C) 6–7 day old six-lobed veliger (600 µm); (D) 14–20 day old six-lobed veliger (900 µm); (E) 14–20 day old six-lobed *S. alatus* veliger with characteristics spots on the end of the lobes (900 µm); (F) 15–21 day old competent veliger (1.0–1.2 mm); and (G) newly metamorphosed conch (1.2 mm). (A–F) courtesy of Megan Davis and (G) courtesy of LeRoy Creswell.

Figure 7.24 For experimental studies conch can be raised in 1 L (0.25 gal) polypropylene tri-corner beakers at 10 veligers per liter. This system was used at the experimental hatchery in The Bahamas. Courtesy of H. Forrest Thomas.

Figure 7.25 Conch larval-rearing tank (1000 L; 265 gal) with conical bottom. Conch veligers are siphoned from the culture tank into a sieve suspended in a bucket of seawater. These were tanks used at the Caicos Conch Farm. Courtesy of Megan Davis.

gal) conical or parabolic larval-rearing tanks with gentle aeration (Fig. 7.25). The conical bottom should be a gentle slope such as 30 to 45 degrees. If the slope is too steep, the conch larvae will continually touch the sides of the cone bottom and this causes them to retract their lobes or damage their shells. To keep the veligers in suspension, a PVC airlift (15 × 15 cm; 6 inches dia. × 6 inches high) with two airlines (without airstones) attached to either side is positioned 0.5 cm (0.2 in) from the bottom of the tank around the standpipe (Fig. 7.26). The air bubbles are small and with the airlift the motion of the water is circular from the bottom to the top of the tank.

It is very important that the conch larvae are stocked initially at relatively low density (150–200/L; 560–760/gal). Once the tanks are stocked, water must be changed every other day to prevent proliferation of bacteria and protozoans. The dead and slow growing veligers pass through the screen, while healthy, fast-growing veligers are retained. As the screen mesh size is increased only the fastest growing veligers are retained. Water changes start by using a 100 µm sieve for new hatch, by day 5 the sieve size is increased to 200 µm, by day 10 to 250 µm, by day 15 to 300 µm, and then 350 µm toward the end of the larval cycle. Progressive culling of the slowest growing veligers should result in a final density prior to metamorphosis of 10–25 veligers/L (95–150 veligers/gal) at metamorphic competence. Optimal culturing temperature is 28°C (82°F) and salinity of 36; however, veligers can be grown at temperatures 24–32°C (75–90°F) and salinities ranging between 26 and 40.

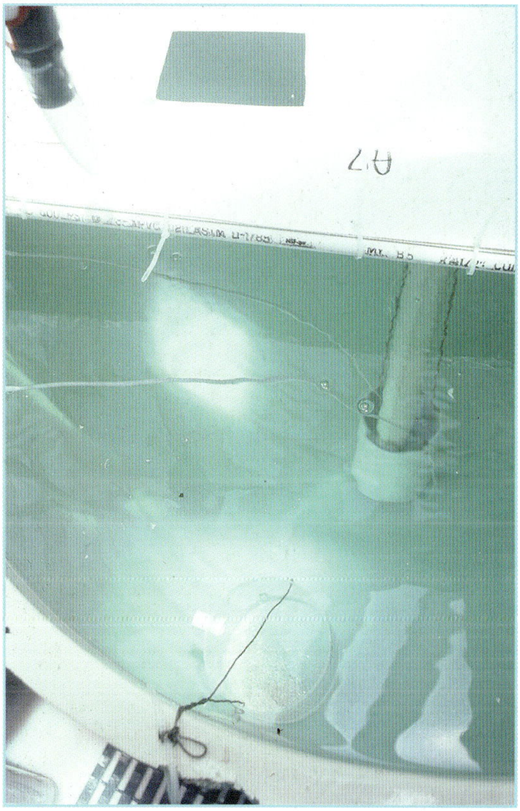

Figure 7.26 Larval-rearing tank showing the airlift surrounding the standpipe toward the bottom of the tank. The egg mass in the container is suspended in the larval tank the evening of hatching. This system was used at the Caicos Conch Farm. Courtesy of Megan Davis.

Larval cultures are maintained either in a flow-through system (1–2 L/min; 0.3–0.6 gal/min) or in static conditions. Higher densities can be maintained in a flow-through system, however, requires more filtered seawater. Water changes begin the day after hatching then occur every 48 hours. Larvae cannot be dry-sieved, therefore, the PVC sieve (20 × 30 cm; 8 inches in dia. × 12 in high) used for water changes is supported in a bucket of seawater (Fig. 7.25 and Fig. 7.27). Smaller or larger diameter sieves can be used for different size tanks or for experimental cultures. For experimental cultures, 5 cm × 5 cm PVC (2 in × 2 in) sieves held inside a container of water is ideal (Fig. 7.28). A 2.5 cm (1 in) diameter siphon hose is used to remove the veligers from the tank (Figs 7.25 and 7.27). To keep the siphon hose from moving around inside the tank, two porcelain weights are secured to the end of the hose with tie wraps. The siphon method is less stressful to the veligers compared to draining them from the bottom of the tank. Often the bottom of the tank has some dead larvae and it is best not to drain the tank completely. After the tank is drained, the veligers are gently hosed off the sieve into the newly filled larval tank and the airlift is turned on to circulate the larvae. Observe the larvae as they come back into suspension. Sometimes it is necessary to turn the air up for a short period of time, then turn the air back down. When the aeration is at the correct amount veligers will be seen with their lobes extended. Once the larvae are fully distributed, the tank is covered and the water is turned on for a flow-through system or the cultures are left static. A special standpipe with Porex material on the outside and an inner standpipe to control water level is needed to retain the larvae during flow-through (Fig. 7.29).

A sample of veligers from each tank needs to be observed daily before the water change. Using a dissecting microscope (20–40×) assists in determining growth, development, and amount of feed in their gut. The criteria of a healthy queen and fighting conch veliger includes: (1) a dark gut that fills the top whorls of the shell, which indicates adequate feeding; (2) clean velar lobes, fully extended, and cilia moving in a wave-like pattern; (3) increase in velar lobe and shell whorl

Figure 7.27 Close up of the sieve and bucket arrangement used to collect the veligers during a water exchange. Courtesy of Megan Davis.

Figure 7.28 A small sieve suspended over a shallow container with an optional funnel was used to change the veliger culture water for small-scale or experimental cultures at HBOI. Courtesy of Tom Smoyer.

development; and (4) growth of 30–50 μm SL/day or faster. A veliger is unhealthy if they are: (1) slow to come out of the shell; (2) bacteria and protozoan in the shell or feeding on the larval tissue; (3) lobes are stunted or show necrosis (indication of a *Vibrio* spp. bacterial infection); (4) erratic movements; (5) debris or protozoa on the outside of the shell (for example, *Vorcella* spp. or *Endoproct* spp.); (6) light gut, indicating lack of feeding; and (7) no or minimal growth and development. If there has been a bacterial infection (for example, *Vibrio* spp.) during the first 5–7 days, the veligers will show normal development, however, there will be 100% mortality of the culture shortly thereafter. If there is a toxicity problem (for example, water quality, copper, new PVC pipes and glue, insecticides) the veligers will not develop and there will be 100% mortality on day 3 to 4. It is important to leach all new culture vessels and supplies that will come in contact with the veligers. Leaching can be accomplished by filling the tanks, containers, tubing with saltwater for 24 hours, draining and filling again for 24 hours. A stressed veliger will excrete mucus strands that trail behind the swimming larvae and cause the veligers to clump together. This occurs from overfeeding, high density, or rough water changes. Usually it is only a temporary condition and can be resolved by decreasing food, lowering density, and improving water exchange methods. Momentary increase in aeration will break the clumps of veligers apart.

Even though veligers hatch with yolk reserves they will begin feeding on single cell microalgae or phytoplankton six to eight hours after hatching and feed on phytoplankton throughout their entire larval-rearing cycle. Therefore, to increase larval survival and vigor the veligers should be fed the morning after hatching. There are two species of phytoplankton that are the preferred species to culture to feed conch veligers: flagellated algae Tahitian *Isochrysis galbana* or Caicos *Isochrysis*, and the diatom *Chaetoceros gracilis*. These batch cultures of phytoplankton are started in test tubes and flasks and then grown in drinking water carboys (20 L; 5 gal) and cylindrical clear fiberglass tubes known as sun tubes (250 L; 65 gal). The desirable cell concentration of *Isochrysis* in the larval tanks is 5×10^3 cells/mL for newly hatched veligers and increases with each water change to as high as 1.2×10^4 cells/mL by the time the veligers are metamorphically competent. *Chaetoceros* is also added to the larval culture starting on day 12–15 depending on when the veligers reach 900–1000 µm SL. The final concentration in the tank for this phytoplankton species is 3×10^3 cells/mL. For static larval cultures, the phytoplankton is added as a batch feed shortly after the water change, whereas in flow-through larval tanks the algae cells are diluted with seawater and are gravity drip-fed into the culture water for approximately 12 hours to keep the desirable density of phytoplankton in the tanks.

Figure 7.29 A standpipe with Porex material on the outside of an internal pipe controls water level and retains the larvae in flow-through culture. This arrangement was used at the Caicos Conch Farm. Courtesy of Megan Davis.

Metamorphosis and post-larval rearing

In the wild, when the veligers are competent for metamorphosis they need a cue to induce settlement in their juvenile habitat. The veligers are attracted to the diatom epiphytes associated with their seagrass nursery habitat. They have an interesting swim-crawl behavior when they are competent for metamorphosis. The veligers at this stage can both swim and crawl on the substrate to 'test' the cue for settlement. In the hatchery queen and fighting conch veligers are metamorphically competent at approximately 14–21 days after hatch (1.0–1.2 mm SL; 0.04–0.05 in SL) depending on density, temperature, water quality, and phytoplankton quality and quantity. The veligers are competent when their eyes migrate outward, tentacles are of equal length, pigments on the foot change from orange to dark green, the ctendium (that is, gill) is elongated and functional, and the buccal mass has developed (Fig. 7.23). Conch in the hatchery also need a metamorphic cue to undergo a series of morphological and physiological changes from a swimming veliger to a benthic snail. Several different metamorphic cues such as the red macroalgae *Laurencia poitei*, potassium chloride, hydrogen peroxide, and epiphytic diatoms have been successfully used to induce metamorphosis of conch larvae in a hatchery setting. Two of the most reliable inducers for large-scale cultures are extracts from the *L. poitei* and small doses of hydrogen peroxide. At a temperature of 28–30°C (82–86°F), these cues should induce 60–75% of the conch veligers to metamorphose. Successful metamorphosis of the batch is going to be based on the uniformity of the veligers in the larval culture. Culling the slow growers during the larval cycle will help keep uniformity. Also testing a small group of veligers first before inducing the whole tank is desirable to improve metamorphic success rate by ensuring that a large percentage (>60%) of veligers in the culture are ready to metamorphose. This test is typically conducted the day before the entire culture of veligers are ready to be exposed to the metamorphic cue.

The *Laurencia* extract is made by hand collecting older thick red-brown stalks of the macroalga from shallow, sandy, grass flats into mesh bags. The younger stalks, yellow-orange in color, are slightly toxic to the veligers and will cause a low percentage of veligers to metamorphose. In the hatchery, the macroalga fronds are saltwater rinsed and sorted to remove coral pieces, sponges, and excess sand. In an industrial blender, a ratio of 2 g of *Laurencia* to 1 mL of seawater is blended for approximately two minutes. The solution is frozen for a minimum of 2 days to lyse the cells and release the molecular algal associated cues. The frozen solution is thawed overnight, filtered through a 200 μm polyethylene screen, and the resulting extract is refrozen. Ten kilograms (4.5 lbs) of collected macroalgae produces 16 L (4.2 gal) of blended slurry, which yields approximately 7.5 L (2 gal) of extract.

Before any new batch of *Laurencia* extract is used on the large scale, the dosage is determined by placing 25 metamorphically competent veligers in three different extract concentrations: 7, 10, 15 mL of *Laurencia* extract/L of seawater for 4 hours. This test setting can be done in small 50–100 mL polypropylene tri-beakers. After removing the veligers from the extract, percent metamorphosis is determined using a dissecting microscope (40× power). Metamorphosis is recognized when the velar lobes are lost, the conch are crawling with propodium (that is, foot), and they are searching for food with their proboscis (Fig. 7.23). A minimum of 60% metamorphosis is considered effective to select a given dosage of *Laurencia* extract for use on the large scale.

A more cost-effective procedure for inducing metamorphosis of queen and fighting conch larvae is to expose the competent veligers to 50 μL of 3% pharmaceutical grade hydrogen peroxide (0.06 mL H_2O_2/1 L of seawater) for 4 hours. The cost per batch of larvae, using the 3% pharmaceutical grade hydrogen peroxide, is considerably less than using *Laurencia.* Using hydrogen peroxide reduces the amount of variability found in natural inducers, while providing a simple, low cost, and safe method for production-scale induction of metamorphosis. A test set of veligers should still be conducted with the hydrogen peroxide cue prior to the entire culture being exposed to the hydrogen peroxide metamorphic cue.

For small-scale cultures or experimental studies, queen and fighting conch can be induced to metamorphose using a natural trophic cue found in juvenile conch seagrass habitats. The detrital blades from turtle grass, *Thalassia testudinum,* can be collected and small pieces (approximately 1 g [0.035 oz] per 80 cm^2 [12 in^2] of area) can be placed in a shallow tray or container with the veligers (Fig. 7.30). These detrital blades are covered in silt and epiphytes including benthic diatoms such as *Navicula* spp. It is important to check the detrital blades for small crabs that will need to be removed because they could be potential predators. As the veligers metamorphose over the 24–48 hours they have a source of readily available food found on the detrital blades. The conch can stay on these blades for several days and can be gently removed with a wash bottle

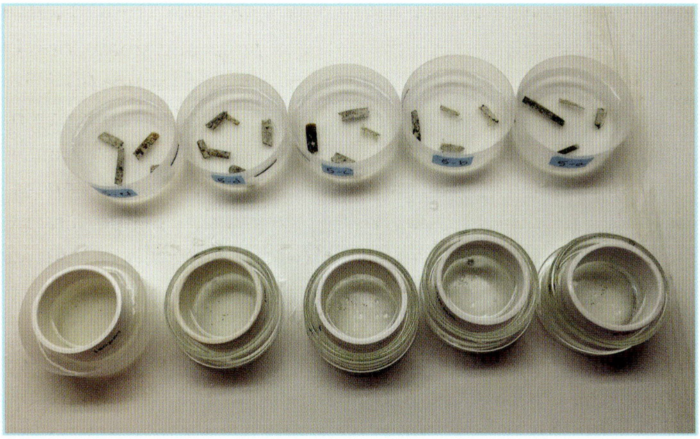

Figure 7.30 Detrital blades from seagrass, *Thalassia* spp., habitat are used for small-scale or experimental cultures to induce metamorphosis of conch veligers. The top row containers have seagrass blades and the bottom row containers have competent veligers to be added to the top row containers. Courtesy of Megan Davis.

or gently hosed off with a saltwater spray. New detrital blades are then added to the containers with the conch.

When a large-scale tank of veligers are metamorphically competent, a shallow tank (for example, 2.4 m L × 0.9 m W × 15 cm D, 2.16 m^2; 8 ft L × 3 ft W × 6 in D, 24 ft^2) is filled with approximately 325 L of seawater and the correct level of *Laurencia* extract (about 3.25 L) or hydrogen peroxide (19.5 mL) cue (Fig. 7.31). The solution is mixed well, then eight shallow screen (250 μm) trays each with an area of 0.28 m^2 (3 ft^2) are supported in the shallow tank. Metamorphically competent veligers are siphoned from the larval-rearing tank, concentrated in a 20 L (5 gal) bucket, and the count is estimated with the mean of three 50 mL subsamples. The veligers are then distributed using a graduated beaker with a handle between the settling trays in the shallow tank. The competent larvae are stocked at 3500/m^2 (325/ft^2) which is equivalent of approximately 7800 veligers/shallow tank. Veligers typically stop swimming 10–30 min after they are exposed to the metamorphic cue. If the veligers continue to swim, the cue concentration needs to be increased slightly. It is recommended that during the metamorphic process the veligers are placed in a subdued light environment. After 4 to 5 hours, water flow is slowly introduced in the shallow tank trays to eliminate the cue from the system. Approximately 50–60% of the veligers will have metamorphosed. Additional veligers may metamorphosis within the next 24 hours. Metamorphosis is recognized when the velar lobes are lost and conch post-larvae are seen crawling across the bottom of the screened trays searching for food with their proboscis.

Figure 7.31 Conch metamorphosis tank used to settle competent larvae with *Laurencia* extract or hydrogen peroxide. The post-larval conch stay on shallow trays in this system for three to four weeks. These tanks were used at the Caicos Conch Farm. Courtesy of Megan Davis.

Newly-metamorphosed conch are maintained on the screen trays for 10–16 days until the conch reach an average size of 3–4 mm SL (0.12–0.16 in SL). Average growth rate is 0.18 mm–0.22 mm SL/day (0.007 in–0.009 in) or higher. These tanks can be maintained either as a recirculating system or a flow-through system. Water enters each shallow screen tray through a small tube (4–6 mm in diameter; 0.15–0.24 inches in diameter) that is angled to create a slight vortex in the downwelling tray. When the post-larval conch reach the desired size, they are sieved on a sorting screen to cull out dead and stunted individuals, and transferred to nursery trays. Typically, about 75% of the induced veligers are recovered as 3–4 mm post-larvae, or 60% survival from the competent veliger stage to 4 mm SL.

Prior to feeding the newly metamorphosed conch, the trays are gently shaken to remove wastes and to dislodge the conch crawling on the sides. The conch are benthic grazers; therefore, they are unable to feed on planktonic algae such as the diatom *Chaetoceros gracilis*. If a flocculation agent is added to this planktonic diatom, however, the algal cells coagulate together and settle out in thick mass that the conch can graze upon. The flocculation procedure works the best with diatoms. The flocculation procedure begins with chitosan, a macromolecule from the exoskeleton of crustaceans. Chitosan is purchased in a powered or flaked processed form. The diatom culture is adjusted to a pH to 6.0–7.0 with the addition of muratic acid (approximately 0.1–1.2 mL of muratic acid/L of phytoplankton). The chitosan solution is mixed well into the microalgae culture at a dose of 5 mL of chitosan solution/L phytoplankton. A stock solution of chitosan (10 g chitosan to 1% acetic acid) is prepared in advance and can be stored for several weeks. Next sodium hydroxide (5% or 5 g NaOH/100 mL freshwater) is added to raise the pH to 8.5–9.0 (approx. 1.0–1.5 mL/L of phytoplankton). The microalgal cells are stirred again for several minutes until flocculation is observed (Fig. 7.32). The flocculated cells settle to the bottom of the vessel and can be separated from the 'clear' water on top. If small quantities of flocculated microalgae are needed on a daily basis, the settled algae can be stored in the refrigerator for up to one week. On the first day post-metamorphosis conch are fed about 6×10^7 cells/conch/day and increases to 20×10^7 cells/conch/day for conch that are 3–4 weeks old after metamorphosis. The concentration and volume of flocculated phytoplankton to feed the conch needs to be determined from the cell count done prior to flocculating the phytoplankton. Once the phytoplankton is flocculated

Figure 7.32 Flocculated diatom, *Chaetoceros gracilis*, using chitosan for feeding post-larval conch; before flocculation on the left and after flocculation on the right. Courtesy of Megan Davis.

cell counts cannot be accomplished. For large-scale feeding needs, this diatom can be grown in 20,000 L (5300 gal) circular tanks located outside with no cover. For smaller hatcheries, *C. gracilis* can be cultured in 20 L (5 gal) carboys or in 250 L (65 gal) clear fiberglass cylinder tubes.

For small or experimental cultures, newly metamorphosed conch can be fed seagrass detrital blades; however, it can be difficult to remove the conch from the blades since the conch are small and become cryptic among the blades. Instead, the detrital blades with their attached silt and epiphytes can be placed in a bucket of seawater and stirred around by hand or a paddle. This dislodges the epiphytes allowing the detrital blades to be removed from the bucket. The silt and epiphytes can then be poured through a 25 µm bag filter to remove large particles and potential predators such as crabs and worms. The silt and epiphytes that go through the bag filter are captured and cultured using standard microalgae techniques. The slurry is added to culture tanks such as carboys (20 L; 5 gal) or clear fiberglass cylinders (1.2 m H × 0.6 m in diameter; 4 ft H × 2 ft in diameter). The diatoms will bloom with the added nutrients and silicate. The cultures need aeration; however, the epiphytes tend to sink because they are mostly benthic diatoms. Therefore, one to two times a day the cultures should be stirred with a paddle. These cultures can be outdoors to utilize natural sunlight and be covered with lids to avoid bugs and rainwater getting into the cultures. It will take about 4 to 5 days for the cultures to bloom and partial harvests can be achieved along the way.

Indoor nursery

The post-larval conch (3–4 mm SL; 0.12–0.16 in) are placed into fiberglass tanks (3 m L × 0.6 m W × 0.6m H, 1.8 m^2; 10 ft L × 2 ft W × 2 ft H, 20 ft^2) at 400 conch/m^2 (316 conch/ft^2) (Fig. 7.33). This system is typically indoors or could be placed outdoors with adequate shade cover to protect the cultures from rain and extreme sun and heat. The tanks are set up to have a thin layer of elevated coral aragonite sand substrate (1–3 mm dia.; 0.04–0.12 in diameter) on top of supported fiberglass window screen. The water downwells through the sand, recirculates through a sump that contains biofilter substrate and returns back to the tanks. This system can also be set up as a flow-through system. Depending on the size of the conch, the sand trays are sprayed

Figure 7.33 Juvenile conch fiberglass rearing-tanks with trays filled with sand substrate. This system was used at the Caicos Conch Farm. Courtesy of Megan Davis.

Figure 7.34 A gel diet comprised of Koi or catfish chow, dried seaweed (*Ulva* spp.), and gelatin is a suitable feed for juvenile conch. Courtesy of Megan Davis.

down every two to four weeks to remove accumulated waste products. During this process, the sand and the animals are left intact and the conch lie dormant.

The conch are fed once per day and are fed a gel-based diet or a dried sinking pellet conch chow that can be produced by the feed companies. The gel feed consists of catfish ground pellets (36%), ground dried seaweed *Ulva* sp. (16%), gelatin or alginate (5%), and freshwater (42%). The gelatin or alginate allows the feed to remain stable for approximately 48 hours. The diet can be either manufactured in a feed mill or made in smaller quantities on site. The dried ingredients are put together and the gelatin is activated with hot freshwater. The ingredients are mixed together into a paste and put onto large aluminum trays, which are placed in the refrigerator overnight. When the gel diet is set it can be cut into cubes and kept in the refrigerator in bags for one week or placed in the freezer for several months (Fig. 7.34).

Juvenile conch (3–4 mm SL; 0.12–0.16 in SL) are fed approximately 10 mg feed/conch/day and this increases over time until the larger juveniles (30–40 mm SL; 1.2–1.6 in SL) are fed approximately 120 mg feed/conch/day. Smaller conch are fed about 15% of wet-meat weight and as the conch increase in size (6 cm SL; 2.4 in SL) they are fed 2–7% of the wet-meat weight of the conch. Typical growth rate on this gel diet is 0.18–0.22 mm/day depending on temperature and stocking density (Fig. 7.35). The feed conversion ratio for conch fed the prepared feed is a ratio of 1.5 : 1 for dry weight of feed to total wet weight of the conch (shell and meat).

Figure 7.35 Small juvenile fighting conch grown on sand with gel diet (1–2 months old). Courtesy of Tom Smoyer.

Figure 7.36 Small juvenile queen conch showing foot and operculum (3 months old). Courtesy of Tom Smoyer.

The thin leading edge of the shell aperture of healthy juvenile conch is usually covered with feed, sand, and feces. The healthy conch animal fills their shell and their foot and operculum is easily seen in the aperture (Fig. 7.36). An undernourished conch recedes back into the shell making it difficult to see the foot and operculum. When a conch is not growing the shell becomes covered with a green or blue-green algae and the shell may get thick and eroded.

Small juvenile conch are most active at night. Typically, during the day conch lay dormant with their aperture toward the sand substrate and tend to be partially buried when not feeding. When underfed, conch can be more active during the day searching for feed. Juvenile conch in poor health are often observed on the sand surface, positioned with their opercula upwards and their bodies partially extending out of their shell. In this state, response to a stimulus is often sluggish. This behavior could be a result of stagnant water flow (for example, low oxygen), overfeeding, or stress from handling. Remedies include flushing the tank, feeding less, and/or adjusting water flow rate. The kicking behavior seen by conch is usually a sign of stress and has been observed in low-oxygen situations or while feeding certain types of artificial feed (for example, *Spirulina*).

Outdoor nursery

After approximately 12–20 weeks at 27–29°C (81–84°F) the conch will reach an average size of (20–40 mm or 2–4 cm; 0.8–1.6 in). Survival from 4 mm to 40 mm is about 75%. The fighting conch can be grown to market size in an indoor nursery system depending on the market size preference. There are two potential commercial markets for the fighting conch: 3 cm conch for the aquarium trade and 6 cm conch for the seafood market as an Ocean Escargot. See 'Market' section below.

The queen conch juveniles are sieved, sorted, counted, and transferred to an outdoor nursery system to be grown to 7–9 cm SL (2.8–3.5 in SL). These juvenile conch can be stocked into shallow, circular fiberglass or cement ponds, or lined raceways at a stocking density of 150 conch/m^2 (14 conch/ft^2). During this nursery stage, the conch will need to be reduced to a density of 75 conch/m^2 (7 conch/ft^2). The Caicos Conch Farm in the Turks and Caicos Islands used circular cement ponds (15.3 m diameter; 50 ft diameter) with a water depth of 10 cm (4 in) and a stocking density of 200 conch/m^2 (19 conch/ft^2) (Fig. 7.37). Other tank configurations to consider include oval shaped ponds or raceways with high density polyethylene (HDPE) liners set in a wooden frame or cinder blocks. A thin layer of sand would also be beneficial for shell growth and would serve as a biofilter to help maintain water quality. It has been determined that conch grown on aragonite sand develop stronger shells and show burial

Figure 7.37 Aerial view of juvenile queen conch Caicos Conch Farm cement ponds for nursery grow-out. Courtesy of Caicos Conch Farm.

behavior. The stronger shell is beneficial for conch that will be grown in sub-sea pens for the seafood market or released for restoration.

The nursery ponds can be set up as a flow-through or recirculating system with a turnover of the full system volume every 2–4 hours. Water should enter the conch ponds or raceways from one or two spray bars to create a circular flow that exits through a central standpipe or two evenly placed standpipes in a raceway system. There are a number of factors that influence the optimal turnover rate, including conch density, feeding rates, and water temperature. To enhance growth and survival, the nursery ponds or raceways should be covered with dense shade cloth (50–75%). Juvenile conch prefer to be in shaded areas, because in the wild they are found buried in the sand or sheltered in the seagrass habitats during their early juvenile stages. The optimum temperature for rearing juvenile conch is 27–29°C (81–84°F). Further research is required to optimize these juvenile conch systems.

Juvenile conch will feed on the naturally occurring diatoms that grow on the sides of the tank and sand; however, they will also need to be fed an artificial pellet diet that can be a blend of catfish and *Ulva* sp. The Caicos Conch Farm used a diet for conch that was manufactured exclusively for their operation. During this stage, conch growth rates range from 0.18–0.25 mm/day and survival rates are as high as 90–95%. At this growth rate the conch cultured at this stage will reach a size of 7–9 cm in 10–12 months in the outdoor nursery area.

Grow-out

Production costs are prohibitive for growing queen conch to sub-adult and adult sizes onshore. Therefore, larger queen conch juveniles can be grown to 20 cm SL (8 in SL) in a sub-sea pasture with a predator-exclusion fence or cages that extend from the substrate to above the high tide line (Fig. 7.38). It is recommended that conch be raised in onshore nursery tanks until they reach a size of 7–9 cm SL before putting them out to pasture to avoid high predation rates.

Figure 7.38 Aerial view of Caicos Conch Farm sub-sea grow-out area with enclosures. Courtesy of Caicos Conch Farm.

Known predators of queen conch include fish, sharks, rays, octopus, sea turtles, spiny lobsters, brachyuran crabs, hermit crabs, and carnivorous gastropods. A variety of fencing or cage material can be used to exclude pelagic and epibenthic predators and the fence also serves to contain the conch stocks. Large circular cages (25 m in diameter, 500 m^2; 80 ft in diameter, 5400 ft^2) with mesh openings (15–20 cm in diameter; 6–8 in diameter) will help to exclude large predators and smaller mesh (5 cm in diameter; 2 in diameter) placed along the bottom will exclude smaller benthic predators and help to keep juvenile conch from escaping. These mesh sizes will avoid fouling and restriction of the water flow. Inexpensive traps such as empty adult conch shells can be used to catch octopus and baited traps can collect lobsters and gastropod (*Fasciolaria tulipa* and *Murex ponum*) predators.

Stocking densities for juvenile conch should reflect natural densities found in the wild (1–2 conch/m^2; 1–2 conch/10 ft^2) or approximately 10,000 conch/hectare (4050 conch/acre). The cages can be stocked at higher density (10 conch/m^2; 1 conch/ft^2) compared to the juvenile conch habitats, however, the natural feed supply may need to be supplemented with a prepared diet. Ideally, feeding should not be necessary if the conch can feed on a wide variety of epiphytes in the seagrass beds. Stock rotation through the sea pasture is necessary to avoid overgrazing. Site selection is critical to successful field grow-out. The ideal grow-out sites are habitats that are historical nursery grounds. These sites have a sandy bottom vegetated with an intermediate density of seagrass (*Thalassia* spp. shoots 600–900/m^2; 56–84/ft^2), high algal/epiphytic productivity, 2–4 m (7–13 ft) deep, and with strong tidal currents (1.03 m/s; 2 knots) to flush the area and import new feed such as epiphytes into the area. There are studies that show that survival rates may be improved by outplanting the juvenile conch in the fall when temperatures begin to decrease. A partial explanation for this phenomenon may be that predators feed less when temperatures are low. Survival in the sea pasture should be 60%–90% and will depend on the effectiveness of the cages, availability of feed and the predator trapping program.

Grow-out from 7 to 20 cm SL (2.8 to 7.9 in SL) will take approximately 1.5–2 years at a growth rate of 0.15–0.20 mm/day (Fig. 7.39). Conch are harvested from the sea pasture by

Figure 7.39 Range of queen conch sizes and development stages from juveniles to fully lipped adult. Courtesy of H. Forrest Thomas.

divers working from small boats. The conch can be processed on site or sold live to restaurants or wholesalers.

Cultured queen conch pearls

The queen conch naturally produces valuable pearls of attractive colors that include pink, orange, white, beige, and many other colors that match the interior conch shell colors. The pink pearls are highest in demand. Natural conch pearls are rare and only one in 10,000 conch produces a pearl of value. Natural conch pearls are valued at US$1500–3000 per carat depending on their color and flame structure. Conch pearls typically weigh less than three carats. Conch pearls are calcareous concretions and are considered non-nacreous pearls. The flame structure is formed due to the concentric layers of fibrous calcium, which is laid down during the pearl formation.

Over the decades many researchers have tried to develop a reliable technique to culture conch pearls, however, the results have been unsuccessful. In 2009, HBOI scientists (H. Acosta-Salmón and M. Davis) made a breakthrough and developed patented seeding techniques to produce both beaded (nucleated) and non-beaded (keshi) culture pearls (Figs. 7.40 and 7.41). Over 200 cultured pearls were produced during their experimental research and several of them were

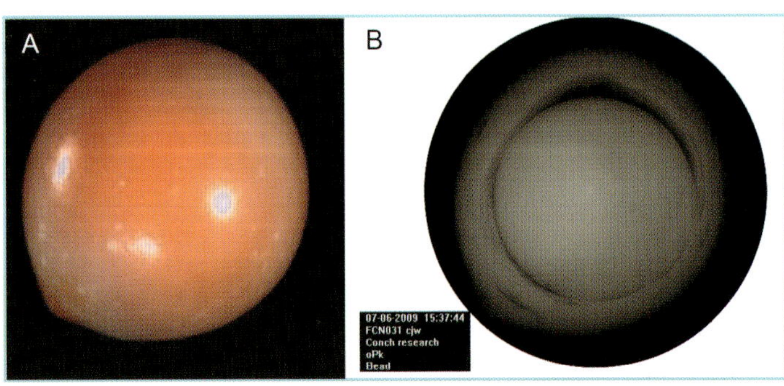

Figure 7.40 Cultured nucleated or beaded queen conch pearl 6 months old from initial seeding: (A) conch pearl photograph; and (B) X-ray of pearl showing bead and growth. Courtesy of Brian Cousin and Gemological Institute of America, respectively.

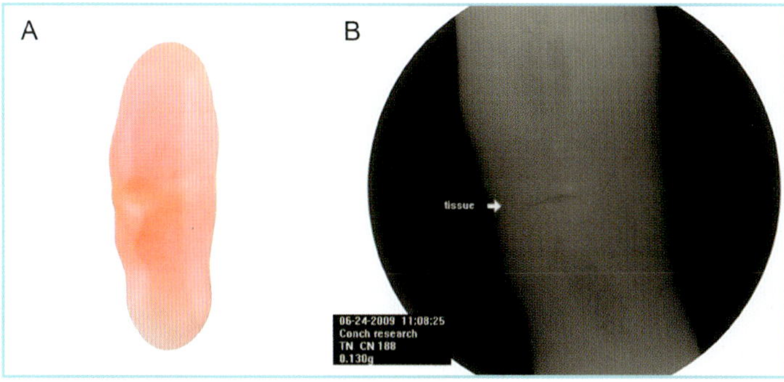

Figure 7.41 Cultured keshi or non-beaded queen conch pearl 12 weeks from initial seeding: (A) conch pearl photograph; and (B) X-ray of pearl showing the mantle tissue. Courtesy of Gemological Institute of America.

top-quality. These scientists recommend that the next step is to test the technology on a pilot-scale in the waters of the Caribbean.

Conch need to be relaxed to perform the pearl seeding and the ideal relaxant for queen conch is magnesium chloride (30 g/L) for 20–30 minutes. This solution can be made in large batches in a tank and the adult conch are placed in the solution. They are typically placed in the solution with their aperture down or on their side with the leading edge of their shell facing up (Fig. 7.42). Although the magnesium chloride causes an increase in salinity in the tank to 47, the short duration of the exposure does not harm the conch. After the seeding is completed the conch are placed back into their culture tank. Survival of conch exposed to magnesium chloride is 100%.

The conch seeding techniques are similar to techniques used in marine pearl oysters where a small piece of mantle is removed from the conch and inserted with the bead into the conch. This would be considered a nucleated or beaded pearl (Fig. 7.40). Keshi pearls or non-nucleated pearls are similar to natural pearls. In the seeding technique only a small piece of mantle is inserted into the animal. The mantle with or without the bead will form a pearl sac, which initiates the pearl formation. It takes approximately 6–18 months to produce a pearl depending on the desired size. The conch is relaxed prior to removing the pearl and does not need to be sacrificed to remove the pearl. After the pearl is removed, the pearl sac remains in the conch and a new keshi pearl will grow at this site. Culturing conch pearls as part of a conch farm would provide a valuable product in addition to the conch meat.

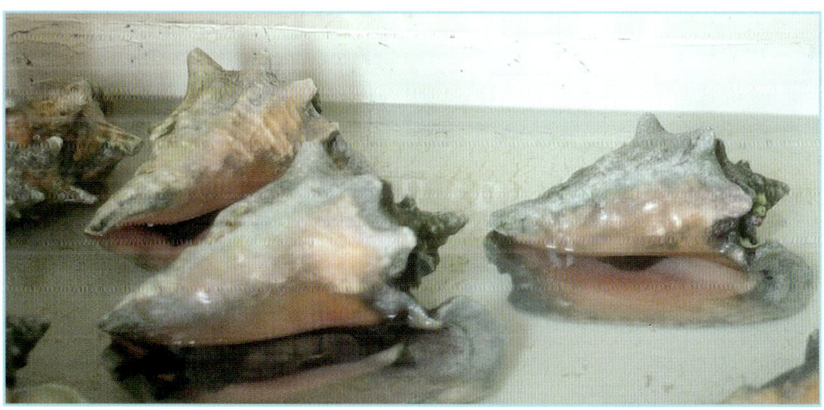

Figure 7.42 Queen conch in magnesium chloride solution, which causes relaxation of the muscle in advance of pearl seeding. Courtesy of LeRoy Creswell.

Markets

Queen conch

Queen conch has long been a favorite item on the menus of restaurants in Florida, The Bahamas, and throughout the Caribbean. Traditional dishes include conch fritters, conch chowder, conch salad, and cracked conch (Fig. 7.43). The tasty, tender meat of the conch is protein-rich and contains no saturated fat. Conch are described as having a crunchy texture similar to geoduck clams and a taste with the sweetness of abalone. Because the market is based 100% on wild harvested conch, the demand is for legal size conch with a fully formed lip (≥15 mm lip thickness; typically >18 cm SL; 4–5 years old). The retail prices for conch sold in the US market is an average of US$21.00 per lb. Fishermen typically receive US$1.25–3.50 per fresh or live conch.

In the 1990s, the Caicos Conch Farm established markets for small farm-raised queen conch that were sold as Ocean Escargot (5–6 cm SL; 2–2.4 in SL) and as Island Princess (8–9 cm SL; 3.1–3.5 in SL) (Fig. 7.44 and Fig. 7.45). Restaurants served the Ocean Escargot in paella, bouillabaisse, gumbos, and traditional escargot dishes. The Island Princess conch were served ceviche-style or were boiled or steamed and served as conch medallions. These farm-raised conch were well received by chefs and their customers note clients. Several chefs stated that the farm-raised conch were sweet and tender.

Figure 7.43 Queen conch salad, a delicacy of the islands. Courtesy of Graeme Davis.

Figure 7.44 Farm-raised juvenile queen conch sold in the shell as Ocean Escargot by Caicos Conch Farm. Courtesy of Tom Smoyer.

Figure 7.45 Farm-raised juvenile queen conch sold with shell removed as Island Princess by Caicos Conch Farm. Courtesy of Megan Davis.

It is very difficult to locate significant numbers of juvenile queen conch measuring <9 cm SL (3.5 in SL) in the wild. Therefore, it is unlikely that wild conch of this size would appear in the market and be passed off as farm grown. Since queen conch are CITES protected, it would be necessary to obtain a CITES permit to sell these smaller sizes of conch for export and other countries may have additional permit requirements.

There are a couple of additional options for ways to build markets for farm-raised queen conch. These include setting up sub-sea pastures in key locations within a country and have fishermen and community members ranch the juvenile conch until they are ready for market. Another option is to have a regional hatchery that uses country specific conch egg masses and raises seed conch that can be shipped back to the country of origin for grow-out or stock enhancement.

Fighting conch

The fighting conch has been known to contribute to the conch fishery more as a subsistence food source of the coastal people of the Caribbean and Gulf of Mexico. It is generally boiled or roasted and has a taste that is similar to the queen conch. The fighting conch may have a potential commercial market as a cultured seafood product, which could help to supplement the queen conch market. The fighting conch are mature at 7–10 cm SL (2.6–3.9 in SL) and could be sold as juveniles for Ocean Escargot or Island Princess type markets (Fig. 7.46). The farm-raised queen and fighting conch have been sold as grazers for the aquarium trade (Fig. 7.47). In the past these conch that were 3 cm SL (1.2 in SL) sold for

Figure 7.46 Florida fighting conch juveniles cultured at HBOI. These conch are a potential seafood market product. Courtesy of Tom Smoyer.

Figure 7.47 Florida fighting conch juveniles cultured at HBOI. These conch were sold for the aquarium trade as herbivorous grazers. Courtesy of Tom Smoyer.

US$2.99 each. The fighting conch are not a CITES species, therefore, a CITES permit would not be needed to sell this species.

Prior to shipment of live queen or fighting conch, they are sorted by product type (for example, food or aquarium market) and purged for 12–24 hours (no feeding) in flow-through or recirculating seawater system. On the day of shipment, the conch are placed on top of layers of moist paper towels inside of plastic bags that are packaged inside of insulated boxes. A small amount of oxygen can be added to the plastic bags prior to closing. In the hot months, cold packs are used to keep the conch cool during shipment (22°C, 72°F). These conch are shipped dry and can be air-freighted live for up to 36 hours with 100% survival on arrival. Some restaurants prefer the small conch shucked or removed from the shell. A specialized tool, 'the conch shucker', is used to release the conch muscle from the columella. The guts are removed and the conch are packaged in 1 lb bags for shipment (Fig. 7.45). When the conch are removed from the shell there are approximately 96 conch/lb at 6 cm SL (2.4 in SL) and 50–60 conch/lb at 8 cm SL (3.1 in SL). These bags of conch are refrigerated or frozen and packaged in coolers for shipment.

References cited and further reading

Davis, M. (1994) Mariculture techniques for queen conch, *Strombus gigas*: egg to juvenile stage. In: Appeldoorn, R.S. and Rodriguez, B. (eds.) *Queen Conch Biology, Fisheries and Mariculture*. Fundacion Cientifica Los Roques, Caracas, pp. 231–252.

Davis, M. (2000) Queen conch (*Strombus gigas*) culture techniques for research, stock enhancement and growout markets. In: Fingermna, M and Nagabhushanam, R. (eds.) *Marine Biotechnology*. Nova Science Publishers, Hauppauge, NY, pp. 127–159.

Davis, M. (2005) Species profile: queen conch, *Strombus gigas*. Southern Regional Aquaculture Center Publication 7203.

Davis, M. and Cassar, V. (2020) Queen Conch Aquaculture: Hatchery and Nursery Phases User Manual. *Journal of Shellfish Research*, 39(3): 731–810.

Randall, J.E. (1964) Contributions to the biology of the queen conch *Strombus gigas*. *Bulletin of Marine Science of the Gulf and Caribbean*, 14: 246–295.

Stoner, A.W. (2019) Forty years of field and laboratory experiments with hatchery-reared queen conch: the case for conserving natural stocks. *Reviews in Fisheries Science & Aquaculture*, 27(4): 490–516.

CHAPTER 8

Culture of the carnivorous marine snail, *Concholepas concholepas*

Patricio H. Manríquez and Juan Carlos Castilla

Introduction

The unique (monogeneric) carnivorous muricid, *Concholepas concholepas*, known in Chile as 'loco' (loko or locabe in Mapudungun, the language of the Mapuche people), and in Perú as 'pata de burro,' 'tolina' or 'chanque,' and in international markets as 'Chilean' or 'false abalone,' is found exclusively along intertidal rocky shores and shallow subtidal reefs along Chile (~18°–53°S) and also central and southern Perú. Particularly in Chile, the species is a traditional sea food resource, even since pre-Hispanic times. During the second half of the 20th century, with the advances of diving technologies and government incentives, under an open access regime, wild loco populations were heavily exploited by artisanal fishers (divers). Since the mid-1970s, the species has been mainly exported to Asiatic markets, and by the end of the 1980s external market demands drove overexploitation. Notwithstanding, during the early 1990s the loco fishery moved to a co-management scheme based on a modern legislation that included Territorial User Rights for Fisheries Program, eventually leading to a fishery regulation where locos can be legally exclusively extracted from the Management and Exploitation Areas for Benthic Resources (several hundred areas along the Chilean coastline), by local small-scale fishing communities.

In Chile, the loco is by far the most scientifically studied invertebrate (>120 peer-reviewed publications) in terms of evolution, genetic, reproduction, larval biology, behavior, physiology, ecology, population dynamics, fisheries, and even regarding its responses to global stressors, such as ocean warming and acidification. During the past 20 years, numerous efforts and substantial economic investments have been made to advance laboratory, pilot, and commercial aquaculture from larval stages; however, no reliable and repeatable commercial-scale aquaculture protocols have as yet been developed. While commercial-scale culture does not yet exist for *C. concholepas* in Chile or Perú, successful juvenile grow-out (>2 cm) under field conditions has been achieved.

In this chapter, the three main biological and technological aspects connected with the future aquaculture of *C. concholepas* are revisited: laboratory-based aquaculture; environmental subsidies, and field-based grow-out (fattening). Several key aspects of the biology of *C. concholepas* are synthesized and illustrated along with the main impediments and challenges to its aquaculture (see Manríquez and Castilla, 2018 and references therein for details). Recent information regarding an economic feasibility analysis for the aquaculture of loco is included.

Laboratory-based aquaculture

Broodstock

The species has separate sexes and under captivity it is possible to differentiate individuals by direct observation of the penis or the female opening. This helps to manipulate the sex ratio for mating and maximize the rate of encounter between males and females. During the reproductive season, *Concholepas concholepas* form reproductive massive mating aggregations in nature and captivity (Fig. 8.1A). Mussels (*Mytilus chilensis*, *Choromytilus chorus*, *Perumytilus purpuratus* and *Semimytilus algosus*), razor clams (*Mesodesma donacium*), and tunicates (*Pyura chilensis* and *P. praeputialis*) are all prey. Sexual maturity in *C. concholepas* is reached at sizes ranging from ~3.0 cm to ~6.0 cm in size and maximal frequencies of sexually mature individuals are achieved at sizes of ~7.0 cm. In captivity, egg capsules of loco can be obtained year round; however, the maximal laying activity of egg capsules under laboratory conditions occurs from May to September. A positive and linear relationship exists between the size of the female individuals and the size of the laid egg capsules. Larger egg capsules have more developing embryos or larvae. Depending on the female size, between 100 and 700 egg capsules can be laid during a single spawning event. In captivity, once mating is completed, males move freely in the tanks in search for food; whereas females search for appropriate sites in the tanks (usually corners) for the laying of the egg capsules. In the first phase, the females use their radula to clean and prepare the surface of the tanks where egg capsules will be laid (Fig. 8.1B). During this process, the female takes about 1 hour to lay one individual egg capsule and the process will not stop (hours or days) until the last egg capsule is laid. Whether the laying process takes place with or without disturbance, females will not search for food during this time period.

Egg-capsule rearing

Recently laid egg capsules are yellowish (Fig. 8.1C) and can be differentiated easily from mature ones which are brownish. Thus, to initiate the aquaculture phase, newly-laid egg capsules can be removed from the matting container to another rearing container with more appropriate seawater flow and aeration conditions for egg capsule rearing. The embryonic development of this species takes place inside the egg capsules and ends with the hatching of small veliger larvae of ~250 μm (Fig. 8.1D; Table 8.1). The veliger growth and development inside each egg capsule is more or less synchronic. The duration of egg development into hatchlings inside the capsule is about 2 months; however, this time period can be shortened or extended if the seawater temperature is increased or reduced, respectively. On average, the number of larvae inside the capsules ranges from 1,000 to 10,000. This suggests that hatching time and the number of hatched larvae are important traits that can be manipulated under aquaculture conditions. To guarantee larval hatching and survival, special care should be taken to maintain the salinity above 25.

The production of purple pigments in *C. concholepas* egg capsules is commonly observed

Figure 8.1 Photographic records of life history stages associated with milestones needed to implement the future laboratory and field-based aquaculture of *Concholepas concholepas*. (A) Mating under laboratory-based conditions; (B) female individual depositing egg capsules on the inside of a glass broodstock container under laboratory-based conditions; (C) group of newly laid egg capsules with the characteristic yellowish color; (D) group of several newly hatched veliger larvae obtained under laboratory-based conditions; (E) competent larva collected from near-shore waters and swimming with an extended tetra-lobated velum; (F) newly settled individual eating a barnacle under laboratory-based conditions; (G) group of small juvenile individuals (~1 cm) grown from newly-settled competent larvae under laboratory-based conditions (dark [left] or light [right] colored shells can be obtained by changing the color of the prey used during rearing); (H) group of juvenile individual grown under laboratory-based culture conditions from individuals shown in (G) and with suitable sizes (~2 cm) to start the field-based culture; (I) group of large juvenile individuals (~5 cm) after being grown from individuals shown in (H) under field based culture conditions; (J) a wild adult individual of legal market size eating mussel beds in an intertidal rocky platform. Courtesy of Patricio H. Manríquez.

under laboratory and field conditions. The apparition of this pigment disrupts normal development and results in unviable eggs or larvae. Under laboratory conditions, purple egg capsules are mainly concentrated in parts of the rearing tanks with reduced water movement or in groups of egg capsules confined by other egg capsules. Attempts to mitigate this condition can be solved by moving capsules from the mating containers to specially designed egg-capsule rearing tanks with heavy aeration. Another common problem observed under laboratory conditions is the growth of bacteria and protozoa on the surface of the egg capsules. As an example, the intra capsular oxygen

Table 8.1 Summary of available information for spawning and grow-up of *Concholepas concholepas*.

Aquaculture phase	Comments	Key references
Broodstock	Adults can be successfully reared in captivity to obtain egg capsules.	Gallardo, 1973*; Bustos and Navarrete, 2001**
Egg capsule rearing	About 30–40 days required for hatching; trait that can be modulated by temperature and salinity.	Gallardo, 1973*
Post-hatching larval rearing	Under laboratory conditions, competence can be reached after about 4 months.	DiSalvo, 1988*; Bustos and Navarrete, 2001**
Larval settlement and metamorphosis	Prey such as barnacles and mussels, or alternatively probiotic biofilms on substrates, can be used as natural triggers for settlement and metamorphosis.	DiSalvo and Carriker, 1994*; Bustos and Navarrete, 2001**; Manríquez et al., 2004*;
Grow-out of small juveniles (up to 1 cm)	After metamorphosis, the mussel *Semimytilus algosus* is the most preferred prey and gives best growth results. Intra cohort cannibalism is common in individuals <1 cm in length and can be minimized with low densities of settlers and high feeding regime.	Bustos and Navarrete, 2001**; Méndez and Cancino, 1990*; Manríquez et al., 2008*
Grow-out of small juveniles (up to 2 cm)	The mussel *S. algosus* supports laboratory-based culture survival and growth. Attainment of 2 cm in length is a milestone size as above this size handling produces minor shell damage. The 2 cm length can be reached after about 4.5–14 months.	Manríquez et al., 2008*
Grow-out to mid-sizes (up to 5 cm)	The mussel *S. algosus* supports laboratory and field-based aquaculture survival and growth. From 2 cm in length, the 5 cm length can be reached after around 1.65–2.64 years.	Manríquez et al., 2008*
Grow-out to legal market size (10 cm)	Depending on the study (direct measurements, modelling, or a mixed study), feeding regime, and prey, between about 2 and 9 years are required to achieve the legal market size of 10 cm.	See Manríquez et al., 2008* for a review

Note: *Peer-reviewed article; ** non-peer reviewed manual.

availability inside the egg capsules of the muricid gastropod, *Chorus giganteus*, can be reduced by the presence of sessile protozoa growing on the outer wall of their egg capsules. This might impair embryonic development of this species and seems to also be present in *C. concholepas*; however, for the loco, this condition can be easily overcome if the outer surface of the egg capsules is carefully cleaned with the aid of a soft brush and 0.45 µm filtered seawater.

Post-hatching larval rearing to competence

The post-hatching culture phase presents the most challenges. In fact, only one published study has fully described the larval development of this species from hatching to competence (DiSalvo, 1988). A period of about 4 months is required for the newly hatched veliger larvae to achieve competence (~1800 µm in shell length, Fig. 8.1E, Table 8.1). A similar larval development period is described in an aquaculture manual for this species (Bustos and Navarrete, 2001). Although this manual represented a potential milestone on the road toward the aquaculture of *C. concholepas*, to the best of our knowledge, no one has yet used it to pilot or set up large-scale aquaculture. During this period, larval survival is low and ranges from ~0.0005% (DiSalvo, 1988) to ~1.5% (author's unpublished data). Despite this important improvement in larval survival rates after hatching, two critical scientific issues still need to be overcome: larval rearing density; and the type and concentration of microalgal food required to reach competent larvae at ~1800 µm in shell length and ~6,000 µm with the full extension of the tetra-lobated velum (Fig. 8.1E).

The appropriate larval densities during rearing are ~1.4 larvae per mL in PVC tanks with a running seawater system at a rate of 1–2 L per min. In this system, larval feeding can be supported by a mix of similar proportions of *Isochrysis galbana* and *Chatocerus gracilis*, with densities of 20,000–40,000 cells per mL initially and on reaching larval competence, respectively. Alternatively, a mix of the microalgae *I. galvana*, *Tetraselmis* sp., *Pavlova* sp., *Chaetoceros* sp., and *Pseudoisochrysis* sp. can be used to culture larvae of *C. concholepas* from hatching to competence. Laboratory observations using natural plankton assemblages have shown that newly-hatched and competent *C. concholepas* larvae consume different plankton cells sizes according to their ability to manipulate the food particles (Vargas et al., 2006). Hatchling veliger larvae are photopositive and geonegative, but during larval development behavioral traits are largely unknown. During this prolonged larval phase, the developing larvae are also exposed to potential bacterial infections. This important threat, along with appropriate larval densities and food provisioning has impeded the successful larval rearing to competence.

Larval settlement and metamorphosis

Prey items that support the very early benthic ontogeny of loco, such as mussels and barnacles (Fig. 8.1F) as well as water turbulence, might induce loco larval settlement and metamorphosis. Under laboratory conditions, competent larvae of loco can also undergo metamorphosis in response to newly-settled conspecifics present in the same experimental arena (Table 8.1). For aquaculture, an unfortunate side effect of settlement near conspecifics is the intra-cohort cannibalism in individuals with sizes below 1 cm. It takes about 3 months to reach 1 cm in shell length (Fig. 8.1G); below this size cannibalism is common if food is not abundant or rearing densities are too high.

Grow-out of juvenile stages

Under laboratory conditions (from settlement), it takes about 5 months for *C. concholepas* to achieve small juvenile sizes of 2 cm (Fig. 8.1H, Table 8.1). During this period, each individual eats between one to five mussels per day with sizes ranging from approximately one-half to three-quarters of the loco body size. The most preferred prey item of those early stages of *C. concholepas* is the mussel *Semimytilus algosus*. It is important to note that *C. concholepas* reaching 2 cm in body size is currently a milestone in aquaculture development; above this size, fattening can be done under field conditions and handling produces minor shell damage. The growing of juveniles in this phase is simply an indoor intensive rearing (fattening) of the individuals with running seawater, aeration at low densities, and abundant food. On reaching 2 cm, the aquaculture laboratory (hatchery)-based phase can be switched to a field-based aquaculture phase (see below) (Table 8.1) (Manríquez et al., 2004).

Environment subsidies

The stickiness of competent loco larvae and the presence of the mucous thread allow for their collection from near-shore foam slicks, where floating substrata accumulates. Collection from near-shore coastal areas could therefore reliably be used to naturally bolster successful aquaculture of the species (Table 8.1). Moreover, new settlers or post-metamorphic (~0.2 cm) and small-size juveniles (< 1.0 cm) can be collected on adult loco shell surfaces encrusted with barnacles (Table 8.1). Presently, locos cannot legally be removed from the shell before trading, resulting in piles of empty shells in the grounds of canning companies. It has been suggested that the return of those shells to the fishing grounds may be a reasonable management strategy to recover both refuges and settlers to the adult habitat. Alternatively, these individuals may be rescued and used to subsidize aquaculture practices. To implement these alternative aquaculture practices, further research is needed into scaling up collection of competent larvae in specially designed collectors, which can both trap them and support their early benthic post-metamorphic life. A potential aquaculture innovation may be the use of empty loco shells incrusted with barnacles kept hanging in the water column as collectors.

Field-based aquaculture

Grow-out to mid-size adult stages (5 cm)

This phase is the first stage that can be carried out as outdoor intensive aquaculture (Table 8.1). Small locos of about 2 cm have been successfully used to implement pilot aquaculture intensive practices under both indoor and field conditions using mono diets of the mussel, *Semimytilus algosus*. Utilizing this practice, individuals can be maintained in two consecutive outdoor aquaculture phases. In the first, individuals of about 2 cm are reared within perforated PVC pipes placed inside plastic and ventilated floating cages until they reach about 5 cm (Fig. 8.1I). During this phase, juveniles need to be fed with the mussel *S. algosus* and take between 4 and 14 months within the pipes to reach 5 cm with 99% survival. Because reproduction is possible in this rearing period, it is suggested that cocktail-size individuals of 5–6 cm in length may be designated as a legal market size for cultured individuals of this species (Manríquez et al., 2008).

Grow-out to legal market size (10 cm)

A few studies have reported field growth rate during the adulthood of loco. According to the projections derived from those studies, and considering the most recent studies since 1990, the time required to achieve the legal market size (10 cm, Fig. 8.1J) of wild population fluctuates from about 3 to 9 years. Using the mussel, *S. algosus,* as prey, the time to reach legal market size, from 5 cm, would range from 1.65 to 2.64 years in field and laboratory-based conditions, respectively. Considering the time required to reach 5 cm from settlement and subsequent projections to achieve the legal market size, the overall time required would be 2–3.8 years. The slow growth rate of this species in the wild corrolates with specific growth rates estimated from tagged and recaptured individuals in nature. This suggests that growth can be significantly improved under intensive aquaculture practices in which feeding and the impact of predation can be manipulated in outdoor caged-housed individuals fed with abundant food. In captivity, loco juveniles and adults can be supported by the razor clam *Mesodesma donacium*, and the tunicate, *Pyura chilensis* that only can be obtained from natural beds. Other prey species such as *M. chilensis* and the tunicate, P. chilensis, however, are the only cultivated and marketed species. The other species can only be obtained from natural beds. This suggests that large-scale and commercial production of *C. concholepas* will require cultivation of its preferred prey items. This is an important issue in need of consideration: the high cultivation cost of a high trophic level carnivorous species such *C. concholepas* which requires large inputs of live prey.

Economic feasibility of commercial aquaculture

Currently, Zuniga-Jara and Soria-Barreto (2018) concluded that the feasibility for the field aquaculture (fattening) of loco, from seed stages of 2–3 cm, is considered to be commercially unprofitable. This assessment is, however, based on a model that calculated 8.76 years for loco to reach 10 cm (legal market size). Given that both field and laboratory fattening efforts reported between 2 and 3.8 years to reach that size, there seems to a realistic possibility that loco can be farmed commercially once the aforementioned impediments are addressed. Moreover, Zuñiga-Jara and Soria-Barreto (2018) reported that in Chile there are several hatcheries that offer stocks of up to 100,000 loco juveniles to fattening centers. To the best of our knowledge, this is not the case. There are numerous reports indicating that the most challenging impediment to loco aquaculture is loco larvae mortality in rearing laboratories, during the nearly 4 months that it takes for the newly hatched veliger larvae to achieve competence (about 1800 µm in shell length). To our knowledge this continues to be the main challenge for the rearing and aquaculture of loco, and there is no research yet suggesting a solution. Indeed, based on a report of a recent loco aquaculture project in Chile, mortality among hatchery-produced larvae is about 99%. Hatcheries providing massive numbers of loco juveniles may thrive in Chile when that problem is solved.

Summary

Aquaculture of *C. concholepas* begins with the conditioning of a broodstock of large and sexually mature individuals. In captivity these individuals are easy to breed. Larvae can be obtained by natural hatching. The main aquaculture challenge is to raise the larvae that live sufficiently long (about 4 months). They are filter feeders and throughout their development need to be raised at appropriate densities and food concentrations. Once larval competence is reached, settlers are obtained either by using natural triggering substrates, such as mussels or barnacles, placed in the bottom of the rearing containers. Alternatively, the use of probiotic biofilms has been suggested

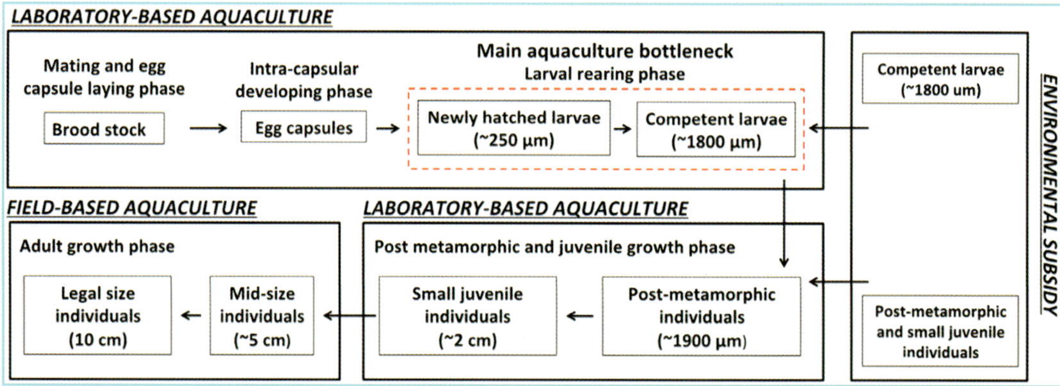

Figure 8.2 Schematic representation of the biological basis for the future aquaculture of *C. concholepas* in Chile.

as an appropriated method to induce settlement of pre-competent *C. concholepas* larvae. To avoid the characteristic intra-cohort cannibalism in small individuals below ~1 cm in length, low settler densities and high feeding regimes are recommended. Production of competent larvae and newly settled *C. concholepas* in the hatchery can be augmented by collection of these stages from natural beds. To sustain settlement to young individuals of about 2 cm in shell length an intensive laboratory-based aquaculture rearing of *C. concholepas*, using mussels as food, is recommended. Under such conditions nearly 140 days are needed to grow-out from settlement sizes of about 2–3 cm in shell length. Regardless of whether the newly settled are hatchery-produced or collected from nature, once reaching ~2 cm and then ~5 cm in shell length, the outdoor intensive aquaculture of small *C. concholepas* using mussel as food is the best method to use. From 5 cm to legal market size (10 cm) in central Chile takes a further 600–960 days. Future laboratory- and field-based *C. concholepas* aquaculture could also help restock wild overexploited populations (Fig. 8.2).

Acknowledgments

We thank Armando Rosson and Cesar Pedrini (Bitecma Ltd) who kindly shared raw data regarding the size of tagged and recaptured adult *C. concholepas* individuals in the wild. María Elisa Jara enthusiastically helped in many of our field and laboratory studies. Our research has been mainly funded by the National Fund for Scientific and Technological Development, FONDECYT of the National Commission for Scientific and Technological Research, CONICYT.

References cited and further reading

Bustos, E. and Navarrete, F. (2001) *Manual: El cultivo del loco (*Concholepas concholepas*)*. IFOP, Puerto Montt.

Castilla, J.C., Espinosa, J., Yamashiro, C., Melo, O., and Gelcich, S. (2016) Telecoupling between catch, farming, and international trade for the gastropods *Concholepas concholepas* (Loco) and *Haliotis* spp. (Abalone). *Journal of Shellfish Research*, 35: 499–506.

DiSalvo, L.H. (1988) Observations on the larval and post-metamorphic life of de *Concholepas concholepas* (Bruguiere, 1789) in laboratory culture. *Veliger*, 30: 358–368.

DiSalvo, L.H. and Carriker, M.R. (1994) Planktonic, metamorphic, and early benthic behavior of the Chilean loco *Concholepas concholepas* (Muricidae, Gastropoda, Mollusca). *Journal of Shellfish Research*, 13: 57–66.

Gallardo, C.S. (1973) Desarrollo intracapsular de *Concholepas concholepas* (Bruguiere) (Gastropoda: Muricidae). Publicación Ocasional No. 16. Museo Nacional de Historia Natural, Santiago.

Lody, M.A. (2013) Optimización de una unidad de cultivo de *Concholepas concholepas* con sistema controlado mediante la implementación de herramientas biotecnológicas fundamentada en la aplicación de probióticos. Informe final. Proyecto FONDEF D07I1138. Available at: http://repositorio.conicyt.cl/bitstream/handle/10533/114586/D07I1138.pdf.

Manríquez, P.H. and Castilla, J.C. (2011) Behavioral traits of *Concholepas concholepas* (loco) competent larvae during ontogenetic plankton-benthic shift. *Marine Ecology Progress Series*, 430: 207–221.

Manríquez, P.H. and Castilla, J.C. (2018) Life history, knowledge, bottlenecks, and challenges for the aquaculture of *Concholepas concholepas* (Gastropoda: Muricidae) in Chile. *Journal of Shellfish Research*, 37:, 1079–1092.

Manríquez, P.H., Navarrete, S.A., Rosson, A., and Castilla, J.C. (2004) Settlement of the gastropod *Concholepas concholepas* on shells of conspecific adults. *Journal of the Marine Biological Association of the United Kingdom*, 84: 651–658.

Manríquez, P.H., Delgado, A.P., Jara, M.E., and Castilla, J.C. (2008) Field and laboratory experiments with early ontogenetic stages of *Concholepas concholepas*. *Aquaculture*, 279: 99–107.

Méndez, M.A. and Cancino, J.M. (1990) Preferencias alimentarias de ejemplares postmetamórficos y juveniles de *Concholepas concholepas* (Bruguière, 1789). *Revista de Biología Marina, Valparaíso*, 25: 109–120.

Vargas, C.A., Manríquez, P.H., and Navarrete, S.A. (2006) Feeding by larval of intertidal invertebrates: assessing their position in pelagic food webs. *Ecology*, 87: 444–457.

Zuñiga-Jara, S. and Soria-Barreto, K. (2018) Economic analysis of loco abalone (*Concholepas concholepas*) commercial cultivation in Chile. *Latin American Journal of Aquatic Research*, 46: 895–908.

CHAPTER 9

Pearl oyster culture and pearl production

Paul C. Southgate

Introduction

Marine pearl oysters (Pteriidae) and freshwater mussels (Unionidae and Margaritiferidae) are uniquely cultured to produce gemstones – cultured pearls – with mother-of-pearl (MOP) (oyster shells) and pearl oyster meat being secondary products. Global cultured pearl production is estimated to be over 2000 metric tons (MT), with more than 98% of this production from freshwater mussels, cultured mainly in China. But because freshwater pearls have considerably less value than cultured marine pearls, the latter generate around 95% of cultured pearl production value, currently estimated at around US$200 million. This chapter focuses on cultured marine pearl production and the reader is directed toward the 'further reading' section for more information on freshwater pearl culture.

Before marine pearls were artificially cultured, they were collected rarely and by chance from oysters gathered for MOP or for food. Large and economically important MOP fisheries existed in areas such as the Gulf of Mannar (India and Sri Lanka), Bay of Bengal, Red Sea, the Persian Gulf, and Polynesia. As pearl oyster stocks became exhausted, many countries with a long history in the MOP fishery became less significant as the world market became increasingly dominated by cultured pearls. The development of a technique for cultured round pearl production in the early 1900s provided the basis for today's cultured marine pearl industry, which now involves more than 30 countries.

Commercially cultivated species

Several pearl oyster species have been extensively exploited for pearl production for over a century. The most important of these are from the genus *Pinctada* from which three major species are used for commercial pearl culture. The Akoya pearl oyster, *P. fucata/martensii*, is the smallest of these, but is the most commonly utilized for commercial pearl production. It provides the basis for Akoya pearl production in Japan and China. The gold- or silver-lip pearl oyster, *P. maxima*, is the largest *Pinctada* species, and is cultured to produce gold and white/silver 'South

Figure 9.1 Cultured pearls produced by Civa (Fiji) Pearls Pte Ltd., Fiji, using the black-lip pearl oyster, *Pinctada margaritifera*. Fiji pearls are well-known for their high quality and unmatched range of colors.

Sea' pearls primarily in Indonesia, northern Australia, Philippines, Malaysia, and Myanmar. This species produces the largest and the most valuable of the cultured marine pearls. The black-lip pearl oyster, *P. margaritifera*, is the second largest *Pinctada* species. It has a broad Indo-Pacific distribution but is particularly abundant in the atolls of Polynesia where it supports significant production of 'black' or 'Tahitian' pearls in French Polynesia. This species is also used for pearl production in the Cook Islands and Fiji, and it produces pearl with the broadest variation in color (Fig. 9.1).

Species from the genus *Pteria* are also used for commercial pearl culture, but to a lesser degree than *Pinctada* species. They are less suitable for round pearl production than *Pinctada* species because of morphological differences between genera, and they are generally used to produce mabé pearls that are also known as half-pearls or blister pearls (Fig. 9.2). The winged pearl oyster, *Pteria penguin* or '*mabé gai*' in Japan, is traditionally used for mabé pearl production. The winged

Figure 9.2 Mabé pearls on the inner surface of a *Pteria penguin* shell will be cut from the shell before value adding. Courtesy of Sophie Gordon.

Figure 9.3 Small bamboo rafts are simple and cheap to construct, and they support small-scale pearl culture in many developing countries. Here used for culture of *Pinctada margaritifera* in Chole Bay, Tanzania.

pearl oyster is restricted to the Gulf of California, and despite also being initially utilized for half-pearl production, has been used for successful production of round and near-round pearls.

Pearl oyster culture – getting started

Pearl oysters are subtidal animals and culture systems must account for this. Pearl oysters are generally cultured using suspended-culture systems, such as longlines or rafts, or bottom culture systems such as fence lines, racks, or trestles. Choice of the system must account for factors such as current speed, water depth, capital costs, operational costs, exposure to wind and

Figure 9.4 Large rafts used for pearl oyster culture in Li'an Bay, Hainan Island, China, are often developed into large structures to provide accommodation for farm staff and a work platform for cleaning, repair, and storage of culture equipment.

wave action, ease of operation, need for direct access from land, security considerations, and tidal variation.

Rafts

Rafts are rigid floating structures composed of a wooden, bamboo, or plastic framework from which pearl oyster culture units can be suspended (Fig. 9.3). Buoyancy is provided by attached floatation devices such as barrels, plastic container, drums, and floats. Rafts can be moored to fixed structures, such as jetties, providing convenient access to culture stock from land, or anchored to the substrate offshore. Rafts are commonly used in relatively sheltered areas with low wind and wave action, where a number of rafts may be moored together. Rafts are relatively cheap and simple to construct, and they provide a work platform from which culture stock can be inspected without the use of divers (Fig. 9.4).

Longlines

Longlines are composed of a large-diameter (>25 mm) rope, or headline, that is supported by floats, and held in position by anchors (Fig. 9.5). Longlines can be deployed on the surface in areas where surface swell and wave action are low, where they pose no navigational hazard, and when security is not a concern. In Australia, surface longlines are generally 100 m long and deployed in block formation with parallel longlines approximately 20–30 m apart. Floats are placed at 1–2 m intervals along the headline and 'dropper' ropes that support individual culture units are attached at 1 m intervals. Sub-surface longlines are less prone to security and navigational risks, and to surface water movement that can affect oyster performance (Fig. 9.6). But they are more difficult to deploy than surface longlines

Figure 9.5 Rows of parallel floats mark surface longlines used for pearl oyster culture in Li'an Bay, Hainan Island, China. Note the large floating raft mid-picture. This is one of few locations in the world where the four major species of pearl oysters (*Pinctada maxima*, *P. margaritifera*, *P. fucata*, and *Pteria penguin*) may be cultured together.

Figure 9.6 Large *Pteria penguin* attached singly to ropes suspended from a sub-surface longline in Tonga. This method of pearl culture is common throughout Polynesia and the Pacific. Courtesy of Sophie Gordon.

Figure 9.7 A benthic rack system used to culture *Pinctada margaritifera* in Manihiki, Cook Islands. Note the oysters are attached singly to ropes that are suspended from the rack. Courtesy of John Lucas.

and rely on scuba divers for regular maintenance and inspection. Sub-surface longlines are widely used in French Polynesia for culture of *P. margaritifera*.

Racks, trestles, and fencelines

Racks and trestles are generally made from timber poles, or metal or plastic pipes to form a platform that is fixed above the seabed (Fig. 9.7). Culture units containing pearl oysters can be placed either on or suspended from such structures. They are simple to construct and relatively low-cost, but, like sub-surface longlines, are more difficult to deploy than surface culture infrastructure and require scuba divers for regular maintenance and inspection. Trestles are the traditional culture system for *P. margaritifera* in Polynesia, but they have now been replaced by sub-surface longlines. Fencelines are composed of a row of galvanized metal posts driven into the bottom substrate and connected by a headline (rope) from which culture units containing pearl oysters are hung. They are used in relatively exposed sites in Western Australia and, like trestles, they have disadvantages associated with deployment and regular maintenance and inspection.

Culture units

The simplest culture method used for pearl oysters is the 'ear-hanging' method, which was adapted initially from the Japanese scallop culture industry. It involves drilling a 2–3 mm hole through the dorsal-posterior region of the pearl oyster shell, then tying the oyster to a rope using monofilament fishing line or wire. Around 10–20 oysters are usually attached to a single rope, either singly or in pairs, to form a 'chaplet', and chaplets are attached directly to a raft or longline or rack (Figs. 9.6 and 9.7). This culture method has low set up and maintenance costs, due to greatly reduced equipment outlay, and because oysters are not enclosed within a culture unit, they have improved water flow and food availability and require reduced cleaning effort. Chaplet-based pearl oyster culture is widely practiced in Polynesia and is well-suited to the relatively protected atoll lagoons where predation is minimal. This culture method is less suited to environments where pearl oysters are exposed to adverse environmental conditions and predation, although threats from predation can be reduced by enclosing chaplets in protective housings (Fig. 9.8).

In areas where there may be significant threats from predators, oysters are generally held in enclosed culture units. In Australia and southeast Asian countries, for example, pearl oysters are predominantly cultured using panel nets that are made from strong steel or galvanized frames supporting mesh pockets that hold single or multiple oysters (Fig. 9.9). The size of the pocket and the size of the mesh can be altered, depending on species and size of the cultured oysters. Nets used for small juveniles may, for example, have 48 pockets composed of 5 mm mesh, while those used for adult oyster are likely to have six pockets made from 30–45 mm mesh. Maximal mesh size is important in considering optimizing water flow and food supply and minimizing fouling of culture units.

For early nursery culture, pearl oysters are often held in large numbers within pearl nets or lantern nets until they reach a size appropriate for deployment to chaplets or panel nets (Fig. 9.10). These culture units are

Figure 9.8 *Pteria penguin* juveniles attached in pairs to chaplets suspended from a sub-surface longline in Tonga. The chaplets are housed within protected mesh cylinders. Courtesy of Sophie Gordon.

Figure 9.9 Panel net used to hold *Pinctada fucata* in Li'an Bay, Hainan Island, China. Note that the mesh is sewn into longitudinal pockets that hold multiple oysters. These pockets are often sewn vertically too, forming pockets that are used to house individual pearl oysters (see Fig. 9.3).

Figure 9.10 Large numbers of *Pinctada fucata* juveniles are held in lantern nets at a pearl farm in Li'an Bay, Hainan Island, China.

Figure 9.11 A polypropylene rope spat collector is harvested at Namarai, Fiji after a 12 month deployment. Three large *Pinctada margaritifera* are visible on the collector.

often used to hold large numbers of smaller oysters during nursery culture. Perforated plastic 'oyster' trays and baskets may also be used to hold pearl oyster juveniles.

Sources of stock

Spat collection

The atolls of eastern Polynesia are relatively small and enclosed. They have good natural stocks of *P. margaritifera* and their larvae are retained within the atoll lagoons which are poorly flushed by oceanic water. Deployment of appropriate materials, or 'spat collectors', to lagoon waters at the right time of year provides settlement substrates for pearl oyster larvae. Some months after settlement, resulting juveniles or 'spat' can be removed from spat collector for further culture. The most commonly used materials for spat collectors include traditional natural substrates such as tree branches, but there has been increasing use of plastic materials such as shade-cloth and frayed rope that offer a three-dimensional structure providing a large surface area and protective spaces for newly recruited spat (Fig. 9.11). Spat collectors are generally suspended from longlines or rafts.

Factors that influence the success of pearl oyster spat collection activities include the type of material used to construct the spat collector, location, depth, season, local currents, and hydrodynamics. For example, recruitment patterns of *Pinctada margaritifera* and *Pteria penguin* to spat collectors across 29 sites in Fiji indicated that *Pteria penguin* has greater tolerance of more turbid inshore sites, while similar research in northern Australia reported improved recruitment of *Pteria penguin* at greater depths. The timing of spat collector deployment is critical; it must coincide with the availability of metamorphically competent larvae of the target species and, where possible, be timed to minimize by-catch of non-commercial fouling species.

The length of time that spat collectors are deployed varies between approximately six months in Polynesia, when *P. margaritifera* juveniles are harvested with a dorso-ventral shell height of approximately 50–65 mm, to around 12–15 months in Fiji, when the shell height of resulting *P. margaritifera* may exceed 100 mm. While increasing spat collector immersion time generally results in larger pearl oysters at harvest, predation and fouling may cause natural attrition of juveniles after recruitment to the collectors. Attempts to protect juvenile pearl oysters by holding spat collectors within mesh bags, was shown to result in high mortality of pearl oyster spat from predation. The larvae of predators also recruit to spat collectors and may grow rapidly once trapped within the mesh bags. Ideally, pearl oyster juveniles should be harvested from spat collectors before the recruits of predatory species become large enough to consume them.

Hatchery production

Broodstock selection

For pearl farms that rely on hatchery produced culture stock, selection of appropriate broodstock is of critical importance because their progeny will influence pearl yield and pearl quality from the farm for years to come. Important factors to consider include rapid growth rate, to minimize the non-productive culture period required for oysters to reach pearl production size, aspects of shell morphometry, which influence the size of nucleus that can be implanted for pearl production, and nacre (MOP) coloration, which influences nacre color in resulting spat, that themselves will be used for subsequent pearl production. Pearl color is a key consideration in assessing pearl quality and value.

Pearl oysters are protandrous hermaphrodites and oyster broodstock of a range of sizes is required to ensure an adequate number of males and females. Reproductive condition is relatively easy to determine in pearl oysters and the various stages of reproductive maturity have been described for the major culture species.

Broodstock conditioning

Two of the three major commercial pearl oyster species, *Pinctada maxima* and *P. margaritifera*, are tropical and have extended spawning seasons that allow hatcheries to rely on broodstock that mature under ambient conditions. As a result, there has been relatively little research on conditioning of these species to bring about reproductive maturation outside normal breeding season. This may also reflect logistic difficulties in providing the large volumes of microalgae that would be needed for adequate conditioning of these large bivalves. A period of at least 5 weeks is required for *P. maxima* broodstock to progress from undeveloped gonads to spawning ripe condition when held on a surface longline under ambient conditions.

Increasing water temperature and food availability are the major cues for reproductive maturation in pearl oysters. Akoya pearl oysters with maturing gonads were brought to spawning condition when held within a water temperature range of 25–28°C and fed a mixed microalgae diet (4 L/oyster/day) supplemented with corn flour (30 mg/oyster/day) for 45 days, and oyster condition can be maintained when held under the same conditions. Elevated water temperatures and an adequate supply of cultured microalgae also brought about successful conditioning of *P. mazatlanica*, with a gradual increase in water temperature, from 20°C to 29°C, over a 2-month period, reported to be more effective than rapid exposure to higher water temperatures.

Spawning induction

After thorough cleaning of the outer shell surfaces of broodstock to remove fouling organisms and potential bacterial contaminants, spawning induction of pearl oysters is normally stimulated by exposure to elevated, or to fluctuating water temperature. Thermal stimulation has been reported to be a successful means of spawning induction for all major commercial pearl oyster species. For example, Akoya pearl oysters from temperate Australian waters spawn at water temperatures above 18°C, whereas those of the same species from tropical Indian waters have been induced by raising water temperature from 28.5°C to 35°C. For the tropical *Pinctada maxima*, an increase in water temperature from 2°C to 5°C above ambient, with a maximum of 33–34°C, was recommended. 'Cold conditioning' has been used to increase the water temperature range to which broodstock of the tropical *P. margaritifera* are exposed to assist spawning induction. Broodstock were held overnight at 20°C (5–6°C below the ambient water temperature), then the following day, placed into another tank containing water heated to 30–31°C to induce spawning. Alternate exposure to elevated water temperature and ambient water temperature, until spawning occurs, is common practice in pearl oyster hatcheries.

Like other bivalves, pearl oysters readily spawn in response to waterborne gametes released by other individuals. A suspension of sperm obtained from a sacrificed male is commonly introduced to the spawning tank to stimulate spawning. Sperm may be washed from shallow incisions made into the gonad of the sacrificed male, using filtered seawater, and the resulting suspension may be added directly to the spawning tank.

Broodstock are usually placed within a communal spawning tank with shallow water that allows spawning individuals to be easily identified, and more immediate impacts of spawning inducing factors (that is, change in water temperature, gametes produced by other individuals, introduction of sperm suspension). Once spawning begins, spawning oysters are removed from the spawning tank and placed individually into separate aquaria where they complete spawning. This allows the gametes of individual broodstock to be examined microscopically, assessed for their quality, and selection of only the better-quality gametes for hatchery production; that is, sperm should be active while eggs should be spherical and negatively buoyant.

Gamete fertilization

Pearl oyster hatcheries generally combine the eggs from a number of individuals prior to fertilization, and the same is done for sperm. The resulting egg suspension may be passed through a nylon sieve mesh to remove debris (for example, feces, and so on) before fertilization. Eggs may also be held on a submerged fine mesh sieve and washed with clean 1-μm filtered seawater. To fertilize the eggs, aliquots of the sperm suspension are initially added to the egg suspension in small volumes (for example, 1 mL of sperm suspension per liter of egg suspension). Microscopic examination is used to determine the number of available sperm per egg and additional sperm can be added if necessary. It is important to minimize the number of sperm used to fertilize the eggs because excessive sperm may lead to 'polyspermy' that can affect survival and development of embryos and larvae and may impact water quality when it decays. The presence of the first polar body resulting from meiotic division of the egg nucleus indicates fertilization.

Successful cryopreservation of pearl oyster gametes has been reported and shown to result in similar larval performance to fresh gametes. This technology is not widely used in pearl oyster hatcheries but has become important in recent years in research programs to develop specific breeding lines for pearl oysters, to improve both biological traits and pearl quality, where the ability to store gametes from genetically superior animals has obvious advantages.

Egg incubation

The fertilized eggs of pearl oysters are generally incubated for around 24 hours to allow embryonic and early larval development to occur. Egg incubation tanks contain gently aerated, 1 µm filtered seawater. Survival to the larval stage is inversely proportional to egg density, with densities of 30–50 eggs/mL used for *Pinctada margaritifera*, ≤30 eggs/mL recommended for *P. maxima*, and 30–50 eggs/mL recommended for *Pteria penguin*. An egg density of 10/mL has been shown to support the highest rates of survival to the veliger larval stage for both *P. margaritifera* and *P. maxima*, compared to higher densities of up to 150 eggs/mL and 100 eggs/mL, respectively.

Water temperature and salinity also affect development and survival of embryos during egg incubation. Optimal development of *P. margaritifera* eggs occurs within the salinity range of 24–39 and water temperature range of 25–30°C; with no development of embryos at 20°C or 35°C. While for *P. fucata* from temperate Australian waters, greatest survival of embryos was recorded within a temperature range of 18–26°C and a salinity range of 29–35. Because of potentially high mortality of eggs and embryos during incubation, some pearl oyster hatcheries may apply a broad-spectrum antibiotic, such as streptomycin sulfate at not as a dose of 10 mg/L, during egg incubation. For *Pteria penguin*, treatment with streptomycin sulfate during egg incubation was shown to improve mean survival to the veliger larval stage by 16% compared to controls without antibiotics.

Embryonic and early larval development of pearl oysters is similar to that of other planktotrophic bivalve molluscs and has been described for the major commercial pearl oyster species. The first motile stage is the trochophore which is ciliated but lacks a shell and is unable to feed. The shelled veliger stage is reached 20–24 hours after fertilization. Early veliger larvae have a characteristic 'D'-shaped shell giving rise to the common names of 'D-stage' or 'D-shaped' veliger (Fig. 9.12). The larval shell provides protection that enables the larvae to be drained from the egg incubation tank after 24 hours and retained on an appropriately sized sieve mesh (for

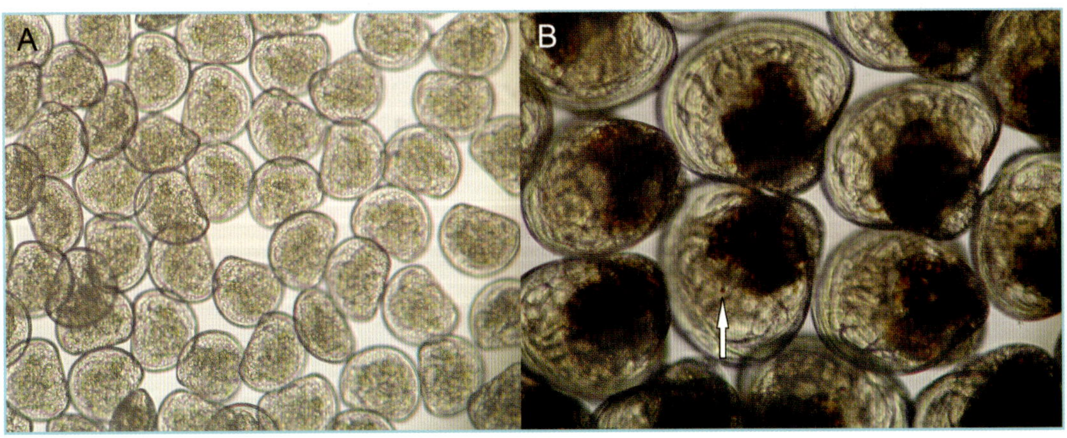

Figure 9.12 Development of *Pinctada maxima* larvae showing (A) 19-hour-old D-stage veliger larvae and (B) 12-day-old eyed pediveliger larvae (eye-spot indicated by arrow). Courtesy of Jens Knauer.

example, 25–53 µm) to separate them from undeveloped eggs and moribund larvae. D-stage veligers are then washed on a submerged sieve mesh with 1-µm filtered seawater, before transfer to larval culture tanks.

Larval culture and water quality

Pearl oyster larvae are generally cultured in 1-µm filtered seawater and require a supply of micro-algae as a food source. It is usual to supply larval culture tanks with gentle aeration although this can vary; early research with *P. maxima* recommended that larvae should be cultured without aeration until just before settlement, and *P. fucata* larvae may be cultured without aeration, which is thought to negatively affect growth and spat-fall of this species. The success of hatchery culture of pearl oysters is influenced by a number of factors including larval stocking density, water temperature, salinity, food density, and food quality.

The initial larval stocking density in culture tanks varies between hatcheries and according to the culture species (Table 9.1). Natural attrition results in reduced larval density as larval culture continues, although grading and disposal of small or weaker larvae, and readjustment of stocking density by mixing larvae from different tanks, may eliminate this. For example, an initial stocking density of 1–8 larvae/mL for *P. maxima* may be reduced to around 0.5–1 larvae/mL at settlement (Table 9.1).

Maximal survival of 6 day old *P. margaritifera* larvae occurs within a salinity range of 26.5–33.5 and a water temperature range of 22.5–26.5°C, with each degree (°C) increase above 29°C reducing survival by approximately 5% to a level of around 20% at 35°C. For older (15 day) larvae, maximal survival occurs at a salinity of 31 and a water temperature of 25°C; however, larval growth at this temperature is very slow. The greatest growth rates of 6 day old and 15 day old *P. margaritifera* larvae occur within salinity ranges of 29–33.5 and 25–36, respectively, and within water temperature ranges of 26–29°C and 24.5–30°C, respectively. Similar information for the larvae of other pearl oyster species is limited, although a temperature range of 27–30°C was recommended for *P. maxima* larvae. Most pearl oyster hatcheries in the tropics use ambient water temperature and salinity for hatchery production. Successful culture of *Pteria penguin* larvae, for example, is undertaken within water temperature and salinity ranges of 24.5–30.1°C and 33.2–34.4, respectively. Culture of *P. maxima* larvae in Indonesia has been successful at a water temperature and salinity of 28±0.5°C and 34, respectively, and in northern Australia at a water temperature and salinity of 28±1°C and 36, respectively.

Table 9.1 Some of the larval stocking densities used successfully for the major pearl oyster species.

Species	Recommended stocking density
Pinctada maxima	1–8 larvae/mL at initial stocking reduced to 0.5–1 larvae/mL at settlement. Initial stocking density of 1–4 larvae/mL.
Pinctada mazatlanica	A 'low' stocking density of 3–4 larvae/mL was reported to be superior to a high stocking density (8–9 larvae/mL).
Pinctada fucata	2 larvae/mL is optimal for *P. fucata* with higher densities resulting in poor growth.
Pinctada margaritifera	Initial stocking density of 1 larva/mL.
Pteria penguin	Initial stocking density of 2.75–3 larva/mL.

Water exchange

Larval mortality, production of feces, and decomposition of uneaten microalgae all contribute to declining water quality in larval culture tanks over time. Because of this, pearl oyster hatcheries must replace the water in larval culture tanks periodically during larval culture. This process involves draining tank water through a submerged mesh sieve that retains the larvae, cleaning of the tank and tank surfaces, refilling the tank with clean 1 μm filtered seawater and replacing the larvae back into the tank. Retained larvae are washed into a vessel of known volume, and microscopic counting of the number of larvae in replicate sub-samples taken from the vessel allow hatchery staff to estimate numbers of larvae and to calculate survival. Samples may also be inspected to assess larval health, or vigor, and may be preserved for subsequent measurement or examination. The frequency at which the water in larval culture tanks is replaced varies. For example, only one complete water change on day 10 was required for *P. maxima* larvae stocked at a density of 1 larva/mL, but four to five water changes were used for larvae stocked at densities of 5–8 larvae/mL. Optimum larval growth and survival of *P. maxima* larvae was supported by a daily water exchange of 25% of tank volume when four different water exchange regimes were compared (25% exchange once every other day and 25%, 50%, and 100% once daily). A more universal approach in pearl oyster hatcheries is to totally replace larval culture water every 2 to 3 days.

Regular 'drain-down' of larval culture tanks and concentration of larvae into a relatively small volume of water during this process, allows larvae to be graded, based on size, by passing them through sieve meshes with varying pore size. This process allows removal of slow growing and moribund larvae, reduces variation in size between larvae, improves the synchrony with which larvae reach settlement, enhances post-settlement survival, and is thought to reduce the period over which settlement occurs. For example, *P. maxima* larvae may be graded on day 14 by passing them through a 100 μm and 60 μm sieve mesh; larvae caught on the 100 μm mesh are retained and placed back into larval culture tanks, while those caught on the 60 μm sieve mesh are discarded. Grading may take place more than once during the larval culture period with *P. fucata* larvae, for example, being graded at 8, 13, and 18 days of age.

Larval nutrition

Hatchery cultured pearl oyster larvae are generally fed live microalgae that are usually cultured on-site at the hatchery. The reader is directed to texts that deal specifically with microalgae culture within a bivalve hatchery context (see Chapter 12).

A diet composed of a mixture of microalgae, generally two of three species, provides a better balance of nutrients for pearl oyster larvae. Many species of microalgae have proved to be appropriate nutrient sources for pearl oyster larvae (Table 9.2). Species from the genera *Isochrysis*, *Pavlova*, and *Chaetoceros*, in particular, have been reported to have high nutritional value and are commonly used in pearl oyster hatcheries. Of seven species of microalgae assessed for *Pinctada margaritifera* larvae, for example, members of the genus *Pavlova* had particularly high nutritional value for both D-stage and later stage larvae, with diatoms, *Chaetoceros* spp., being of high nutritional value for later stage larvae only.

Hatchery culture of live microalgae is labor and resource demanding. It contributes significantly to hatchery establishment and running costs and requires specialized technical skills. On this basis, it is not well-suited to the developing countries where pearl culture predominantly occurs, because they often lack required resources. A simplified and cheaper larval culture protocol,

Table 9.2 Some of the species of microalgae assessed as a food source for pearl oyster larvae and their reported nutritional values.

Species	Micro-algae	Nutritional value
Pinctada margaritifera	*Pavlova pinguis* *P. salina* *Pavlova* sp. (PS 50) *Isochrysis* (T-Iso) *Chaetoceros muelleri* *Chaetoceros* sp. *Skeletonema* sp. *Micromonas pusilla*	*P. salina* and *Pavlova* sp. supported greatest growth of D-stage and umbone larvae. *Chaetoceros* sp. supported poor growth of D-stage larvae. Larval growth correlated with carbohydrate, lipid, protein, and fatty acid (HUFA) content of microalgae.
	P. salina *Pavlova* sp. (PS 50) *Isochrysis* (T-Iso) *Micromonas pusilla* *C. muelleri*, *Chaetoceros* sp., and *Skeletonema* sp. fed in binary and ternary combinations.	*Pavlova* sp. and *Pavlova* sp./*Chaetoceros muelleri* recommended for D-stage and umbone larvae, respectively.
	P. salina *Isochrysis* (T-Iso) *Chaetoceros simplex*	*Isochrysis* (T-Iso) supported greatest growth rate and survival.
	C. muelleri *C. simplex* *Isochrysis* (T-Iso) *P. salina* *P. lutheri*	Clearance rate (CR) and ingestion rate (IR) lower with diatoms than flagellates. CR and IR of *P. salina* were about five-times higher than those for *Chaetoceros* spp.
Pinctada maxima	*Isochrysis* (T-Iso) *Chaetoceros calcitrans* *C. gracilis* *Nannochloropsis oculata*	*N. oculata* not easily digested. Spines of *C. gracilis* impeded ingestion by small larvae.
	Isochrysis (T-Iso) *C. calcitrans* *C. gracilis* *N. oculata* *Tetraselmis suecica*	*Isochrysis* (T-Iso), *C. calcitrans*, and *C. gracilis* were most nutritious.
	Isochrysis zhanjiangensis, *Chlorella* spp., *Nannochloropsis oculata*, *I. zhanjiangensis* with *Chlorella* spp. *(1:1)* *I. zhanjiangensis* with *N. oculata (1:1)*	*Chlorella* spp. and *N. oculata* could replace 50% *I. zhanjiangensis* ration without affecting growth and survival
	Brachiomonas submarina *Chlorococcum* sp. *Dicrateria* sp. *Dunaliella tertiolecta*, *I. galbana* *M.* (*Pavlova*) *lutheri* *Nannochloris* sp. *Nitzschia closterium*	*Isochrysis galbana* and *Monochrysis lutheri* were the most nutritious species. At the larval rearing temperature of 30°C both species lost motility. The greater suitability of tropical micro-algae species was suggested.

Species	Micro-algae	Nutritional value
Pinctada maxima	*Pyramimonas grossi* *Chlorella* sp.* Green flagellate*	
Pinctada fucata	*I. galbana*	High.
	Pavlova sp.	Good.
	Chromulina sp.	Good.
	Dicrateria sp.	Good.
Pinctada mazatlanica	*Isochrysis* (T-Iso) and *P. lutheri* (1:1) mixture	Supported production of spat.
Pteria sterna	*Phaeodactylum tricornutum* *Chaetoceros calcitrans* *C. muelleri* *Thalassiosira weisflogii* *Dunaliella salina* *Nannochloris* sp. *Tetraselmis tetrahele* *T. suecica* *Isochrysis* (T-Iso) *Pavlova lutheri*	Only *Nannochloris* sp., *Isochrysis* (T-Iso) and *P. lutheri* were ingested. Only *Isochrysis* and *P. lutheri* were digested.
Pteria penguin	Instant Algae® micro-algae concentrates# Isochrysis 1800® and Pavlova 1800® (1:1)	Successful replacement of live microalgae with micro-algae concentrates which supported production of high numbers of spat.

Notes: *Local isolates; #Reed Mariculture Inc., San Jose, CA, USA.

using commercially available microalgae concentrates as a replacement for cultured live microalgae, was reported for *Pteria penguin* (Table 9.2). The products used, composed of concentrated *Isochrysis* and *Pavlova*, proved nutritious for *Pteria penguin* larvae and supported normal growth and development through settlement. Microalgal concentrates have been used as the sole nutrient source in routine hatchery production of *Pteria penguin* in Tonga since 2008. Replacing live microalgae with microalgal concentrates supports establishment of simpler, cheaper hatchery facilities, and larval rearing protocols that are more appropriate for developing nations.

Food supply

Successful hatchery culture of pearl oysters, like that of other bivalves, requires larvae to be provided with a ration that supports high growth rates, but one that does not compromise water quality or cause undue increase in hatchery running costs. Feeding schedules used for the larvae and early post-larvae of pearl oysters (Table 9.3). Research to determine optimal rations for pearl oyster larvae is scarce. Growth rates of D-stage *P. margaritifera* larvae fed a 1 : 1 mixture of *Isochrysis* (T-Iso) and *Pavlova salina*, at rations greater than 5×10^3 cells/mL, were significantly greater than those of larvae fed lower rations, while maximal survival was shown by larvae fed 10×10^3 cells/mL. This shows that optimal algal ration varies with larval age and indicates the likely relationship between microalgae density and larval density.

Subsequent research reported that D-stage *P. margaritifera* larvae showed high survival and greatest growth rate when cultured at a density of 3/mL and fed a ration of $4.5–11.5 \times 10^3$ cells/mL. For 13-day-old *P. margaritifera* larvae, however, greatest growth rate was shown within a

Table 9.3 Feeding schedule (cells/mL/day) used for larvae and early post-larvae of some pearl oyster species. Values in parentheses are mean antero-posterior shell sizes of the larvae. Source: Southgate (2008).

Age (days)	Species (*Pinctada*)			
	P. maxima[a]	*P. margaritifera*[b]	*P. fucata*[c]	*P. mazatlanica*[d]
1	5,000 (80)	1,000 (82)	5,000 (67.5)	10,000
2	6,000	1,000	5,000	10,000 (80)
3	7,000	1,000	5,000	10,000
4	8,000	1,000	5,000	10,000
5	10,000 (98)	2,000	5,000	10,000
6	12,000	2,000	5,000	10,000
7	14,000	2,000	5,000	10,000
8	18,000	4,000	5,000	10,000
9	20,000	8,000	5,000	10,000
10	25,000 (115)	8,000	5,000 (135)	10,000
11	28,000	8,000 (138)	10,000	10,000
12	30,000	8,000	10,000	10,000 (139)
13	32,000	10,000	10,000	10,000
14	34,000	10,000	10,000	10,000
15	38,000 (144)	12,000	10,000 (180)	10,000
16	40,000	12,000	10,000	10,000
17	42,000	12,000	10,000	10,000
18	44,000 (165)	12,000	10,000 (200)	10,000
19	48,000	12,000	15,000	15,000
20	50,000 (185)	10,000 (214)	15,000 (220)	15,000 (193)
21	50,000	10,000	50,000*	15,000
22	50,000	10,000	50,000*	15,000
23	50,000	10,000	50,000*	15,000
24	50,000	10,000	50,000* (300)	15,000 (223)
25	50,000 (237)	10,000	50,000*	15,000
26	50,000	12,000	50,000*	15,000
27	50,000	12,000	50,000*	30,000
28	50,000	18,000	50,000*	30,000
29	50,000	18,000	50,000*	30,000
30	50,000 (281)	25,000	50,000*	30,000

Notes: *50,000 cells/day/spat. [a]Diet composed of *Isochrysis* (T-Iso), *C. gracillis*, and *N. oculate* (~50–10%). [b]Diet composed of *Isochrysis* (T-Iso), *P. salina*, and *C. simplex*. [c]Diet composed of *I. galbana* only to 15 days after settlement when *Chaetoceros* is added to the diet. [d]Diet composed of 1 : 1 mixture of *Isochyrsis* (T-Iso) and *P. lutheri*.

microalgae ration range of 15–32 × 10^3 cells/mL, but greatest survival occurred at a much lower microalgae ration (2.5 × 10^3 cells/mL), which supported poor growth. Based on these results, it was recommended that *P. margaritifera* larvae should be cultured at an initial density of 3 larvae/mL up to 8 days of age and fed a microalgae ration of around 8 × 10^3 cells/mL, while older (13–20 day) larvae should be cultured at a density of 2 larvae/mL and fed a microalgae ration of around 25 × 10^3 cells/mL.

The relationship between larval stocking density and microalgae ration was also investigated for *Pteria penguin* with results indicating that hatcheries should adopt a stocking density of 6

Table 9.4 Timing of the stages in embryonic development reported for five pearl oyster species.

	Pinctada fucata	*Pinctada maxima*	*Pinctada margaritifera*	*Pteria sterna*	*Pteria penguin*
Temperature (°C)	27–29	27–29	28	24	26–28
Polar body	—	—	24 min		28 min
1st division	45 min	40 min	—	90 min	60 min
2nd division	60 min	60 min	2 h	4 h	75 min
4th division	108 min	—	—	4 h	90 min
Morula (hours)	—	3	—	4–5	2.5
Blastula (hours)	6	–	—	5–6	4.5
To gastrula (hours)	6.75	5	5	6–12	5–6
Trochophore (hours)	11	7	8–12	12–16	7–14
D-stage (hours)	20–21	18–24	24	21–24	18–22

larvae/mL and a ration of 10 × 10^3 cells/mL for D-stage larvae, but use a lower stocking density of 1 larva/mL and an increased ration of 10 × 10^3 microalgae cells/mL for umbone larvae. Survival of *P. maxima* larvae was similarly reported to be significantly affected by stocking density and ration, but not by water exchange, yet all three factors significantly affected larval growth.

Settlement

Like those of other commercial bivalves, pearl oyster larvae pass through a sequence of recognized developmental stages before settlement. The distinctive D-shape of early veligers is lost as they increase in size and the umbo area above the hinge line becomes increasingly prominent in umbone larvae toward settlement (Fig. 9.12). Larvae develop a pigmented spot, known as the 'eye-spot' on both sides of the base of the foot primordium indicating metamorphic competence (Fig. 9.12). The larval foot becomes increasingly active as the larvae transition between swimming and crawling when settlement is imminent. The larval life of most pearl oysters is similar (20–23 days after fertilization), and settlement and metamorphosis occur at a similar larval shell size (230–266 µm) among species (Table 9.4).

Once a prominent eye-spot is visible in some of the larvae, these larger metamorphically competent larvae are separated from smaller larvae using appropriately sized sieve meshes. For *Pinctada maxima* and *P. margaritifera*, appropriate mesh pore sizes are 140 µm and 150 µm, respectively. Larvae retained on the sieve are transferred to settlement tanks, while those that pass through the sieve are returned to larval culture tanks for further development. This procedure may be repeated a number of times.

Settlement tanks for pearl oysters must contain a settlement substrate or substrates that are hung into the settlement tanks. Appropriate substrates include PVC slats or polypropylene rope positioned horizontally or vertically within panel net or PVC frames, and shade-cloth, frayed ropes or plastic strips that may or may not be housed in mesh bags (Fig. 9.13). Settlement tanks are generally filled with 1 µm filtered seawater and they are more vigorously aerated than larval culture tanks. This provides good circulation of water through the settlement substrates to encourage settlement onto the substrates, not the tank bottom, where mortality can be high

Figure 9.13 Settlement substrates prepared for *Pinctada margaritifera* larvae at a hatchery in Fiji. Collectors are composed of a PVC frame with bundles of polypropylene ribbon fastened vertically between the frame.

because of accumulated debris. The density of metamorphically competent pearl oyster larvae introduced to settlement tanks influences subsequent growth and survival and a setting density of ≤1 larva/mL has been recommended for *P. maxima* and is considered a good guide for all species. Settlement and byssal attachment of larvae onto substrates in settlement tanks occurs over a number of days. Once settlement has occurred, a flow-through water system can be established in settlement tanks.

Conditioning of PVC slat collectors with epifloral bio-film has been shown to improve recruitment of *P. maxima* spat but did not result in a growth advantage. Extension of the larval settlement period because of insufficient stimulus resulting from inappropriate conditioning of spat collectors has been suggested. Laboratory experiments showed that to optimize pearl oyster settlement, both appropriate chemical and physical cues (that is, substrate) are required. The color of settlement tanks has also been shown to be an important factor in successful settlement of pearl oyster larvae. For example, improved settlement of *P. fucata* larvae was reported in dark-colored tanks compared to lighter ones, and in black tanks, compared to white and blue tanks.

Early nursery culture

After settlement, pearl oyster spat are retained on collectors in settlement tanks until they are transferred from the hatchery to an ocean-based culture system. Culture tanks are generally supplied with a 1 μm filtered flow-through sea water supply during this period that provides a 50–100% water exchange per day. Spat density influences growth and survival with higher growth rates and survival associated with lower stocking densities because of competition for space and food at higher densities. A density of around 70 spat/100 cm^2 was recommended to maximize growth and survival of *P. maxima* spat.

The ration of microalgae provided to settlement tanks containing pearl oyster spat is considerably higher than that used for larval culture (Table 9.5). Growth rates of spat are influenced by the nutrient composition of the microalgae used to feed them and, as with larvae, species from the genera *Isochrysis*, *Chaetoceros*, and *Tetraselmis* are considered to be of high nutritional value (Table 9.5). In particular, *Chaetoceros muelleri*, *T. suecica*, and *Isochrysis* (T-Iso) have been shown

Table 9.5 Feeding schedule for post-larvae and spat of *Pinctada maxima*. Source: Rose (1990).

Day	Shell length (mm)	Algal density (cells/mL)		Percentage of ration		
		AM	PM	T-Iso	*Chaetoceros*	*Tetraselmis*
43		26,000	34,000	30	60	10
48	1.60	30,000	35,000	30	60	10
53		30,000	40,000	30	60	10
58		30,000	50,000	35	60	5
63		35,000	50,000	35	60	5
68	5.10	40,000	50,000	35	60	5
73		45,000	50,000	40	55	5
78		45,000	55,000	40	55	5
83		45,000	50,000	40	55	5
88		45,000	55,000	45	50	5
93		50,000	50,000	45	50	5
98	14.80	52,000	53,000	45	50	5
103		53,000	55,000	50	45	5
108		55,000	55,000	50	45	5
113		55,000	55,000	50	45	5
118		55,000	60,000	55	40	5
123		55,000	60,000	60	35	5
128		60,000	60,000	65	30	5
135	41.70	60,000	60,000	70	25	5
138		60,000	60,000	75	20	5

to have high nutritional value for 75 day old *P. maxima* spat, whereas *C. calcitrans* and *P. lutheri* had relatively poor nutritional value.

Pearl oyster juveniles are also influenced by salinity. Despite a broad tolerance to salinities within the range 45 to 25, which did not affect survival, *P. maxima* spat showed reduced growth rates at 45, 40, and 25, with best growth at a salinity of 30. Similarly, *P. imbricata* juveniles (17 mm) held within a water temperature range of 14–26°C showed greater survival at salinities of 32 and 35, and high mortality at salinities of ≤23. Rapid changes in salinity are effective in inducing detachment of *P. maxima* spat from settlement substrata with, for example, 92.3% detachment following exposure to hypersaline (45) seawater for 1 hour. Exposure of *P. maxima* spat to low pH (pH4) is also effective in inducing significant detachment (86%) of *P. maxima* spat.

Nursery culture and grow-out

Transfer from the hatchery

Pearl oyster spat are usually transferred to field culture conditions on the spat collectors used in the settlement tanks; however, many spat may settle directly onto the inner surface of settlement tanks, move off the collectors or become dislodged from them. They can be removed from tank surfaces using thin plastic film or by gentle brushing, and they will readily reattach to a more appropriate substrate prior to transfer. It is common practice for collectors to be placed into protective housings when transferred to the field, to protect spat from predators. These may include mesh trays and fine mesh sleeves. Protective units must have fine pore sizes to retain spat, but this leads to biofouling and reduced water flow and food supply to the spat. Mesh sleeves have been shown to improve survival of 1.8 mm *P. maxima* spat attached to PVC slats, although spat growth improved significantly with increasing mesh pore size.

Early nursery culture systems, like hatchery systems, provide optimized conditions to maximize survival and growth rates of pearl oyster juveniles; water quality, water temperature, food availability, and food quality being among the major considerations. But this changes when spat are transferred to field culture systems, where the operator has very limited influence over the culture environment, and efforts must be made to minimize resulting stress. This is done in a number of ways, including:

- gradual reduction of the level of water filtration in culture tanks, from 10 to 20 µm, to 30 µm, prior to transfer to exposed spat to natural phytoplankton and silt
- retaining spat until they reach a minimum size or age before transfer to minimize mortality; 3 mm for *P. fucata*, 2–3 months after settlement (6 mm) for *P. maxima*, 3–7 weeks after settlement for spat of *P. margaritifera*
- transferring spat to a sheltered site with minimal wave action and protection from predators
- timing the transfer of spat to account for seasonal conditions (that is, most appropriate water temperature at field site and minimal potential impacts from seasonal storms and resulting run-off).

Despite what are often considered optimized conditions within land-based culture systems for pearl oysters (that is, adequate and reliable food supply, no predators, and relatively high-water temperature), research with *P. margaritifera* showed that, when graded at 3.5 months of age, spat transferred from the hatchery 3 weeks after settlement were larger than those transferred 5, 7, and 9 weeks after settlement. This is thought to reflect superior nutrition in the ocean (that is, a broader range of potential food particles and nutrients) and suggests that pearl oyster spat should be transferred from the hatchery as soon as possible after settlement to maximize growth rates.

Figure 9.14 *Pteria penguin* juveniles attached to a plastic mesh collector in Tonga. Competition for food and space may inhibit oyster performance and these oysters will soon be graded and placed into early nursery culture units. Courtesy of Max Wingfield.

At some point following the transfer of spat collectors to field culture conditions, the juvenile pearl oysters they contain must be removed and transferred to other culture apparatus (Fig. 9.14). Larger juveniles may be selectively and sequentially removed from spat collectors over time with smaller spat retained on collectors. Alternatively, all spat may be removed from collectors at a certain age and graded. For example, when graded at 3-months of age, mesh sizes of 15 mm, 10 mm, and 5 mm retained 0.2%, 8.9%, and 67.3% of *P. margaritifera* spat, respectively. Grading allows pearl oyster spat to be held in culture units with the largest possible mesh size, supporting improved food availability.

Nursery culture

The well-established cultured pearl industries in Japan, Australia, and French Polynesia traditionally source their culture stock from the wild; either as adults or large juveniles from spat collection. In French Polynesia for example, *P. margaritifera* juveniles are usually held on spat collectors until they reach a size of around 60–90 mm, and in Australia, adult *P. maxima* are collected within a managed fishery, and may be used directly for pearl culture. Because of this, early nursery culture of pearl oysters, focusing on spat and younger juveniles, had, until relatively recently, received little research attention. But increasing emphasis on hatchery production of pearl oysters generated considerable research interest into the most appropriate methods for ocean-based nursery culture of pearl oysters. Nursery culture of pearl oysters is a key part of the production cycle as it represents a 'non-productive' culture period prior to pearl production. Maximizing growth rates of pearl oysters during nursery culture reduces the time required for them to reach a size appropriate for pearl production, and this has economic implications for pearl farms. Furthermore, uniform growth rates within a given cohort of pearl oysters ensures that a greater proportion of them reach operable size at a given time, thereby maximizing financial returns.

As described above, adult pearl oysters are commonly cultured on ropes, either singly or in pairs (Figs. 9.6 and 9.7), or in panel (pocket) nets (Figs. 9.3 and 9.9). The small size and relatively large numbers of pearl oyster juveniles that must be managed during nursery culture, however, provide logistical difficulties, and small juveniles, at least, are commonly grown in communal culture units such as pearl nets, lantern nets (Fig. 9.10) and baskets or trays. Some of the nursery culture and grow-out methods used for pearl oysters are shown in Table 9.6.

When housed communally within a culture unit, pearl oyster juveniles tend to aggregate to form clumps and the innermost individuals within the clumps may become stunted or die (Fig. 9.15). This phenomenon results in reduced growth rates, shell deformity, and a higher proportion of smaller oysters. A number of studies have shown that when aggregation is prevented by separation of individual oysters in a nursery culture unit, then these problems do not occur and culture units which hold individual oysters separately, such as panel (pocket) nets, generally support greater growth rates (Fig. 9.16). Many of the studies that have assessed the performance of different culture units for nursery culture of pearl oysters, however, have worked with *Pinctada* spp. In contrast, *Pteria penguin* was reported to show relatively poor growth rates when cultured in panel nets compared to communal culture in perforated trays or pearl nets, and three-dimensional culture units are generally recommended for juvenile culture of *Pteria penguin* in order to maximize water flow. It should be noted here that pearl culture occurs in many developing countries where access to desired culture equipment may be limited or prevented by cost. As well as growth and survival of pearl oysters, choice of culture unit under such circumstances may also consider economic and practical factors, such as equipment cost, ease of construction, degree of fouling (and therefore maintenance inputs) and the materials that are locally available.

Table 9.6 Some of the nursery culture methods used for pearl oysters.

Species	Culture unit	Oyster size (mm)	Findings
Pinctada mazatlanica	Lantern nets or pocket net suspended at a depth of 10 m Plastic cages attached to an iron platform at a depth of 10 m	32.6	• Growth did not vary significantly among treatments over 22 months. • Survival was better (99%) in cages than in suspended culture (65%).
Pteria sterna	Pocket nets compared to cages	37.8	• Significantly greater growth rates in pocket nets compared to cages over 18 months.
Pinctada margaritifera	Lantern nets, panel nets and plastic mesh trays in suspended culture at 5 m	16, 24, 33	• 16 mm juveniles showed significantly greater growth and survival in lantern nets than in panel nets over 3 months. • No difference in growth or survival of two larger size classes between culture units. • Oyster growth and wet weight were greater in mesh trays when oysters were fixed to prevent aggregation.
	Ear-hanging (chaplets)	58–78	• High survival (~87%, 12 months) and an annual growth rate of 60–64 mm. • Algae removal by brushing every 3-4 weeks significantly improved growth.
	Juvenile panel net (24-pocket), adult panel net (8-pocket), plastic mesh tray and 'ear' hanging; in suspended culture at 4 m	41.5	• Ear-hanging and 24-pocket (panel) nets were superior (5 months). • Small 5 mm mesh size in some culture units promoted rapid fouling.
Pteria penguin	Panel nets (1 cm^2 mesh), perforated plastic trays (1 cm^2 mesh) and pyramidal pearl nets (0.5 cm^2 mesh); suspended at 3 m and 6 m at an inshore and off-shore site	28.0	• Oyster survival did not differ significantly between culture units at either site over the 6-month study. • Trays and pearl nets supported the best growth rates at the off-shore site. • Trays supported the best oyster growth rates at the inshore site.

Stocking density

Stocking density is an important factor influencing pearl oyster performance during nursery culture. When 63-day-old (6.2 mm) *P. maxima* juveniles were cultured on PVC slats within mesh protectors at four densities (1.3, 6.7, 13.3, and 20 juveniles/100 cm^2), oyster growth and survival were significantly greater at the lowest stocking density, and growth rates were significantly smaller with each increase in stocking density above 1.3 juveniles/100 cm^2. Oysters held at higher stocking densities also showed greater rates of shell deformity which affected ≥25% of juveniles at the highest stocking density tested. This relationship between oyster performance and stocking density, during nursery culture, has also been shown for *Pteria sterna* (6.0–8.2 mm) cultured at four densities in perforated culture trays (25, 50, 75, and 100 individuals/tray); again, best

Figure 9.15 *Pteria penguin*, like other pearl oyster species, aggregates to form clumps if house in communal culture units resulting in a broad size range, shell deformity, stunting and mortality.

growth and survival were shown by oysters held at the lowest stocking density. The lowest possible stocking density should generally be chosen for nursery culture of pearl oysters but with consideration of economic viability.

Figure 9.16 *Pinctada maxima* juveniles being place individually into a panel net. Note the fine mesh of juvenile panel nets. These larger juveniles were removed from spat collectors that were transferred to the ocean from the hatchery two to three months earlier. Courtesy of Jens Knauer.

Control of fouling

Pearl oyster culture adopts suspended-culture methods, where both culture equipment and culture stock are held subtidally. As a result, culture equipment and culture stock are readily fouled by a wide array of fauna and flora. The most common fouling organisms at a pearl farm in West Paua, Indonesia, for example, were barnacles, other bivalves (*Pinctada* spp., *Pteria* spp., *Crassostrea* spp.), and polychaete worms. Removal of biofouling is a routine and labor-intensive component of pearl farming that has been estimated to make up 25–30% of pearl farm operating costs. Biofouling has a number of negative impacts on pearl oyster culture. Direct impacts on culture stock include reduced water flow through culture units potentially reducing food availability, reduced food availability because of direct competition from filter feeding organisms within the fouling community, reduced growth rates of pearl oysters as a result of food limitation, and an increase in shell deformities caused by increased competition for space within culture units, potentially influencing whether an oyster can be used for subsequent pearl production. As well as the direct impacts on culture stock, biofouling increases the weight and drag of suspended-culture equipment, increasing maintenance costs and the potential for

Figure 9.17 A team of young female workers pull pearl culture nets into small wooden boats and manually remove fouling from them using brushes and knives in in Li'an Bay, Hainan Island, China.

Figure 9.18 A cleaning boat servicing a longline on a *Pinctada margaritifera* farm in Fiji. The longline is raised out of the water by hooks on the boat providing access to the farm stock that are help on chaplets.

storm damage to the farm, and it also stimulates further fouling and potential recruitment of pearl oyster predators.

Although chemical-based (biocides) means are available to prevent biofouling, most pearl farmers focus on regular removal of biofouling from oysters and culture equipment instead. The methods used to do this including exposure to air, freshwater dips, hypersaline dips, high temperature, manual removal by hand, mechanical and high pressure water cleaning, and even the grazing activity of herbivorous animals such as rabbitfish. Probably the most common methods are hand-cleaning of oysters and culture equipment using bushes and knives and the use of high pressure water cleaning machines (Fig. 9.17). The degree of mechanization varies between farms and larger, better resourced, pearl farms may have dedicated cleaning boats which progress along a longline, bringing oysters onto the boat for cleaning as they go (Fig. 9.18). In Western Australia cleaning boats may winch attached culture units (panel nets) onto a dedicated cleaning platform on the boat, where they are cleaned using high pressure water jets. The frequency of cleaning varies between sites and according to local conditions. In West Papua, Indonesia, a monthly cleaning frequency was recommended for *P. maxima* because longer cleaning cycles (8 and 16 weeks) resulted in slower growth of oysters and an increased rate of shell deformity resulting from direct contact with biofouling organisms.

Pearl production

Cultured marine pearls can be divided into two major categories: (1) half-pearls, that are also known as blister pearls or mabé pearls; and (2) beaded or nucleated 'round' cultured pearls.

Half-pearl production

Cultured half-pearls are produced by adhesion of semi-spherical nuclei to the inner shell surface of a pearl oyster (Fig. 9.19). The oyster is pegged to hold the shells apart allowing the operator access to the inner

Figure 9.19 Mabé pearl production involves insertion of hemispherical nuclei onto the inner shell surface of *Pteria penguin*. It can be achieved by local people following appropriate training.

shell surfaces. Alternatively, anesthetics may be used to relax oysters prior to nucleus implantation; this reduces stress to the oyster and improves operator access for nucleus placement. After nucleus insertion, the oyster is cultured for a further 10–12 months to allow implanted nuclei to be covered with successive layers of nacre to form the pearl (Fig. 9.2). A minimum commercial nacre thickness of 0.25 mm is recommended for mabé pearls produced using *Pteria penguin*; however, the culture period required to achieve this varies according nucleus position within the shell, culture conditions and season. After harvest, the resulting half-pearls are cut from the shell for value-adding. It is usual for three to five nuclei to be implanted into a single oyster although the height (profile) of the nuclei used and number of nuclei implanted, as well as their positions, greatly influence pearl production and quality.

Mabé pearls are not as valuable as high-grade round pearls, but they are simpler and cheaper to produce, and they require around half the culture time of round pearls. Unlike round pearl production, where an oyster generally produces a single pearl, multiple mabé pearls are usually produced from a single oyster, and their collective value may exceed that of a single round pearl. Other advantages of mabé pearl culture, compared to round pearl production, are that mabé pearl farms can be established on a small scale, require much less capital in the establishment phase and, perhaps most importantly, mabé pearls can be produced by local people with appropriate training (Fig. 9.19). Not surprisingly, given the significant socio-economic and livelihood opportunities it can provide, mabé pearl culture is a growing livelihood activity in Pacific island countries.

Round pearl production

Round pearl production requires two individual oysters; a suitable 'donor' to provide the mantle graft tissue for the implanting procedure, and another recipient or 'host' oyster within which the pearl will be cultured. Donor oysters are selected on the basis of favorable nacre characteristics and coloration because the graft tissue removed from it will proliferate to form the 'pearl-sac' that will secrete the nacre for pearl formation. Selection of appropriate donor phenotypes has been shown to be a key factor influencing the quality of pearls produced by *Pinctada margaritifera*.

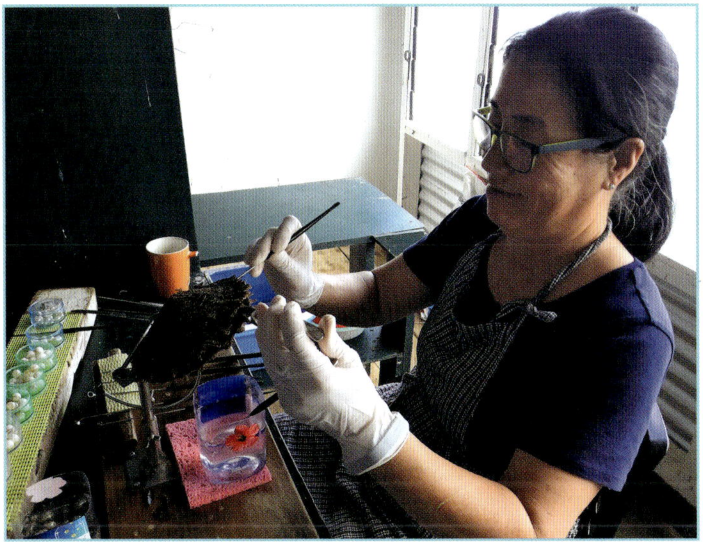

Figure 9.20 A pearl grafting technician nucleates a *Pinctada margaritifera* for round pearl production. Note the various sizes of nuclei (in containers on the left side of the photo) that can be chosen by the technician depending on oyster condition. Courtesy of Claude Prevost, Civa (Fiji) Pearls Pte Ltd.

Host oysters must be able to accommodate the largest possible implanted nucleus (within its gonad) and have reduced ability to eject the nucleus. To facilitate this, potential host oysters may be conditioned prior to the implanting procedure, to stimulate gamete resorption, and to weaken the oyster to minimize the potential for nucleus rejection. Conditioning can be achieved by overcrowding oysters within a culture unit or restricting water flow to reduce food availability, by increasing culture depth to reduce food availability, or by moving oysters to warmer waters in shallower depths to induce spawning.

An appropriate section of mantle tissue is removed from the donor oyster, trimmed, and sectioned into small (approximately 3 × 3 mm) pieces that will form single grafts, called 'saibo' in Japanese. Each piece of saibo is implanted with a spherical nucleus into the gonad of a host oyster. This procedure, called 'grafting,' 'seeding,' 'implanting,' or 'nucleation,' is highly skilled and the experience and skill of pearl grafting technician(s) is directly related to the quality and value of the resulting pearl crop (Fig. 9.20). After the nucleus is inserted, implanted oysters are carefully nurtured for at least two weeks to reduces stress and minimize rejection of the nucleus after implantation.

Culture of nucleated pearl oysters

Special care of nucleated oysters is required to maximize pearl yield (by minimizing nucleus rejection) and pearl quality. Nucleated oysters are held in calm conditions with appropriate and constant water quality parameters, and care is taken to minimize stress. In Australia, nucleated *P. maxima* are placed in a holding area for 7–8 days to facilitate recovery; in Japan, nucleated *P. fucata* are held in deep, calm waters for 2–3 weeks to allow recuperation; and in China, nucleated oysters are held in a resting zone for at least 2 weeks before being returned to an area of calm ocean water at a depth of 2–3 m. It is common for pearl farmers to assess the rate of nucleus retention among nucleated oysters. This may be done effectively by placing nucleated oysters into fine mesh bags during the recuperation period (Fig. 9.21). Rejected nuclei are retained within the bag identifying oysters that have rejected the graft. Nucleus retention can also be checked using X-ray. Survival of implanted oysters and nucleus retention are influenced by factors such as size and age of the oysters, size of the nucleus, grafting method, season and environmental conditions at the culture site. Season was reported to have a highly significant effect on nucleus rejection from *P. margaritifera* in French Polynesia, with nucleus rejection rate being significantly greater in spring than in autumn.

Culture of nucleated *P. maxima* in Western Australia may also include a 'turning' program which is thought to assist pearl-sac development, improve nucleus retention, and increase the likelihood of obtaining spherical pearls. Oysters are held in panel nets and laid horizontally for the first few days after nucleation; then turned to rest on the opposing shell valve at 2 to 5 day intervals. The turning frequency decreases over time and the turning program lasts for 6–10 weeks when nucleated oysters join the main farm stock.

Figure 9.21 A nucleated *Pinctada margaritifera* is placed into a fine mesh bag then attached to a rope before transfer back to the ocean. Should the implanted nucleus be rejected, this will be immediately evident. Courtesy of John Lucas.

Pearl production and harvest

Graft saibo tissue proliferates to create a 'pearl-sac' that completely covers the implanted nucleus. This occurs within 14 days of grafting in *P. margaritifera* but requires up to 30 days in *P. maxima*; periods between 30 days and 65 days have been reported for pearl-sac formation in *P. fucata*. Water temperature has a major influence on the rate of pearl-sac formation. The first layers of nacre are secreted onto the nucleus by epithelial cells within the pearl-sac around 32 days after grafting in *P. margaritifera* and 45 days after grafting in *P. maxima* cultured at 25–30°C.

Nucleated oysters are generally cultured for a further 1 to 2 years before a commercially acceptable nacre thickness of approximately 0.35–0.8 mm is achieved, and the resulting pearl may be harvested. Cultured marine pearls are generally harvested in the cooler months when nacre secretion rate slows to produce a more lustrous pearl surface. An incision is made in the pearl-sac and the pearl removed. If the harvested pearl is of high quality, a new larger nucleus may be placed into the existing pearl-sac, and the oyster returned to the water for another 1 to 2 years to produce a second pearl. This technique is called 're-seed' or 'surgreffe' and it may be used two to three times with the same oyster assuming pearl quality is maintained. Compared to first harvest pearls, improvements in both pearl size and shape have been reported in second-graft pearls produced by *P. margaritifera*.

Pearl grading

Pearl oysters are farmed for cultured pearl production. This chapter would therefore be incomplete without a brief overview of the factors, or 'five virtues' that are used to determine the quality, and therefore value, of cultured pearls; size, surface, shape, color and luster. For pearls, size does matter, and generally the larger a pearl, the greater its value. The largest cultured round pearls are produced by *P. maxima* and they range from around 8 mm to more than 20 mm in size. Of the major culture species, the Akoya pearl oyster, *P. fucata martensii*, produces the smallest cultured pearls of around 6–11 mm. Pearls are natural objects and subject to surface imperfections such as spots, wrinkles, and grooves. Generally, the fewer surface imperfections a pearl has, the higher its value. Cultured pearls vary in shape from perfectly round, to drop and irregular 'baroque' shapes. Shape does not affect the quality of a pearl, but it does influence pearl value because particular shapes have higher demand. Perfectly spherical pearls are rare and generally command higher prices. Pearl color is often species-specific and reflects differences in shell nacre color between species. The colors of pearls produced by *P. maxima* include gold, silver, white, silver-rose, and white-rose, while the natural colors of Akoya pearls include white, cream, and rose. The broadest range of colors within cultured marine pearls is seen in those produced by *P. margaritifera* (Fig. 9.1). Rare colors, including those in the green and blue spectra, may command particularly high prices. The last of the five virtues is luster, which assesses shine, reflection and luminescence of a pearl's surface. Pearls with high luster have much greater value than less-lustrous pearls with comparatively dull surfaces. Pearl luster is influenced by nacre thickness, with finer, thinner nacre on the surface of a pearl generating improved luster. Finer, thinner nacre secretion occurs at cooler water temperatures and, for this reason, pearls are usually harvested during cooler months.

References cited and further reading

Gervis, M. and Sims, N.A. (1992) The biology and culture of pearl oysters (Bivalvia Pteriidae) International Center for Living Aquatic Resources Management (ICLARM) Studies and Reviews, Vol. 21. Overseas Development Administration, London.

Rose, R.A. (1990) *A Manual for the Artificial Propagation of the Silverlip or Goldlip Pearl Oyster,* Pinctada maxima *(Jameson) from Western Australia.* Fisheries Department, Perth.

Southgate, P.C. (2008) Pearl oyster culture. In: Southgate, P.C. and Lucas, J.S. (eds.) *The Pearl Oyster.* Elsevier, Oxford, pp. 231–272.

Southgate, P.C. and Lucas, J.S. (eds.) (2008) *The Pearl Oyster.* Elsevier, Oxford.

Strack, E. (2006) *Pearls.* Rühle-Diebener-Verlag, Ludwigsburg.

Zhu, C., Southgate, P.C., and Li, T. (2019) Production of pearls. In: Smaal A., Ferreira J., Grant J., Petersen J., and Strand Ø. (eds.) *Goods and Services of Marine Bivalves.* Springer, Cham, pp. 73–93.

CHAPTER 10

Site selection for molluscan shellfish aquaculture

Lynne Falconer and Trevor C. Telfer

Introduction

Site selection is one of the most important decisions that any aquaculture producer will make. For molluscan shellfish, the conditions and features of a site influence almost every aspect of culture so it is vital that a suitable site is selected during the planning stage of a development. There are many factors to consider so site selection can be a challenge. For molluscs, the criteria that determine a suitable site will vary depending on the species being cultured and the system used for production. Farmed molluscs include bivalves, gastropods, and cephalopods, and these classes vary considerably, from their fundamental physiology to the farming practices required. These species-specific factors must be considered when selecting a site.

Bivalves, which include clams, oysters, mussels, and scallops are the most farmed mollusc and are key contributors to global aquaculture production. The group is characterized by two shells that are connected via a hinge, protecting the soft invertebrate body within. As filter-feeders, they are dependent on food from the environment and are a form of extensive aquaculture as no supplementary feed is provided, which has important implications for site selection. For bivalves, production techniques include Bouchot culture, longlines, rafts, and bottom-culture. There are advantages and disadvantages for each culture system, and some work better for certain bivalves than others.

Aquatic gastropods are cultured in many parts of the world, with abalone, conch, and trochus as the most commonly farmed. Gastropods are prized for both their meat, but also their shells which are popular with collectors and tourists in many locations. They are mostly grown in cages in the marine environment or onshore in tanks or raceways. For cage systems, there may be a need to provide additional feed in addition to the natural food supply. For tank systems, the feed must be provided by the aquaculture producer.

Only a few species of cephalopods are cultured at present and this is a small-scale industry, largely still in the research and development phase, with further work needed to address many of the knowledge gaps associated with production. Furthermore, though there is high market demand for cephalopods, there are a number of ethical concerns about the culture of these animals as cephalopods are complex organisms, with high levels of intelligence, large brains, and well-developed nervous systems.

This chapter focuses on bivalves and gastropods and overarching themes for mollusc culture.

Site selection process

Many of the undesirable issues associated with aquaculture have been as a result of poor planning or inadequate site selection. At the farm level, site selection is fundamental to the success of a farm while region-wide planning should be implemented to minimize cumulative impacts and avoid conflict with other resource users, activities, and protected areas. Prior to establishing a site, it is therefore essential to determine if a location is available for development, it has the ecosystem services necessary to support aquaculture, and to assess the overall suitability for the chosen species.

It is important to note that within governance of aquaculture, site selection is usually the first step in the planning process and once a site has been identified the aquaculture producer will then have to embark on the more formal licensing process to obtain the relevant permits and permissions to establish a farm (see 'Licensing and regulation', below).

Site availability

Not all sites will be available for aquaculture as there may be permanent, long-term or short-term restrictions that prevent the establishment of a farm. These restrictions are known as constraints and should be identified as part of the site selection process. The restrictions may be formal and outlined in policies and regulations for a county or region. This is particularly the case for protected areas and locations where there are navigational issues or high volumes of marine traffic. There will also be constraints that are not supported by policy or regulation and these are sometimes more difficult to recognize and identify. If assessing a new area for development, then a prospective aquaculture producer may find it useful to speak to local stakeholders and representatives from the planning authority to get an insight into the location and the activities and restrictions that are in place.

A site may be unavailable because it is already in use for other activities. Coastal locations are under a lot of pressure from many different users and there is often conflict over space. The constraints may be permanent structures that physically prevent a system being installed. There may also be temporary activities, such as fisheries or tourism, which can be seasonal, but would still prohibit establishment of a farm in an area as the production cycle can overlap with the time used for the other activity. Furthermore, ownership of the land or water-rights can also restrict potential areas for development.

In some cases, there may be the opportunity for co-location of activities and there has been some research into multiple use of space, for example co-location of mussel culture with offshore windfarms. Most such studies have focused on the theoretical concept or pilot scale assessment so there is a lack of research on feasibility at a commercial scale. One of the key challenges is finding available locations that are suitable for both activities, as areas that are used for offshore wind may not necessarily be appropriate for mussel culture and vice versa. Furthermore, there are logistical issues that need to be overcome as each activity will access and use the space in a different way.

An area may be suitable for a species but unavailable for a specific technology or system as the constraint may be the physical conditions of the environment. For example, a coastal location may be too exposed to wind and waves, which would prevent installation of trestles, or the area may not be deep enough to deploy longlines. For land-based culture of gastropods the limiting factor may be water availability, restricting the availability of areas for development.

Some jurisdictions may use aquaculture zones to coordinate development within specific areas. Aquaculture zones are areas where aquaculture development is prioritized over other activities and are a useful way of allocating space within multi-user environments. Zones can play a key role in marine spatial planning (see 'Marine spatial planning', below).

Site suitability

The suitability of a site will depend on a number of factors including the biological and physiological requirements of the species, the physical characteristics of the production system and accessibility and associated transport infrastructure. There may be trade-offs between the different factors when selecting a site to establish a farm, however, there will be some fundamental conditions that are essential for production and cannot be compromised.

Biological and physiological requirements of the species

Environmental conditions must be within the tolerance ranges of the particular species to support growth and good health and welfare. The actual ranges and threshold values will depend on the species, but key physicochemical parameters that influence production are temperature, salinity, oxygen, and pH levels. The site selection process will involve assessing existing data or collecting new data to identify the environmental conditions at the site and determine if they are suitable for the species. For intertidal, coastal, and open ocean conditions, the molluscs are directly exposed to the environment, whereas in land-based tank systems producers have much more control over the environment if required.

Food availability is essential in determining the overall suitability of a site for the culture of bivalve molluscs. As filter-feeders, bivalves rely on the environment for food so must be located in an area where they will be exposed to a sufficient source of food. Phytoplankton is the primary food source, but zooplankton and detritus are also dietary components, especially in seasons when phytoplankton levels are low. During the site selection process, it is important to identify locations with high levels of food availability. Furthermore, depending on the production system there may be a reliance on natural spat, making it important to identify locations that can support spat collection.

Physical characteristics of the production system

For intertidal, coastal, and open ocean systems, physical characteristics of the environment that will affect the suitability of the area for different production systems include depth and hydrodynamics. Suspended culture methods, such as longlines, require sufficient depth and space to be installed. They also must be located in an area where current speeds allow sufficient flushing and access to food. Recently, there has been increased interest in moving bivalve production further from the coast to more offshore locations. Moving to new areas where production has previously not been established will require more thorough investigation and perhaps pilot-scale trials to assess production potential and suitability of the technology.

Physical structures in the intertidal, coastal, and marine environment can have navigation conflicts for other users. When identifying a potential site, it is important to assess how existing users are actively occupying and travelling within the area and attention should be paid to this in site design and demarcation. Multiple users and activities can co-exist, but there are some measures that may be required to ensure minimal risk, such as clear identification of where farm structures are and consultation with the relevant navigational authorities to ensure the site is in an appropriate location.

For land-based systems, the key physical characteristics will be the suitability of the terrain for construction of the facilities. Most land-based facilities will be tanks or raceways, therefore there is less focus on the soil characteristics than there would be for earthen pond culture. The key issue will be the water supply, both in terms of water availability and water quality. For sites located in coastal areas, using water pumped ashore, it is important to identify where the extraction will take place and what filtration and control measures are required prior to entering the tanks. The amount of water required will depend on the scale of operations and the setup of the tank system, notably whether it is flow-through, partial recirculating or fully recirculating systems.

Accessibility and infrastructure

Accessibility and the associated infrastructure that enables producers to travel to and work at a site, as well as allowing the transport of the products, are key considerations for site selection. For coastal systems such as longlines there is a need to ensure the site can be accessed by boat and the land-based infrastructure is available to support that. Depending on the scale of the operation there may be a need for large boats with winches and specialized technical equipment, so the site must be able to accommodate this. For systems in the intertidal zone land-based vehicles and machinery such as tractors and trucks will be required for installing and harvesting the stock so the site must be accessible, albeit off-road conditions will prevail. Land-based farms, such as the tank facilities used for gastropods, will usually be located in areas connected to a suitable road network.

Wider transport infrastructure and road networks are also important as this affects not only access to a site, but also the potential routes to further processing, seafood buyers, or direct access to market. Access to depuration facilities may be an important consideration, especially for small-scale producers who do not have the resources available to develop their own systems. For fresh produce there will be a desire to have good transport links to distribution points to ensure the best possible product arrives with the buyer.

Ease of access may be a disadvantage in some areas if poaching or theft is a concern. In these circumstances it may be desirable to locate bivalve farms in more remote or hard to access locations. Surveillance systems can be installed to monitor the stock and act as a deterrent. In some jurisdictions, bivalve poachers have been prosecuted based on surveillance footage. Theft is also an issue for gastropods that are grown in tank facilities and abalone farms in particular have been subject to armed robberies due to the high value of the species. Therefore, at these sites there may be a need to install security systems and hire guards to protect the stock.

Another consideration is the risk posed by predators and the ease of access they may have. This is particularly important for systems such as longlines which are open to the environment and easy to access. Effects of benthic predators, such as crabs, can be reduced by using off-bottom culture, but pelagic predators are hard to avoid. Mesh bags and nets can be used, but some predators such as otters may still be able to gain access. Even for land-based systems there is a need to consider if predators may be an issue at a site and perhaps there is a need to implement physical control measures such as nets. Knowledge of the animals and birds that occupy the area of a proposed site is important as this can be used to assess potential predation problems.

Carrying capacity

Concepts of carrying capacity

The concepts of carrying capacity are complex and can be confusing but are important when considering the sustainable use of aquatic resources for shellfish production and also in selecting suitable sites. The general definition of carrying capacity for natural biological systems, such as shellfish culture that obtains its food from the environment, relates to a population size that is naturally constrained by using resources in equilibrium with the surrounding environment. When using this to manage sustainable exploitation of a shellfish, ideally the carrying capacity should not be achieved but only approached within pre-defined limits.

In the case of aquaculture, the concept of carrying capacity can be divided into four categories; physical, production, ecological, and social. These are discussed in detail in Smaal et al. (2019). These are often considered independently and consequently there are often trade-offs between the different categories, where in some locations one category will be considered more important than another.

Physical and production carrying capacities are generally applied at a local scale for a particular site or location. Ecological and social carrying capacities are often applied at a larger, regional scale to ensure overall site suitability and sustainability in a wider context. Ecological carrying capacity considers the Ecological Approach to Aquaculture, as promoted by the Food and Agriculture Organization of the United Nations (FAO), and is used as a basis for wider environment management and sustainability of aquaculture development. Ecological and social carrying capacities are based on acceptability of environmental change or social impacts, and so depend on social values and defined environmental quality standards for the particular regions or locations being considered for culture. Often consideration of the categories of carrying capacity is in stages with consideration of the site first, physical, and then the production, environment and social factors (Fig. 10.1).

There is often feedback from social carrying capacity to ecological as environmental variables of interest are largely defined by society, as an iterative process, before determining the ecological carrying capacity. Assessment of carrying capacity and interactions between the different categories can be difficult and complex involving environmental measurement, ecosystem and production modelling, questionnaires, and interviews. To translate this into something meaningful, robust indicators and approaches to measurement are required.

Physical carrying capacity can be considered the initial identification of sites, taking into account the environment and physical features required for the production system. In other words, the question is asked "Can the location physically accommodate and sustain the type of aquaculture development being suggested?" Variables of interest include bathymetry, currents,

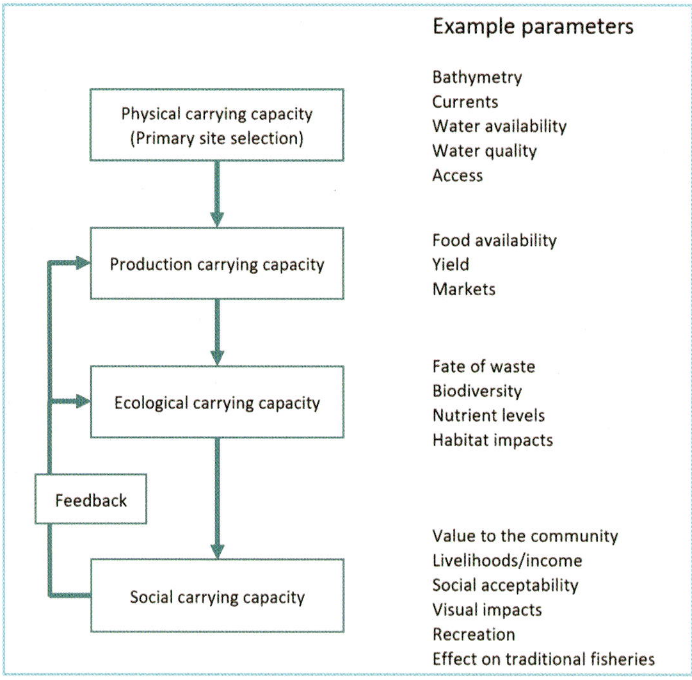

Figure 10.1 Hierarchical structure to determine carrying capacity of a given area for shellfish aquaculture. Re-drawn and modified from McKindsey (2006).

waves, and water quality, but other site suitability factors are taken into account. This often has a number of indicators and approaches which can assess the suitability of locations. These can be combined to aid in decision making at this stage using spatial modelling with GIS-based methods (see 'Geographic Information Systems (GIS) for site selection').

Production carrying capacity is the maximum aquaculture output that can be generated in a selected location. It is based on the potential intensity of production or yield and is normally evaluated at farm level but may be assessed over a wider area or region. For bivalves, this will often depend on the availability of natural resources (normally for food, as assessed for physical capacity) and how their utilization is related to ecosystem function. Modelling approaches to define these functions in relation to capacity have been widely used but tend to focus on hydrodynamics, food availability/production, feeding physiology, and husbandry. Though the approaches are consistent and widely used they are limited through their restricted ability to determine ecosystem responses to their activity and account for seed collection or harvesting of shellfish. Production is also limited by economic aspects such as investment potential and market value, which may act as limiting factors. Though important to shellfish producers, production capacity can be used by environmental regulators and decision makers to set limits on production for a given area to ensure that there is no excessive stress on natural resource for other users of the environment. It can ensure sustainability of shellfish production over the longer term, and be instrumental in licensing decisions.

Ecological carrying capacity is the magnitude of aquaculture production that can be achieved without changes in ecological processes, species, populations, and communities in the environment. Ecological carrying capacity may have adverse connotations for some forms of aquaculture development and is often linked to the impacts on the immediate environment through waste or chemotherapeutant release. Shellfish aquaculture is usually considered to have limited direct impacts on the local environment due to release of wastes, and those that occur are mainly through fecal and pseudofecal materials settling to the seabed. Over-production of shellfish in any location may deplete the plankton community and affect the lower trophic levels of the food web, with consequent knock-on effects throughout the ecosystem (see Shumway, 2011).

A key concern for any aquaculture development is the impact on protected areas or sensitive species. Prior to establishing a site, it is important to identify if there are potential species or habitats that may be affected by the development. In bivalve culture, for example, there are concerns that if not planned and managed appropriately there can be impacts on seabed communities and priority marine features that are important in local and regional conservation. If this is a concern for a particular site, then that area may not be suitable for development.

Social carrying capacity is the level of aquaculture that can be developed without adverse social impacts. This includes consideration of what is acceptable to the community and wider public. The social dimensions of mollusc aquaculture are complex, and social carrying capacity, which may also be referred to as social license, will vary depending on the area and the community in which it is practiced. Furthermore, within a community there may be individuals or groups with differing opinions and perceptions of mollusc farming and the acceptability of the activity. During site selection it is important that aquaculture producers recognize potential concerns from the local community as this could affect the sustainability of the operation or could prevent the development from occurring in the first place.

Conflict with other resource users and activities will affect social carrying capacity. If space and resources are limited, then there could be competition with individuals or groups over access and use of an area. It is important to have a constructive dialogue with the community and other activities to ensure optimal use of the location. In the planning stage, potential conflicts can be identified through stakeholder engagement and this may influence site selection. Mollusc producers will have their fundamental requirements for establishing a site but there may be other factors which have some leeway. For example, to avoid conflict with recreational boat users, the producer may select a site further from a jetty than originally planned. Such issues will vary between sites and communities; engaging with the local community at an early stage is key to improve social acceptance.

Loss of visual amenity is often a concern for coastal communities, particularly those that rely heavily on tourism. This can lead to objections in the planning process which may delay or prevent a development. As part of the planning process, producers can identify sites that are less visible and are not located in areas of outstanding beauty. Other mitigation factors may be selecting specific types of material for the farm system, so they are visible to other water users but perhaps not as noticeable from viewpoints on land. Some jurisdictions will have guidelines or regulations in place concerning impacts on visual amenity from developments. It is also important to note that it is not only the farm structures that is an issue for communities, but also waste and discarded items. Therefore, it is important that producers tidy up any waste items that are collected at the farm or along the shoreline such as ropes or buoys and store them in appropriate locations.

Mollusc producers may also wish to engage with and build relationships within the area they operate. Some companies embrace their position in a community and sponsor local events or host visits and tours at the farm site. Although this involves some effort, the rewards may be beneficial as there can be improved public perception as well as increased marketing opportunities and brand or company recognition. Aquaculture companies can work with local businesses, such as restaurants and bars, to promote the locally grown produce that is often highly attractive to tourists visiting an area. This means there are mutual benefits for the mollusc producers and community. Working with schools, producers can also provide tours and information for educational purposes and demonstrate how the industry operates and contributes to food production. Such activities may increase acceptability of the farm in the community and also more widely.

Some of the most important societal benefits from mollusc aquaculture are its opportunities for employment and income generation. The wider benefits of this will depend on the scale of the farm as some operations may be single-handed and only require seasonal support for seeding or harvesting. Larger operations may have a larger workforce, and this may be amplified if there are processing facilities as well. For site selection the labor force may also be a key consideration as aquaculture producers may need to locate in an area where there is a supply of people that could be employed either permanently or temporarily.

Application of carrying capacities to shellfish aquaculture

The different categories of carrying capacity have different implications in different regions around the world. These are often considered as trade-offs for site selection and management. The major consideration for setting up of aquaculture in one region, globally, is not necessarily the same in another. For example, for the US, Europe, and Canada the greatest limiting factor is site availability and production, whereas for Southeast Asia and China the greatest limitation is social and livelihoods considerations (Fig. 10.2).

Carrying capacity and application of the categories are an essential part of site selection for shellfish aquaculture. How this is applied and assessed depends on a number of factors, such as system, scale, and the location, which can be used to define the trade-offs in models for site selection and subsequent management. Assignment of carrying capacities is often achieved through modelling of indicators, particularly for initial site suitability assessment (physical capacity) and the impacts and levels of production available for a given location (ecological and production capacity, respectively). This is discussed in different sections of this chapter.

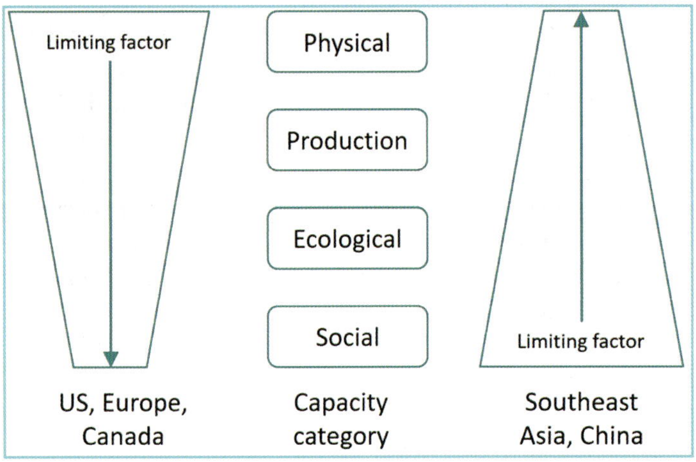

Figure 10.2 The relative importance of the four categories of carrying capacity. The narrowest segment for the regional areas corresponds to the greatest constraints on carrying capacity. Re-drawn and modified from Byron and Costa-Pierce (2013).

Decision support

Site selection involves consideration of many different factors and often there will need to be trade-offs between multiple components and concerns. Decisions may be subjective and based on value judgements as different stakeholders will have varying priorities. This presents a challenge for mollusc producers as well as the relevant planning authorities and other resource users. To help make a more informed and objective choice, decision support tools can be used. Decision support is any guidance, procedure, or analytical model that helps the decision-making process, usually allowing easier interpretation, simplification, and/or communication of complex information.

A key element of decision support is that it helps the decision process and does not make it more complicated. Sometimes there is no need for complex computer-based modelling approaches and written guidance may be enough. In other cases, computer-based modelling approaches will be required to support site selection. Models take many different forms; approaches are diverse and structures vary. There is often more than one way to develop a model for a particular area or species. When selecting an approach, it is important to consider the resources that are available, and the questions being asked. In some cases, a relatively simplistic approach with minimal data requirements will be sufficient but other occasions will require more complex and data-intensive modelling approaches. A key advantage of using models for decision support is that they allow simulation of past, present, future, and alternative scenarios, and they can provide information that would otherwise be difficult to obtain.

Written guidance

Regulators and planning authorities can support aquaculture development by providing guidance on the planning process and information required to select a site and obtain the relevant licenses or permits to establish a farm. Aquaculture producers benefit from this as they know the information that is required as part of a licensing process, allowing them to conduct a thorough assessment of all relevant parameters and prepare a full application with all the necessary detail. This then benefits the authorities by reducing the administrative burden of processing incomplete applications or spending time on applications for sites that would not be appropriate for development. See 'Planning process' section for examples of the planning process for shellfish licensing and regulation.

To be effective, guidance should be written in clear and unambiguous language. The decision-making process should be outlined clearly, and advice should be given at the level of detail and information required to support an application. Visual aids and diagrams may also be useful to explain the process. If there are different requirements depending on a particular species or system, then flowcharts that follow a structured and logical process can guide a user to an answer. For example, a planning authority can prepare a flowchart that can be used to help producers identify if they would need to perform an environmental impact assessment (EIA) as part of the license application. Another way of guiding site selection is to have a checklist that can be used by producers to identify potential issues that may need to be investigated further. A simple yes/no checklist can allow for a quick assessment and determine if it is worth investing time and resources in more detailed analysis (Table 10.1).

Table 10.1 Example of a yes/no checklist that could be used to help mollusc producers identify suitable sites.

Potential issue	Yes	No	Comments
Is the site available for development?	X		n/a
Is aquaculture allowed in the area?	X		n/a
Are there potential conflicts?	X		Fishermen, sailing club, hotel overlooking the bay
Suitable depth?	X		
Is the site sheltered?		X	The site is exposed to the west, need to establish if there is a risk of waves and wind.
Is the site easily accessible?	X		

Geographic information systems (GIS) for site selection

For site selection, models developed using geographic information systems (GIS) are powerful way of assessing the availability and suitability of areas for aquaculture. GIS is a combination of computer hardware, software, organizational infrastructure and data that is used to interpret, analyze, and/or model geographic data to gain further understanding of spatial issues and support decision making. GIS-based models have been used to identify potential sites for aquaculture species, including molluscs, for many years and the potential applications of GIS are increasing each year as more data becomes available and technological capabilities improve.

There are several stages in development of a GIS-based site selection model (Fig. 10.3). First there is a need to define the scope and scale of the assessment. This is then used to identify the data requirements based on existing knowledge and stakeholder opinion. Once identified, there is a need to search for existing datasets that may be suitable for use. If data are not available, or not of sufficient quality, then sampling campaigns must be used to generate new data for use in the assessment (see 'Data requirements' section). Once data are available, they must be processed into an appropriate format for the GIS model. If data are not already in a spatial format, then this will require geo-referencing and development of spatial layers. Other procedures such as cropping to the desired study area may also be required. Spatial analysis may be required to transform data, for example calculating distance from a key feature or estimating the costs to travel along a specific transport network. Once all the spatial layers have been prepared, they are then combined within a site selection model.

For GIS-based models, the most popular site selection approach is the use of multi-criteria evaluation (MCE). Within this methodology, data layers are reclassified to a common suitability scoring system which represents different categories of suitability (for example, highly unsuitable, unsuitable, suitable, or highly suitable) or a scale of suitability (often from 0 to 1, where 1 is most suitable). Each layer is then assigned a weight based on its relative importance. Weights can be established based on experience or through consultation with stakeholders and focus groups. For the model, one single output may not be enough, and a series of scenarios may be explored. For example, in some cases biological performance may be prioritized, whereas in others it may be environmental impact.

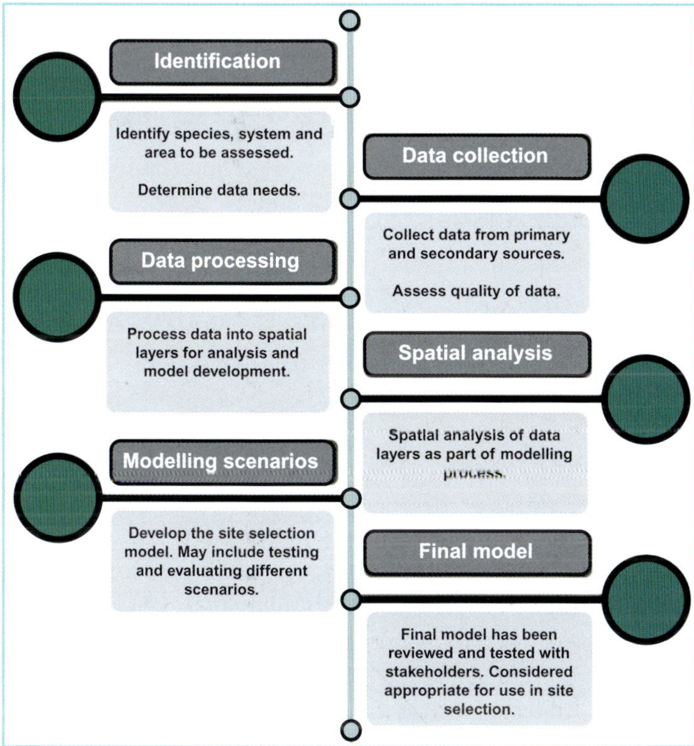

Figure 10.3 Process of developing a GIS-based site selection model.

GIS model outputs can be presented in several different ways. They can be used in reports and documents, or they can be hosted on web-based servers that allow end-users to explore the output. When presenting the model outputs, it is also important to show the model structure in a clear and transparent manner to allow end-users to understand how the model outputs were developed. Furthermore, if the models are hosted online then it is important to maintain and update the website to ensure accessibility.

Modelling potential production and ecosystem interactions

Biological models can be used to predict the growth of molluscs at a particular site based on the characteristics of the species and local environmental conditions. This allows assessment of different locations and producers can identify areas which have more desirable characteristics such as faster growth. Growth models are commonly used for bivalves and require data on temperature and food availability to simulate physiological processes and overall growth of an individual animal. If required, the models can be scaled up to population level, which allows assessment of overall production potential. A key advantage of investigating the potential of a site using growth models is the ability to simulate different production strategies and assess the costs and benefits of choosing one approach over another.

As bivalves are filter-feeders an understanding of the potential ecosystem interactions may be useful when selecting a suitable site for production. Models can be used to simulate the food web and the dynamics between the different species in an area. This allows assessment of ecological carrying capacity and producers and regulators can then evaluate the potential impacts of establishing a new site within that area. Such an approach is useful for areas where there may already be a number of existing bivalve farms and concerns over cumulative effects or in areas where there may be sensitive species.

Data requirements

The effort, resources, and time required to collect and process data to run a model is often underestimated. This can be a substantial task and it can take longer than the actual model development. The amount and type of data required will depend on the model being used. In some cases, it will be data availability that determines the choice of the model in the first place.

There is an increasing number of online data portals that are easy to access and provide regularly updated datasets. These may be specific to an area or jurisdiction or they may be international. Examples include EMODnet (http://www.emodnet.eu/) and Copernicus Marie Environment Monitoring Service (http://marine.copernicus.eu/). These are very useful data sources and have an important role to play in aquaculture planning and management. The scale and coverage of data provided in such portals can be used to identify areas of interest for more detailed analysis and fieldwork. Fieldwork and sampling campaigns are used to generate new in-situ data or for validation purposes.

Data storage limitations may restrict the size of databases. As computer capabilities and technology has improved in recent years this has become less of an issue for datasets collected through fieldwork or monitoring programs. But it could be an issue if using data from earth observation or large complex hydrodynamic and biogeochemical models which may need more specialized computing systems and large data storage facilities.

Regardless of the data source, it is essential to check if the data is fit-for-purpose as the quality can vary considerably and there may be errors which would be propagated in any modelling exercise. Furthermore, the resolution and frequency of data points may not be appropriate for decision-making purposes. For example, yearly averages can miss the variable conditions of interest so more frequent data points may be necessary. The importance of assessing the quality of input data should not be ignored as this determines the final output of the analysis and modelling. If unsuitable or erroneous data are used, then this could result in misidentification of suitable sites.

There is often a trade-off between data that should be included and data that can be included. Some of this will be due to data availability and data quality, but also the time and resources available to perform the analysis. It may be desirable to perform extensive field campaigns and develop highly complex and sophisticated models, but if there are restrictions on time and resources then there may need to be concessions and perhaps use more simple approaches that are less data intensive. In other cases, simple approaches may not capture the complexities of a system, therefore it may be necessary to collect new data and use more comprehensive modelling approaches. The advantages and disadvantages of different approaches may have to be considered on a case by case basis and will depend on the answers being sought and purpose of the analysis.

Licensing and regulation

There are two aspects in which the location of molluscan shellfish culture is regulated; their interactions with other users and the surrounding environment, and their food safety. Both are important in considering site selection and suitability of culture sites.

Planning process

Usually mollusc producers will select their own sites, but they should allow themselves to be guided by the planning process, regulations, and licensing requirements. A healthy environment is key to a productive aquaculture sector. Aquaculture must be planned and managed appropriately to ensure that negative impacts are minimized while benefits are optimized, avoiding conflict with other users and maintaining biodiversity. Most countries have some method of licensing and consenting aquaculture that incorporates national and international law. The number and nature of licenses required, and time taken to get through the planning process can vary between countries and sometimes even within countries, as different regions may have different requirements. The licensing process can be a bottleneck for aquaculture development. A good example of the planning and licensing process for shellfish aquaculture developments, and interactions between developer, authorities, and consultees, comes from Scotland (Fig. 10.4).

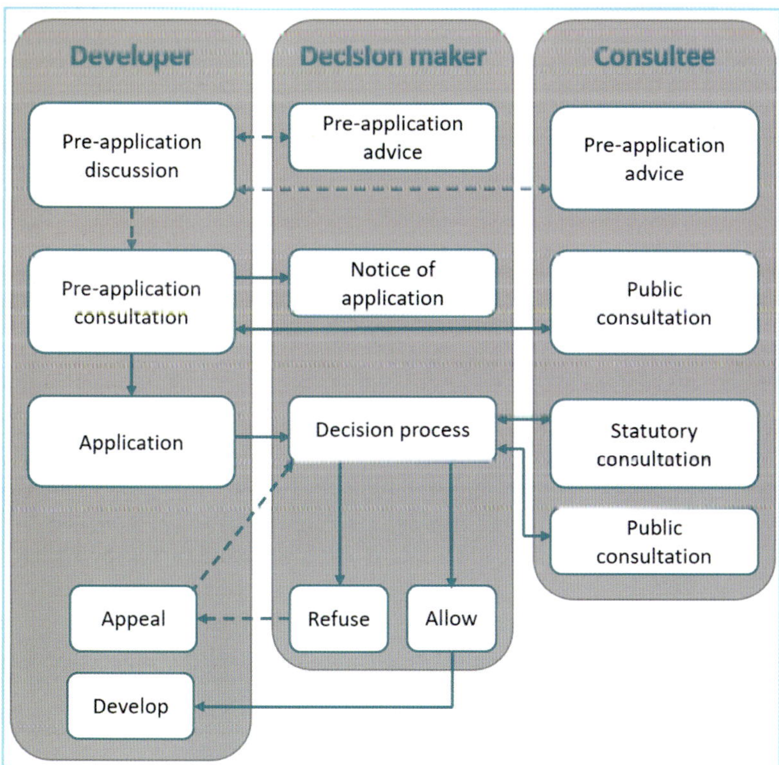

Figure 10.4 Flow chart of an example of planning application stages and determination process for shellfish aquaculture developments in Scotland, UK. Dotted arrows are optional pathways, solid arrows are required pathways.

The process has four stages from pre-application to appeal and involve the developer, the planning authority, and statutory and public consultees. This is enacted in order to set up a new site for mollusc culture along the Scottish coast. Ultimately the decision is made by the local planning authority (district council) under guidance from the consultees.

The first stage in the process is a pre-application consultation where a report is produced with a description of the development and a location plan. Evidence must be provided of consultation with the relevant community council and at least one public event. The developer must also publish notice of the proposed development in a local newspaper. Part of the planning determination process involves compliance with national and local policies such as the Scottish Planning Policy, the National Marine Plan, and any local marine spatial plans or local development plans (see 'Marine spatial planning' section for further information). Proposals are also assessed by the statutory consultees against planning criteria shown in Table 10.2. All criteria are considered equally important.

The criteria are assessed by the consultees in order to advise the local planning authority, however, there is no requirement in Scotland for a formal EIA to be completed as part of the planning application.

Table 10.2 Assessment criteria with the relevant responsible statutory organizations and consultees in Scotland.

Assessment criteria	Responsible statutory organisation/consultee
• Impact on, and benefits for, local communities • Economic benefits of the sustainable development of the aquaculture industry • Other marine sectors and users (commercial and recreational activities, including existing consented aquaculture sites) • On-shore facilities • Effects of construction, operation and decommissioning (including waste, noise, light and odor) • The availability of any necessary infrastructure and potential impact on existing infrastructure where relevant	Local government (local planning authority)
• Biodiversity • Biological carrying capacity (of the waterbody)	National government (Marine Scotland Science)
• Local water quality	National environmental regulator (Scottish Environment Protection Agency)
• Landscape, seascape and visual amenity • Seabed habitats • Nature conservation designated sites (impacts on and requirements for appropriate assessment) • Designation of Priority Marine Features • Predator control	Conservation non-governmental organization (Scottish Natural Heritage)
• Historic environment (e.g. archeology; listed buildings; wrecks etc.)	Archeology non-governmental organization (Historic Environment Scotland)
• Navigation (ensuring safe navigation is maintained with respect to any military activities, navigational routes, ports and harbors, anchorages, tourism, recreational and leisure activities)	Ports and Operations Authority

Scotland is unusual in using assessment of carrying capacity in its decision-making process. A spreadsheet-based model is used by Marine Science Scotland as part of its statutory consultation to estimate production carrying capacity of coastal shellfish culture (Gubbins et al., 2008), and especially mussels. The model estimates the relative risk of new and existing shellfish culture sites within areas of restricted water exchange. The model estimates the maximum 'clearance potential' of all the cultured shellfish from the potential standing biomass, assuming 5 kg mussels m^{-1} on suspended ropes and a clearance rate of 3866 L d^{-1} kg^{-1}. Whole area clearance potential is estimated using the flushing rate of the system (total quantity of water exchanged per annum). Direct comparison of percentage clearance potential estimates and flushing rate allows comparison of the relative risk of food becoming limited for cultured mussels. In Scotland, local planning authorities are advised to take carrying capacity into account in licensing where developments result in a clearance potential exceeding 50% of the flushing rate.

Once granted, consent is specific to the area of foreshore or seabed and not a specific person or company. At present, planning consent is permanent and not time limited, though the development must commence within 3 years of permission being granted, and equipment should be removed if the development is not in continuous use for 3 years. If planning permission is refused there is a right to appeal to the local authority review body or further to Scottish ministers.

Food safety

For any food production system, food safety and potential risks to human health are a key concern. As bivalve molluscs are filter-feeders they can accumulate contaminants, bacteria, viruses, and toxins, from natural or anthropogenic sources, which can affect human health. Many countries have implemented standards and monitoring procedures based on fecal indicator bacteria such as *Escherichia coli* (*E. coli*), and, in some cases biotoxins. These are used to classify shellfish harvesting areas and depending on the levels certain conditions will apply prior to, or after harvest regarding treatment. If coliform or biotoxin loading exceed acceptable levels, or the area is not classified for production, then molluscs harvested are usually prohibited for sale for human consumption.

Different countries have different regulatory systems for food safety. In the UK food safety and hygiene comes under the remit of the Food Standards Agency (FSA) in England and Wales, and Food Standards Scotland in Scotland. Waters are classified according to their coliform loading as A, B, or C. Where in class A waters (80% of samples <230 *E. coli*/100 g flesh) shellfish can be harvested directly for human consumption, class B waters (90% of samples ≤4600 *E. coli*/100 g flesh) shellfish require purification relaying or heat treatment, or class C waters (all samples <46,000 *E. coli*/100 g flesh) shellfish require relaying for at least 2 months in class B followed by treatment relevant to class B. Areas are reclassified annually based on collection of eight to nine samples spread over the year (FSA, 2020).

The safety of food for human consumption in a specific location, and how it varies over time and season, is clearly relevant to site suitability and selection. This can affect both the economics and marketability of the shellfish produced and so the viability of production. Each location from which the shellfish are cultured or collected must have a license based on its water classification before the products can be sold or used for human consumption. Culture and collection areas may be also open or closed to shellfish exploitation at different times depending on the loading of biotoxins, caused by plankton blooms, in the shellfish tissues. Occurrence of biotoxins in cultured shellfish in relation to algal blooms is complex. This is reviewed in Shumway et al. (2018).

Marine spatial planning

The coastal and marine environments that are used for mollusc culture are under increased pressure from multiple activities and users. With more demands on natural resources, and potential interactions between multiple activities it is important to consider how to optimize use of multi-user environments through more strategic planning, involving all stakeholders and decision makers. Marine spatial planning (MSP) is a strategic and integrated approach that supports policy and planning decisions within the marine environment. Implementation requires use of analytical approaches, information and tools. MSP is usually considered at a strategic scale, but for the MSP process to work effectively the needs and issues associated with individual activities should also be considered (Fig. 10.5).

The initial stage will involve determining the spatial area and defining goals and objectives. Once this has been decided, it is necessary to collect and analyze spatial data about the area (for example, maps detailing important ecological areas, existing activities, and potential conflict areas). This data is then used to prepare a spatial management plan to achieve the goals and objectives that were set at the start of the process. If the information is insufficient there may be a need to return and collect more data or refine the aims and objectives. When available, the plan should

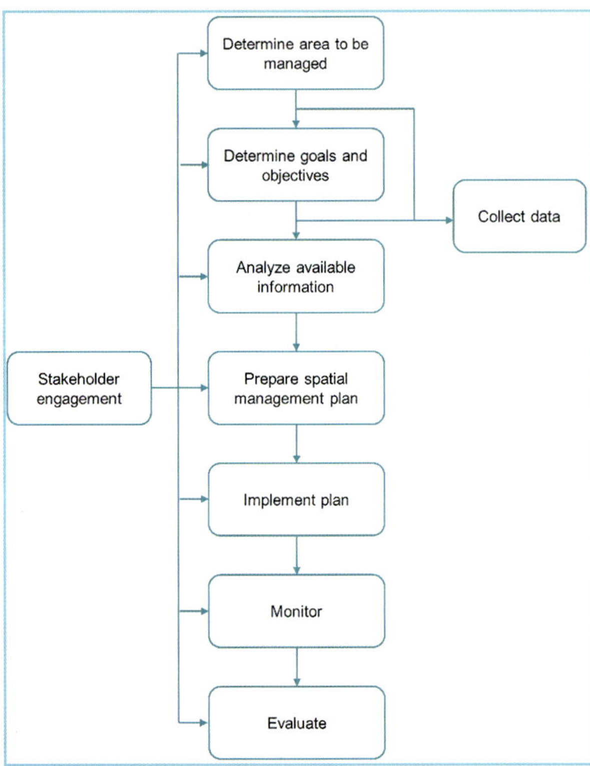

Figure 10.5 Diagram outlining different stages within the marine spatial planning (MSP) process.

be implemented, and each user and activity in the area should know their role and how they can and should make optimal use of the locations. The plan should be regularly monitored and evaluated to ensure it is effective in meeting the aims and objectives and that unintended negative consequences are minimized. MSP is an iterative process so can be adapted and improved based on the evaluation and changes in circumstances. For MSP to be effective, it is important to involve all stakeholders throughout the process, and mollusc producers should play a key role in MSP to ensure their needs and requirements are understood. Furthermore, any MSP strategy must also conform to existing national and international regulatory frameworks, which governs a number of aspects of marine management (for example, United Nations Convention on the Law of the Sea, European Union Marine Strategy Framework Directive).

As part of the MSP process, aquaculture zones may be established which prioritize areas for aquaculture production. Zones are usually established by national or local governments and supported by relevant policy and regulation. Depending on the process, zones may be dedicated for exclusive use by aquaculture or aquaculture may be prioritized but other activities may not necessarily be prohibited. Zoning enables more integrated planning of aquaculture development, allows for better regulation, and helps avoid conflicts with other sectors. This is important for bivalve mollusc production in crowded areas as they are dependent on water quality that is highly influenced by other activities.

A communication platform that encourages coordination of multiple activities within an area can support MSP for aquaculture. The advantage of integrating all users within a group is that it provides a strong communication network, there is a sense of ownership among all users and all stakeholders are engaged in the process. The goal being that conflicts are avoided or resolved at an early stage. For mollusc site selection this is useful as it allows the development of farms in the most appropriate locations and considers potential interactions with other activities and the needs of stakeholders in the area. This process is also useful for planning and regulatory authorities as it can streamline the planning process by identifying potential issues for new or existing development quickly. Furthermore, as the community have their say in how the environment is used and managed there can be increased acceptance of the activities that take place.

Climate change and ocean acidification

When selecting a site for aquaculture it is also important to consider the long-term future of the area and potential changes that may affect the suitability for mollusc production. Climate change and ocean acidification are altering environments where mollusc aquaculture takes place and will impact the suitability of areas for some species. Some areas will be affected more than others as the rate of change is not uniform throughout the world. Local factors will also play a key role in how sites are affected and the impact on mollusc culture. For extractive species, climate change and ocean acidification will not only affect the physiology of the cultured animal, but also the food source, larval recruitment, and ecosystem structure.

Effects will vary depending on the species of mollusc, the system used for production and the location of the farm. Some examples for potential impacts are discussed here. Increasing temperatures beyond the thermal limits of a species may affect the suitability of a site for existing species. There may be the potential to introduce new species to areas where they have not been cultured before. But any introduction of a new species must be carefully planned and managed to avoid an unwanted invasion and negative impacts on the natural ecosystem. Change in temperatures may also affect pathogens and predators, increasing their numbers or supporting new species, which could affect the suitability of some areas for production. Another potential impact from climate change is an increase in storminess, which could affect farm infrastructure and damage stock. Extreme weather events are becoming more frequent and intense. Coastal areas that are commonly used for bivalve production may be susceptible to cyclones, hurricanes, coastal flooding, and extreme water levels. It is important to note that the effects of climate change on mollusc culture are complex, as the animals will be subject to a combination of multiple stressors within the environment.

Ocean acidification is an issue for mollusc culture as the resulting changes in the carbonate chemistry of seawater affects the formation of exoskeletons. While the general trends of open ocean acidification are understood, there are uncertainties about the rate and magnitude of change in coastal areas. Human activities and coastal runoff also play an important role in changes to pH levels in coastal waters. Therefore, there is a key need for research at farm level to understand how different local conditions may affect pH levels and carbonate chemistry in order to identify potential risks to mollusc production.

There are many uncertainties and knowledge gaps over how climate change and ocean acidification may affect all kinds of aquaculture. One of the challenges is the difficulty disentangling climate change impacts from other factors that influence production. There is also a lack of long-term continuous datasets that can be used to assess the changing environmental conditions and the implications for mollusc shellfish. Predictive models can be used to assess the trends in climate change and ocean acidification but much of this research is still at an early stage. Most climate models are ocean models and focus on long-term averages across large areas, but higher resolution models are required that capture the farm environment (Falconer et al., 2020a). Research is urgently needed into the impacts from climate change and ocean acidification, the affects this will have on site selection and production, as well as identification of potential adaptation responses.

Summary

Site selection for aquaculture is influenced by many factors. A suitable location for a mollusc farm will have to meet certain biological, environmental, technical, and socio-economic criteria. It is important to note that these criteria are not universal among all molluscs and will be species and system specific. Site suitability may also change over time, so it is essential to identify the short- and long-term issues that may affect production and/or the health of the stock prior to establishing a farm site. Potential sites will also experience seasonal variability in environmental and production conditions, which should be evaluated through analysis of existing data or from new sampling campaigns, before the site is declared suitable for production. Decision support tools such as computer-based modelling can be used to help identify suitable sites, by combining multi-factor datasets and evaluating carrying capacity for areas or regions. They can also support regulators and planners make licensing decisions and to establish potential zones as part of the MSP process.

References cited and further reading

Byron, C.J. and Costa-Pierce, B. (2013) Carrying capacity tools for use in the implementation of an ecosystems approach to aquaculture. In: Ross, L.G., Telfer, T.C., Soto, D., Aguilar-Manjarrez, J., and Falconer, L. (eds.) *Site Selection and Carrying Capacity for Inland and Coastal Aquaculture*. FAO Fisheries and Aquaculture Proceedings, No. 21. FAO, Rome, pp. 87–101.

Falconer, L., Hjøllo, S.S., Telfer, T.C., McAdam, B.J., Hermansen, Ø., and Ytteborg, E. (2020a) The importance of calibrating climate change projections to local conditions at aquaculture sites. *Aquaculture*, 514: 734487.

Falconer, L., Middelboe, A.L., Kaas, H., Ross, L., and Telfer, T. (2020b) Use of geographic information systems for aquaculture and recommendations for development of spatial tools. *Reviews in Aquaculture*, 12(2): 664–677.

FSA (2020) Shellfish classification. https://www.food.gov.uk/business-guidance/shellfish-classification (accessed 10 June 2020).

Gubbins, M.J., Greathead, C., Amundrud, T., Gillibrand, P., Tett, P., Inall, M., Hawkins, A.J.S., and Davies, I.M. (2008) Towards determination of the carrying capacity of Scottish sea lochs for shellfish aquaculture. ICES CM 2008/H:13. Available at: http://www.ices.dk/sites/pub/CM%20 Doccuments/CM-2008/H/H1308.pdf.

McKindsey, C.W., Thetmeyer, H., Landry, T., Silvert. W (2006) Review of recent carrying capacity models for bivalve culture and recommendations for research and management. *Aquaculture*, 261:451–462.

Sanchez-Jerez, P., Karakassis, I., Massa, F., Fezzardi, D., Aguilar-Manjarrez, J., Soto, D., Chapela, R., Avila, P., Macias, J.C., Tomassetti, P., Marino, G., Borg, J.A., Franičević, V., Yucel-Gier, G., Fleming, I.A., Biao, X., Nhhala, H., Hamza, H., Forcada, A., and Dempster, T. (2016) Aquaculture's struggle for space: the need for coastal spatial planning and the potential benefits of Allocated Zones for Aquaculture (AZAs) to avoid conflict and promote sustainability. *Aquaculture Environment Interactions*, 8: 41–54.

Shumway, S.E. (2011) *Shellfish Aquaculture and the Environment*. John Wiley & Sons, Chichester.

Shumway, S.E., Burkholder, J.M., and Morton, S.L. (2018) *Harmful Algal Blooms: A Compendium Desk Reference*. Wiley-Blackwell, London.

Smaal, A.C., Ferreira, J.G., Grant, J., Petersen, J.K., and Strand, Ø. (2019) *Goods and Services of Marine Bivalves*. Springer International Publishing, Cham.

CHAPTER 11

Water and shellfish quality concerns for commercial bivalve shellfish aquaculture

Gregg W. Langlois and Jorge Diogène Fadini

Introduction

A number of bivalve shellfish species are harvested and sold commercially throughout the world. These may be grown and harvested using a variety of aquaculture technologies, harvested from natural populations, or harvested via mixed strategies, for example, planting natural substrate with juvenile shellfish (referred to as seed) and returning to collect the adults. Those interested in the commercial aquaculture of bivalve shellfish for human consumption face a number of challenges, including rigorous environmental quality standards and numerous, detailed food safety requirements, and regulations that follow the product from the growing area to the consumer. There is considerable government oversight of commercial shellfish aquaculture worldwide. There is good reason for this oversight and for the existing environmental and food safety requirements. The efficiency of filter-feeding bivalve shellfish allows them to concentrate a number of contaminants to levels that are dangerous to human health. Pollutants of concern include bacterial, viral, and protozoal pathogens, marine toxins, and anthropogenic chemicals such as pesticides.

To protect the consumer, conservative limits are established for contaminants of concern in commercial aquaculture areas to ensure acceptable water and shellfish quality. Some prospective growing areas may be free of contamination, while others may experience intermittent impacts that can be managed by closing the area to harvesting when necessary. Still other areas may be so heavily impacted that direct harvesting for human consumption is not allowed, with the shellstock required to go through a cleansing stage either by relaying to a clean growing area or by depuration in a certified plant. In some extreme cases, the impacts to an area may be so severe that it would not be safe to harvest and consume shellfish from them under any circumstances. To address these different scenarios, the agencies responsible for regulating commercial shellfish production have classification systems for growing areas that define how they may be operated and managed to ensure a safe product. This chapter is focused on the US system but includes discussion of an alternative system adopted by European Union (EU) countries. Countries with a memorandum of understanding (MOU) or a trade agreement with either the US or EU will have either adopted the respective classification system or developed a modified system that has been demonstrated to be equivalent and is acceptable to the importing country.

A tremendous amount of information and data are required to classify a new shellfish growing area properly. An assessment must be made of the potential contaminants present, the risk they pose, where they originate, and if they can be managed in a way to ensure that harvesting only occurs when the product is safe for consumption. This information can also be of value to resource agencies responsible for correcting sources of water quality impairment. A portion of this chapter focuses on the task of surveying a prospective shellfish growing area, including the surrounding watershed, the various procedures and tools used in this effort, and how the information and data generated is interpreted and used to classify the area.

How this assessment is conducted will vary for different types of contaminants. Testing of some pollutants requires sampling the growing waters, whereas testing of shellfish tissue may be needed for other contaminants. In addition, the analytical turn-around times for the various contaminants differ considerably, ranging from hours to days. For example, the tools and methods exist to monitor for the presence of marine toxins in near real time, a possibility that does not exist for routine testing of the large number of microbial pathogens of concern. The laboratory methods for marine toxins can often be completed the same day the sample arrives, whereas the approved microbial assays require 24 to 72 hours. Shellfish samples can also be qualitatively tested in the field for the presence of one or more toxins, allowing additional opportunistic sampling when toxicity is detected; this, in turn, can speed up the survey process for assessing the risk from these toxins. There are no such options for the routine monitoring of microbial pathogens, which require samples to be tested in the laboratory, typically involving time-consuming culture methods to grow enough cells to allow detection.

These practical differences and limitations in how risk can be assessed for the different contaminants also have a significant impact on how that risk can be managed. The strategy for limiting risk from microbial disease organisms is focused on eliminating, or at least managing, the sources of contamination, whereas the strategy for marine toxin management is focused on early detection and increased monitoring, with proactive harvest closures prior to toxin levels becoming unsafe.

While focusing on the various public health concerns associated with bivalve shellfish and acknowledging the trepidation that they can create, it should be kept in mind that safe and nutritious bivalve shellfish are being grown for human consumption worldwide. Areas exist with clean water and better, more conscientious, sewage handing (for example, improved human sewage treatment systems and best management practices in agriculture), allowing aquaculture to expand into regions that were previously unsuitable. There will always be significant challenges as different and sometimes conflicting uses compete for the same resources. It is essential, therefore, to keep abreast of the existing and potential impacts to aquaculture water and shellfish quality and develop control strategies proactively. Success involves understanding the risks, avoiding them when possible, managing them properly when feasible, and knowing how to navigate a number of regulatory processes and requirements intended to ensure the production of a safe and wholesome seafood product. While water quality is the focus of this chapter, it is important to present this discussion in the context of the overall process for evaluating and classifying a new area for shellfish aquaculture. The following sections will provide: (1) an orientation to the regulatory world that a shellfish aquaculturalist will become a part of; (2) a review of the more common microbial pathogens and marine toxins than can contaminate shellfish growing waters and shellfish tissue; (3) an overview of the various efforts required for classifying a shellfish growing area, which include shoreline and sanitary surveys and studies of the area's hydrography and hydrology; (4) a discussion of the factors involved in designing and conducting the necessary water quality studies in the growing area and watershed; (5) examples of typical fecal coliform (FC) data sets and how they are used to determine the appropriate classification for the proposed growing area; and (6) a description of common management control measures that may be necessary to keep the growing area, once classified, in compliance with the classification requirements. Throughout these discussions there will be suggestions on how the prospective shellfish grower can participate in this process. While the focus is on commercial shellfish aquaculture in the US, information will be provided on alternative approaches and requirements in the EU. Several of the topics discussed in this chapter are far too complex to cover thoroughly; in these instances, suggestions are provided for obtaining more detailed information, which can be found in Appendix I, links and references. A glossary is provided in Appendix II to provide more detail on terms and acronyms used in the field of shellfish growing area classification and management.

Overview of regulatory programs and requirements

For a person with no prior experience in the shellfish industry, the regulatory infrastructure may be a bewildering entanglement of local, state, and federal rules. Viewed in top-down fashion it can actually make some sense. In the US, the minimum requirements for commercial shellfishing for human consumption exist within the National Shellfish Sanitation Program (NSSP), which is administered by the federal US Food and Drug Administration (FDA). Each state with a shellfish industry agrees to adhere to these guidelines, which helps ensure uniform practices among the states. Uniformity among states is further ensured by the annual review of each state program by the FDA. Local health agencies are also involved either directly or indirectly via such activities as inspection of retail shellfish facilities and providing the initial response to shellfish-related illness reports.

Given the prominent role of the NSSP in establishing the guidelines that each state must follow, it is in the best interest of each shellfish aquaculture business to understand how the program functions. It may be one of the few federal/state programs in which the regulated community can participate and provide input. Before defining and describing the NSSP it is necessary to first introduce another organization called the Interstate Shellfish Sanitation Conference (ISSC). The ISSC is recognized by the FDA as the primary voluntary national organization of state shellfish regulatory officials that provides guidance on matters related to the sanitary control of shellfish. Participants in the ISSC/NSSP include state agencies responsible for regulation of commercial shellfish production (growing and marketing), non-producing states that receive shellfish, tribal representatives, FDA and other federal agencies, and the shellfish industry. The ISSC meets biennially, serving as the mechanism for discussing and implementing changes to the NSSP. So, by definition, the NSSP is a federal and state cooperative program that is recognized by both the US FDA and the ISSC for the sanitary control of shellfish produced and sold for human consumption. The purpose of the NSSP is to promote and improve the sanitation of shellfish (oysters, clams, mussels, and scallops) moving in interstate commerce.

At the state level, the agency, or agencies, responsible for ensuring the state shellfish program meets NSSP guidelines is referred to as the shellfish control authority (SCA). A prospective shellfish grower will work directly with the SCA in establishing a growing area suitable for shellfish production intended for human consumption. New shellfish growers should also be aware of regional shellfish organizations that typically meet annually (for example, the Pacific Rim Shellfish Sanitation Association or the Gulf and South Atlantic Shellfish Conference). These organizations include participation from member state shellfish program managers, the local shellfish industry, tribal representatives, FDA and other involved federal agencies and university researchers. The regional meetings provide a forum for local topics to be shared and discussed, for development and debate of proposed changes to the NSSP to be submitted for consideration at the next ISSC meeting, and for review of proposals submitted by ISSC members from other regions. These regional associations can be contacted by way of the member states (see Appendix I). Although a new grower may find the initial business start-up incredibly time consuming, it is not too early to also become involved in these regional discussions. Active participation in the regional shellfish associations and the ISSC committees and meetings is the most effective way for the shellfish industry to have input on the requirements that affect their business.

Changes to the NSSP that are approved by the ISSC must first receive FDA concurrence, then they are adopted by way of periodic revisions to a document called the NSSP Guide for the Control of Molluscan Shellfish. This document includes the Model Ordinance, which contains the requirements that each shellfish-producing state (and therefore each shellfish producer) is bound to follow, a section on public health reasons and explanations for the various requirements, and several other supporting documents (US Food and Drug Administration, 2019; see also Appendix I).

The Model Ordinance consists of both pre-harvest and post-harvest requirements for commercial shellfish growing and harvesting. The information in this chapter focuses on the various pre-harvest requirements with which prospective and existing aquaculture companies must understand and comply, all of which relate to ensuring acceptable water and shellfish quality for public health protection.

In the EU, member states that have shellfish harvesting areas need to comply with EU regulations (for example, regulations 852, 853 and 854 /2004). Depending upon the state administration, these responsibilities can be transferred to regional governments within each member state. This is the case in Spain, for example, where regional governments are in charge of implementing the current regulations. Production areas need to be clearly defined and member states have to put in place monitoring programs covering all legal administrative and technical requirements in these areas. Laboratories implementing official controls need to be accredited (ISO 17025) and follow specific regulations set to conduct official controls.

Food contaminants of concern in shellfish growing waters

Microbiological contaminants

Some historical background on shellfish-related illness, along with a basic understanding of the more common shellfish-related illnesses, will help a new grower come to grips with the myriad NSSP requirements that are intended to ensure a safe product and prevent such outbreaks. It is also important to have a basic awareness of the types of pathogens that can be present, the possible sources, and how they are monitored. Of the numerous bacterial, viral, and protozoal pathogens that can be concentrated by shellfish and infect humans, the majority are associated with human sewage. In addition to these enteric pathogens, there are also free-living pathogenic bacteria that are not associated with contaminated water, such as some species in the genus *Vibrio*. This lack of association between pathogen and sewage for *Vibrio* spp. presents a unique problem for managing their presence, a topic that will be addressed later in this section.

Historically, the first significant pathogen associated with shellfish consumption was the typhoid bacterium, *Salmonella typhii*, which was responsible for outbreaks in the early 1900s that resulted in numerous illnesses and deaths. The typhoid epidemic was eventually associated with raw sewage and the fecal–oral transmission of the pathogen. Over time, with the understanding of the association between sewage-contaminated waters and shellfish-borne diseases such as typhoid fever, steps were taken to expand and improve wastewater treatment, to remediate pollution sources, and to use more intelligent aquaculture and harvest site selection processes based on thorough shoreline surveys and sanitary surveys of the watershed. As a result, by the 1950s, there were far fewer shellfish-related illnesses associated with enteric bacteria. There are many other enteric bacterial pathogens that are also associated with shellfish consumption, including non-*typhii Salmonella* species, *Shigella* spp., *Campylobacter jejuni*, *Plesiomonas shigelloides*, and *Listeria* species; at least some of these pathogens may be introduced during post-harvest handling, and are not necessarily a growing water problem.

While typhoid illness from US shellfish declined decades ago, there has been an increase in illness outbreaks associated with enteric viruses and the non-enteric, free-living *Vibrio* bacteria. The most common viruses involved in illness outbreaks associated with shellfish consumption are the noroviruses (NoV; previously referred to as Norwalk-like viruses), other small round structured viruses (SRSV), and the Hepatitis A virus (HAV). Of these, NoV is the most common pathogen associated with foodborne illness. As with the enteric bacterial pathogens, NoV and HAV are associated with sewage contamination. Illness onset times differ, with NoV symptoms appearing rapidly (for example, 24–48 hours), HAV symptoms taking longer to appear (for example, 15–50 days), and bacterial pathogens having a range of symptom onset times (for example, 6–30 days for typhoid and 12–72 hours for other bacterial pathogens). NoV and bacterial pathogen infections result in various symptoms associated with gastroenteritis (for example, nausea, abdominal cramps, vomiting, and diarrhea). Infection due to *S. typhii* also results in a high fever and requires medical treatment. Bacterial pathogen infections generally last for 2–7 days and are self-limiting. NoV-related illnesses can last from 24 to 60 hours. Hepatitis-A virus is a very serious infectious liver disease that can potentially last between 3 to 6 months, although the majority (66%) of patients have recovered within 2 months. In rare cases HAV can also be life-threatening. A particular concern is that, unlike bacterial pathogens, the enteric viruses are resistant to standard wastewater treatment processes, including disinfection, and can survive in the marine and estuarine environments for a long time, with estimates ranging from months to over a year in the case of HAV. The NoV are also highly virulent compared to bacterial pathogens, requiring very few virus particles to cause illness.

In addition to these bacterial and viral agents of foodborne disease, a number of protozoal pathogens have also been linked to human illness outbreaks associated with contaminated water and food, but not specifically with shellfish. It seems that there is at least the potential for illness from these microbes, however, as several disease-causing protozoans associated with human and animal sewage have been identified in shellfish. In fact, shellfish have been used in biomonitoring studies of these pathogens because of their ability to concentrate these organisms, coupled with the ability of the protozoal oocysts and cysts to remain viable inside the shellfish. While the lack of illness associated with these microorganisms may indicate that they are somehow inactivated inside the shellfish, their presence does raise concern. With this uncertainty in mind, it is still useful to be aware of the potential risk and understand a little bit about these protozoans. Some of the more common protozoan parasites include *Cryptosporidium parvum*, *Giardia lamblia*, *Cyclospora cayetanensis*, *Toxoplasma gondii*, and several species of microsporidia (for example, *Encephalitozoon* spp.). The transmittable forms of some of these pathogens (for example, *C. parvum* oocysts and *G. lablia* cysts) can survive for long periods of time in the environment as well as in concentrated fecal wastes (for example, livestock sewage ponds). With the exception of *T. gondii*, infections caused by these protozoan pathogens result in mild to severe gastroenteritis and can last up to six weeks depending on the specific pathogen. Infections from *T. gondii* do not typically cause noticeable symptoms in people with healthy immune systems, but can cause serious health problems in people with weakened immune systems and in pregnant women. Immunocompromized persons can also suffer life-threatening illness from microsporidial infections. Similar to the enteroviruses, the protozoal pathogens are resistant to standard water treatment processes and may occur in very low densities in domestic wastewater, making sampling and detection difficult. Conversely, infected humans and animals can produce tremendous densities of oocysts or cysts, so any direct discharge of sewage would significantly impact the receiving waters.

There are numerous free-living bacteria in the genus *Vibrio* that occur in marine and estuarine waters. These bacteria prefer warmer water, with the May to October period associated with the greatest number of illnesses in the US. The chances for ingestion and concentration of *Vibrio* by shellfish increase because of its ability to attach to particles. The most important *Vibrio* species that are associated with illness from consumption of raw or undercooked oysters are *V. cholera* non-01, *V. parahaemolyticus*, *V. vulnificus*, and *V. alginolyticus*. Serious illness from *V. vulnificus* can also occur by infecting wounds that are exposed to brackish and saltwater containing this bacterium. For clarity it should be noted that the 01 serotype of *V. cholera*, a more virulent strain than the non-01 serotype, is associated with sewage contamination and is therefore not included in this discussion of free-living bacterial pathogens. The majority of *Vibrio* illnesses are attributed to *V. parahaemolyticus* and result in gastroenteritis with a 24 to 48 hour onset time, typically lasting up to 3 days. The exception to this is *V. vulnificus* that, in addition to gastroenteritis, can also cause septicemia from wound infection (symptoms occurring within 36 hours) and has the highest mortality rate of the pathogenic *Vibrio* species. Of greatest concern to immunocompromized persons, especially those with preexisting liver disease from alcohol consumption or viral meningitis is *V. vulnificus*.

Given the numerous microbial pathogens that can contaminate shellfish, it is important to know when they are present in the growing area. Monitoring for the presence of all possible microbial pathogens is, unfortunately, problematic. In fact, it is considered an impossible task to routinely test for every potential bacterial or viral pathogen that could be present in a shellfish growing area. Some may occur in numbers that are too small to be reliably detected, but are still sufficient to cause illness. The cost, personnel time, and length of time required to complete the assays for each pathogen are also limiting factors to this approach. To address this problem, the use of an 'indicator organism' has been adopted as a proxy for the many microbial pathogens of fecal origin that may be present. There are notable differences in the indicator organisms used for various applications. While the NSSP focuses on detection and quantification of thermotolerant FC in shellfish growing water, for example, official control in the EU focuses on *E. coli* in shellfish tissue.

Microbial indicator organisms

Because of the diversity of bacterial and viral pathogens and the impracticality and high cost of testing for every known shellfish-born pathogen, the use of an indicator organism for fecal contamination has been adopted as the most practical approach for water quality monitoring. The characteristics of an ideal indicator organism for fecal contamination include the following: (1) the indicator should be associated with the intestinal microflora of warm-blooded animals; (2) the indicator organism should be present whenever enteric pathogens are present; (3) the indicator organism should have a longer survival time than the pathogens of concern; (4) the indicator organism should not multiply in the environment; (5) the density of the indicator organism should have some direct relationship with the magnitude of fecal pollution; (6) the testing method should be easy to perform, provide rapid quantitative results, and be inexpensive; and (7) the indicator organism should be reliable and perform uniformly across all waters or shellfish species. The presence of the indicator species at a specified threshold does not predict the presence of one or more pathogens, but is an indication of fecal contamination that increases the risk of a pathogen being present.

The indicator bacteria that have been used for detection of fecal contamination of surface waters in the US are total coliforms (TC), fecal coliforms (FC), *Escherichia coli* (*E. coli*), *Streptococcus*, and *Enterococcus*. In the US, scientists in federal agencies differ with respect to the indicator(s) of choice. The US Environmental Protection Agency (EPA) currently recommends the use of *E. coli* for freshwater systems and *Enterococcus* for marine waters. The FDA, via the NSSP guidelines, requires the state SCA to use TC or FC for water quality studies, including freshwater and estuarine-marine waters, classification of shellfish growing areas, and for routine verification of shellfish growing water quality. The EU regulates shellfish growing areas on the basis of *E. coli* numbers in shellfish tissue. In contrast, the NSSP does not contain a threshold standard for TC or FC in shellfish meat in the growing area, but FC limits do exist for depurated shellstock. Although a quantitative FC limit does not exist for shellfish in the growing area, there is a NSSP requirement for reopening a growing area closed to harvest only when "the emergency situation or condition has returned to normal and sufficient time has elapsed to allow the shellstock to reduce pathogens or poisonous or deleterious substances that may be present in the shellstock to acceptable levels" (US Food and Drug Administration, 2019a). The criterion of 'acceptable levels' for pathogens is not defined in the Model Ordinance, but rather is expected to be established by the SCA based on studies in the affected area that determine the decay time required for the contaminant, typically FC, to return to normal background levels. This topic will be revisited in the 'Data analysis and interpretation' section.

The most common indicators of fecal contamination of surface waters and shellfish growing areas meet many of the desired characteristics listed above, but also have some common weaknesses. They are all bacteria that are assumed to serve adequately as indicators of pathogenic bacteria, however, some bacterial pathogens (for example, *Salmonella*) are known to survive for longer periods of time than FC and *E. coli*. It has also been known for some time that fecal indicator bacteria are not reliable for determining the risk from enteric viral pathogens or free-living bacterial pathogens, both of which have been dominant sources of shellfish-related illness outbreaks in recent decades. As mentioned earlier, viral and protozoal pathogens are more resistant to wastewater treatment processes and persist in the environment for long periods of time. Considerable research has been conducted over the past several decades on alternative indicator organisms for monitoring enteric viruses more effectively to improve the assessment of public health risk in shellfish growing areas. One of the more studied indicator groups are the coliphage, which are viruses that infect *E. coli*. In fact, the NSSP has started to use coliphage as a regulatory tool, a topic that will be discussed in the section on 'Shoreline and sanitary surveys.'

Marine biotoxins

There are a number of naturally occurring marine toxins that can accumulate in bivalve shellfish, potentially causing illness or death when consumed. These toxins are produced by certain species of phytoplankton; some toxins may only pose a threat when there are large numbers of phytoplankton cells present (for example, a 'bloom' or 'harmful algal blooms' [HAB]), while others are so potent that even low cell numbers can pose a serious public health threat. That these toxins and shellfish aquaculture are not mutually exclusive is a testament to the effectiveness of marine biotoxin monitoring programs, coupled with the fact that these toxic events are not constantly occurring. Despite the episodic nature of these events, they are largely unpredictable and require routine and frequent monitoring for early detection.

Phytoplankton can be monitored for the presence of toxin-producing species and shellfish can be sampled regularly and tested for the toxin(s) of concern, with results usually available on the same day that the samples arrive at the laboratory. There is an obvious advantage to selecting a potential new shellfish aquaculture lease in a region already under commercial production, as the risk from marine toxins will have been evaluated and a routine monitoring program established as needed. If the proposed growing area is located in a completely new region with no past history of biotoxin monitoring, however, then significant effort will be required to determine which toxins may occur. Before discussing the basic requirements for marine toxin monitoring and risk management, some general information on the various toxins will be helpful (refer to Shumway et al., 2018, for more detailed information).

Marine toxins of concern include the potentially lethal Paralytic Shellfish Poisoning (PSP) toxins and domoic acid (for example, Amnesic Shellfish Poisoning or ASP), and the nonlethal Neurotoxic Shellfish Poisoning (NSP) toxins, Diarrhetic Shellfish Poisoning (DSP) toxins, and Azaspiracid (AZA) Shellfish Poisoning (AZP). The most well-known and studied of this group are the toxins responsible for PSP, consisting of saxitoxin and a large number of analogs. PSP toxins are the most potent of the known marine shellfish toxins and have been responsible for a large number of illnesses and deaths in different parts of the world. The federal action level for PSP is 80 micrograms of saxitoxin (STX) equivalents per 100 grams of shellfish meat (80 µg STX eq/100 g = 0.8 ppm). The toxins are produced by several genera of gonyaulacoid dinoflagellates (*Alexandrium*, *Pyrodinium*, and *Gymnodinium*), which are normal members of the marine phytoplankton community in different parts of the world. Eating shellfish that contain dangerous levels of PSP toxins typically leads to an acute neurologic disturbance within a few minutes to a few hours. The PSP toxins are sodium channel blockers, inhibiting neural transmission. Symptoms begin with tingling and numbness of the lips, tongue, and fingertips, followed by disturbed balance, a floating sensation, lack of muscular coordination, slurred speech, and difficulty in swallowing. In severe poisonings, complete muscular paralysis and death from asphyxiation can occur if breathing is not maintained by artificial means. There is no antidote for PSP and anyone who suspects they have been poisoned must seek medical support immediately. Until recently these toxins could only be detected by a live animal assay, the PSP mouse bioassay (MBA). The NSSP now recognizes two additional methods for detecting the PSP toxins: a high performance liquid chromatographic (HPLC) method and a receptor binding assay (RBA). The latter method is approved for use with mussel tissue for the purposes of closing and reopening harvest areas; the RBA is currently designated a 'limited use' method for screening clam and scallop samples, with any positive sample resulting in a precautionary harvest closure, requiring analysis by an approved method to confirm closure and for reopening the harvest area. Additional matrix validation studies are underway that will hopefully permit the use of the RBA with oyster tissues, an urgent need given the focus of shellfish aquaculture on oyster production in many regions. In addition to the development of new quantitative methods of analysis, qualitative tests for the PSP toxins have been developed commercially, one of which (Scotia Rapid Test for PSP, Scotia Rapid Testing Systems, Ltd.) is NSSP-approved as a screening test. The use of a qualitative screening test can be an important addition to the regulatory laboratory, as it eliminates the need to test negative samples with one of the quantitative regulatory methods. The use of qualitative toxin test kits for field screening shellfish samples will be discussed in more detail in the 'Water and shellfish sampling' section. In addition, new quantitative methods of analysis continue to be developed and considered for adoption as a regulatory method. Recently, for example, an LC-MS/MS (liquid chromatography coupled to mass spectrometry) method for determination of PSP toxins has been proposed for use in the EU.

Amnesic shellfish poisoning (ASP) is caused by a chemical compound called domoic acid, which is produced by several species of diatoms in the genus *Pseudo-nitzschia.* Domoic acid was first identified as a shellfish toxin in 1987 when it was linked to an outbreak in Prince Edward Island (PEI), Canada. In this event the consumption of domoic acid-contaminated mussels was linked to three deaths and over 100 illnesses. Domoic acid has subsequently been detected in shellfish from a number of countries around the world. The federal action level for domoic acid in shellfish is 20 μg/g (20 ppm). Although there have not been additional ASP outbreaks in human consumers, domoic acid has had significant impacts on marine seabirds and marine mammals, particularly along the California coast. Domoic acid is a neuroexcitatory amino acid that causes over-stimulation of certain nerve cells, called glutamate receptors, in two regions of the brain; the effects range from minor and temporary to potentially permanent or fatal, depending on the dosage. Case studies of the PEI outbreak reported that the most common symptoms were gastrointestinal, followed by neurologic symptoms including headaches, loss of balance and/or dizziness, memory loss, varying degrees of confusion, disorientation, changes in the level of consciousness, and in some cases seizures. The only method accepted by the NSSP for domoic acid analysis is an HPLC method coupled with UV detection. A qualitative method (Reveal 2.0 ASP, Neogen Corporation) has also been approved for use as a screening tool, with negative samples not requiring further testing by HPLC.

Diarrhetic Shellfish Poisoning (DSP) is associated with a number of toxins (okadaic acid, dinophysistoxins, pectenotoxins) produced by the dinoflagellates *Dinophysis* and *Prorocentrum*. The first shellfish-related illnesses positively associated with the DSP toxins and linked to *Dinophysis fortii* occurred in Japan in 1976, although similar illnesses from shellfish consumption had occurred a decade earlier in the Netherlands. These toxins and the associated dinoflagellates occur worldwide and are considered an emerging problem, with the first-documented DSP outbreaks in North America occurring in British Columbia, Canada and in Washington State in 2011. The federal alert level for DSP is 160 µg okadaic acid eq/kg (0.16 ppm). Onset of symptoms can occur within 2 hours of consuming DSP-contaminated shellfish. Health effects, while not fatal, include severe gastrointestinal symptoms including diarrhea, vomiting, and abdominal pain, which can last several days. The NSSP recently approved an LC-MS/MS method of analysis for the toxins associated with DSP, but is currently only applicable to clam samples. In comparison, an LC-MS/MS method for the DSP toxins has been approved by the EU as the reference method and a protein phosphatase inhibition assay (PPIA) is a supplementary method for the okadaic acid (OA) group of lipophilic toxins.

Azaspiracid (AZA) Shellfish Poisoning (AZP) is caused by a lipophilic toxin produced by dinoflagellates of the genus *Azadinium*. Another relatively new marine toxin, AZA was identified in 1998 as the cause of an earlier shellfish poisoning event in the Netherlands that was associated with mussels from Ireland. Since that time there have been numerous additional AZA analogs identified, many the product of shellfish metabolism. AZA produces similar symptoms to DSP toxins, and in the EU regulations fix a maximum permitted level of 160 µg AZA eq/kg (0.16 ppm), a threshold also adopted by the NSSP.

Chemical contaminants from anthropogenic sources

Chemical contaminants (for example, pesticides, herbicides, polycyclic aromatic hydrocarbons, metals such as mercury) represent a significant challenge for assessing the potential risk for consumers when present above regulatory levels. The analytical methodology is usually complex and costly and beyond the capacity of most government shellfish programs. A survey of the area in proximity to the aquaculture site and a review of past land use may reveal the presence of a source of one or more chemical contaminants or reveal the presence of legacy pollutants. Often this is enough to reject the site for bivalve shellfish aquaculture, redirecting the survey to areas that are more pristine. Conversely, knowledge of the lack of industry (for example, mining, manufacturing, agriculture), both historically and currently, can allow the sanitary survey to proceed without the need for unnecessary, and prohibitively expensive, chemical analyses. Given the large number of potential existing and legacy pollutants, the cost of evaluating their impact, and the impracticality of aquaculturalists attempting to conduct these studies on their own, these contaminants will not be addressed in detail in the following discussions.

The ability to characterize the risk from known or suspect chemical contaminants in the growing area requires considerable expense for laboratory analyses. Special sample collection and handling procedures are also required for many of these poisonous substances. The question also arises as to what constitutes an 'acceptable' level of chemical contaminant in shellfish intended for human consumption. There are action levels or guidance levels for some chemical contaminants, but not all, in several FDA guidance documents; additional guidance is available in several EU documents. Appendix I contains citations and links for several of these guidance documents.

In the EU, regulations require monitoring of several chemical contaminants in marine products, including shellfish: heavy metals, dioxins, dioxin-like polychlorinated biphenyls, non-dioxin-like polychlorinated biphenyls and polycyclic aromatic hydrocarbons. Regulations set the maximum permitted levels for all these contaminants and the evaluation of data in a given area may be considered in order to foresee eventual risks associated with these compounds.

Classification of shellfish growing waters

It is clear from the previous sections that there are a number of factors involved in determining if a given area is suitable for shellfish aquaculture. Selecting a site for an aquaculture lease is one of the most important decisions a shellfish grower will make. The distance or proximity to pollution sources, the predictability and magnitude of those impacts, and the history of biotoxin activity in the region, will be key factors in determining when, and how frequently, the lease may be closed to harvesting. As an example, a common situation involves a growing area subjected to water quality impacts from rainfall events. The watershed runoff transports pollutants to the growing area via waterways such as rivers and streams. Overland runoff from the nearshore area of the lease can also contribute contaminants. In rural areas the major concern may be fecal bacteria and viruses from agricultural operations (for example, grazing land, confined animal operations such as dairies), failing septic systems, herbicides and pesticides, and increased nutrient loads in coastal waters that may increase eutrophication. In more urban landscapes the list of pollutants can expand to include impacts from wastewater treatment plants (WWTP) and a variety of chemical contaminants from human activities (for example, pesticides, herbicides, petroleum products from roadway runoff). Additional to these sources of contaminants, local coastal hydrography must be taken into account to understand the dispersion model of the contaminants in coastal areas, and if possible, identify the best locations with lower risks. If the causative factors of the contamination are known and predictable, the growing area can be managed by closing harvest activities until water and shellfish quality return to acceptable levels. Although no product is lost during a harvest closure, the lack of sales represents an impact to the business' cash flow, a consideration for any prospective shellfish grower. There are options for the shellfish grower to buffer the impacts from harvest closures, including wet storage and depuration, and availability of product from different, non-impacted locations to counteract the unavailability of product during closures of one specific area.

It is advisable that, prior to obtaining an aquaculture lease for growing shellfish commercially, the operator contacts the SCA to discuss the location. If other shellfish leases are operating in the vicinity then there will be documentation for the known and potential pollution sources, as well as the current control measures employed to ensure that harvesting does not occur until the impact has ceased and water and shellfish quality have returned to acceptable levels. There will also be a history of the number of harvest closure days per year that can be expected. This knowledge can help the prospective shellfish grower develop their business plan and determine the most suitable site for their operation.

If the proposed lease is in an area with no preexisting shellfish aquaculture, then a thorough shoreline survey and sanitary survey of the watershed must be conducted, as detailed below. A sampling program must also be developed to characterize water and shellfish quality in the proposed lease area and throughout the watershed that drains into the area. Seasonal, ongoing, and unpredictable impacts to water quality must be assessed and quantified.

When enough information on pollution sources has been gathered and the water quality of the area has been thoroughly evaluated, the lease can be classified in one of several categories that define how the leaseholder must operate. The water quality data consists of fecal coliform (FC) levels, expressed as a most probable number per 100 mL of sample (MPN/100 mL); for simplicity this will be abbreviated to MPN. While a detailed explanation of laboratory methodology is beyond the scope of this chapter, it is important to have a basic understanding of the regulatory data and how the threshold compliance values are derived. The current acceptable methods for determining FC levels are referred to as culture-based methods: a portion of the water sample is added to an enrichment media that allows the target indicator (for example, TC, FC, *E. coli*) to grow and be enumerated. To complicate matters there are several variations of acceptable methods and compliance standards for both TC and FC. A common assay currently employed is called a multiple tube fermentation (MTF) method, in which serial dilutions (for example, 1 : 10, 1 : 100, 1 : 1000, also referred to as decimal dilutions) of sample are added to test tubes of enrichment media, with replicate tubes for each dilution, and incubated (for example, for 24 hours). The long-standing American Public Health Association (APHA) MTF method for FC requires that, at the end of the initial incubation period, each positive tube is used to inoculate another media specific to FC, which then incubates for another 24 hours. An alternative, the A-1 decimal dilution method, is formulated to be completed in 24 hours. At the end of the final incubation period the number of positive tubes at each dilution is determined and used to find the MPN value from a table. It is important to recognize that the MPN is not a quantitative value of FC density, but is rather an index value representing a probability based on the number of positive tubes in each dilution. There are several variations of the MTF method, each involving a different number of tubes and dilutions, which are accepted under the NSSP for different applications. Each of these variations has a different level of precision, increasing with the number of tubes, and therefore each has a unique regulatory threshold compliance MPN value for either TC or FC. The standard MTF method can take from 48 to 72 hours to complete and is therefore not a valuable tool for responding quickly to a pollution event. There is also a membrane filtration method, called the membrane-thermotolerant *Escherichia coli* (mTEC) method, which is approved for use in shellfish growing waters (US Food and Drug Administration, 2019). Briefly, the mTEC method involves filtering a specified volume of sample, placing the filter on a petri dish of agar media, incubating for 24 hours, and counting the number of colonies to produce a value for colony forming units per 100 mL (CFU/100 mL). In an effort to simplify an already complicated explanation, the following discussions will assume that a five-tube APHA decimal dilution MTF method is being employed and only the related water quality standards will be referred to. The reader can consult the NSSP Model Ordinance for details on the other FC assay formats and their respective water quality standards.

Fortunately, the explanation of FC testing for shellfish tissue is greatly simplified compared to the variety of growing water methods. There is only one FC method approved for use by the NSSP for the survey and classification of growing areas, although other methods are approved for different purposes (for example, depuration). The NSSP recognizes the five-tube, four-dilution MTF method for shellfish tissue as described in the APHA manual. This shellfish assay produces a value reported as MPN/100 grams of tissue. The EU method for shellfish *E. coli* is referred to as the Horizontal method for the enumeration of beta-glucuronidase-positive *E. coli* (ISO 16649–3:2015) and values are reported as MPN *E. coli*/100 g.

It should be noted that, for any of the FC assays, additional dilutions should be performed when contaminated samples are suspected. The testing may otherwise produce an indeterminate result that is of limited use in characterizing the impact to water or shellfish quality. For example, a highly contaminated water sample may cause all the tubes in each dilution to be positive, preventing the determination of a specific MPN value. Instead, an indeterminate result, such as ≥1600 MPN/100 mL, will be reported. In actuality, had additional dilutions been performed to produce some negative tubes at the higher dilutions, the actual MPN could be much higher (for example, 16,000 or 1,600,000 MPN/100 mL).

In the US the NSSP classification system consists of five categories: *Approved*, *Conditionally approved*, *Restricted*, *Conditionally restricted*, and *Prohibited*. These categories are distinguished by the degree of water quality impacts in the growing area based on the FC indicator data, the types of control measures required based on the cause and magnitude of the elevated FC levels, and other factors such as the presence of one or more marine biotoxins that must be monitored and managed. Each classification has a water quality standard based on the FC geometric mean (the average of the log FC MPN values, converted back to arithmetic scale by calculating its antilog) of the 30 samples collected during the sanitary survey, as well as a secondary water quality standard that differs depending on which of two accepted sampling strategies are used: adverse pollution condition (APC) or systematic random sampling (SRS). The APC strategy is appropriate for classification and monitoring of growing areas impacted by point (for example, a wastewater treatment plant) and nonpoint sources of contamination. This approach focuses on sampling during events known or suspected of causing impacts to water and shellfish quality in the growing area. Rainfall is a common adverse condition. The secondary APC-specific water quality standard is based on a maximum allowable FC MPN, which cannot be exceeded by more than 10% of the sample data. The latter criteria is intended to account for method variability, not environmental variability. The actual FC MPN for this secondary standard differs for *Approved* and *Restricted* classifications, as detailed below. The SRS strategy requires a predetermined schedule of dates, essentially randomizing sampling with respect to environmental conditions and events that can impact water and shellfish quality. SRS sampling is appropriate for classification and monitoring of growing areas impacted by nonpoint sources of contamination but cannot be used for areas that are impacted by point sources of pollution. The SRS-specific secondary water quality threshold is based on the estimated 90th percentile of the sample FC MPN data, which differs for *Approved* and *Restricted* classifications. A brief explanation of each classification category is as follows.

1 An *Approved* growing area is one that does not experience impacts to water quality as measured by the FC indicator, nor does it experience impacts from marine biotoxins or other poisonous or deleterious substances. Direct harvesting for human consumption is permitted. The FC geometric mean for a minimum of 30 water samples cannot exceed a value of 14 MPN and a value of 43 MPN cannot be exceeded by either (a) more than 10% of the samples (APC), or (b) the estimated 90th percentile (SRS).

2 A *Conditionally approved* area is one in which the *Approved* criteria are met the majority of the time, but periodically fails these criteria due to significant impacts to water quality or unacceptable levels of a marine biotoxin. As a result, the area must be closed to harvesting for a specified period of time. The impacts must be predictable and manageable to ensure that contaminated product is not harvested and sold to the public. A management plan must be developed to document the conditions under which the area must be closed to harvesting (for example, a rainfall-related closure when a certain threshold is exceeded, such as 23 mm [1 in] within 24 hours). If the cooperation of other agencies or facilities are considered essential for the successful administration of the management plan, those entities must agree to the conditions of the plan. Failure of any involved party to agree would disallow the use of the *Conditionally approved* classification. Direct harvesting for human consumption is permitted when the area is in the 'open' status. The water quality criteria are the same as the *Approved* classification standards.

3 An area classified as *Restricted* does not meet the *Approved* or *Conditionally approved* classification criteria and therefore direct harvesting for human consumption is not allowed. The sanitary survey report documents a certain degree of pollution but water quality still meets the following minimum requirements: The FC geometric mean for a minimum of 30 water samples cannot exceed a value of 88 MPN and a secondary limit of 260 MPN cannot be exceeded by either (a) more than 10% of the samples (APC), or (b) the estimated 90th percentile (SRS). The *Restricted* classification may only be used if (a) the shellfish grower has an SCA-approved relay plan for moving shellstock to an *Approved* area for cleansing (see NSSP Model Ordinance, Chapter V, US Food and Drug Administration, 2019), or (b) the operator has an approved depuration facility for cleansing of shellstock (see NSSP Model Ordinance, Chapter XV, US Food and Drug Administration, 2019). Both of these options require specific permitting, intensive sampling, and thorough recordkeeping.

4 A *Conditionally restricted* area is one in which the *Restricted* criteria are met the majority of the time, but periodically fails these criteria due to significant impacts to water quality or unacceptable levels of a marine biotoxin. As a result, the area must be closed to harvesting for a specified period of time. The impacts must be predictable and manageable, and a management plan must be developed to document the conditions under which the area must be closed to harvesting, as detailed in section (2) above. When the area is open for harvesting it is subject to the restrictions discussed in section (3) above. The water quality criteria are the same as the *Restricted* classification standards.

5 A *Prohibited* area has either been determined by the sanitary survey process to be unsuitable for shellfish harvest intended for human consumption, or does not have a current sanitary survey. An area classified as *Prohibited* may be impacted by unpredictable pollution events or other sources of fecal contamination such that minimum water quality standards cannot be met.

The EU has three classification zones based on *E. coli* content (MPN *E. coli*/100 g meat and intravalvular liquid) in shellfish from the harvesting areas (REG UE 853/2004 i REG UE 2073/2005). These areas are classified each year into A, B or C zones (alternatively referred to as Class A, B, and C) according to the values obtained during the last 3 years, with a minimum of 24 values. If needed, values over a maximum range of 5 years can be used. A brief explanation of each EU classification category is as follows.

1 *A* zones should have at least 80% of MPN values ≤230 MPN *E. coli*, allowing a maximum of 20% of values between 230 and 700 MPN, with no values exceeding 700 MPN. In these areas, when values are below 230 MPN, shellfish can be put in the market without depuration.
2 *B* zones are those with at least 90% of the values ≤4600 MPN *E. coli*, with none exceeding 46,000 MPN. In these areas, when values are below 4600 MPN, shellfish can be put in the market after depuration in an approved facility, after relaying in an approved relaying area in the A classification, or after a European Community (EC) approved heat treatment process.
3 *C* zones are those whose *E. coli* values may reach, but not exceed, 46,000 MPN *E. coli*. In these areas, shellfish can only be put in the market after relaying for at least 2 months in an approved relaying area in the A classification, or after relaying for at least 2 months in an approved relaying area in the B classification followed by treatment in an approved purification center, or after an EC approved heat treatment process. If a value exceeds 46,000 MPN *E. coli* harvesting is not allowed and the area may be classified as Prohibited.

In addition to the three classification zones, the EU provides for five types of classification: preliminary, provisional, annual, seasonal, and part-year. Details on the different types of classification are available in the 'Sanitary survey' links in Appendix I. The EU also defines the 'reclassification level,' which is the value at which the zone automatically changes the classification value when *E. coli* levels in shellfish are above this limit. For example, an *A* zone has a classification level of 230 MPN and a reclassification level of 700 MPN. Should one single value be obtained above 700 MPN, the area automatically changes into a *B* zone, and this zone will only come back to an *A* zone after further revision of data as explained above (values obtained during 3 years, with a minimum of 24 values).

Shoreline and sanitary surveys

A common misconception of those new to aquaculture is that obtaining a commercial shellfish aquaculture permit is as simple as obtaining a fishing license. A prospective company can compromise its ability to get started in business by not understanding the timelines involved in surveying and classifying a potential growing area before it can be used for commercial shellfish production. As mentioned earlier, if the proposed growing area is not in the vicinity of currently certified aquaculture operations, then a thorough shoreline survey and sanitary survey must be conducted to determine the proper classification. This survey work is time consuming and encompasses a number of efforts, including: (a) a thorough records review of the region to establish what is already known and what information gaps need to be filled; (2) a field survey of the immediate shoreline and surrounding watershed to identify all potential pollution sources; (3) studies and information gathering on the hydrographic characteristics of the growing area and watershed; (4) information gathering and data collection to understand the meteorological conditions that affect the proposed growing area; (5) water sampling for fecal coliform (FC) throughout the area to characterize the baseline water quality and the magnitude of impacts to water quality under different adverse conditions (for example, rainfall-related runoff); (6) collection of water samples and stream flow measurements throughout the watershed to determine baseline and adverse condition FC levels and to estimate FC loadings for the major tributaries; and (7) collection of shellfish samples to determine baseline levels of FC and document the magnitude of impact under adverse conditions and the time required to return to acceptable levels. The nature of the sanitary survey work is similar in the EU, with the exception of monitoring *E. coli* levels rather than FC. This work is typically conducted by the professional staff of the SCA, who may allow or even require the participation of the prospective shellfish grower. The shellfish industry can in fact provide valuable support throughout the sanitary survey process in a number of ways, helping to expedite the process. If they have knowledge of the water body to be used for shellfish aquaculture, as well as the surrounding land uses, this information can help focus the sanitary survey efforts. Familiarity with navigating the water body, particularly shallow areas often used for shellfish cultivation, can be a tremendous assistance for expediting the collection of samples. The potential shellfish grower can also be trained to collect water and shellfish samples to provide additional support beyond what the SCA can contribute. If shellfish aquaculture already has a position in the community as a local business, contributing jobs and tourism, it may help encourage cooperation among the community in helping to identify and correct problems. Proactive community action to solve problems can avoid regulatory action by government agencies, which often becomes a divisive and drawn out process.

Pollution source survey

When conducting a sanitary survey, a thorough records review should be implemented to obtain all of the existing information on potential pollution sources and coastal hydrography. Valuable information can be obtained from local communities, corporate associations, tourist offices, food safety and environmental agencies, government and university laboratories, research centers and other sources that can greatly reduce the time and effort needed to conduct the sanitary survey. Obtaining current information on such things as the location of known pollution sources, performance of permitted waste handling facilities (for example, septic systems, WWTP, dairies, industrial operations), and records of complaints can help the SCA set priorities for the sanitary survey. Having accurate maps showing the exact location of the proposed lease location and configuration, the nearby shoreline, bathymetry, watershed topography, land use, and soil types is essential to the sanitary survey process. Fortunately, the availability of geographic information system software and inexpensive, but accurate handheld global positioning system units, allows most of the SCA to easily create digital map layers for each type of geographic data. A consultation and review of records with other regulatory agencies can provide information and perhaps digital map layers on potential pollution sources. For example: the location of all residences and businesses can be identified and mapped; the location of agricultural animal waste ponds may be mapped so the proximity to tributaries can be determined; homes with septic systems can be located and inspection reports on performance or reported problems and complaints reviewed; and WWTP discharge location, performance standards and actual performance via monitoring reports can be assessed.

The survey will characterize any seasonal impacts from rainfall-related runoff or human activity (for example, seasonal marinas or live-aboard anchorages, WWTP and septic system performance), as well as ongoing impacts (for example, wildlife). This investigation will determine a number of factors, including: (1) triggers for water and shellfish quality impacts that exceed control limits; (2) the relationship between the magnitude of the pollution event and the relative degradation of water and shellfish quality; (3) the duration of the impact; and (4) the time required for shellfish to cleanse themselves once the impact is over.

Armed with a general knowledge of the shoreline and watershed pollution sources, the survey team can then conduct the necessary field evaluation to verify these sources and identify any others that may impact the growing area. The latter could include a number of unregulated sources such as concentrated grazing operations, areas of concentrated wildlife in the watershed as well as on the water (for example, marine mammal activity and seasonal influxes of waterfowl), and illegal boat moorings representing potential sources of human waste discharged overboard. The presence of a marina requires an inspection of the operation and dilution modeling to determine if the growing area could be impacted by overboard discharge of human waste. The SCA may establish a *Prohibited* area to prevent harvesting in the region potentially impacted by the marina, based on the simple dilution modeling documented in the NSSP Model Ordinance (US Food and Drug Administration, 2019). Likewise, a *Prohibited* area must be established for each WWTP outfall in the vicinity that could potentially impact the growing area as per the NSSP dilution modeling criteria. More detail on effluent plume modeling is available at the links in Appendix I. In some cases, the records review may have identified studies conducted on the fate of the wastewater plume that can replace, or fine tune, the modeling. The information gathered from a thorough pollution source survey will guide the design of the water quality sampling studies needed to characterize the actual impacts to water quality in the growing area, which is discussed in the 'Water and shellfish sampling' section below.

Hydrography, hydrology, and meteorology

The physical features, or hydrography, of the water body in and around the proposed growing area must be documented and understood. The depth profile, or bathymetry, of the area may already be mapped (for example, by the National Oceanic and Atmospheric Administration). Knowledge of shoreline features, water body depths, and bottom composition and topography are all important for understanding how pollutants may affect the growing area: do these features allow the rapid transport of water through or around the area, or do they reduce flow, increasing the residence time of contaminants in the growing area? Information on the magnitude of tides and currents, and how they vary throughout the year, are also factors that affect how contaminated water moves through the growing area. Salinity and temperature profiles in different seasons are also valuable information. Routine salinity measurements can help assess the impact of freshwater flow, and therefore the level of contamination from the watershed, into the growing area. Frequent temperature data can reveal the presence of a thermocline in warmer months that can trap contaminants either above or below a certain depth; knowing this can guide sampling efforts and inform the SCA on relative risk of growing area contamination.

Hydrologic data on movement of water through the watershed must be documented. For each tributary that transports waterborne pollutants to the growing area, either directly or indirectly, the seasonal volume and flow velocity must be determined. Some tributaries may only flow seasonally following some threshold of cumulative rainfall, a condition that needs to be accounted for in the sampling program design. Estimates of stream or river flow velocity and volume can be made using a variety of tools and procedures, from simple to sophisticated. Larger tributaries may contain permanent flow gauges for flood control and these data may be readily accessible online or from the managing agency. The agency should also have data on the cross-sectional area at the gauging site, information that will be needed along with the velocity measurements to estimate flow, usually expressed in cubic feet per second (ft^3/sec). Depth and flow velocity measurements for larger, ungauged tributaries can be conducted from bridges by lowering a handheld flow meter at different points along the cross-section of the river; depths can also be sounded at each sampling point. Smaller streams can be evaluated by wading to collect depth and velocity measurements. More detailed explanation of these procedures is provided in the Appendix I links. The stream flow data will be combined with the water quality data to estimate pollutant loadings, that is, the quantity or mass of pollutant delivered to the proposed aquaculture site, as discussed in the next section.

Of the various meteorological conditions, rainfall is probably the most common event related to water quality impacts in a shellfish growing area. The SCA must monitor the magnitude and duration of precipitation in conjunction with water quality sampling throughout the watershed and growing area. Combined with the hydrologic and hydrographic data discussed above, a relationship can be established between the rate and amount of rainfall, or in some cases river stage, and the timing and magnitude of water quality impact in the growing area. Examples of these relationships will be provided in the 'Water and shellfish sampling' section, below. Suffice to say that an understanding of the relationship between rainfall rate/intensity and water quality degradation in the growing area is essential for managing these impacts.

Water and shellfish sampling

Water and shellfish quality must be characterized in the proposed aquaculture site for the variety of environmental conditions that may impact the area. Each of the known or suspect pollution sources identified in the shoreline and sanitary survey must be evaluated for its potential to impact the proposed growing area. These different sources of contamination are categorized as point or nonpoint sources. The EPA (Clean Water Act, Section 502) defines a point source as "... any discernible, confined and discrete conveyance, including but not limited to any pipe, ditch, channel, tunnel, conduit, well, discrete fissure, container, rolling stock, concentrated animal feeding operation, or vessel or other floating craft, from which pollutants are or may be discharged. This term does not include agricultural stormwater discharges and return flows from irrigated agriculture." In contrast to these single sources of contamination that can be identified, nonpoint sources are characterized as being diffuse and not attributable to a single source and are related to different land uses. Nonpoint sources are often a combination of pollution sources that enter waterways via runoff from rainfall. Examples of rural nonpoint source pollutants include fecal pathogens from animal grazing land and failing septic systems and chemical contaminants such as pesticides or herbicides from agricultural croplands. Types of urban nonpoint sources include storm drain water that can contain a mix of domestic waste, including fecal pathogens from pets and urban wildlife (for example, opossums and raccoons) and common chemicals used in landscaping (for example, fertilizers, herbicides), as well as runoff from paved surfaces that contribute a variety of chemicals.

The impacts from point and nonpoint sources of contamination must be assessed to document differences in pollution sources due to season. Areas that experience well-defined wet and dry seasons will probably experience significant differences in water quality between seasons due to rainfall-related impacts (WWTP overloads, septic system failures, agricultural runoff) that are negated during the dry months. There may be a large increase in marina and boat activity during summer months that affects the growing area and results in closures.

Water quality sampling in the growing area and watershed can be conducted under either the APC or SRS strategy, the latter assuming there are no point sources of pollution. Under the APC strategy, sampling will be focused on those environmental conditions likely to impact water and shellfish quality in the growing area (for example, rainfall), while the SRS approach will require a predetermined (for example, systematic), randomized sampling schedule that is intended to capture all conditions experienced in the growing area, including the adverse conditions. The goal is to adequately characterize the magnitude, duration, and variability of the sources of water quality degradation that occur under all environmental conditions. A sampling plan will be developed for the growing area and the watershed that documents all aspects of sampling, including methods, equipment, maps and geographic coordinates of sampling stations, sample handling and transport procedures, and data quality control (QC) measures.

Shellfish growing area sampling

A good starting point in the design of a sampling program for the proposed lease is to determine how many sampling stations are needed to represent the growing area. Stations must be located closest to the known or suspect pollution sources to evaluate any impact. It may be desirable to locate additional stations at incremental distances away from these 'worst case' stations: should water quality be unacceptable in the portion of the growing area closest to the pollution sources, these secondary stations may be less impacted and able to meet the requirements of a higher classification. To determine the time of sample collection each day, the SCA may need to consider tides, at least for stations in the growing area and other sites under tidal influence. An ebb tide may carry pollutants to the growing area and/or away from it. The former case represents the adverse condition and therefore needs to be factored into the sampling schedule. The size of the watershed, the relative amount of permeable versus impermeable surface area, and transport times to the growing area will affect the time it takes to impact the growing area and the time of maximum impact of water and shellfish quality. To address the magnitude and duration of the impact to water and shellfish quality, samples must be collected frequently enough to determine the lag time between the start of the event (for example, rainfall) and the period of maximum impact, as well as to adequately characterize the decay time of fecal coliform and the time required for water and shellfish quality to return to an acceptable level. Sampling the growing area stations on consecutive days of an adverse condition is important for determining when the maximum impact to water and shellfish quality occurs. A minimum of 30 samples per station are required for classifying the growing area. If there are significant impacts to the area it may be necessary to collect additional samples to better characterize water quality under both baseline (for example, dry season) and adverse conditions.

There are many environmental variables reflected in this type of data, as each adverse condition sampled will be unique (for example, in magnitude, duration, antecedent conditions, and so on). Impacts from a rainfall event can be immediate and short-lived, or there may be a lag of 1 or more days before maximum pollutant loadings are reached. The magnitude of impact may increase throughout the rainy season as soils become saturated, with less precipitation needed to generate a large amount of runoff. If the source(s) of contamination are renewable (for example, a WWTP that becomes overloaded due to infiltration of the sewage collection system with each storm event; animals continually present on grazing lands and in confined animal facilities) then there may actually be an increase in the level of fecal indicator bacteria throughout the rainfall season. The data from these adverse condition sampling events will establish if the growing area water and shellfish quality are impacted and, if so, will be used to determine the length of time the growing area must be closed to harvest. Given all the variables involved, it bears repeating that there is no substitute for sampling as frequently as possible over the variety of environmental conditions that can be expected.

Shellfish samples will also need to be collected to characterize baseline conditions when there are no water quality issues, as well as the magnitude and duration of impacts under adverse conditions. These data, together with the complementary water quality data, provides the basis for establishing rainfall closure thresholds and durations. A proposed aquaculture site for shellfish growing is presumably absent of shellstock. Sentinel bags or cages of mussels or oysters can be strategically placed in the lease area to provide a source of samples. Restocking should be done as needed, allowing enough time for the shellfish to acclimate before sampling so they represent ambient conditions. The use of a sentinel species will allow an initial evaluation of the impact from adverse conditions. The extent and duration of contamination will vary among species; however, therefore it will be necessary to evaluate each species intended to be aquacultured or present on the proposed site.

Watershed sampling

The sampling program will also incorporate watershed samples to document the magnitude and duration of contamination contributed by each of the known or suspect pollution sources. Sampling stations should be established as close as possible to each point or nonpoint source to assess the contribution of each. There is not an NSSP requirement for a minimum number of watershed samples. It can be helpful to collect samples on consecutive days at each station, together with the tributary flow and volume estimates discussed earlier, to help understand the response of the watershed to the rainfall event and how, and when, it impacts the growing area water quality. Ideally, stream flow measurements will at least be conducted at representative, if not all, watershed sampling stations. This information, combined with the fecal coliform (FC) data, will allow calculation of the pollutant loading and transport time to the growing area for each freshwater input. Estimates of FC loading in different tributaries can help identify portions of the watershed responsible for the greatest contribution of fecal waste. This is a tremendous help to the agencies responsible for remediation of pollution sources, allowing them to prioritize clean-up efforts in the areas with the highest FC loadings. The residence time of a contaminated water mass transported to the growing area must also be determined for a range of environmental conditions to help determine the length of time an area must be closed to harvest.

The sampling effort is another opportunity for involvement of the prospective shellfish grower. The SCA can train the applicant in the proper collection and handling of samples so they can participate as a member of the SCA sampling team. The prospective grower may be able to sample at times when the SCA is not available. In the case of a SCA with limited resources, the applicant may take on a larger share of the sampling activities and share in the cost of laboratory testing. The actual collection of water samples is relatively quick and easy, although care must be taken to avoid contamination of the sample or mishandling between collection and delivery to the laboratory. The SCA will have protocols for proper sample collection and handling; Appendix III contains an example water sampling protocol.

Additional sampling logistics

There are additional considerations for the sampling program if resources allow. As part of the QC effort, collection of replicate samples can help assess different sources of variability in the data. Collection of multiple discreet samples at a certain percentage of the sampling stations (for example, 10%), referred to as field replicates, can provide a measure of environmental variability. Field replicates can also help account for an occasional outlier datum that, taken alone, could confuse the interpretation of the data set. To assess laboratory variability a large volume of water can be collected and split into two or more replicate samples for a certain percentage of stations; this may be particularly valuable when more than one laboratory is used.

Because samples must be transported to the lab in time to allow the assays to be set up, it is not always possible to account for all adverse conditions on a given sampling event (for example, tides). As a result, it is often necessary to conduct additional sampling events to adequately represent each of the adverse conditions. No two sampling events will represent identical environmental conditions due to the inherent variability within each adverse condition (for example, different intensities and magnitudes of rainfall, varying tidal prisms and current velocities, variable tributary runoff throughout the season, different antecedent conditions). This resulting patchwork of data is not ideal for characterizing water quality in the growing area but is often the best approach possible. The variability in sampling conditions, combined with the variability of the FC assay, make data interpretation challenging. For these reasons it is advisable to conduct as many sampling events as possible to make data analysis and interpretation feasible with a high level of confidence in the resulting classification.

There are a number of important factors involving laboratory support that can impinge on the sampling design. A laboratory must be identified that is certified by the state for testing water and shellfish samples. If a state laboratory is used, then it must be certified by the FDA. Typically, the SCA agency will have a certified laboratory to conduct all of the water and shellfish sample assays. If the state laboratory is a considerable distance from the growing area it may not be feasible to collect and ship samples to the lab within the required holding time of the assay. In that circumstance a possible solution is to engage the support of a local health department lab that is, or is willing to become, certified by the state for the appropriate methods. A local laboratory is often more invested in the community and willing to accommodate the need for overtime and weekend work. If this is not an option then the SCA, or even the prospective shellfish grower, may need to contract with a commercial lab in the vicinity that is certified by the state, or capable of being certified, to conduct the necessary sample processing and testing. The capacity of the laboratory can also influence the design of the sampling plan with respect to the number of samples collected at one time. The standard MTF serial dilution method for FC is laborious, time consuming, and requires a significant amount of physical space and incubator capacity. Some laboratories may not work on weekends, limiting sampling options to just a few weekdays. Samples must be delivered to the laboratory early enough in the day to allow time for processing and starting the assay. Shellfish samples require considerably more time than water samples to process. A minimum of 12 individuals per sample, or more to provide at least 200 g of tissue, must be shucked aseptically and homogenized before the assay can begin; 50 or more individuals may be required for smaller species such as littleneck clams. To meet the needs of the laboratory it may be necessary to modify the sampling plan, requiring more events to be captured to compensate for fewer samples being collected for each event, or for collection of some samples under less than ideal adverse conditions (for example, sampling under incoming or slack tide when ebb tide is the adverse condition). The complexity of the sampling plan will be a reflection of the size of the growing area and watershed, the diversity of pollution sources that must be assessed, and the more pragmatic concerns of staff and lab resources. All of these variables need to be addressed early in the design of the sampling program to ensure that the work is conducted as thoroughly and efficiently as possible.

While the basis for the classification decision is the FC water quality data, this information does not provide insight into the original source (for example, human, cattle, birds, domestic pets) of the indicator bacteria or the presence of enteroviruses from a WWTP or other human sources. Thus, the need for a thorough sanitary survey to identify all potential sources and refine the determination of risk, particularly if human sources are possible. There may be circumstances, however, where the information provided by the survey work does not adequately explain the patterns of water and shellfish quality impact. These disparities can potentially be addressed by employing one or more of the microbial source-tracking (MST) methods discussed below.

Microbial Source-Tracking (MST)

Simply stated, the goal of MST is to determine the host specie, or species, associated with fecal contamination. This information would be of value in the identification and remediation of pollution sources associated with the host species of concern and would improve the assessment of public health risk. At present, MST is an evolving and complicated science that presents numerous challenges to its use in sanitary surveys. Unlike the traditional, more general indicator organism assays (for example, fecal coliform [FC]), there is not yet a standard method for MST or one agreed upon approach that is available for use. The various experimental methods have used a variety of microbiological species as the indicator organism, including FC, *E. coli*, fecal streptococci, enterococci, anaerobic enteric bacteria in the *Bacteroidales* order, adenovirus, and coliphage, a virus that infects *E. coli*. A variety of molecular and biochemical detection methods continue to be investigated as well, including several forms of polymerase chain reaction (PCR), antibiotic resistance, biochemical fingerprinting, and ribotyping. Microbial source-tracking (MST) methodologies further divide into library-dependent and library-independent methods. The former relies on the development of a large collection of genetic fingerprints for the various potential fecal sources in the area, against which the genetic information from samples are matched using sophisticated statistical techniques (for example, discriminant analysis). The library-independent methods rely on the use of indicator organisms that are thought to be specific to a particular animal host (for example, cattle); method comparison studies have shown the library-independent methods to perform better than the library-dependent methods and the former are now the focus of most MST research. The current MST approaches are essentially qualitative, lacking the ability to determine the relative pollution load associated with each animal or human source. Although still in the research and development phase, there may be some useful applications of a qualitative MST method as an adjunct to the traditional sanitary survey process. Ultimately, however, there will be the need to determine the relative contribution of each source if the goal is to identify the major pollution sources and achieve a significant reduction in fecal contamination through prioritized remediation activities.

Microbial source-tracking is also a costly technology and often well beyond the financial resources of the SCA or a broader stakeholder group interested in determining sources of fecal contamination. In most cases a thorough shoreline and sanitary survey will be successful in discovering and assessing the various pollution sources, allowing the SCA to estimate their relative contribution to the overall pollution load reaching the proposed growing area. Conversely, there are occasional circumstances when the sanitary survey process identifies water quality impacts in the growing area that may not be consistent with observations and water quality data for the watershed and surrounding shoreline. In these cases, there is an argument for exploring the use of MST to potentially gain a better understanding of the host organisms contributing fecal contamination to the area. There is a steep learning curve involved with MST and it is advised that the strengths and weaknesses of the chosen methodology be fully understood by all involved parties to avoid misinterpretation of the results and overstated conclusions. The design of a MST study should incorporate a rigorous QC program to limit false positive and false negative results and ensure the validity of the data generated. Guidance on this topic can be found in Stoeckel and Harwood (2007) and the links in Appendix I. At present MST is another tool in the toolbox, an experimental and expensive tool at that, and not the first to be reached for when conducting a sanitary survey that requires defensible data on pollution sources.

Male-specific coliphage (MSC)

To address the inadequacy of bacterial indicator species to assess the risk from enteric viral pathogens such as NoV, one avenue of research has focused on the use of bacteriophage, which are viruses that infect bacteria. Certain phages that are host-specific to *E. coli*, referred to as coliphage, have been of particular interest. The terms F+ coliphage and MSC are used interchangeably in the literature: they refer to coliphage that infect the *E. coli* cell via structures called pilli, which are filamentous appendages on the cell surface. Methods for culturing and enumerating the individual enteric viruses of concern (for example, NoV, HAV) have not yet been standardized for regulatory use, with current experimental methods requiring both a high level of expertise and expensive equipment. In any event, as with the bacterial pathogens, it is more practical and cost-efficient to assay for a single representative indicator species rather than all viruses of concern. Fortunately, there are culture-based assays for coliphage that are simple, inexpensive, do not require sophisticated equipment, and can be accommodated in a typical microbiology laboratory. There are several methods approved by EPA (for example, EPA Methods 1601 and 1602), the International Organization for Standardization (for example, ISO 2000, 2001), and the NSSP. The method approved by the NSSP in 2009, referred to as a modified double agar overlay method, was limited for use with eastern oysters (*Crassostrea virginica*) and soft-shelled clams (*Mya arenaria*); this approval was modified in 2015 to include northern quahogs (*Mercenaria mercenaria*). Presumably the ISSC will require additional validation studies to be conducted with other commercial shellfish before this method can be used with them for regulatory purposes.

The premise for MSC as an indicator organism is that it is present in high numbers and will parallel NoV, HAV, and other enteric pathogenic viruses in their resistance to treatment processes, environmental survival rate, and uptake and residence time in bivalve shellfish. MSC is in fact found in very high numbers in human sewage originating from a large number of people, such as a WWTP. The results in the research literature on the correlation between MSC and enterovirus are somewhat confounding, however, as some studies demonstrate a positive relationship and others do not. A 2015 EPA review of a number of MSC-related studies concluded that there is a better association between enteric pathogens and indicator organisms in brackish and marine environments compared to freshwater. Studies conducted by FDA in a number of shellfish growing areas around the country reported a strong correlation between the detection of MSC and NoV in WWTP effluent. In addition, there has been considerable success in demonstrating that WWTP influent and effluent contains predominantly human sources of MSC, further supporting the use of this indicator for evaluating the impact of human sewage contributed by a WWTP or other large sewage collection system.

Male-specific coliphage (MSC) has also been investigated as a MST tool for sanitary surveys. A variety of PCR methods exist that allow the determination of four recognized MSC genogroups (GI, GII, GIII, and GIV). Each group is associated with a specific source, although not exclusively so. GI and GIV have been associated with animal fecal sources, while GII and GIII are associated predominantly with human sources. So MSC has the potential to be of tremendous value in evaluating human and non-human sources of fecal contamination, although currently with some limitations. While MSC is present in very high numbers in WWTP influent and effluent, it is often not found in individuals or in small, on-site treatment systems (for example, septic systems). At present, therefore, MSC is not a suitable source-tracking tool when surveying rural areas that rely on individual on-site sewage disposal and no central sewage collection system. On the other hand, if the sanitary survey has identified a WWTP as a cause of concern it would be useful to monitor for MSC in the growing area and perhaps with additional sentinel shellfish stations moored at incremental distances from the effluent discharge.

As a result of these findings, the NSSP has adopted MSC as an indicator for the presence of enteroviruses under specific circumstances. To consider reopening a growing area that has been closed to harvest due to a release of untreated sewage from a WWTP or other large community collection system, representative samples of commercially harvested species cannot exceed a level of 50 MSC per 100 grams of tissue. The source of sewage contamination will have to have been corrected, and a minimum of 7 days elapsed, before the samples can be collected. The other application of the MSC standard is for evaluating shellfish in a growing area classified as *Restricted* prior to relaying to an *Approved* area.

Free-living bacterial pathogens: *Vibrio* spp.

The new shellfish grower will be faced with a conundrum when it comes to determining the safety of their growing area and shellfish from pathogenic *Vibrio* spp. This is because these bacteria cannot be monitored reliably in the environment and there are no other reliable indicator organisms to help assess risk. The presence of *Vibrio* spp. can occur in unpolluted waters; therefore the fecal bacterial and viral indicators are of no use. The pathogenic *Vibrio* (for example, *V. parahaemolyticus* and *V. vulnificus*) are also typically present in low numbers, making detection in seawater samples difficult and unreliable. Existing NSSP methods rely on detection of *Vibrio* in shellfish tissue, focusing not only on the total numbers of *Vibrio* present but also on genetic markers for virulence. Unfortunately, there has been a poor relationship between illnesses and the density of *Vibrio* in shellfish samples from the harvest areas linked to the illnesses. As a result, routine shellfish monitoring is not a reliable control mechanism and, in fact, the NSSP does not require it as part of an initial risk evaluation during the sanitary survey or as a routine testing requirement during harvesting. Particularly under warm conditions, *Vibrio* can continue to multiply in shellfish. The focus of the NSSP control mechanisms, therefore, has been on the need to get harvested shellstock under proper temperature control in a short period of time when *Vibrio* is reasonably likely to occur (that is, in summer and fall when water temperatures are at their highest). The NSSP has put significant effort into the control of vibriosis, with 13 separate entries addressing *Vibrio* control in the Model Ordinance table of contents (US Food and Drug Administration, 2019). It is clear that a new shellfish grower must educate themselves on this most common source of bacteria-related shellfish illness and the various post-harvest temperature control requirements. The grower can also expect to participate in a *Vibrio* Control Plan developed by the SCA for *V. parahaemolyticus* and/or *V. vulnificus* as per NSSP requirements.

Marine biotoxin sampling

A proposed aquaculture lease will likely be located in an existing shellfish growing area, as most bodies of water suitable for shellfish growing are already in production. As a result, the SCA will have knowledge of the toxins of concern in the proposed area, with insight into seasonality of occurrence based on the accumulated historical data. If this is the case, then it may not be necessary for the new grower to conduct additional sampling as part of the sanitary survey. The SCA may still require some level of monitoring, however, if the new area is not contiguous with existing leases or if there is some other reason to suspect a difference in toxin exposure at the new lease. For example, if the new lease is located closer to the mouth of a bay or estuary it may be subject to higher levels of toxin-producing phytoplankton that are carried in on currents from the open ocean. The prospective grower should become familiar with the history of toxin-related growing area closures, as this information combined with the record of adverse condition harvest closures will provide an estimate of the number of open and closed harvest days to anticipate in an average year.

Alternatively, if little is known about marine toxin occurrence in the proposed new growing area, routine sampling of phytoplankton and shellfish can be integrated into the sanitary survey process. This will begin to provide baseline data on the seasonality, frequency, and distribution of toxin-producers and shellfish toxicity. Sampling frequency will be guided by the SCA and will be partly dependent on the capacity of the laboratory. Weekly sampling is a good place to start, increasing in frequency if toxicity is detected so the toxic event can be tracked from beginning to end. Shellfish samples will usually be collected and submitted to the SCA laboratory; although it is rare to find a non-SCA laboratory that is certified to perform the necessary NSSP-approved analytical methods for toxin analysis, this is an option for the grower to augment the SCA samples. A sample will consist of at least 12 organisms and may require 50 or more for smaller species to provide enough tissue (100 g) for the Association of Official Analytical Collaboration International (AOAC) extraction method used in the PSP bioassay; more tissue will be required if additional toxin analyses (for example, ASP, DSP) are needed, so a minimum of 200 g is recommended. Some laboratories will accept shellstock and others, lacking the staff to do the time-consuming work of opening each shell and removing the tissue, will require that the shucking be done by the grower. The NSSP laboratory checklists provide guidance on the proper handling of samples that are partially processed in the field and shipped to the lab under proper temperature control (see 'FDA-accepted laboratory methods' section, in Appendix I, for link to checklists).

Phytoplankton samples can also be collected weekly, preferably at the same time and location as the shellfish sample(s). With training the grower can participate in the phytoplankton sampling and identification; this is a cost-effective approach for cataloging the presence of toxin-producing species and can help guide the decision-making on which toxin(s) should be analyzed in shellfish samples. Phytoplankton sampling and identification by the shellfish grower also facilitates the collection of more frequent samples, increasing the likelihood of detecting the start of a toxic bloom. Data can also be collected on the other phytoplankton species present, which may be of use to the shellfish grower in understanding certain aspects of their shellfish farm, such as variability in the rates of growth, mortality, and seed survival. Another tool for the biotoxin toolbox is the use of qualitative field test kits for various shellfish toxins, some of which were mentioned earlier. The prospective grower can use these kits to screen samples in the field, providing an immediate warning of toxin presence that can be of tremendous value to the SCA for increasing and expanding sampling. Detailed guidance on procedures and options for phytoplankton and shellfish sampling and testing can be found in *Harmful Algae: A Compendium Desk Reference* (Shumway et al., 2018).

Data analysis and interpretation

Water and shellfish fecal coliform (FC) data

During the sanitary survey initial data summaries should be prepared after each sampling event. Evaluation of the geometric mean and the percentage (APC), or estimated 90th percentile (SRS), of samples exceeding 43 MPN can help guide subsequent sampling. For example, as data are collected on adverse conditions it may be necessary to make changes in sampling stations, adding new ones to better focus on impacts or dropping existing ones that are not contributing to the study. In a worst case scenario, the initial data will not meet the water quality criteria for an *Approved* area and does not appear to be associated with a predictable impact event such as rainfall. In that circumstance it is clear that the area is unsuitable for shellfish aquaculture and there is little point in carrying out the remainder of the study. This is a rare occurrence and it is more common that the data reflects some degree of correlation between water quality impacts and the adverse condition(s) of concern. The following discussion will provide several simple examples of FC data sets to highlight common situations, assuming that the water and shellfish quality impacts are not associated with the performance of a WWTP or other sewage collection system; there are separate requirements associated with the latter point source impacts that are not covered in these examples. While not a thorough treatment of all possible scenarios, the following information should serve as a primer for a new shellfish grower, providing an understanding of how the FC data are used to guide the classification process.

Following the collection of at least 30 samples under the APC strategy, a data set with a geometric mean less than 14 MPN and with less than 10% of the samples exceeding 43 MPN would meet the *Approved* classification. Significant rainfall events have been captured in the APC sampling (Table 11.1).

While some of the higher rainfall rates show an increase in FC level, none of the values exceed 43 MPN and the geometric mean is well below the 14 MPN limit. In contrast, the example

Table 11.1 Fecal coliform (FC) most probable number (MPN/100 mL) data and classification statistics for 30 water samples, collected under adverse pollution conditions, which meet the *Approved* classification. Rainfall data represents cumulative inches of rain over 24 hours. The geometric mean is the antilog of the mean of the log FC values.

Sample number	24 hour rainfall (in.)	FC MPN	
1	0.00	2	
2	0.00	2	
3	0.00	2	
4	0.60	23	
5	0.00	2	
6	0.20	7.8	
7	0.00	2	
8	0.10	2	
9	0.75	13	
10	0.00	2	
11	0.00	2	
12	0.00	2	
13	0.10	2	
14	0.00	2	
15	0.20	2	
16	0.00	2	
17	0.55	2	
18	0.10	7.8	
19	0.00	2	
20	0.40	2	
21	0.00	2	
22	0.25	13	
23	0.00	2	
24	0.00	2	
25	1.10	13	
26	0.20	2	
27	0.00	2	
28	0.00	2	
29	0.00	2	
30	0.00	2	
		2.9	= Geometric mean
		0.0	= % > 43 MPN
		30	= *n*

data set in Table 11.2 contains elevated FC values associated with significant rainfall, exceeding both the 14 MPN and 43 MPN benchmarks. In this case the growing area represented by these data does not meet the *Approved* classification criteria because more than 10% of the samples (13.3%) exceed 43 MPN. The geometric mean is well within the limit, however, suggesting the area may be suitable for shellfish growing if there is a predictable relationship between rainfall amount and FC MPN such that the area can be managed using the *Conditionally Approved* classification. A closer look at the rainfall data reveals a strong relationship between the magnitude of 24 hour rainfall totals and FC MPN (Table 11.2). A wide range of 24 hour rainfall totals (0.10 to 1.10 inches) is represented in the data set, helping to clarify this relationship.

Rainfall of 0.55 inch or greater within 24 hours results in FC densities greater than 43 MPN. In addition, increasing 24 hour rainfall totals above 0.55 inch correspond with higher FC levels, providing some level of confidence in this apparent relationship. Eliminating these values from the data set resolves the unacceptable percentage of samples exceeding 43 MPN and the geometric mean is lowered further from its already acceptable value (Table 11.3). If the relationship between rainfall and water FC levels is ambiguous, it would be advisable to collect additional samples to confirm the correlation, replace the violative sample data that was removed from the data set, and bolster the total number of samples. This is most efficiently done as part of the ongoing review of each data set as it is produced. It can be anticipated that samples exceeding 43 MPN might result in the failure of the data set to meet *Approved* criteria, so additional sampling can be scheduled, presumably under slightly less adverse conditions, to refine the relationship between rainfall and water quality degradation.

The data in Table 11.3 support the idea that the growing area has acceptable water quality for a reasonable period of time, only being subject to seasonal impacts. Because a predictable relationship exists between measured rainfall amount and FC density, this adverse condition can be managed by closing the growing area to harvest when the trigger rainfall threshold has been reached or exceeded. The Table 11.2 data suggest this threshold is 0.55 inch of rain within 24 hours, however there is a data gap for rainfall between 0.40 (acceptable FC MPN) and 0.55 inch (unacceptable FC MPN). The more conservative approach would be to set the closure threshold to greater than 0.40 inch of rain: when this threshold is exceeded within any 24-hour interval, the area will be closed to harvest.

Table 11.2 Fecal coliform (FC) most probable number (MPN/100 mL) data and classification statistics for 30 water samples, collected under adverse pollution conditions, which do not meet the *Approved* classification. Rainfall data represents cumulative inches of rain over 24 hours. The geometric mean is the antilog of the mean of the log FC values.

Sample number	24 hour rainfall (in.)	FC MPN	
1	0.00	2	
2	0.00	13	
3	0.00	2	
4	0.60	49	
5	0.00	2	
6	0.20	23	
7	0.00	2	
8	0.10	2	
9	0.75	79	
10	0.00	2	
11	0.00	2	
12	0.00	2	
13	0.10	13	
14	0.00	2	
15	0.20	2	
16	0.00	2	
17	0.55	49	
18	0.10	7.8	
19	0.00	2	
20	0.40	23	
21	0.00	2	
22	0.25	7.8	
23	0.00	2	
24	0.00	2	
25	1.10	130	
26	0.20	13	
27	0.00	2	
28	0.00	23	
29	0.00	2	
30	0.00	2	
		5.4	= Geometric mean
		13.3	= % > 43 MPN
		30	= n

The next question that must be addressed is: how long does the impact persist on both water and shellfish quality in the growing area before each return to an acceptable level? Several time series of data will be required to make these determinations, which emphasizes the earlier recommendation for sampling water and shellfish on consecutive days during and after an adverse condition. The assumption is that shellfish, having concentrated contaminants during the adverse condition, will lag behind the growing water in returning to an acceptable level. It will therefore be necessary to collect additional shellfish samples beyond the time when water quality returns to an acceptable level. The first step is to determine when the growing waters have returned to normal, which requires that the meaning of an 'acceptable level' must be defined. In the examples above the FC level of 43 MPN was used to differentiate acceptable from unacceptable data. In *Conditionally approved* areas it is often this criterion that fails before the geometric mean; however, an elevated FC value slightly below 43 MPN does not represent the baseline for good water quality in a *Conditionally approved* growing area. So what

Table 11.3 Fecal coliform (FC) most probable number (MPN/100 mL) data from Table 11.2, with values removed when associated with rainfall that exceeds 0.40 inch. The classification statistics now meet the *Approved* classification. Rainfall data represents cumulative inches of rain over 24 hours. The geometric mean is the antilog of the mean of the log FC values.

Sample number	24 hour rainfall (in.)	FC MPN	
1	0.00	2	
2	0.00	13	
3	0.00	2	
5	0.00	2	
6	0.20	23	
7	0.00	2	
8	0.10	2	
10	0.00	2	
11	0.00	2	
12	0.00	2	
13	0.10	13	
14	0.00	2	
15	0.20	2	
16	0.00	2	
18	0.10	7.8	
19	0.00	2	
20	0.40	23	
21	0.00	2	
22	0.25	7.8	
23	0.00	2	
24	0.00	2	
26	0.20	13	
27	0.00	2	
28	0.00	23	
29	0.00	2	
30	0.00	2	
		3.7	= Geometric mean
		0.0	= Percent > 43 MPN
		26	= *n*

is an appropriate target for improving water quality, beyond which it becomes 'acceptable'? There are several benchmarks that can be considered for this purpose, using the *Conditionally approved* criteria in this example: (1) FC values below 43 MPN; (2) the baseline FC geometric mean during open harvest conditions; or (3) FC values below 14 MPN, the upper limit for the geometric mean. Considering any value less than 43 MPN as acceptable carries some risk of driving the geometric mean upward until it exceeds 14 MPN, with the area then failing its classification criteria. This risk may be reduced by the fact that additional time will be added to the closure to account for shellfish cleansing, as discussed below. A much more conservative approach would be to use the calculated geometric mean of the growing area under open harvest conditions as the target (for example, Table 11.3). This represents the baseline water quality of the growing area, however, there can be significant variability associated with an acceptable geometric mean, as seen in the earlier data sets. Waiting for the FC MPN to return to the baseline geometric mean may be onerous given this variability. A compromise approach would be to use the maximum allowable geometric mean (14 MPN) for the classification as the target, below which water quality would be considered acceptable. This option is more protective of the geometric mean of the growing area than the 43 MPN target, although it may still result in a slight elevation of the geometric mean over time. This approach is also more realistic than using the growing area's baseline geometric mean, as it accommodates some variability in the data. For consistency with the earlier example, the 14 MPN target will be applied to the next data set discussed below.

The data set in Table 11.4 illustrates one example of the relationship between water and shellfish fecal coliform (FC) values during adverse conditions (for example, rainfall) and will be used to examine both the classification determination and the development of a management strategy

Table 11.4 Fecal coliform (FC) most probable number for 30 water (MPN/100 mL) and shellfish (MPN/100 g) samples, collected under the adverse pollution condition sampling strategy. The associated classification statistics illustrate that the data do not meet the *Approved* classification. Rainfall data represents cumulative 24 hour rainfall in inches. The geometric mean is the antilog of the mean of the log FC values.

Sample number	Sampling date	24 hour rainfall (in.)	FC MPN (water)	FC MPN (shellfish)	
1	10/7	0.50	49	2300	
2	11/12	0.40	13	200	
3	12/10	0.60	450	1100	
4	12/11	1.10	1700	4600	
5	12/12	0.00	790	9200	
6	12/13	0.00	130	7900	
7	12/14	0.00	79	4900	
8	12/15	0.00	11	3000	
9	12/16	0.00	7.8	2300	
10	12/17	0.00	2	230	
11	12/18	0.00	2	130	
12	2/1	0.00	7.8	80	
13	2/17	0.00	4.5	130	
14	2/28	0.40	23	230	
15	3/3	0.20	11	170	
16	3/11	0.65	1100	7900	
17	3/22	0.10	7.8	49	
18	4/6	0.80	1300	2300	
19	4/7	0.00	2300	11000	
20	4/8	0.00	1100	9200	
21	4/9	0.00	450	7900	
22	4/10	0.00	110	4600	
23	4/11	0.00	7.8	3300	
24	4/12	0.00	4.5	2300	
25	4/13	0.00	7.8	220	
26	5/24	0.00	2	49	
27	6/17	0.00	49	20	
28	7/2	0.00	2	230	
29	8/21	0.00	2	130	
30	9/14	0.60	79	4900	
			37.3	908	= Geometric mean
			46.7	n/a	= Percent > 43 MPN
			30	30	= *n*

for the adverse condition. These data include water and shellfish FC data from a hypothetical study carried out at one growing area sampling station as part of a sanitary survey. The data was collected under the APC strategy and includes a time series of samples for two significant rainfall events, as well as single samples associated with rainfall and samples collected in the absence of an adverse condition. Focusing first on the water quality data that is the basis for classification, it is obvious that there are significant water quality impacts associated with rainfall, a clear indication that the growing area represented by this sampling site will not meet the *Approved* classification.

There is a positive relationship between the 24 hour rainfall totals and the FC level, with increasing rain totals resulting in increased FC levels. This suggests the possibility that rainfall can be managed to ensure the area is closed during these predictable adverse conditions, allowing consideration of the *Conditionally approved* classification as an option. As with the earlier example, it appears that significant water quality impacts are likely to occur after a 24 hour cumulative rainfall greater than 0.40 inch. Examining the first time series of adverse condition data, beginning on December 10, there is a rapid increase in growing water FC with the first 0.60 inch of rain, and a further increase on the second day as a result of an additional 1.10 inches of rain. By the third day there has been no more rainfall and the FC MPN has started to decrease, but remains well above the allowable limit (that is, greater than the target of 14 MPN). Water quality returns to an acceptable level 4 days after the cessation of rainfall (11 MPN on December 15). The second rainfall event that was sampled on consecutive days occurs on April 6. The 1 day storm event has a 24 hour rainfall total of 0.80 inch, resulting in a significant impact to water quality. There is no subsequent rainfall and the growing area station returns to an acceptable FC level 5 days after the end of the storm (7.8 MPN on April 11).

At the bottom of Table 11.4 are the summary statistics used to evaluate APC data. It can be seen that the geometric mean (36.6 MPN) is well above the 14 MPN maximum and that 46.7% of the 30 water samples exceed 43 MPN, violating the *Approved* classification limit of less than 10%. If the unacceptable growing water FC levels associated with significant rainfall (Table 11.5) are excluded, the geometric mean is reduced to an acceptable level (7.3 MPN) and 3.3% of the remaining values exceed an MPN of 14; the remaining errant value (49 MPN on 6/17) is not associated with an adverse condition and therefore remains a part of the data set for open harvest conditions. The second rainfall event resulted in greater water quality impacts than the first event despite a lesser amount of rain, and an additional day was required for water quality to return to an acceptable level. These observations might be explained by saturation of the watershed during the rainy season, with less precipitation needed to generate greater runoff and higher pollutant loads. The water quality response to the two storm events was similar, despite the noted differences in FC levels, increasing confidence that rainfall closure rules can be developed that are reliable and protective of public health. This satisfies one of the major tenants for a *Conditionally approved* classification: that the condition(s) responsible for the area failing the *Approved* classification criteria are predictable and not so complex as to be unmanageable.

The next step is to evaluate the shellfish FC data relative to the magnitude and duration of impact to shellfish tissue quality. This will hopefully provide some guidance for determining a baseline FC MPN value for shellfish. Returning to the December 10 rainfall event (Table 11.4), an immediate increase in shellfish FC MPN can be seen in response to rainfall. Unlike the water FC level that began declining on the first day following the cessation of rain (December 12), the FC level in shellfish continues to increase, a not uncommon occurrence, and then begins to decline 2 days after the cessation of rain. Similarly, for the second rainfall event on April 6 the shellfish exhibited a 1 day lag in the FC maximum, which was higher than observed in the first event. There is a marked improvement in the FC geometric mean for the shellfish samples (Table 11.5) when the data associated with unacceptable water quality during adverse conditions is removed (245 MPN versus 908 MPN for the adverse condition data set in Table 11.4).

Before the shellfish data can be interpreted, the issue of what constitutes an acceptable level of FC in shellfish must be addressed. If it is assumed that the shellfish are adequately cleansed when

Table 11.5 Fecal coliform (FC) data for water (MPN/100 mL) and shellfish (MPN/100 g) samples from Table 11.4, with values removed for rainfall-related impacts to water quality. The classification statistics illustrate the area now meets the *Approved* classification. Rainfall data represents cumulative 24 hour rainfall in inches. The geometric mean is the antilog of the mean of the log FC values.

Sample number	Sampling date	24 hour rainfall (in.)	FC MPN (water)	FC MPN (shellfish)	
2	11/12	0.40	13	200	
8	12/15	0.00	11	3000	
9	12/16	0.00	7.8	2300	
10	12/17	0.00	2	230	
11	12/18	0.00	2	130	
12	2/1	0.00	7.8	80	
13	2/17	0.00	4.5	130	
14	2/28	0.40	23	230	
15	3/3	0.20	11	170	
17	3/22	0.10	7.8	49	
23	4/11	0.00	17	3300	
24	4/12	0.00	4.5	2300	
25	4/13	0.00	7.8	220	
26	5/24	0.00	2	49	
27	6/17	0.00	49	20	
28	7/2	0.00	2	230	
29	8/21	0.00	2	130	
			6.5	245	= Geometric mean
			5.9	n/a	= Percent > 43 MPN
			17	17	= n

water quality returns to an acceptable level, the 245 MPN shellfish geometric mean from Table 11.5 can be used as the target. There is some concern, however, that the FC levels in shellfish continue to decline even after the growing water FC had returned to an acceptable level. So, there is uncertainty as to what the true shellfish FC baseline really is. As a starting point, if the tentative target of 245 MPN is applied to the shellfish data in Table 11.5, there are several values associated with the rainfall impacts that exceed this maximum. A closer look reveals that the shellfish require 2 additional days to decline below this target once the growing water has reached an acceptable level. If the adverse condition shellfish data that exceeds 245 MPN is removed, the geometric mean of the remaining data is further reduced to 117 MPN (Table 11.6).

While this may be a more accurate estimate of the baseline shellfish FC levels, using this value as the target 'acceptable level' carries the same drawback as discussed for the water quality data. Recalling the discussion of the difficulty in using the baseline water quality geometric mean as a benchmark for acceptable water quality, the variability of the baseline shellfish data would likewise make this an onerous target to apply to the shellfish FC data: only four of the 13 shellfish values in Table 11.6 would be considered acceptable, despite the obvious lack of water quality impacts. While the 245 MPN target for shellfish may seem reasonable, this geometric mean is influenced by values that are obviously impacted by the adverse condition. In addition to the concerns raised for each of the two efforts to establish a shellfish FC baseline value, there

Table 11.6 Fecal coliform (FC) data for water (MPN/100 mL) and shellfish (MPN/100 g) samples from Table 11.5, with additional records removed for rainfall-related impacts to shellfish quality. Rainfall data represents cumulative 24 hour rainfall in inches. The geometric mean is the antilog of the mean of the log FC values.

Sample number	Sampling date	24 hour rainfall (in.)	FC MPN (water)	FC MPN (shellfish)	
2	11/12	0.40	13	200	
10	12/17	0.00	2	230	
11	12/18	0.00	2	130	
12	2/1	0.00	7.9	80	
13	2/17	0.00	4.5	130	
14	2/28	0.40	23	230	
15	3/3	0.20	11	170	
17	3/22	0.10	7.8	49	
25	4/13	0.00	7.9	220	
26	5/24	0.00	2	49	
27	6/17	0.00	49	20	
28	7/2	0.00	2	230	
29	8/21	0.00	2	130	
			5.9	117	= Geometric mean
			7.7	n/a	= Percent > 43 MPN
			13	13	= *n*

is a separate concern that growing areas in separate bodies of waters will probably have different baseline values: from a public health perspective it would be difficult to justify allowing shellfish to be harvested from a bay that had a considerably higher baseline FC (for example, 790 MPN) than that for the growing area in the present example. As an aside, the use of the estimated 90th percentile of the baseline data (for example, 311 MPN for shellfish data in Table 11.6) might be more practical as it accounts for the inherent variability in the non-adverse condition data set. This approach is still area-specific, however, and will probably create an un-level playing field among non-contiguous growing areas.

One piece of information that is lacking from this process is the relative public health risk associated with different FC levels in shellfish. Although the NSSP lacks a growing area standard for shellfish FC, the Model Ordinance does establish critical limits for FC in depurated shellstock: for oysters, mussels, and clams (other than soft-shell clams), the geometric mean of the most recent 10 process samples cannot exceed 20 MPN and the estimated 90th percentile cannot exceed 70 MPN. This is a special application for depuration process control and is not a good fit for growing area monitoring, leaving us to look elsewhere for guidance. Earlier editions of the Fish and Fishery Products Hazards and Controls Guidance (US Food and Drug Administration, 2011) contained a standard for removal of suspect product from the marketplace: one or more of five subsamples exceeding either FC or *E. coli* levels of 330 MPN, or two or more subsamples exceeding 230 MPN for either indicator. The latest edition of this guidance document (US Food and Drug Administration, 2020) has removed this guidance, making it consistent with the lack of a shellfish meat standard for FC or *E. coli* in the Model Ordinance. The EU classification system, discussed earlier, requires that shellfish that can be direct marketed after harvest

(*A*-zone) must have less than 230 MPN *E. coli*, with a maximum of 20% of values between 230 and 700 MPN and no value exceeding 700 MPN. *E. coli* is a subset of FC, so applying the 230 MPN as a target FC baseline in the study data would potentially be more conservative, or at least equivalent; in some cases, particularly with 'fresh' sources, all of the FC can be in the form of *E. coli*. Having a uniform standard to apply to all growing areas is preferable to unique site-specific targets that are likely to result from the sanitary survey data for growing areas in different bodies of water. For the following discussion a shellfish FC standard of 230 MPN will be applied. Examining the shellfish FC data in Table 11.4 for the two rainfall events, it can be seen that the FC in shellfish declines below 230 MPN 2 days after water quality returns to an acceptable level. For the December 10 event the water quality returned to an acceptable level 4 days after the cessation of rain, with an additional 2 days required for shellfish FC to decline below the 230 MPN target. This results in a 6-day harvest closure following greater than 0.40 inch of rain in any 24 hour period. The April 6 rainfall event resulted in higher FC levels than the first event, requiring 5 days for water quality to return to an acceptable level and an additional 2 days for shellfish to do so. The data from this second rainfall event supports a 7-day harvest closure, 1 day more than indicated by the data from the December sampling event. The differences in FC magnitude and cleansing rates between the two rainfall events studies suggest that greater water and shellfish quality impacts occur with less rain as the watershed becomes saturated. Unfortunately it is probably beyond the capacity of the SCA to develop a reliable 'sliding scale' for rainfall closures based on time of year or cumulative rainfall throughout the wet season. The most conservative approach will therefore be taken, which means a 7-day closure in the current example. In practice the SCA will periodically collect samples to verify the closure and reopening criteria for classified growing areas, ensuring that the reopening rules remain protective of public health without being unnecessarily restrictive.

In the current discussion, assuming that any additional sampling stations in the growing area responded similarly, the recommendation would be to classify the area as *Conditionally approved*. The growing area met the *Approved* water quality criteria for a significant period of time in the absence of an adverse condition, but failed to meet acceptable water and shellfish quality when significant rainfall occurred. The periods of impact were determined to be predictable and manageable by implementing a harvest closure when rainfall exceeds a 24 hour cumulative amount of 0.40 inch. The harvest area would be reopened 7 days after the cessation of rain. A final consideration in this example is the occurrence of extreme environmental conditions, for example, rainfall events that exceed the totals represented in the sanitary survey data set. In such cases the harvest closure would remain in place until sufficient sampling data was collected to demonstrate that water and shellfish quality had returned to acceptable levels.

This example has been a simplistic treatment of a complicated process and is primarily intended to provide a basic understanding of FC data and some insight into the challenges of interpretation. In reality there may be numerous sampling stations in the growing area and additional APC variables, such as tidal direction and current velocity, or an upstream treatment plant that fails after a certain intensity of rainfall-related runoff, which were not considered in the current examples. Initially it may seem a burdensome and time-consuming task to collect 30 samples per station for classification. It is, in fact, a very limited data set for developing an understanding of the complex relationship between adverse conditions, pollution sources, and growing area water and shellfish quality. The mantra for most studies involving growing water and shellfish quality is the need for more and more samples to help resolve the 'noise' of environmental variability. Although the example data set contained the necessary minimum of 30 samples, only 13 samples remained after filtering out the adverse condition data. There are also only two rainfall events that had a time series of growing water and shellfish data, a scant amount upon which to base an understanding of rainfall impacts in the area and to develop criteria for harvest closure and reopening. In reality it may take several more sampling events to develop a somewhat accurate assessment of the relationship between rainfall rate, magnitude, and duration and the resultant FC levels in growing area water and shellfish tissue. This relationship may vary over the course of the rainy season as initial rainfall is absorbed by the soil, resulting in lesser impacts to the growing area; after saturation it may take less and less precipitation to produce runoff that is increasingly contaminated. The rules developed for closure and reopening will have to account for the worst case scenario for the sake of public health protection, which may result in some early season closures when water or shellfish quality were not impacted significantly. An alternative approach to a fixed length harvest closure is to collect growing area water samples during the closure to determine when fecal coliform has returned to an acceptable level: if water quality was acceptable the area could be reopened once the additional time necessary for shellfish to return to an acceptable level had passed. This approach may prove useful for the first few seasonal storms and less so once the watershed is saturated. Of course, rainfall is not the only possible adverse condition, and a different approach will be needed for the different sources identified in the shoreline and sanitary survey. For example, a *Conditionally approved* area impacted by WWTP, whose performance degraded after significant rainfall, might have a harvest closure based on a loss of adequate disinfection at the plant (for example, determined by a declining chlorine residual). Regardless of the pollution source, it would benefit the decision-making process to gather more adverse condition data, particularly time series studies, as well as much more baseline data for both growing water and shellfish samples. An alternative to the APC approach used in the previous example would be to apply the SRS strategy, establishing a set schedule of random sample collection dates for the growing area stations (for example, every other Monday) and conducting separate intensive studies of adverse conditions to determine thresholds for harvest closures and reopening criteria.

As detailed in the 'Classification of shellfish growing waters' section, the EU classification system is based on the levels of *E. coli* in shellfish tissue. Two shellfish growing areas on the northwest Mediterranean coast of Spain will be used to illustrate the application of this microbiological classification system. The Cases d'Alcanar growing area has 49 shellfish samples collected between February 2013 and December 2017 (Fig. 11.1). All of these samples have *E. coli* levels below 230 MPN, therefore this area would be classified as *A* zone, suitable for direct marketing of harvested shellstock.

The Fangar Bay growing area has 40 shellfish samples collected between January 2013 and December 2017 in each of three regions of the bay (Fig. 11.2). While the majority of *E. coli* levels in shellfish are below the 230 MPN limit, one or more samples from each region (outer, middle, and inner bay) exceeds the 700 MPN limit, violating the requirements for classification as *A* zone. None of the samples exceeds the 4600 MPN limit for *E. coli* in shellfish tissue, so Fangar Bay meets the classification criteria for *B* zone, requiring that all shellfish be cleansed in an approved depuration facility before being placed in the market.

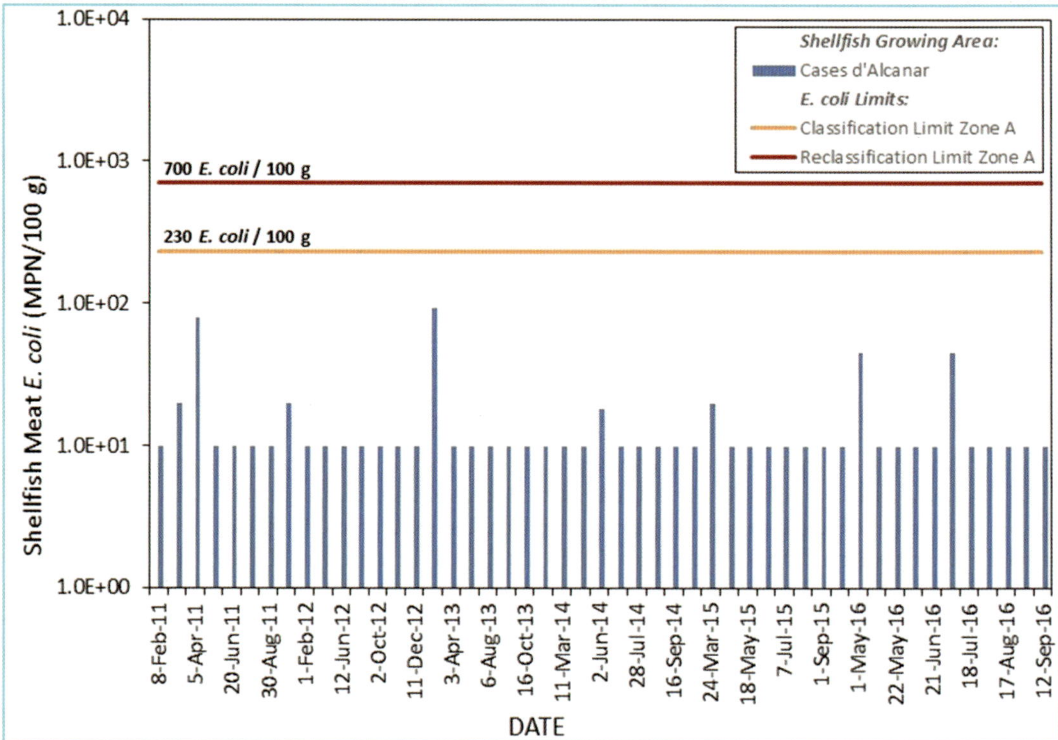

Figure 11.1 Shellfish tissue *E. coli* data that meets the EU microbiological classification requirements for an *A*-zone shellfish harvesting area. *Núria Sagristà*, Programa AMiC, IRTA-Sant Carles de la Rapita.

Marine biotoxin data

In contrast to the somewhat ambiguous matter of 'acceptable levels' for water and shellfish FC data, the marine toxin action levels described earlier are definitive: a growing area is closed when a given toxin threshold is met or exceeded (for example, 80 µg/100 g for PSP) and cannot reopen until shellfish toxin concentrations decline, and are demonstrated to remain, below this level. An area that experienced frequent and consistently high toxin levels would not be considered for commercial growing, as there are no proven treatment processes for reducing toxin levels in shellfish. Assuming dangerous toxin levels occurred in growing area shellfish for limited periods of time, the sanitary survey process would concern itself with developing management strategies for the various toxins of concern. The SCA may determine that additional toxin data was needed as part of the sanitary survey; this additional information would provide insight into the toxins, and toxin-producing phytoplankton, of concern, and help guide the development of a routine monitoring strategy once the area was classified. The management considerations for marine toxins are discussed further in the next section.

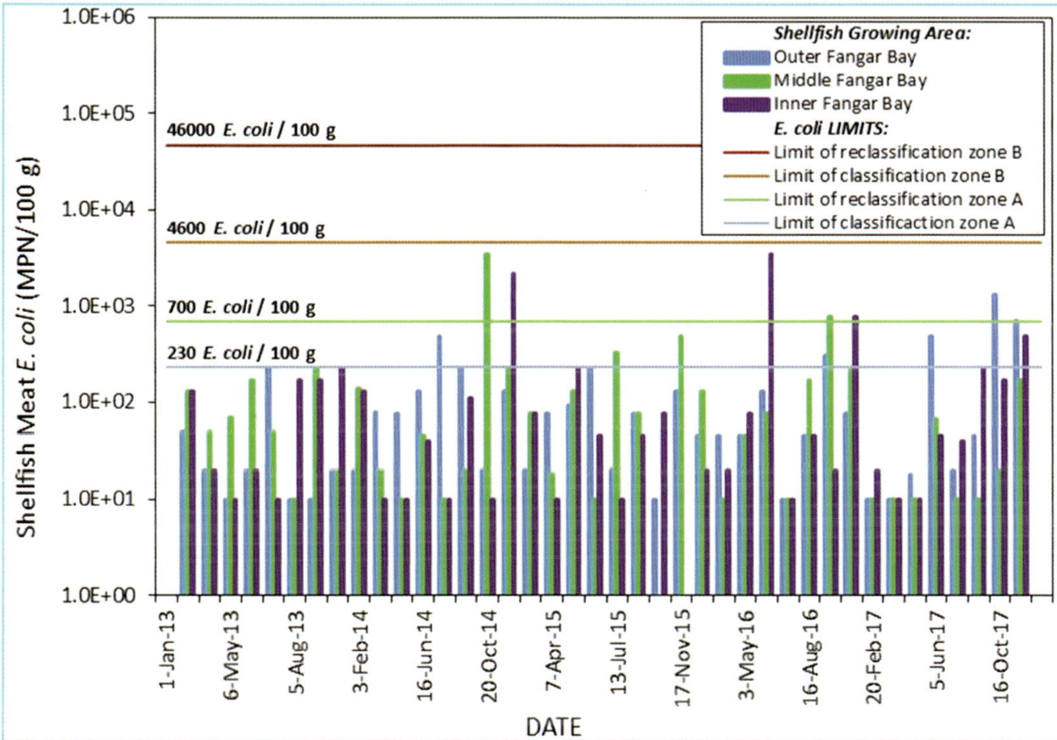

Figure 11.2 Shellfish tissue *E. coli* data that meets the EU microbiological classification requirements for a *B* zone shellfish harvesting area. Núria Sagristà, Programa AMIC, IRTA-Sant Carles de la Rapita.

Next steps

Finally, once the shoreline and sanitary survey work, special studies, and data analysis have been completed, the SCA will prepare a report of the findings, including all of the elements discussed earlier (for example, hydrography, hydrology, water and shellfish quality studies), with a recommendation for the determined classification. The sanitary survey report will typically be subject to a review and comment period, allowing input from all of the stakeholders. Once a final report is distributed there are some remaining decisions to be made and some additional requirements that must be met before the classification of the area can be finalized. These topics are addressed in the next section on 'Classification considerations'.

Classification considerations

While the water quality data collected during the sanitary survey guides the classification of the proposed area, there are a number of additional considerations that also affect this decision. An *Approved* classification is the simplest for the SCA to manage, with a requirement for routine monitoring of FC levels at the designated growing area sampling sites. This classification is also the least restrictive for the grower, with no harvest restrictions other than what may arise from an emergency event (for example, oil spill). The *Conditionally approved* classification, on the other hand, requires a tremendous amount of oversight and resources on the part of the SCA, as well as a significant commitment on the part of the grower. Because the *Conditional* classification is so resource-intensive, the Model Ordinance considers this classification optional. The SCA must develop a management plan for each *Conditionally approved* area that describes each of the conditions (for example, biotoxins, rainfall, WWTP performance, seasonal marina) that may require a harvest closure and all of the related requirements and procedures for closing and reopening the growing area. The area covered by the plan may include a number of different leases in a common body of water, as appropriate. The management plan must also have provisions for patrol of the growing area by law enforcement personnel (for example, local sheriff or fish and wildlife warden) during harvest closures to prevent illegal harvesting; a separate patrol policy document must also be developed as per NSSP requirements. The management plan is a cooperative agreement between the SCA, the involved shellfish industry members, any WWTP operators in the watershed, and other state, federal, tribal, or local agencies with jurisdiction in the growing area or surrounding watershed (for example, natural resource and water quality agencies). All of the involved parties must be signatories to the management plan; failure of any one party to agree to the conditions of the plan will prevent the use of the *Conditionally approved* classification. In support of the management plan the SCA must monitor rainfall frequently each day, including weekends and holidays. If the growing area is potentially impacted by marine toxins then the SCA must conduct a routine monitoring program, documented in a biotoxin control plan. Likewise, a *Vibrio* control plan must be prepared if shellfish in the growing area are potentially affected by this free-living pathogen, as determined by the Model Ordinance risk assessment criteria for *V. vulnificus* and *V. parahaemolyticus*. All of the assorted documents must be updated at least annually and are subject to audit by FDA as part of their routine evaluation of state shellfish programs. In addition, the sanitary survey process must be repeated every 12 years, or sooner if there has been a significant change in the growing area or watershed that can potentially impact water and shellfish quality in the growing area.

The *Conditionally approved* classification also creates a significant demand on the shellfish grower, as the conditional area management plan will have numerous requirements that pertain to their operation. The grower will probably be required to keep records of all harvest activity, including location, date, time, species, and quantities; records of closure and reopening times may also be required. The records will be subject to audit by the SCA, so it is in the grower's best interest to develop a reliable and accurate system. The SCA may require the grower to have some means of contact (for example, cell phone, answering machine) 24 hours a day, 7 days a week, to ensure they can be notified of harvest closures, reported illness, and emergency situations that could impact the growing area. There are additional *post-harvest* record keeping requirements in the Model Ordinance (for example, tracking the time required to get harvested shellstock under proper temperature control) that the grower must also become familiar with. There is also a responsibility on the part of the grower to immediately notify the SCA of potential illnesses reported to them or illegal activities that they may observe (for example, illegal harvesting during a closure). While the *Conditionally approved* classification provides for the use of growing waters that are occasionally impacted by microbial contaminants or marine toxins, the management of these areas involves a significant commitment on the part of the shellfish grower and the SCA.

If the growing area cannot be classified as *Conditionally approved* due to lack of agreement by one or more stakeholders or poor water quality that does not meet the requirement, there is the possibility of using the *Restricted* classification. As explained in the 'Classification of growing waters' section, this classification can only be used if the grower has either a certified depuration facility for cleansing the contaminated shellfish or a separate growing area in the *Approved* or *Conditionally approved* classification to which contaminated shellfish can be transferred and held for cleansing, a process called 'relaying' in the NSSP. Relayed shellstock must remain in the destination area for greater than 60 days, assuming only microbial contaminants must be reduced. The portion of the *Approved* or *Conditionally approved* growing area used for receiving relayed product must remain in the closed harvest status for the duration of the relay process. This receiving area must be physically marked so that relayed product remains separated from other product raised in the growing area; individual relay lots must also be clearly marked and distinguishable from one another. Shorter cleansing times may be used only if a contamination reduction study is conducted and demonstrates the reduction of microbial pathogens to background FC levels of the shellstock grown in the receiving lease. If the contaminant reduction study supports a cleansing time period shorter than 14 days then shellfish tissue samples must be collected and tested for FC for each lot before and after the relay process. If the growing area is impacted by WWTP discharge then MSC may be used to monitor the reduction of viral pathogens in shellfish.

Use of the *Restricted* classification for relaying carries a number of requirements for both the SCA and the grower. For example, the SCA must have the legal authority to issue a relay license, which is only valid for a specific relay activity. The SCA or other authorized patrol agency must oversee each relay operation from start to finish. The grower must keep detailed records for each relay, for example the date and time relayed from the *Restricted* area, the date and time the shellstock entered the destination growing area, the species and quantity relayed, the exact location in the destination growing area with a description of how different lots are marked and kept separate, and the date and time harvested after the end of the relay period. Although the requirements of relaying may be beyond the capacity or interest of most prospective growers, for those with the resources it can be a viable option by using the *Restricted* areas for increasing production, with cleansing in *Approved* or *Conditionally approved* waters. It should also be noted that shellstock in a *Conditionally approved* area can be relayed from that area when it is closed to harvest, provided the water quality always meets the *Restricted* criteria during closures. So, there are obvious advantages of having more than one growing area, particularly if at least one is in the *Approved* classification and has room to receive relayed shellfish from the others. The requirements for approval and operation of a depuration facility are well beyond the scope of this chapter and a prospective grower considering use of a *Restricted* area is encouraged to review the complete requirements for relaying and depuration in the Model Ordinance, Chapters V and XV, respectively (US Food and Drug Administration, 2019). The last iteration for classification is for a *Restricted* area that cannot meet the water quality criteria for this classification due to some adverse condition such as rainfall. In this case the *Conditionally restricted* classification may be considered. All of the requirements discussed for the *Conditionally approved* classification would also apply in this circumstance.

Regardless of the classification, there will be the need for routine water quality monitoring in the growing area. In some states the SCA conducts all sampling activities, while other states have trained and certified specific shellfish growers to collect samples following a sampling protocol developed by the SCA. These data will be evaluated as they are generated to ensure ongoing compliance with the classification requirements. The growing area water quality data are also used for the annual reevaluation of the classification; this reevaluation includes a consideration of the cooperation and compliance with the management plan by all involved parties, as well as grower compliance with all harvest closures and the various post-harvest requirements they are required to meet. The SCA may also conduct field inspections of pollution sources as warranted. Every third year the SCA conducts a more detailed evaluation of pollution sources along with a review of the past 3 years of data and the various elements described for the annual reevaluation. As noted earlier, a complete sanitary survey must be conducted at least every 12 years for each growing area.

Marine biotoxin management

If the SCA determined the need for marine biotoxin data collection as part of the sanitary survey, this information would be used to develop a management strategy once the area is classified. If data collected during the survey indicated that the proposed growing area was more heavily impacted, or more quickly impacted, than other growing areas in the vicinity, a separate sampling plan would be necessary. For example, this plan may involve more frequent sampling of phytoplankton and shellfish to ensure early detection of toxicity, the maintenance of a sentinel mussel station to detect toxins before they accumulate in less susceptible aquacultured species, and/or the use of a field test kit for toxin presence. Conversely, if the new area did not appear to represent a greater risk from marine toxins than adjacent growing areas, the current sampling effort of existing shellfish lease(s) may serve to protect the new area, avoiding unnecessary duplication of effort and saving laboratory resources until they are most needed. An existing marine biotoxin monitoring program will have effective, proactive control mechanisms, ensuring that harvesting can be suspended before contaminated product is in the marketplace. In some cases, it may be possible to restrict harvesting to certain safe areas in a body of water while the impacted areas are closed. An example would be a long embayment that experiences shellfish toxicity in growing areas near the mouth, with toxin impact lessening the farther inside the bay you moved. In some areas it may be feasible to move shellfish from a high-risk area to a safer area before the former is closed. Another option is to allow batch harvesting (referred to as an end product testing program in the Model Ordinance) from high-risk areas (for example, a growing area with increasing toxin levels that are still below the action level, or an area adjacent to a recently closed lease), with each batch being held in cold or wet storage until representative samples are analyzed and the safety of the batch determined. Management strategies will vary from region to region based on local knowledge of toxicity events and the resources of the agencies involved. Fortunately, the NSSP provides some flexibility for regional differences among states or coasts instead of a 'one size fits all' approach. This allows the SCA to use their regional knowledge, experience, and professional judgement to manage shellfish toxins in a way that protects public health and avoids costly and unnecessary harvest closures.

Eventually, when the growing area is classified, the new grower will need to be aware of their responsibility during a biotoxin event. The NSSP requires the SCA to develop and maintain a marine biotoxin management plan for areas potentially impacted by one or more marine algal toxins. The biotoxin management plan defines the administrative procedures and resources necessary to: (1) maintain a routine shellfish sampling and testing program; (2) close growing areas and embargo shellfish; (3) prevent harvesting of contaminated species; (4) provide for product recall; (5) disseminate information on the occurrences of toxic algal blooms and/or toxicity in shellfish meats to adjacent states, shellfish industry, and local health agencies; (6) coordinate control actions taken by the SCA and federal agencies; and (7) establish reopening criteria. The grower will likely be required to collect shellfish samples throughout a biotoxin event, perhaps multiple times per week, to track toxin levels and to determine when the earliest possible time is to reopen safely. Conversely, for labs operating with limited resources, sampling frequency may be reduced until toxin concentrations begin declining and are near the action level. The collection and identification of phytoplankton samples and use of the qualitative toxin-testing kits are ways in which the grower can generate more frequent information on the status of the event. Although the NSSP does not require phytoplankton monitoring, the Model Ordinance does recommend an early warning system that includes phytoplankton sampling and it is to the grower's advantage to collect this information. The NSSP also requires the SCA to develop and adopt a marine biotoxin contingency plan for all shellfish growing areas in response to the emergence of a toxin-producing phytoplankton that has not historically occurred or an illness outbreak caused by marine biotoxins. The contingency plan addresses the management of PSP, ASP, NSP, DSP, and AZP, defining the administrative procedures and resources necessary to "(a) initiate an emergency shellfish sampling and assay program; (b) close growing areas and embargo shellfish; (c) prevent harvesting of contaminated species; (d) provide for product recall; (e) disseminate information on the occurrences of toxic algal blooms and/or toxicity in shellfish meats to adjacent states, the shellfish industry, and local health agencies; (f) coordinate control actions taken by authorities and federal agencies, and (g) establish reopening criteria including the number of samples over what period of time (US Food and Drug Administration, 2019). More information on the requirements for marine biotoxin monitoring and control can be found in Section IV, Chapter II of the Model Ordinance.

The EU has also established the need for a sampling plan for the required monitoring of both phytoplankton and toxin levels in shellfish, taking into consideration that an appropriate risk evaluation is conducted if frequencies are to be reduced. The EU regulations also provide guidance on a uniform procedure for enumeration of phytoplankton (European Committee for Standardization, 2011), and a specific procedure for the enumeration of phytoplankton with the Utermöhl method (European Committee for Standardization, 2006), which describes acceptable laboratory procedures for microscopic algal analyses. EU member countries develop their region specific management plans for marine toxins that contain details on sampling sites, sampling frequency, sampling methodology, the reference and official analytical methods for marine toxins, and all of the other region-specific information necessary to manage effectively for the occurrence of marine toxins in shellfish to protect public health.

Summary

There are many challenges faced by those intending to start a shellfish aquaculture business, not the least of which is coming to grips with the regulatory infrastructure they will have to operate within. To the uninitiated, the world of shellfish regulation may seem unfathomable, overwhelming, and maybe a little discouraging. Every shellfish company, new or established, is encouraged to participate in their regional shellfish sanitation organization and the ISSC: these are forums for learning about advancements in the science of shellfish sanitation and for providing input on the development or modification of NSSP guidelines.

In addition to overseeing commercial shellfish operations to ensure compliance with NSSP guidelines, the SCA is a valuable resource for the potential new shellfish grower. The SCA can provide information and guidance on the overall process for obtaining the required license(s) to grow and market bivalve shellfish. Prior to investing significant time and expense, the prospective shellfish aquaculturalist should contact the SCA in their state to discuss what is known about prospective growing areas before obtaining a lease. For example, are there existing aquaculture operations in the area and, if so, how are they classified? Are there records of water and shellfish quality data for the specific area of interest that can be reviewed? What are the marine biotoxins of concern and what control strategies currently exist? Is it possible to estimate the annual number of harvest closure days expected for the area based on water quality and biotoxin impacts experienced by any current aquaculture operations? This is a good time to also discuss the sanitary survey and classification process; the applicant should gain an understanding of the timelines involved for designing and conducting this work and determine what level of involvement is expected of them. The prospective grower should also discuss the various post-harvest requirements for their anticipated operation that were not discussed in this chapter; these may involve additional time and expense for training (for example, Hazard Analysis Critical Control Point [HACCP] certification), purchase and verification of monitoring equipment, and so on.

Given the preference for raw consumption of some bivalve shellfish (for example, oysters and some clam species) there is an increased risk of illness to the consumer from a variety of pathogens compared to cooked seafood, even under the best of practices. Some consumers may not be aware of these risks, so the shellfish company has a role in educating their customers and ensuring they are aware of proper handling procedures (for example, proper refrigeration). This of course means that a new shellfish grower must first educate themselves on the various public health risks associated with their product. This chapter has introduced the more common concerns that exist in the growing area prior to harvest, including a variety of microbial pathogens and marine toxins. Additional risks associated with other food pathogens occur throughout the processes of harvesting, handling, shipping, and marketing. It is essential that the grower understand these risks, as well as the control strategies required to manage them. More information on these topics can be found in the links and references in Appendix I and the Model Ordinance. Ultimately it is the responsibility of the growers to ensure that a safe and wholesome product is delivered to their customer. Understanding the various public health concerns, managing for them with the best available control strategies, and being an active participant in the regional and national shellfish organizations will provide the best opportunity to have a sustainable business producing wholesome and healthy shellfish.

Acknowledgments

The authors would like to gratefully acknowledge the Monitoring Programme for the Quality of Waters in Shellfish Harvesting Areas in Catalonia, funded by the Direcció General de Pesca i Afers Marítims of the Catalan Government and Núria Sagristà, Programa AMIC, IRTA-Sant Carles de la Rapita for providing graphics of *E. coli* numbers in shellfish. We also want to thank Vanessa Zubkousky-White, California Department of Public Health Preharvest Shellfish Program, for review of the fecal coliform example data and associated helpful comments, and for providing the water quality sampling plan that was modified slightly for use in this chapter.

References cited and further reading

AOAC International (2019) *Official Methods of Analysis of AOAC International.* 21st edition. Association of Official Analytical Chemists, Arlington, VA.

EPA (2015) Review of coliphages as possible indicators of fecal contamination for ambient water quality. Office of Science and Technology, Health and Ecological Criteria Division, 820-R-15-098.

European Committee for Standardization (2006) *Water quality. Guidance standard on the enumeration of phytoplankton using inverted microscopy* (Utermöhl technique), EN 15204:2006.

European Committee for Standardization (2011) *Water quality. Guidance on quantitative and qualitative investigations of marine phytoplankton.* European Committee for Standardization, CSN EN 15972:2011.

Food and Agriculture Organization of the United Nations (FAO) (2014) Assessment and management of seafood safety and quality. Ryder, J., Karunasagar, I., and Ababouch, L. (eds.) FAO Fisheries and Aquaculture Technical Paper – TP 574.

Gerba, C.P. (2009) Indicator microorganisms. In: Gerba, C. (ed.) *Environmental Microbiology.* Elsevier, Oxford, pp. 485–499.

Greening, G.E. and McCoubrey, D.-J. (2010) Enteric viruses and management of shellfish production in New Zealand. *Food and Environmental Virology*, 2(3): 167–175.

Horner, R. (2002) *A Taxonomic Guide to Some Common Marine Phytoplankton.* Biopress, Bristol.

Livingstone, D.R. and Gallacher, S. (2003) Shellfish: contamination and spoilage of molluscs and crustaceans. In: Trugo, L. and Finglas, P.M. (eds). *Encyclopedia of Food Sciences and Nutrition.* Academic Press, Oxford, pp. 5228–5245.

Oliveira, J., Cunha, A., Castilho, F., Romalde, J.L., and Pereira, M.J. (2011) Microbial contamination and purification of bivalve shellfish. Crucial aspects in monitoring and future perspectives – a mini-review. *Food Control*, 22: 805–816.

Shumway, S.E., Burkholder, J.M., and Morton, S.L (eds.) (2018) *Harmful Algae: A Compendium Desk Reference.* Wiley Science, Chichester.

Stoeckel, D.M., Mathes, M.V., Hyer, K.E., Hagedorn, C., Kator, H., Lukasik, J., O'Brien, T.L., Fenger, T.W., Samadpour, M., Strickler, K.M., and Wiggins, B.A. (2004) Comparison of seven protocols to identify fecal contamination sources using *Escherichia coli. Environmental Science and Technology*, 38: 6109–6117.

Stoeckel, D.M. and Harwood, V.J. (2007) Performance, design, and analysis in microbial source tracking studies. *Applied and Environmental Microbiology*, 73(8): 2405–2415.

US Food and Drug Administration (2011) *Fish and Fisheries Products Hazards and Control Guidance.* April 2011. FDA, Washington, DC.

US Food and Drug Administration (2019) *National Shellfish Sanitation Program Guide for the Control of Molluscan Shellfish.* FDA, Washington, DC.

US Food and Drug Administration (2020) *Fish and Fisheries Products Hazards and Control Guidance.* March 2020. FDA, Washington, DC.

Work, T.M., Barr, B., Beal, A.M., Fritz, L., Quilliam, M.A., and Wright, J.L.C. (1993) Epidemiology of domoic acid poisoning in briown pelicans (*Pelecanus occidentalis*) and Brandt's cormorants (*Phalacrocorax penicillatus*) in California. *Journal of Zoo and Wildlife Medicine,* 24(1): 54–62.

World Health Organization (WHO) (2010) *Safe Management of Shellfish and Harvest Waters.* WHO, Geneva.

Appendix I Links and references for more information

Pathogens and health effects

https://www.cdc.gov/norovirus/about/index.html
https://www.cdc.gov/parasites/food.html
https://emedicine.medscape.com/article/999282-clinical
https://www.cdc.gov/foodborneburden/2011-foodborne-estimates.html

Fecal indicator organisms

https://www.epa.gov/sites/production/files/2016-05/documents/tech_notes_9_dec2013_pathogens.pdf
www.who.int/water_sanitation_health/dwq/iwachap13.pdf
https://www.foodsafety.gov/poisoning/causes/bacteriaviruses/ecoli/index.html
https://archive.epa.gov/water/archive/web/html/vms511.html
https://www.epa.gov/sites/production/files/2016-07/documents/review_of_coliphages_as_possible_indicators_of_fecal_contamination_for_ambient_water_quality.pdf

Marine toxins

https://www.cdc.gov/habs/illness-symptoms-marine.html
https://wwwnc.cdc.gov/travel/yellowbook/2018/the-pre-travel-consultation/food-poisoning-from-marine-toxins
PSP: http://www.fao.org/docrep/007/y5486e/y5486e05.htm
DSP: http://www.fao.org/docrep/007/y5486e/y5486e0e.htm#TopOfPage
ASP: http://www.fao.org/docrep/007/y5486e/y5486e0n.htm#TopOfPage
NSP: http://www.fao.org/docrep/007/y5486e/y5486e0o.htm#TopOfPage
AZP: http://www.fao.org/docrep/007/y5486e/y5486e0p.htm#TopOfPage

Chemical contaminants

https://fsa-catalogue2.s3.eu-west-2.amazonaws.com/fsa-chemical-contaminant-shell2015.pdf

FDA-accepted laboratory methods

https://www.fda.gov/Food/FoodScienceResearch/LaboratoryMethods/ucm2006949.htm
http://www.issc.org/2019-nssp-guide

State shellfish control authorities

http://www.issc.org/ssca

Management of shellfish growing areas

http://www.issc.org/2019-nssp-guide
https://www.inspection.gc.ca/food/requirements-and-guidance/preventive-controls-food-businesses/fish/cssp/eng/1563470078092/1563470123546
https://dpipwe.tas.gov.au/biosecurity-tasmania/product-integrity/food-safety/seafood/shellfish-quality/australian-shellfish-quality-assurance-program
https://www.cefas.co.uk/icoe/seafood-safety/designations/fao-reference-centre-for-bivalve-mollusc-sanitation/
http://whqlibdoc.who.int/publications/2010/9789241563826_eng.pdf?ua=1
https://www.seafish.org/article/bivalve-mollusc-safety

Sanitary surveys and classification

https://www.fda.gov/food/federalstate-food-programs/national-shellfish-sanitation-program-nssp
https://www.cefas.co.uk/data-and-publications/sanitary-surveys/
http://www.who.int/entity/water_sanitation_health/emerging/sanitary_profiling.pdf?ua=1
https://www.foodstandards.gov.scot/downloads/Shellfish_-_Classification_Protocol_-_final_22-01-2019.pdf
https://www.food.gov.uk/sites/default/files/media/document/protocol-for-classification-of-shellfish-production-areas-in-england-and-wales.pdf

Methods for estimating river and stream flow

https://water.usgs.gov/edu/streamflow2.html
http://www.appropedia.org/How_to_measure_stream_flow_rate
https://archive.epa.gov/water/archive/web/html/vms51.html
https://fortress.wa.gov/ecy/publications/documents/0510070.pdf

Wastewater plume modeling

http://www.cormix.info/applications.php
https://www.epa.gov/exposure-assessment-models/visual-plumes

Microbial source-tracking

https://www.epa.gov/sites/production/files/2015-07/documents/mst_for_tmdls_guide_04_22_11.pdf
https://pubs.usgs.gov/tm/2005/tm2a3/
https://fortress.wa.gov/ecy/publications/summarypages/1203010.html
https://fortress.wa.gov/ecy/publications/summarypages/1103038.html
https://iwaponline.com/jwh/article/1/4/181/1562

Male-specific coliphage

http://www.who.int/water_sanitation_health/dwq/iwachap13.pdf
http://www.wcponline.com/2006/09/27/coliphage-better-water-quality-indicator/

Shellfish nutritional information

http://www.berkeleywellness.com/healthy-eating/food/article/shellfish-delicate-briny-treats-sea
https://wsg.washington.edu/wordpress/wp-content/uploads/publications/Nutritional-Value-of-Shellfish.pdf
https://onlinelibrary.wiley.com/doi/10.1111/1541-4337.12312

Climate change

http://www.climateknowledge.org/Food_Water_Illness_Models/Marques_Climate_Change_Seafood_FoodResInt_2010.pdf

Appendix II Glossary

adverse pollution condition (APC) – Any meteorological, hydrological or seasonal event (for example, rainfall) or point source discharge that has historically resulted in elevated FC levels in a particular growing area.

Approved – The highest level of growing area classification, an *Approved* area is free of water and shellfish quality impacts, allowing harvesting for direct marketing to the consumer.

colony forming unit (CFU) – A measure of bacterial (for example, FC) concentration. Usually expressed as colony forming units per 100 milliliters of water sample (CFU/100 mL).

Conditionally approved – A growing area classified as *Conditionally approved* meets the *Approved* classification requirements except during periods when water quality is negatively impacted by events that are predictable and manageable and documented in a conditional growing area management plan.

Conditionally restricted – A growing area classified as *Conditionally restricted* meets the *Restricted* classification requirements except during periods when water quality is negatively impacted by events that are predictable and manageable and documented in a conditional growing area management plan.

depuration – The process of reducing the pathogenic organisms that may be present in shellstock by using a controlled aquatic environment as the treatment process.

E. coli – A bacterium that grows in the intestines of humans and other animals. *E. coli* is a subset of FC and is one of several bacteria that are used as indicator organisms in water quality monitoring.

fecal coliform (FC) – A group of bacteria originating in the intestinal tract of warm-blooded animals. FC is a subset of total coliform and includes *E. coli*, and is used as an indicator for the presence of pathogenic bacteria (for example, *Salmonella*).

geometric mean – A summary statistic used in the classification of commercial shellfish growing areas. A geometric mean is the antilog of the mean of the log values of FC data.
milliliter (mL) – one thousandth of a liter.
most probable number (MPN) – a measure of bacterial (for example, FC) concentration. Usually expressed as most probable number per 100 milliliters of water sample (MPN/100 mL).
relay – The process of transferring shellstock from a *Restricted* or *Conditionally restricted* area to an *Approved* or *Conditionally approved* area for removal of pathogens or other poisonous or deleterious substances.
Restricted – A growing area classification signifying an area where water quality is routinely impacted such that direct harvesting for marketing is not allowed. Harvesting from a *Restricted* or *Conditionally restricted* area is only allowed under special license, with the shellfish being cleansed via relaying or depuration for a specified period of time before marketing.
Prohibited – An area classified as *Prohibited* cannot be used for shellfish harvesting for human consumption.
systematic random sampling (SRS) – A sampling strategy for collecting growing area water samples in the absence of point sources of pollution. Sampling collection occurs based on a pre-established, randomized schedule intended to represent all meteorological, hydrographic, and any other pollution event that may impact the growing area.
total coliform (TC) – An indicator bacterial group that includes FC and other, nonenteric organisms.

Appendix III Example sampling protocol for shellfish growing waters

Sampling can be as simple as a hose clamp on a pole to hold a sterile sample bottle for surface sampling, or it can involve more sophisticated equipment if sampling at depth is required (for example, Van Dorn or Niskin water samplers). In the latter case a certain amount of the collected volume of water should be flushed through the exit port, and the remaining volume subsampled by decanting aliquots into a sterile sample bottle. Any equipment used for sample collection must be cleaned and sterilized between each sample to prevent cross-contamination. Thorough rinsing with water at the current sampling location prior to sample collection may be sufficient when there is no or minor contamination present, or submersing in a disinfectant solution may be required when more significant contamination is suspected.

A. Equipment and supplies needed by sampler

1 Sterile water sample bottles (approximately 110 mL or larger size) or equivalent (for example, Whirl-Pak® bags), with numbered labels, 1 per sample plus several extras.
   a Optional: large volume (for example, 500 mL) sterile bottle for preparing laboratory replicates.
2 Sterile plastic bags and twist-ties for shellfish samples.
3 Laboratory sample submission form(s) provided by a certified shellfish laboratory.

4 Water sampling pole for surface sampling (growing area, tributaries, culverts, and so on): an extendable aluminum pole (for example, a boat pole) with an attached hose clamp for holding a sample bottle is sufficient.
5 Depth sampler (for example, Niskin or Van Dorn sampler): only needed if aquacultured shellfish will be grown at depth.
6 Salinity meter or refractometer.
7 Temperature meter or water thermometer capable of reading to nearest 1/2 degree.
8 Water bucket. Can be used for grab samples to determine temperature and salinity if using thermometer and refractometer, respectively.
9 Cooler or insulated shipping container with frozen gel packs.
    a NOTE: Do NOT use wet ice: potential source of sample contamination.
10 Clipboard.
11 Field data form printed on waterproof paper.
    a Form should contain columns for sample station # or name, date and time of sample collection, sample bottle identification number, salinity, water temperature, and notes (for example, environmental conditions, observations).
12 Waterproof marker and pen or pencil capable of writing in wet weather.
13 Watch for recording time of sample collection.
14 Sample station map.
15 Boat in safe operating condition with required safety and personal flotation equipment.

B. Water sampling procedures (instructions to sampler)

1 General requirements
    a When an adverse condition (for example, rainfall) is anticipated or has occurred, determine the appropriate day/time to collect water samples at all stations representing the growing area for classification purposes.
        i Consider all relevant adverse conditions (for example, rainfall intensity, tides).
        ii Notify the laboratory to confirm their ability to receive the estimated number of samples for FC assay.
    b Assemble sampling equipment in section A. above.
    c If weather or other conditions are hazardous, postpone sampling until the next safe opportunity (appropriate tide when growing area is open). If a sampling is postponed, notify the laboratory as soon as possible.
2 Sample collection procedures: surface samples
    a Insert empty sample bottle into holder of sampling wand.
    b Carefully remove cap and hold in one hand so inside surfaces of cap and bottle are not touched or otherwise contaminated.
    c Avoiding visible debris or floating material, dip bottle underwater, mouth down, and with a slow sweeping motion to one side, turn bottle right side up to fill. Take sample beneath the surface about six inches and no deeper than one foot.
        i It may not be possible to invert the sample bottle when sampling shallow tributaries or culverts. For these situations collect the sample in fast-moving water. Avoid disturbing the upstream area prior to sampling.

- d Bring bottle to surface and tip out a little water to produce a small head space of air, approximately ¼ inch. Ensure that the minimum volume of sample required by the laboratory remains.
- e Carefully replace cap, without contaminating the sample, and screw on tight. If a sample bottle accidentally becomes contaminated, discard and resample with a clean, sterile bottle.
- f Immediately place the sample bottle in a cooler with frozen ice packs. It may be necessary to provide some insulation between the bottle and ice pack to prevent freezing.
- g Record the station number/name, sampling time, and bottle identification number on the field data form.
- h Collect an additional sample at the first station and label it "temperature blank." This sample should be handled identically to all other samples. Upon receipt of the samples, the laboratory will check the temperature of the temperature blank to ensure it is within the required holding temperature range (1–10° C).
- i Record water temperature and salinity on the field data form. If using a thermometer and refractometer, a grab sample can be used by collecting an adequate volume of water with the bucket.
- j Repeat steps 2.a through 2.i for each station.

3 Sample collection procedures: depth samples

- a Ensure that the interior of the sampler has been sterilized to avoid contamination of samples.
- b Prepare the instrument for sample collection as per instrument instructions.
- c Lower sampler to appropriate depth, activate mechanism for closing the sampling chamber (usually a weighted messenger sent down the line), and retrieve the instrument.
- d Carefully remove cap and hold in one hand so inside surfaces of cap and bottle are not touched or otherwise contaminated.
- e Flush drain spout of sampler thoroughly. If water is needed for measuring temperature and salinity with a thermometer and refractometer, collect water in bucket. Immediately record water temperature.
- f Fill sample bottle, leaving a small head space of air (approximately ¼ inch). Ensure that the minimum volume of sample required by the laboratory remains. If particulates are present such that material is settling out in the sampler, keep the sampler volume agitated and thoroughly mixed while collecting subsamples of its entire volume.
  - i Reserve unused sampler volume if needed for measuring salinity.
  - ii For the first sample of the day also collect a temperature blank sample (see 2.h).
- g Carefully replace cap, without contaminating the sample, and screw on tight. If a sample bottle accidentally becomes contaminated, discard and resample with a clean, sterile bottle.
- h Immediately place the sample bottle in a cooler with frozen ice packs. It may be necessary to provide some insulation between the bottle and ice pack to prevent freezing.
- i Record the station number/name, sampling time, and bottle identification number on the field data form.
- j Record water temperature and salinity on the field data form.
- k Repeat steps 3.b through 3.j. Sterilize sampler prior to collection of next sample.

4 Complete a laboratory sample submission form for each sample.
5 Transport or ship samples so they are delivered to the laboratory as soon as possible, and no more than 24 hours after first sample was collected. Samples should be kept in a cooler with frozen gel packs or placed in a refrigerator adjusted to a temperature of 4° Celsius (39°F). Samples must be held at 1–10°C (do not use wet or dry ice).
6 In cases where the elapsed time between sample collection and delivery to the laboratory is minimal, the sample temperature must be at or below the ambient water temperature measured in step 2.i. Under this circumstance, when the ambient water temperature at the time of collection is significantly greater than 10°C, there may be inadequate time for the sample to chill enough to reach the proper holding temperature range. Regardless of this scenario, the samples must be held under refrigeration at all times between collection and delivery to the laboratory.
7 Provide a copy of the field data form to the appropriate person (for example, data custodian) as soon as possible.

Appendix IV Example shellfish sampling protocols for *E. coli* assay

Although the NSSP does not stipulate a growing area standard for FC levels in shellfish, there are requirements for the handling of shellfish samples intended for fecal coliform testing. Part III of the Shellfish Laboratory Evaluation Checklist, within the Guide for the Control of Molluscan Shellfish, requires that: (1) A representative sample of shellstock is collected; (2) Shellstock samples are collected in clean, waterproof, puncture resistant containers loosely sealed; (3) Shellstock samples are labeled with collector's name, type of shellstock, the source or harvest area, sampling station, time, date and place (if applicable) of collection; (4) Immediately after collection, shellfish samples are placed in dry storage (ice chest or equivalent) which is maintained between 0 and 10°C with ice or cold packs for transport to the laboratory. Once received, the samples are placed under refrigeration unless processed immediately; and (5) is initiated as soon as possible after collection.

Shellfish samples are not tested if the time interval between collection and analysis exceeds 24 hours.

The following links provide detailed descriptions of the EU procedures for collecting, handling, and preparing shellfish intended for testing for the presence of *E. coli* or other indicator organisms:

- Centre for Environment Fisheries and Aquaculture Science, Food Standards Agency, Protocol for the Collection of Shellfish under Microbiological Classification Monitoring Programme (EU Regulation 854/2004): https://www.cefas.co.uk/media/kgadkybh/202005-cefas-classification-sampling-protocol-for-local-authorities-version-10-final.pdf
- UK Centre for Environment Fisheries and Aquaculture Science. National Reference Laboratory for monitoring bacteriological and viral contamination of bivalve molluscs. Generic protocol - Final. Enumeration of Escherichia coli in bivalve molluscan shellfish by the most probable number (MPN) technique (based on ISO 16649-3):https://www.cefas.co.uk/nrl/methods/bacteriological-contaminants-of-bivalve-molluscs/

CHAPTER 12

Design and construction considerations for a molluscan hatchery

Samia Sarkis, Richard Karney, and R. LeRoy Creswell[†]

General principles in hatchery construction

The design of hatcheries varies with site characteristics, target species, production level, culture methodologies, and available funds. Because of these variables, no two hatcheries are the same, yet fundamental principles and key criteria remain common to all. Molluscan hatcheries range from small, technically simple family operations producing a limited amount of seed of one species for the local market, to large corporate multi-species production facilities producing billions of seed and utilizing computer controlled automated systems.

An operating hatchery is a life-support system for the organisms cultured within it, and it is designed with the purpose of providing reliable conditions for the optimal survival, growth, and production of the cultured organisms. The hatchery produces 'spat' or 'seed' for transfer to grow-out sites or 'farms' in the natural environment. In order to achieve this, a molluscan hatchery includes both phytoplankton food production and target species seed production systems.

This chapter discusses the design criteria for a successful hatchery operation, incorporating the associated key production techniques. It is intended to be an informative guide, outlining options and strategies applicable to a range of species and/or to a multi-species hatchery.

There are major stages to the development and construction of a hatchery which involve an inter-disciplinary team; identifying the phases and timeline of the project at the outset facilitates collaboration and implementation.

Phase 1: Project details

- Identifying target species and goals.
- Identifying production techniques.
- Identifying market-related strategy, and product type.

[†] We dedicate this chapter, one of his last completed publications, to our friend and colleague R. LeRoy Creswell. Larger than life, with a boundless enthusiasm for aquaculture, he was a doer and an optimist. His advice always rang true and his passing leaves a great void in the aquaculture world.

- Data-gathering related to culture techniques and market demand.
- Identifying risks and risk-management approaches, including the output of an alternative final product or target species.

Phase 2: Site selection

- Data-gathering for environmental, economic, and regulatory considerations.
- Permit regulations and application processes.

Phase 3: Hatchery complex design

- Identify partners: engineers, builders, technical support.
- Timeline for design, construction, start of culture operation.
- Hatchery design.
- Budget: design, construction and operation (capital and operating costs).
- Business plan.

Phase 4: Construction and implementation

- Planned aquaculture production targets over first 3 years.
- Year 1 dedicated to testing for entire system and training team onsite.

First and foremost, the hatchery production target and long-term goals must be clearly defined (Phase 1). This will avoid limitations in production levels, species cultured, and operational challenges in the long term. The system needs to have the capacity to adapt to the continual improvement in production techniques, and to changes in hatchery strategy; this often leads to the diversification of the final product and/or species produced. The end result should be a reliable, efficient facility with the capacity to be flexible, allowing for expansion and/or modifications in culture techniques.

This chapter addresses practical considerations in site selection, and in hatchery design (Phases 2 and 3), once production output and operation goals have been defined.

Site selection

Proper siting of both the land-based hatchery, for seed production, and farm, for grow-out, is key to a successful and sustainable aquaculture operation. The site selected determines to a large extent the final design of the building, production systems, and operational logistics.

Ideally, the hatchery will be located where incoming seawater provides for the requirements critical to seed production. More specifically, an ideal hatchery site is located in proximity to a water body providing clean seawater for all sectors: live microalgal culture, broodstock holding or conditioning, embryo development, larval culture, settlement, and post-set culture (see Appendix I for definition of stages as referred to in this chapter).

Environmental considerations, namely high-quality seawater is in truth, the first consideration and the single most important factor in siting a hatchery. A hatchery with a reliable source of high-quality seawater facilitates successful algal and larval culture, and reduces daily operational complexities in protocols and sustained challenges. This in turn leads to a more cost-effective operation.

Temperature control and filtration requirements are dependent on the incoming water quality, and species produced. Temperature and filtration will vary for the various sectors of the hatchery. Although broodstock and seed benefit from incoming seawater high in natural phytoplankton composition, this is not the case for early life stages. Water treatment in a hatchery operation ranges from coarse filtration for broodstock and seed to fine filtration and sterilization for algae, with ambient, chilled, and heated seawater needed for larval and spat stages.

Hatcheries can be operated in sub-optimal water quality conditions; however, this will impose additional costs in filtration and sterilization in the hatchery/nursery/algal complex, and often lead to sub-optimal seed production, with respect to both quantity and quality. For these reasons, it is advantageous to gather as much data as possible on the biological, physical, and chemical characteristics of the water at the proposed hatchery and farm sites.

If possible, it is prudent to operate a small pilot facility at the site to test its suitability for shellfish culture before making a large investment. Particular attention should be paid to the following characteristics.

Water exchange and tidal currents need to be sufficient to provide an adequate supply of good quality seawater, including appropriate depth for intake. In areas of high tidal exchange, care must be taken where pumps are located; if these are on a floating dock, the head on the delivery side may increase markedly at low tide, and affect the efficiency of the pump in delivering water to the hatchery complex; this can especially restrict operations during times of high water demand in the larval and algal sectors.

The more constant *temperature, salinity, and oxygen fluctuations* are in the incoming seawater supply for the hatchery, the easier operation will be, and the lower the risks will be in larval and post-larval production. Most fluctuations can be counteracted in a hatchery, but require additional labor, monitoring, and operational costs. Salinity variations, possibly in sites with heavy river or rainfall runoffs, can be problematic for many marine molluscs if not monitored and their effects mitigated.

Solids and dissolved substances, in general, should be absent; however, it is also recognized that dissolved organic material may be utilized by bivalve larvae, and that some bacteria may be an important part of the larval diet; this proves especially true post-set, as food demand increases and production benefits from supplements in incoming seawater.

Silt input (turbidity) is increased by periods of heavy rainfall, and associated runoff, and under storm conditions, which suspend bottom sediments creating high silt loads in source water; addressing the need for frequent removal will require the installation of an additional and costlier filtration system, or the need to attend to the frequent clogging of the filtration system. Selecting a site with access to deep water will minimize the intake of suspended sediment, as compared to a shallower site more exposed to sediment re-suspension.

In areas of high *maritime traffic*, assessment of the potential increase in re-suspension of sediments, occurrence of toxic fuel spills, and the impact of toxic bottom paints is required.

Effluents from nearby industrial plants, marinas, boat yards, agricultural land, and sewage outfalls impact water quality and need to be considered.

The occurrence of algal (and possibly bacterial) blooms needs to be understood. If an alternative site is not an option, historical data and a good understanding of the types of bloom will assist in developing and implementing a monitoring program and subsequent strategy for production and harvest. In some cases, and depending upon the algal species composition, high levels of primary productivity in water may be advantageous for intermediate nursery stage or juvenile culture and will supplement the food source to tanks or ponds; this may reduce the demand on the algal production sector, but can also result in increased settlement by other marine organisms and fouling of the supply lines.

The occurrence and frequency of natural disasters is a serious consideration; sites protected from hurricanes, earthquakes, or strong storms are recommended.

It is often the case that the best sites from an environmental standpoint may be unavailable due to societal constraints, usually codified as governmental regulations. Practically, considerations other than environmental ones often play a major role in the location of the hatchery, such as, regulations and legislation of the country, state, or municipality, involving effluent restrictions, and the compliance of operational protocols for production, transport, and documentation with local environmental concerns and regulations.

It is essential early in the planning phase to make enquiries of local, state, provincial, or federal authorities to investigate all the laws relevant to the installation, construction, and operation of a hatchery. It is also critical to research the permitting requirements for the target species; for example, culturing non-indigenous species is usually against regulations. There may be further prohibitions, such as the construction of commercial operations close to the shoreline, or that of a dock and placement of sea water lines, and limits to the size, height, and future expansion of buildings. Any of these could thwart construction at an otherwise ideal location.

An aquaculture operation should comply to best management practices (see also Chapter 17), which dictate that aquaculture operations have minimal environmental and social impacts on their surroundings. For example:

- the location of intake pipes should not present an impediment to navigation
- the physical plant should be aesthetically consistent with adjacent properties
- erosion during construction should be minimized
- areas with exposure to the sun and wind should make use of alternate energy to lower operating costs.

Economic considerations, such as labor and utility costs, accessibility and expenses incurred for transport to 'farm sites,' 'processing' plants, and market need to be evaluated. These are specific to the strategy of the commercial operation, dependent on the product sold (seed or market size; level of processing for final product).

Ideally, a hatchery should have, or be in close proximity to, a dock; or alternatively, easy access to alternative transport (road or air) for the shipment of seed and/or market product to facilitate transfer from the hatchery, and to optimize seed survival rate post-transfer.

A prime location will have the added benefit of access to external technical support and the availability of a local labor force with suitable skill sets.

Criteria for sizing a hatchery

The size of a hatchery is dependent on the target species, the production output, and the financial investment. Production output also relates to the size of seed cultivated; a hatchery that aims to produce larger size seed will require additional space for nursery rearing and increased microalgal production, than a hatchery focused on remote setting of eyed larvae or transferring young seed (<2 mm) to farm sites.

Determining the size of the land-based hatchery requires a good understanding of its functional requirements; more specifically, the steps involved in the culture process, and the seawater requirements for each stage must be understood. The economic efficiency of the future hatchery is also key; incorporating a flexible, modular concept to the design will enable future product diversification and improvement of techniques if needs be. Hatcheries can be categorized as small, medium, and large scale.

In a first instance, pumping capacity, tankage capacity, and scale of algal production have to be estimated for achieving hatchery target production. This can be done by determining the following.

1 Volume of seawater required for maximal production – The volume needed defines the size of the pumps, and the diameter of the intake and supply lines to rearing tanks. It includes water requirements for each of broodstock, larval, nursery, and algal sectors, and takes into consideration the seawater system used for each – continuous, semi-continuous, or static.
2 Yearly production of larvae required to achieve target spat production – Targets define the size and numbers of larval tanks. Several biological factors need to be taken into consideration at this time, including frequency of spawning, duration of larval life, and optimal larval culture density (see 'Hatchery: larval and post-set rearing systems'). It is often the case that several larval cycles are required within one season to achieve larval production goals.
3 Species-specific culture techniques for spat rearing and optimal seed size and time of year for transfer at sea – these are the defining factors for the type and size of nursery tanks, and for the level of intermediate culture required, such as a floating upweller systems (FLUPSY) or pond (see 'Intermediate nursery'). This is based in part on the commercial goal of the aquaculture operation and its business plan. A hatchery may act strictly as a seed producer selling seed ready for transfer to growers operating a farm for market production, or selling/providing seed for stock enhancement purposes; a hatchery can also be both a producer and a grower, keeping all seed for its own market production; or, a hatchery can adopt a mixed strategy where a portion of the seed is sold to growers and/or for stock enhancement, and a portion kept for grow-out to market size. If seed is sold, the size at which it is supplied is largely dictated by the requirements of the grow-out industry. This will in turn define the facilities for the nursery sector of the land-based operation.
4 Extent of broodstock conditioning required to achieve larval production – conditioning is dependent on the candidate species and on their natural reproductive cycle. It will define the number of broodstock tanks and the size of temperature-control units.

All of the above will define the algal volume and algal species cultured. This will translate into a selection of the algal culture system(s) used, ranging from batch, to semi-continuous, and continuous. It will also identify the requirements for commercially produced feed as a supplement to live microalgae.

There are many things to consider when sizing a hatchery, including the reproductive capacity of the target species and the survival rate in the early life stages (Worksheet 12.1). Please note that Worksheet 12.1 is intended as a guide only, and does not address all questions specific to an operation. System specifications vary from site to site, dependent upon culture methods and environmental site characteristics, especially on incoming seawater quality (Example 12.1 illustrates a real-life bivalve hatchery case study).

General seawater system layout

There is no rigid design for a shellfish hatchery, yet there are a few underlying principles in designing a functional and efficient molluscan hatchery.

1. The physical plant of a land-based hatchery must be large enough to house the main hatchery and the ancillary buildings. These include the following inter-related sectors: (a) algal culture production; (b) broodstock quarantine (if imported); (c) broodstock conditioning; (d) larval production; (e) settlement and nursery for spat/young seed production; (f) intermediate nursery for larger seed production (including pond culture if available); (g) dry laboratory; and (h) workshop and storage areas.
2. Each production sector named above must have dedicated seawater supply lines, to avoid cross-contamination, with temperature-control and filtration levels specific to each sector.
3. Back-up systems are a prerequisite to ensure continuity in production, in the event of mechanical or electrical failure.
4. Biosecurity must be ensured, production sectors for each life stage should be isolated; this assists in disease control and prevents contamination, reducing the level of loss in case of a disease outbreak.
5. Available open space adjacent to the hatchery is useful for storing materials and equipment, expansion of a production sector at a later date, and building new tanks or other equipment.
6. Ease of cleaning pipes and all components of the seawater system; presence of clean-out ports located at the lowest points in the lines, and/or unions in critical areas, should be built in for ease of flushing and dismantling for thorough cleaning.
7. Access to and the size of drains returning water from the hatchery to the source for routine cleaning should be planned. The lower the flow rate in these lines, the greater the setting of fouling organisms (especially in horizontal lines). Open troughs that allow for easy cleaning are best by far to remove difficult clogging organisms, such as barnacles, mussels, and limpets. At the very least, clean-out ports that are easily accessible with big diameter chimney cleaning brushes are needed. The size of drains depends on the capacity of the tanks to be emptied.
8. The hatchery floor should be well drained, contributing to cleanliness in the hatchery.

In general, the seawater system for a commercial hatchery includes a primary pumping system with a dedicated intake line, and a secondary (or back-up) system in case of malfunction (Fig.

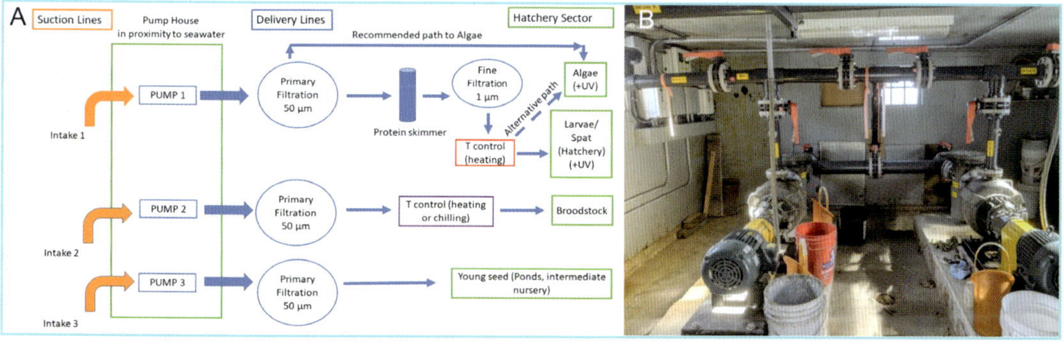

Figure 12.1 Pumping seawater to a hatchery facility. (A) A schematic diagram illustrating dedicated pumps to three main culture sectors in facility, with filtration level and temperature control specific to each sector: Algae and hatchery, broodstock, and intermediate nursery. (B) Main pumps installed in parallel and supplying entire hatchery complex. (B) Courtesy of John T. Hughes Hatchery and Research Station.

12.1). In a large-scale operation, the primary pumping system consists of three intake lines, each with a dedicated pump, supplying (1) the hatchery (larvae and spat, and algae), (2) the broodstock, and (3) the intermediate nursery (Fig. 12.1). Water treatment differs among these. The algal and 'hatchery' (larvae and spat) require the highest level of filtration and sterilization, with often the capacity to heat incoming water. Most hatcheries have a dedicated independent water treatment for algal production, preferred to the alternative shown in Fig. 12.1A. The broodstock sector receives coarsely filtered seawater, with the need to control the temperature dependent on the conditioning regime; this can be either heating or cooling or both. In this hypothetical large-scale operation, the third pump delivers coarsely filtered seawater to the nursery sector – comprising a dedicated intermediate nursery, where young seed are transferred from the 'hatchery' sector; in this case, there is no need for additional temperature control. If seed are not acclimated to ambient seawater conditions, the intermediate nursery may require flexibility in its filtration and temperature-control system, dependent on the size of spat initially reared in this sector, ranging from relatively fine filtration (10 µm) to coarse filtration, with the added capacity to acclimate young seed to natural environmental temperatures.

A reliable supply of fresh seawater is critical to the production success and sustainability of the operation. For this reason, several smaller hatcheries duplicate the entire seawater supply system (intakes, pumps, seawater lines, and storage tanks). Should any of these components fail, a readily available back-up ensures the sustainability of the operation by facilitating the switch to a functioning system and isolating broken components for repair. For a larger-scale hatchery, all parts of the seawater supply system are interconnected, acting as a back-up to any sector in the event of a failure (Fig. 12.1). Please note, the interconnection enables the use of either intake or pump to the sectors, but there are no interconnections between delivery lines. This prevents contamination of water between sectors, such as between coarsely filtered broodstock seawater, and fine filtered (or sterilized) larval seawater. The two most important points are that the system should be designed such that there is no interruption of seawater flow in the hatchery complex, despite potential malfunction of the primary pump, and no cross-contamination between sectors.

A seawater system will include:

- main and secondary intakes of seawater with dedicated pumps.
- a continuous supply of coarsely filtered seawater, independently plumbed, but with the option to interconnect in case of failure of any one component, or blockage of intake.
- a fine-filtration system and sterilization system for larvae, early spat, and algal cultures.
- a temperature-control system to support broodstock conditioning, algal culture, larval culture, and early spat rearing, if needed.

One of the key operational principles is the maintenance of a clean seawater system. Good practice calls for the cleaning of the entire system from mouth of intake to rearing tanks at the beginning of every hatchery season; a dry-out period in-between seasons is also beneficial to eliminate bacteria or other harmful organisms from the pipelines.

Throughout the season, having duplicate seawater lines is an easy means of keeping the lines free of fouling, by alternatively shutting down the lines. Slightly different methods are used by hatcheries for the removal of fouling organisms or pathogens. Some aquaculturists leave water in the closed line, creating an anaerobic environment; this kills any fouling organisms before they reach a size requiring mechanical brushing to remove. When the lines are switched, the anaerobic water is flushed from the closed line for several minutes before the line is put back in service. Other aquaculturists drain the lines using drains located at the lowest point in the lines, sterilize the line, and let it dry before re-use. In locations of heavy fouling, or at times of heavy fouling possibly during the spring and summer months, alternating shut down of lines may be as frequent as every two weeks.

Intake and main pumping station details

The hatchery, or at the very least the pump house, should be located in close proximity to the water source, in order to avoid pumping water over great distances, and close to sea level, avoiding the additional demand of pumping vertically (Fig. 12.1). It is more efficient to push water than pull it, so it is advantageous to locate the pump as close as possible to the water source.

Operating a hatchery using well water is another possibility, depending on the nature of the geological strata, and the water quality of the seawater aquifer. Such a water source often exhibits more constant temperature and salinity, and is often pre-filtered by percolation through the strata. Well water can be a desirable source of water with a minimal load of bacteria, and viral or parasitic pathogens. On the other hand, this phytoplankton-poor water will not benefit operations with a long nursery phase; seed >2 mm grow rapidly requiring an exponential demand of food, which is often supplemented by coarsely filtered incoming seawater, induced algal blooms or commercially available diets. A site with incoming seawater rich in phytoplankton will support seed growth and reduce costs. In addition, well water may require oxygenation before use and it may also contain undesirable chemical characteristics, such as hydrogen sulfide, carbon dioxide, and ammonia, unsuitable for larval and post-larval rearing – the presence of iron in certain well water supply is known to also result in a rapid fouling of filters, requiring increased labor and costs for filter replacement.

Intake

The first step in assessing the location of the intake and the source of seawater pumped is to obtain data on the seawater parameters. An evaluation on the seawater quality will determine intake location, as well as the subsequent production strategy, especially with regards to the nursery section. For example, clean, pre-filtered well water has advantages in culturing phytoplankton and algae; on the other hand, it will require increased phytoplankton production for the culture of spat and seed which have higher food demands. At least two separate intakes and two main pumps are recommended for a reliable supply of seawater. Both intakes should be connected such that either pump can be used (Fig. 12.1). If more than one intake and more than one pump is used, the same principle applies, and all intakes and pumps should be interconnected, providing a back-up in case of failure or blockage. The layout should also be such that either pump can supply any main sector of the facility.

Depth of the intake

The vertical profile of the water quality dictates the depth at which the mouth of the intake is situated. In the case of high tidal amplitudes or high surface seasonal fluctuations in temperature, salinity, or other factors, the mouth of the intake should be located at a depth where variations may be less. Intakes at depth may also help in avoiding algal blooms, some of which may be harmful to the early life stages. In addition, if thermoclines or haloclines are known to occur in the area, the mouth of the intake should be below them whenever possible. At a minimum, the mouth of the intake pipe should be at least 1 m off the bottom, to avoid an intake of sediment churned from the bottom, especially during storm events. Intakes attached to floating platforms that rise and fall with the tides allow for positioning at a constant depth from the surface regardless of water surface fluctuations.

Materials

Intake pipes are usually PVC (polyvinylchloride); schedule 40 and 80 are used, with the latter often favored for plumbing pumps, or for higher pressure areas and larger systems. Pipes can be rigid or flexible PVC tubing dependent on the physical characteristics of the site; pipe diameter increases with the scale of the hatchery, ranging for the most part from 100 mm to 150 mm (see Appendix II for pipe size conversion chart).

Mouth of intake

The open end of the intake pipe needs to be protected by a screen to prevent large objects and animals (finfish or shellfish) being drawn in or from swimming in and settling, thus clogging the suction pipe and pump. Various means of avoiding this are used; on smaller 100 mm diameter pipes, the equivalent of plastic roof drains can be fitted; for larger pipes, rebar frames lined with 20 mm to 25 mm polyethylene mesh are fixed to protect the intake, yet still allow water flow; existing frames as used for crab or lobster traps can also be modified and secured around the mouth of the intake. The mesh will foul over time, reducing flow into the pump, and a maintenance schedule for the changing of the intake mesh is needed (Fig. 12.2). The frequency of cleaning and/or changing of the mesh is site-specific dependent on the productivity of the area and the level of fouling. Dependent on locations and the level of fouling, this can range from a few weeks during peak blooms in high nutrient areas to once a year in oligotrophic waters. Thought must be

Figure 12.2 Fouled filter screen on intake (left); intake without screen (right). Courtesy of John T. Hughes Hatchery and Research Station

given to securing the protective mesh for ease of disconnection and replacement. Removing the protective mesh for a thorough clean and dry out on land, while replacing it with a new protective mesh is the most efficient.

Length of intake pipe

A long intake pipe will increase the chances of cavitation in the pump, and will make it more difficult to pinpoint clogging issues, air leaks, or other malfunctions. The location of the pumping station is key in minimizing the length of the intake pipe, and the suction lift or head.

Securing intake pipe

If the intake pipe runs along the sea bottom, it needs to be secured in place to avoid excessive movement due to currents, storms, and tides.

Intake pipes running vertically alongside a pier or floating platform, may consist of a flexible PVC hose, hooked over the dock with the section exposed at low tide secured to the pier using clamps. The mouth of the vertical pipe needs to be weighted.

For intake pipes in shallower areas, concrete blocks and sandbags can be saddled over the pipe laid on the sea bottom for this purpose. Alternatively, in areas relatively protected and with low currents, a rigid intake pipe can be maintained midwater by the use of floats (Fig. 12.3); care must be taken that there are no undulations in the pipe, especially for one running horizontally, to avoid any air traps, or air leaks, within the intake pipe, which would affect the operation of the pump, resulting in cavitation and possible loss of pump prime. Further, air leaks on the intake side of the pump can create air bubbles in the water delivered to the shellfish that if not degassed can concentrate in the shellfish killing them, or causing them to float out of flow-through nursery systems.

Main pumping station

A well-ventilated pump house that protects pumps and parts from weather and seawater spray is recommended. Pump size and type depend on the scale of the operation, the volume of seawater

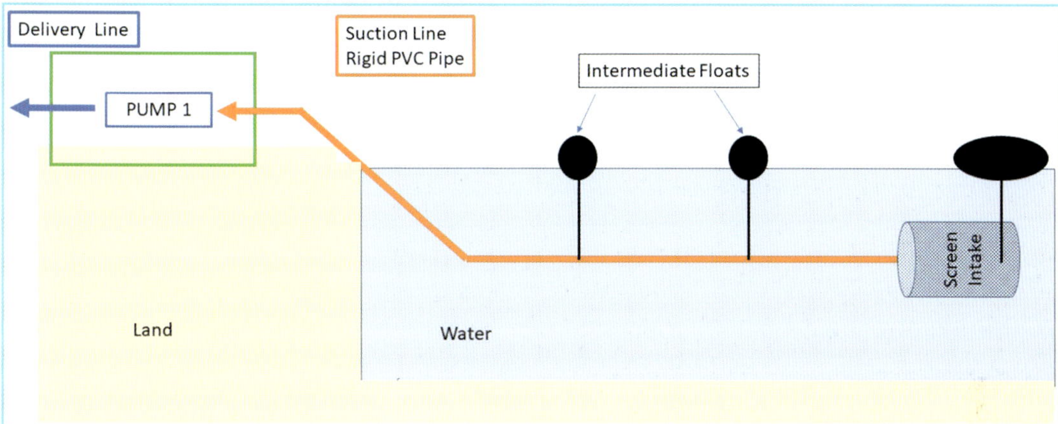

Figure 12.3 Schematic diagram showing rigid intake pipe suspended midwater using floats.

supply required, incoming seawater quality, location of the pumping station, and seawater system used. Pump size for molluscan hatcheries varies, ranging from 1.5 HP to 15 HP, as does the mode of operation, running continuously on a 24 hour basis over the hatchery season, or on demand, if a header tank system is used. The volume of seawater required at peak times for each section of the hatchery is calculated for accurate pump sizing. In addition, information on suction head and delivery head is required to assess resulting flow rate and available water volume. There are new pump designs which provide for automatic variable flow rates adjusting to the amounts needed. These can substantially reduce energy consumption reducing costs and carbon footprint.

Pumps

Submersible pumps are immersed in the source water, and draw water directly without the need for an intake pipe. Although efficient in pumping directly from the source, they are more expensive as they need to be made from non-corrosive metal and are generally more difficult to access for maintenance. They may, however, prove useful for some sites, and/or as an independent pump for ponds, providing flow rates of up to 600 L/min.

Surface-mounted pumps are ideally installed on a floating platform or pier. This reduces the suction head and optimizes the efficiency of the pump by being positioned very near to source water. A pier mounted pump may also allow for the placement of the intake higher in the water column helping to avoid the suspended sediments that are more likely to contaminate seawater drawn from closer to the bottom. Self-priming centrifugal pumps are efficient and reliable pumps for continuous water pumping. It is recommended that a routine pump maintenance schedule and log be established for optimal and long-term operation.

Primary and secondary pumps

Both pumps should be connected and can be used simultaneously in case of high seawater demand or, alternatively, in case of breakdown. In a small-scale hatchery, one main pump may prove sufficient to supply seawater to the entire facility, with a secondary pump available in case of failure of the primary pump. As the demand for seawater volume increases with production size or hatchery scale, several pumps may be needed to operate simultaneously for a sufficient flow. The type of pump used depends also on the type of system used – that is, 24 hour open-flow system, or storage in header tanks. It is critical that pumps used do not have any copper, brass, or bronze components; stainless steel components should be required when ordering. In Worksheet 12.1, the main 2 HP pump was backed up by a secondary pump (self-priming 1.5 HP pump), which proved reliable in case of primary pump failure.

Materials

All materials and surfaces coming into contact with seawater need to be non-toxic. Copper and its alloys are particularly toxic to bivalve shellfish and extreme care must be taken to ensure that none (including internal pump parts) come into contact with the culture water. Commonly used non-toxic materials include most plastics, cast iron, titanium, and certain grades of stainless steel. All valves and piping are usually PVC.

Priming and check valves

A check valve installed between intake and pump is a precautionary measure that will assist in priming the pump and will avoid loss of water in the pump in the event of a power failure. Unions on either side of the check valve facilitate the removal and cleaning of the check valve, and/or replacement in case of damage due to blockage. Check valves are often placed at the low tide mark (on the suction side of the pump), and the size reduced from the intake pipe diameter to the inlet size of the pump. Prior to starting the pump, the suction pipe between the pump and check valve should be filled with water.

Strainer baskets

Some pumps have a built-in strainer on the suction side, acting as a second filtration prior to the pump, and often retaining small shells; pumps used commonly for swimming pool applications often have built-in strainers. Other pumps can be purchased with a strainer basket or pot, or an independent strainer basket can be fitted in-line prior to the pump (Fig. 12.4). Minimizing the amount of hard matter passing through the impeller of the pump increases the pump life, as it limits any unnecessary laboring of the pump due to an impeller blockage. Even for heavy-duty pumps, with impellers capable of handling small shells, a strainer basket is recommended on the suction side.

Figure 12.4 Strainer baskets installed on suction side of pump (small- and large-scale operations). Courtesy of R. Karney (left) and S. Sarkis (right).

Primary filtration

The primary filtration, or pre-treatment filtration system on the delivery side of the pump, serves to remove most larger particles prior to water entering the main supply lines to the hatchery (Fig. 12.1). This assists in maintaining the clean supply lines, and reduces the demand on the finer filtration system required for larval rearing and algal cultures.

Dependent on the characteristics of the incoming seawater, various means of pre-filtration are used in molluscan hatcheries. Information on maximum total particulate load, and in some cases on the composition of the particulate load, is useful in assessing the type of filter best suited to the site.

Sand filter

For small-scale operations and a limited budget, sand filters as used for swimming pools, are a low-cost, easy maintenance coarse filter option. This is especially true for areas with low levels of turbidity, where vulnerability to clogging of filter media is reduced, and efficacy of filtration ensured. Lighter and more efficient media than the standard sand is the better option. An initial operating period of 3 to 4 weeks before use in the hatchery is advised to allow for the establishment of a microbial community contributing to the filtering function. Thereafter, routine backwash, the frequency of which depends on the particle load of the incoming seawater, will ensure that seawater is coarsely filtered to approximately 50 µm, prior to entry in the hatchery complex. The use of clear PVC piping for the incoming and outgoing sections connected to the sand filter assists in monitoring the clarity of the water. It is recommended to plumb a drain or bypass so that if the filter is sitting off-line for extended periods of time new water can be pumped through the sand filter and associated plumbing without sending it into the hatchery. This will maintain the active filter media until the sand filtered seawater is needed. At no time should there be any stagnant seawater in the sand filter, as this will create a substrate for undesirable bacteria, and become unsuitable for hatchery use. Cleaning of the sand filter and replacement of the filter media or sand can be done during the dry-out period. It is recommended that a log sheet be completed daily, noting backwash completion and any other comments pertaining to the main seawater pump operations. If possible, the sand filter should be installed immediately after the pump, housed in the same complex as the pumps. For small-scale operations, both the primary and secondary pump can be plumbed to the sand filter, alternating between the two if and when needed (see Chapter 1 in Sarkis and Lovatelli, 2007).

Drum filter

Drum filters are more costly than sand filters, but can filter larger volumes of seawater; the volume filtered depends on the level of filtration required and the total suspended solids (TSS) load. For example, for a TSS equal to 25 mg/L, drum filter maximum capacity is in the range of 1480 liters per minute (372 gallons per mnute) to filter to 20 µm; if the TSS is lower (15 mg/L), and required filtration less (30 µm), capacity increases (1600 liters per minute). The drum has fine screens mounted to its periphery. Water flows through the screens, and the solids adhere to the screen surface (Fig. 12.5). The advantages of a drum filter are the option to install screens of various mesh size (10–100 µm), dependent on the level of filtration required, and the automatic activation and shutting down of the backwash system once filters are cleaned of solids. In areas where the TSS load is exceptionally high, there have been reports of the drum filter overflowing.

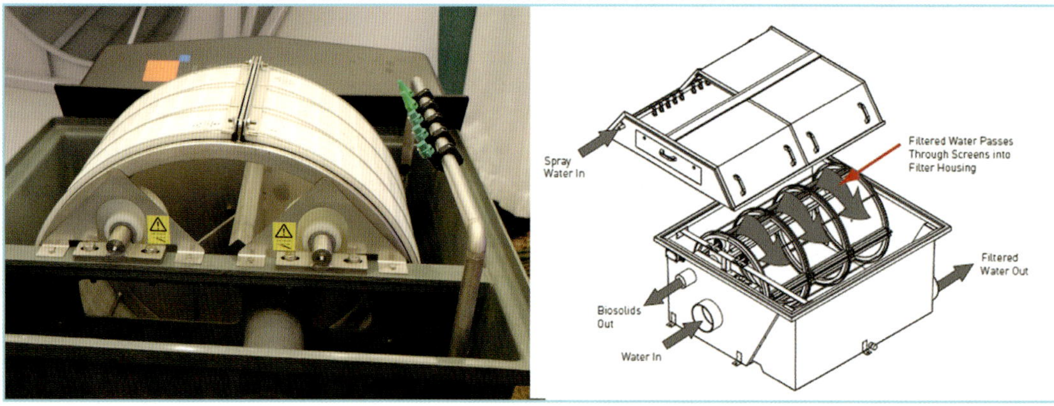

Figure 12.5 Open drum filter. Photo courtesy of S. Sarkis; schematic courtesy of Pentair Aquatic Ecosystems.

A first step in selecting the micro-screen size is determination of the TSS load, and the particle size distribution in the incoming seawater. The particle size distribution relative to the pore size of the microfilter is the single most influential factor in the mechanical efficiency of a micro-screen filter.

Whether a drum filter or a sand filter for primary filtration is chosen depends on the budget available, scale of operation, and/or physical characteristics of incoming seawater. Comparing the two for a commercial scallop hatchery, Magnesen et al. (2013) report that the combined use of drum filter and protein skimmer was most effective in reducing dissolved organic carbon and *Vibrio* spp. bacteria in the seawater and increased larval activity, compared to the combined use of a sand filter with a protein skimmer. Incoming water for the study consisted of total organic carbon (TOC) consisting mainly of dissolved organic carbon (DOC), peaking at 4.9 mg/L. The backwashing action of the drum filter was thought to assist in the reduction of the DOC content for water passed through the protein skimmer, removing solids of organic material. There was, however, no significant difference in egg development or microalgal growth between both systems. In regions with high phytoplankton blooms, leading to increased DOC in incoming seawater, or high initial levels of suspended solids, the removal of suspended solids of organic material will assist in limiting bacterial growth and maintain a constancy in incoming seawater quality for the early life stages of bivalve larvae and microalgae. In regions with lower phytoplankton production, or less pronounced seasonal fluctuations, a cheaper sand filter is a reliable pre-filtration method for microalgae and early life stages of scallop species (Sarkis and Lovatelli, 2007). Alternatively, and dependent on the scale of the operation, drum filters may be used for larval, post-larval, and algal cultures; in this case, broodstock may be independently supplied by a sand filter reducing capital costs for filtration.

If an aquaculturist selects a drum filter to treat high seawater demand, it may prove more economical to purchase several small filters rather than one large unit. Typically, smaller filters operating at continuous backwash mode are found to operate at maximum efficiency, as this ensures an unblocked screen. Intermittent backwashing of an oversized drum filter is suitable when there is extra flow capacity within the filter; if not, the filtering process may be affected due to the formation of too dense a 'cake' in-between backwashes. Test methods for optimal micro-screen drum filter selection are available (Dolan et al., 2013).

Figure 12.6 (A) Arkal filter system showing a series of cartridges and (B) disk change during maintenance. (C) Arkal system installed on palette for ease of subsequent safe storage in case of storm events. Courtesy of Michael C. Voisin Oyster Hatchery.

Arkal filtration system

The Arkal system consists of a series of disk filters with the reported capacity of filtering incoming raw seawater to 20 μm. It is a costly system used in large-scale operations, for supply to all sectors of the facility. In areas with high TSS load in incoming seawater, the Arkal system can be high maintenance. In order to ensure optimal operation, a strict maintenance schedule is required (Fig. 12.6). Failure to adhere to proper maintenance will result in decreased flow rates, and a lack of filtration leading to potential bacterial contamination.

Disk filters

Filters using disks, other than the Arkal system described above, initially designed for irrigation can be purchased individually and installed as required; this is a cheaper version adequate for small-scale operations. Disk filters are promoted for use with water containing high amounts of organic or inorganic matter. They can be installed at an angle, horizontally or vertically. A two-sided disk system increases the surface area resulting in longer periods between cleaning and less filter maintenance. The lowest level of filtration is 75 μm, which is relatively coarse and will need to be supplemented with subsequent finer filtration units to achieve similar water quality to that obtained using primary filtration systems above.

General layout of a molluscan hatchery facility

Biosecurity in the facility is especially critical if more than one species are cultured and/or if broodstock or young seed is imported. In case of import of animals, effluent water must be treated; otherwise broodstock holding tanks, larviculture systems and nursery facilities should be separate and isolated from each other to minimize the likelihood of cross-contamination. A schematic diagram of a general hatchery layout is given in Figure 12.7; the facility includes the following.

- Main seawater treatment area – where coarsely filtered seawater supplied from the pump is treated to accommodate larval and post-set culture requirements.
- Hatchery – for larval cultures receiving heated, filtered, and sterilized seawater; this houses a series of tanks, a modular design precluding any transfer of veliger larvae, bacteria or

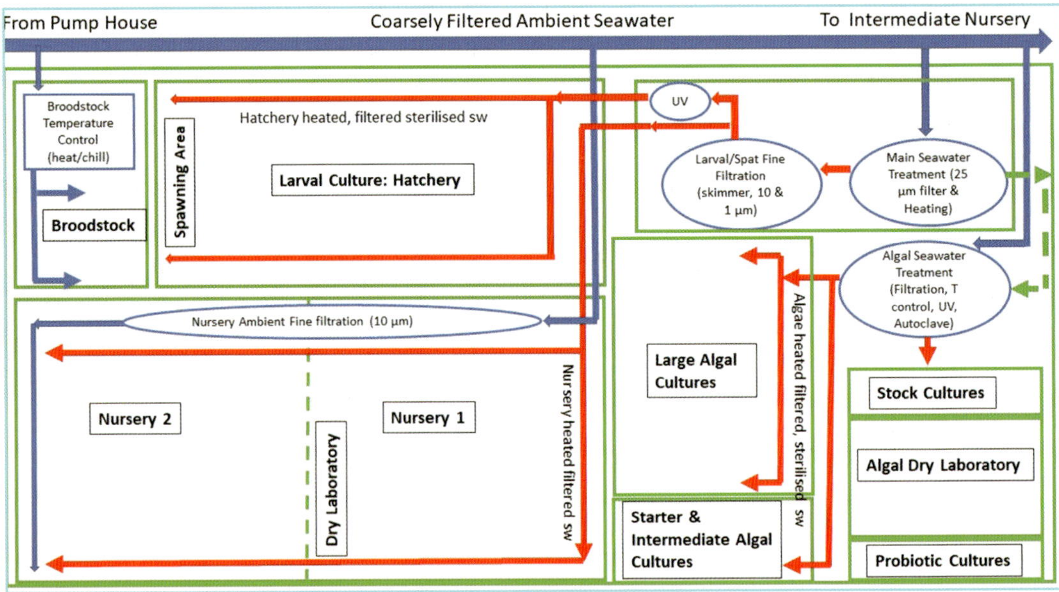

Figure 12.7 Schematic of a general hatchery complex, excluding external areas such as storage, dock, and pond.

other contaminants from one unit to the next. Climate control of this sector is often required.

- Algal culture system – indoor cultures are housed in climate-controlled rooms, including one for stock cultures, probiotic cultures, intermediate, and large algal cultures. All algal culture areas receive heated (if needed), filtered, and sterilized seawater. A separate seawater treatment area is recommended for algal cultures; alternatively, pre-treated water can be obtained from the main seawater treatment unit (shown by green dashed arrows). An additional greenhouse or outdoor algal culture area may be necessary for large-scale operations (not shown here).
- Broodstock sector – ensuring isolation from other sectors with its dedicated temperature-control system, if conditioning is planned.
- Nursery sector – this is divided into Nursery 1 and Nursery 2 on the diagram. Nursery 1 is for settlement of pediveligers and rearing of early spat. Nursery 2 is for young seed (6 weeks and older after settlement), grown to a larger size prior to transfer to the intermediate nursery or farm. Nursery 1 receives similar heated and filtered seawater to that of the hatchery; Nursery 2 requires access to both heated/filtered seawater, and ambient seawater filtered to no more than 10 μm; this allows for acclimation of seed prior to transfer to intermediate nursery or farm.
- Dry laboratory – a dry laboratory for the algal culture facility, and a second one for the hatchery/nursery sectors are shown here.
- Storage area for filter cartridges, sieves, and other larval and pre-nursery materials (not shown in diagram but an important space allocation in every sector of the facility).
- Intermediate nursery – this sector is not shown in the figure; it receives ambient, coarsely

filtered seawater from a dedicated pump. The need for an intermediate nursery is dependent on the size of seed targeted for transfer at sea, and time of transfer. The system used may be in the form of external raceways, ponds and/or a dock-related system, such as FLUPSY or suspended cages.

The layout of the hatchery, if carefully designed, will facilitate efficient daily operations. For example, easy access from the phytoplankton production to the larvae and nursery sectors facilitates the daily distribution of algal food to the tanks. The broodstock sector should be separate, but in proximity to the larval sector, with a spawning area close to larval tanks for distribution of eggs. Similarly, having the larval production sector adjacent to the nursery reduces the operational load during settlement. This becomes increasingly relevant to the smooth operation and cost-effectiveness of the hatchery, as the scale of production increases. Wherever possible, the use of gravity, especially for feeding tanks, is recommended.

Larval production: fine filtration, sterilization and temperature control

The level of sterilization, fine filtration, and type of temperature control varies with quality of incoming seawater, species cultured, and stage of culture. Fine filtration and/or sterilization is necessary for algae, larvae, setting, and post-set stages of culture. Fig. 12.8A illustrates the potential components of the hatchery seawater fine-filtration system, fine filtration is explained below. *Please note:* Not all hatcheries may require every component.

The level of treatment depends on the incoming seawater quality and the target species. In cases of a high bacterial load in the incoming seawater, a protein skimmer may be required, as a first step to finer filtration following the coarser drum or sand filter unit (Fig. 12.8). This is followed by a gradual increase in filtration, often through the step-down of filters from 10 μm

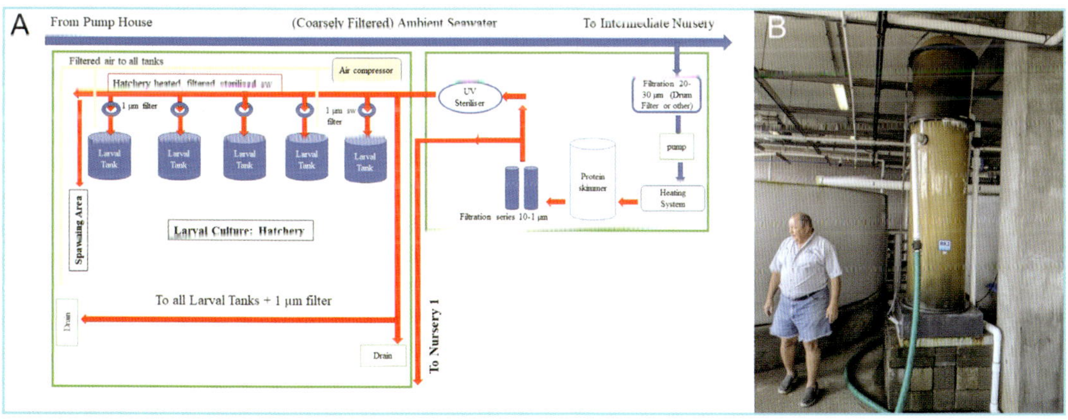

Figure 12.8 Seawater treatment for the hatchery sector. (A) Schematic details of hatchery (larviculture) seawater treatment and distribution, in case of poor water quality requiring both protein skimmer and sterilization for the hatchery tanks. (B) Protein skimmer in use in a commercial bivalve hatchery. Courtesy of (B) Michael C. Voisin Oyster Hatchery.

to 1 μm. In cases of poor water quality or very sensitive target species, a UV sterilization unit, prior to the larval tanks, is installed. Several bivalve species benefit from sterilized seawater for embryonic development to D-larvae, but will perform well in 1–5 μm filtered seawater for the remainder of their larval life. The extent to which water is filtered is species dependent; for example, oysters are often reared in 5 μm filtered seawater, and scallops in 1 μm. Larvae are usually reared in double-filtered seawater, achieved by installing a cartridge or bag filter in line with the incoming flow of the larval tank, in addition to the fine filtration provided by the larger units (Fig. 12.8A). This double filter can be threaded to the incoming seawater pipe, and easily removed when not required, thus keeping the system clean.

Considerations for the use of sterilized seawater for larval cultures are discussed below; although sterile water offers some advantages to embryonic development, it also creates the opportunity for any introduced pathogen to bloom due to the lack of competition and may have subsequent devastating effects.

Fine filtration

The main line of finely filtered seawater is designed to supply fertilized eggs, larvae, setting larvae, and post-larvae. Algal cultures, or the greenhouse sector, often have a filtration and sterilization unit in addition to that delivered by the main supply lines. Sterilization is always a must for the greenhouse.

Protein Skimmer

A protein skimmer, fitted prior to the series of fine-filtration units assists in removing dissolved organic matter and many waste products, including carbon dioxide, animal feces, ammonia, and other complex organic substances from decomposing feed, metabolic by-products, algae, and phenols. Protein skimmers are commonly used in recirculating systems (RAS), depuration of bivalves, or aquaculture ponds; however, in cases of high concentrations of suspended particles in the incoming water source, protein skimmers are used in open-flow or semi-recirculating bivalve systems for larval and post-larval rearing (Fig. 12.8).

Protein skimmers in marine hatcheries have proven to be efficient in reducing DOC and fine particles in seawater. By using filters and protein skimmers in the hatchery, a larger amount of the organic material will be removed from the incoming seawater, at the same time reducing the threat of opportunistic bacteria.

The capacity to filter water beyond 50 μm is usually achieved through a series of filtration units. Cartridge filters, bag filters, and disk filters are commonly used in hatcheries (Fig. 12.9).

Cartridge filters

Pleated cartridges, such as those supplied for swimming pools, are relatively easy to maintain and can filter to approximately 25 μm for water flows up to 600 L/min. Cartridges are cleaned using a jet of freshwater and a bactericidal agent (such as Chlorox or Virkon) bath routinely; the frequency of which is dependent on the incoming seawater particle load. If cartridges are not cleaned, clogging will occur, and filtration capacity will be reduced, resulting in a low outgoing flow. Such a filter can be the first in-line of the filtration series, followed by 10 μm, 5 μm, and 1 μm cartridge filters.

Smaller wound polypropylene or pleated cartridges are supplied in 10" and 20" housings, and are useful in smaller hatcheries, where tanks are less than 1000 L in volume. They are easily cleaned and can be used repeatedly over a period of several weeks or months depending on the incoming water quality. Wound polypropylene cartridges will eventually shred, at which time they have to be discarded and replaced. The small filter housings are also useful as a double filtration to 1 μm immediately prior to the larval tank; they can be easily plumbed in-line and fitted with a flexible hose.

Bag filter

These are an alternative to cartridge filters, with several models suitable for low and high flow rates (greater than 140 L/min). Several sizes of housings are available, with a range of bags filtering from 800 μm to 1 μm. Bag filters are fitted with a ring that seals directly against the filter housing, eliminating any chance of bypass. Ten levels of filtration bags are offered. These can also be fitted with a differential pressure gauge that indicates when it is time to change the bag. Smaller models are also used as the double filter prior to larval tanks, similarly to the cartridge filters above.

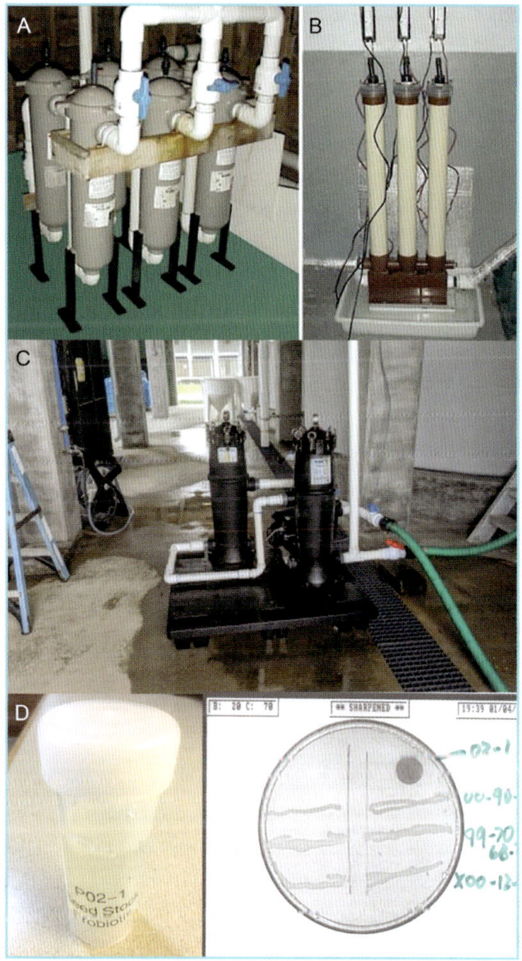

Figure 12.9 Filtration and sterilization units. (A) A multiple bag filtration unit, showing a series of three filters in use while second bank is cleaned (Helm et al., 2004). (B) UV sterilization unit consisting of three individual lamps, more reliable than treatment by a single bank unit (Helm et al., 2004). (C) Tertiary filtration to 1 μm for larval and algal cultures on trolley for safe storage during storm events. (D) Probiotic bacteria starter culture (left) and inhibition plate (right) with 02–1 being the probiotic bacteria. Courtesy of (A) and (C) Michael C. Voisin Oyster Hatchery, (B) R. Karney, (D) R. Elston, Aquatechnics.

Sterilization and bacterial control

UV sterilizer

Installing a sterilizing unit as part of the general seawater supply to larvae and pre-nursery sectors may not always prove the best solution. Sterilization of larval culture water removes both pathogenic and non-pathogenic bacteria and often creates a bacterial 'vacuum' that can result in any introduced pathogenic species (for example, *Vibrio* sp. introduced with the eggs) replicating quickly in the absence of other harmless bacteria that would otherwise keep them in check. If sterilization is used, the culturist must institute additional sanitation protocols to ensure that no pathogenic bacteria enter the culture. The management of pathogens rather than their total removal may reduce the need for difficult and expensive protocols to maintain totally axenic cultures. As a guide, the acceptable level of bacterial contamination in bivalve hatcheries is that seawater must contain less than 100,000 colony forming units (CFU)/mL for safe use; more species-specific data on vulnerability of larvae to *Vibrio* species are available in the literature.

UV sterilizers can be purchased as a single lamp unit, or for larger scale operations, as a multi-lamp system with a flow of capacity of up to 109 L/min (Fig. 12.9). Single lamp units can be installed in parallel by the aquaculturist to satisfy hatchery-specific requirements – namely embryonic development stage of target species.

Alternative bacterial control

The use of probiotics is to be considered in any new hatchery operation, as a biocontrol agent and as an alternative to the use of anti-microbial drugs. This shift away from anti-microbial drugs is an expanding trend in the shellfish industry, and at the time of writing, is incorporated in oyster culture. There has been little reported in the literature for scallops at a commercial scale, however, unpublished preliminary trials have shown promising results.

As a brief overview, larval and post-larval cultures are a common environment shared by larvae and bacteria, including both beneficial bacteria – enhancing larval development – and potentially pathogenic bacteria – resulting in the loss of entire batches. Recent research reflects the importance of bacterial balance in maintaining healthy cultures, and the role of probiotics (especially of probiotic bacteria) as a means of controlling pathogens.

Probiotics can be defined as an entire microorganism or its components that are beneficial to the health of the host; a probiotic can benefit the host either nutritionally or via a change in its immediate environment. This could apply to certain phytoplankton species – capable of producing substances toxic to harmful bacteria and potentially acting in a beneficial manner – and to certain bacteria species.

More specifically, some bacterial species are known to have beneficial effects on the digestive system of bivalve larvae. Probiotic bacteria can be cultured and introduced to larvae and post-larvae directly into larval tanks or with microalgal food as the entry point into the larviculture system. Co-cultivation of probiotic bacteria with phytoplankton species used in aquaculture operations has been demonstrated to not affect algal growth but also, in several cases, to improve the ingestion of the probiotic bacterium by larvae. This yields an improvement of the larval culture itself. Most studies pertain to the response of oyster and scallop species, to *Pseudomonas* spp., and other marine bacterial isolate strains, under experimental conditions.

Practical considerations for probiotic use

For a new hatchery operation, commercially available probiotic bacteria need to be identified. Bacteria of the genus *Pseudoalteromonas* are commercially available, and are known to produce both antibacterial and antivirulence compounds, making them interesting from an applied point of view as bivalve probiotic. A probiotic starter culture in a simple vial inhibits three strains of virulent *Vibrio tubiashii* (*corallyticus*) (Fig. 12.9C and D). Care must be taken, however, prior to the application of probiotic to the entire culture operation, as this genus has been reported to have toxic effects on *Artemia* sp. (Neu et al., 2014).

The first step in setting up a probiotic-enriched larval and post-larval culture is to confirm the compatibility of the selected probiotic bacteria with live microalgae cultured for food through co-cultivation trials. The second is the testing of larval and post-larval performance when exposed to probiotic bacteria.

Regardless of the method of integration in larval/post-larval tanks, the culture of probiotics requires additional equipment and space to that of the algal facility for the stock cultures of probiotics (Fig. 12.7). Equipment and materials necessary to prepare the nutrient media probiotic culture, and to harvest them include: a fume hood, autoclave, magnetic heated stirrer plate, shaker table, incubator, centrifuge, and bacterial count materials. Monitoring of probiotic growth is conducted using microbiological plating techniques. It is recommended that a rigorous evaluation and monitoring of both probiotic, and co-cultivated algae is conducted initially to establish the suitable dosage. The initial research and development phase for probiotic use is not to be underestimated. It should also be clear that although the use of probiotic bacteria has most likely long-term benefits for an aquaculture operation, the initial assessment and daily culture requires additional labor to the daily culture operations, and more specifically to the greenhouse personnel with the required laboratory skill set.

Temperature control: broodstock, larvae, seed, algae

It is often the case that one or more of the life stages of the targeted species require temperature control – either heating or cooling – in order to extend the production season, and/or to enable timely transfer of seed to the farm in the natural environment.

The first consideration is the reproductive cycle of the broodstock, its duration, and frequency of spawning within one season. If the natural reproductive cycle enables several reliable spawnings in one season, it may not prove necessary to condition the broodstock. If the natural cycle does not enable several spawns in one season, broodstock can be conditioned either to accelerate the latter stages of gametogenesis and produce ripe gonads, or in extreme cases, to undergo the entire gametogenic cycle repeatedly over a production period. The shorter the conditioning regime required, the the lower the effort and costs involved. Broodstock are conditioned in great part via exposure to temperature cycles mimicking the natural cycles. Dependent on the target species and geographical location of the hatchery, broodstock will either be chilled or heated for conditioning and spawning.

Larval rearing may also benefit from increased temperatures throughout the larval life. This may shorten the duration of the larval life, and also enables a higher frequency of larval cycles throughout the production season, maximizing and ensuring seed production output.

Post-larvae will need to be cultured under similar temperature conditions as the larvae, for at least the first month. Gradually, spat and/or seed will be acclimated to the natural temperatures specific to the farm site, in preparation for transfer to the sea.

Similarly, for optimal production of algal cultures, the temperature of incoming ambient seawater will most likely be too low in northern temperate regions, and too high in equatorial regions. Algal cultures are generally grown at an optimal temperature range between 20°C and 24°C. Species-specific optimal temperature data can be obtained either from the supplier or in the literature, and it is recommended to carefully select algal species, focusing on four or five species with similar optimal temperatures that are easily maintained in the facility. High temperatures have a more deleterious effect than lower temperatures; temperatures below 16°C will slow down growth, temperatures higher than 35°C are lethal for several species. For small and intermediate algal cultures (up to 20 L), temperature may be maintained in a climate-controlled room; however, for larger algal vessels, and continuous systems, incoming seawater needs to be heated or chilled accordingly. Although most microalgal species are capable of carrying out photosynthesis and cellular division over a wide range of temperatures (15–30°C), cultures fare better at a constant optimal temperature and result in a higher quality algal culture necessary for the early life stages of molluscan species.

Heating seawater

A range of heating methods are used, from submersible aquarium heaters to oil or gas boilers, from static, to the use of header tanks, and flow-through systems. The method and unit used depends on the volume of heated water required, and whether the need for heated seawater is continuous or intermittent.

Submersible aquarium heaters

A basic and low-cost approach is the use of static systems and submersible aquarium heaters. This method, used by some aquaculturists for conditioning northern quahogs (hard clams) and oysters, consists of two conditioning tanks (Fig. 12.10). One tank is prepared for transfer by gradually heating seawater over 24 hours using aquarium heaters; algal food ration is added prior to the transfer of broodstock suspended in pearl nets. The process is repeated alternating the broodstock between tanks on a daily basis, cleaning thoroughly each tank post-broodstock holding and prior to preparation. During transfer to the new heated tank, care is taken that there is no thermal shock, otherwise entraining spontaneous spawning. Initial ambient seawater temperature approximates 6°C, which is raised 1°C per day, until it reaches pre-spawning temperature – this temperature is usually 1–2°C below that triggering spawning. Animals are conditioned in this manner for up to 30 days.

Submersible heaters can also be used in small semi-recirculating tanks. The last stage of the

Figure 12.10 Small-scale and low-tech conditioning tanks for broodstock, regulating temperature with submersible aquarium heaters. Courtesy of Richard C. Karney Solar Shellfish Hatchery.

gametogenic cycle of the calico scallop, *Argopecten gibbus*, was completed by heating a 300 L tank, with two aquarium heaters. A temperature differential of 4°C, gradually increased over a 10 day period was achieved, with temperature remaining at ±0.5°C. The semi-recirculating system within this tank compensated for the small heating capacity of aquarium and enabled a continuous flow of incoming ambient seawater (Sarkis and Lovatelli, 2007).

Storage heating tank system

When larger volumes of heated seawater are required, incoming seawater can be heated and stored in a header or reservoir tank. From there, it is distributed to tanks using a delivery pump. There are several ways of achieving this, one of which being the use of titanium heaters with digital temperature controllers. Sarkis and Lovatelli (2007) illustrate the use of three titanium heaters of 4000 W each, sufficient to heat a differential of 8°C, stored in a 1000 L tank and supplying a total tankage of 5000 L over an 8 hour period. Incoming ambient water is finely filtered and flows at a rate equivalent to the outgoing rate (outgoing flow includes partial flow to hatchery and partial recirculation to storage tank). The entire storage heating tanks system is turned off when not required, and can be used, as in this reported case, during water change of static larval tanks. This system is not advised for supplying a continuous (or 24 hour) flow of heated seawater, unless it is modified to shut off in case of pump failure. The advantage of this small-scale heating unit is that it is easy to keep clean, in a similar fashion to protocols used for larval and algal tanks.

Boiler systems for continuous heating

Heating seawater becomes increasingly costly with larger scale operations. Traditional methods for the continuous supply of heated seawater for algal, larval, and nursery sections of the hatchery facility include the use of boiler systems – oil or gas – in conjunction with heat exchangers (Fig. 12.11). Oil boilers entrain large electricity bills; gas boilers are more cost-effective depending on the hatchery site. For both, the challenge is ensuring that heat exchangers are maintained clean; this is an important component to maintaining the efficiency of the heat exchangers and

Figure 12.11 A gas boiler system, showing associated heat exchanger, used in larger scale operations. Courtesy of Michael C. Voisin Oyster Hatchery.

controlling energy consumption. Boiler heating systems do offer a compact way to provide centralized heating to multiple points in the aquaculture facility, used to heat any body of water, such as algae, broodstock, larvae, and post-set. As a guide, a 1 million BTU (British thermal unit) gas boiler system is capable of increasing temperature by an 8°C differential for a flow of up to 600 L/min. Efficiency of boiler heating systems can be improved by considering heat recovery, discussed below.

Water source heat pump and heat recapture systems

Water source heat pumps are found by some aquaculturists to be more cost-effective than boiler-reliant temperature-control systems; especially if heat pumps are used in conjunction with a heat recovery system. The transition from boiler to heat pump with heat recovery system is highly recommended. In brief, the heat pump removes heat from the water and transfers it into new water entering the hatchery. A heat recovery system makes use of the effluent water leaving the hatchery, as it flows through a heat exchanger. This recovered heat assists the heat pump in raising the temperature to the temperature differential required by expending less energy. For example, in the case of incoming ambient seawater of 0°C to be heated to 10°C, without heat recovery, the heat pump would need to heat to a 10°C differential; with heat recovery, the heat pump only needs to heat from 9°C to 10°C, or 1°C differential (see Fig. 12.12A). Note that the effluent water leaving the culture tank only flows through the heat exchanger to recover heat, it is not re-used to flow back into the hatchery. Incorporating a heat recapture method in the temperature-control system of a newly designed hatchery is worthwhile in the long term.

Heat recapture can also be installed in an operating facility, and the types of modifications needed to the temperature-control system are described in the literature. Some aquaculturists completely replace previous boiler systems with a heat pump and heat recovery system, with a reported result of 70% savings in utility costs. Others rely on a ready-made compact heat pump system plumbed to the outflow. As a real-life example, an oyster hatchery estimates an energy cost reduction of 85% (accounting for the electrical costs to operate the pump), with the displacement of up to 95% of propane previously used. This hatchery added a collector pipe for waste water, and a 13-ton heat pump system plumbed to the outflow (Fig. 12.12B); heated waste water (T = 14–22°C) is passed through a heat exchanger and the MULTISTACK heat pump, reclaiming heat which would otherwise be lost to the hatchery. The principle is similar to that illustrated in Fig. 12.12A, in that the heat pump reclaims heat and adds it to the incoming ambient seawater (7–12°C) before it passes through the main gas boiler.

In the case of replacing an existing system, it is advisable to keep the former system as a back-up, or until such time that it is considered obsolete. In all cases, maintaining the heat exchanger clean remains key for sustained efficiency; namely, free of bacterial growth and solids, which easily foul these components. Monitoring the pressures and temperatures at the heat exchanger inlets and outlets will provide an early warning sign of fouling level, before it becomes an issue and significantly affect energy cost. Regular backwashing of a heat exchanger will assist in minimizing fouling.

Renewable technology: geothermal and solar energy

The application of renewable energy technology is increasing, driven by the trend to adopt clean technology and to reduce reliance on non-renewable fossil fuel.

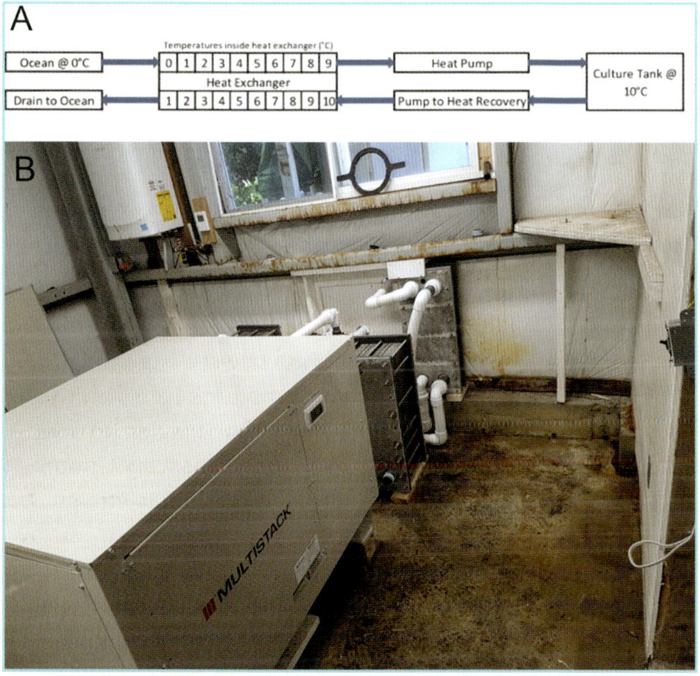

Figure 12.12 (A) Heat recovery schematic; using counter-current exchange maintains a gradient throughout the heat exchanger for optimum capacity. (B) In-line MULTISTACK heat pump, modifying existing heating unit to a heat recapture system. Courtesy of J.P. Hastey, Nova Harvest.

Geothermal heat pumps are commonly used for swimming pools and can accommodate flows approximating 250 L/min. Such pumps rely on the stored solar energy in natural bodies of seawater, and utilize this thermal energy by removing the heat from the ambient seawater to provide the required heated seawater in culture tanks. This energy movement is driven by electricity, at an output reported to be substantially greater than that obtained using the boilers. The specifics and cost-effectiveness of such pumps depends on incoming ambient temperature and temperature differential required.

Solar energy for heating seawater in small-scale hatcheries has been tested and proves energy efficient, supplying a reliable source of heated seawater when used in conjunction with electrical heating. The combination of both a solar and electrical power source ensures that constant temperature control is achieved; during times of high sun exposure, solar energy is the primary heat source, and during times of low exposure, electrical heating elements are relied upon. A solar-assisted hatchery was built on Martha's Vineyard, Massachusetts, during a time of energy crisis in 1980, and incorporated both passive and active solar components (Fig. 12.13). The hatchery is south facing on the shore of a 535-acre embayment; it encompasses all hatchery sectors.

Phytoplankton is cultured in a well-insulated greenhouse on the upper floor; starter cultures are grown in a hood under continuous fluorescent lights, whereas intermediate and larger volumes are grown mostly under natural light, with fluorescent and metal halide lamps available as back-up lighting during cloudy periods. This is relatively successful for the standard species of *Isochrysis* sp., *Chaetoceros* sp., *Tetraselmis* sp., and *Thalassisosira weisflogii*; *Pavlova* sp., appears to be less tolerant of natural light and temperature fluctuations.

Larvae are also cultured on the same level. The primary source of heated seawater comes from four active solar plate collectors. Freshwater is pumped from a well, and heated by solar plate

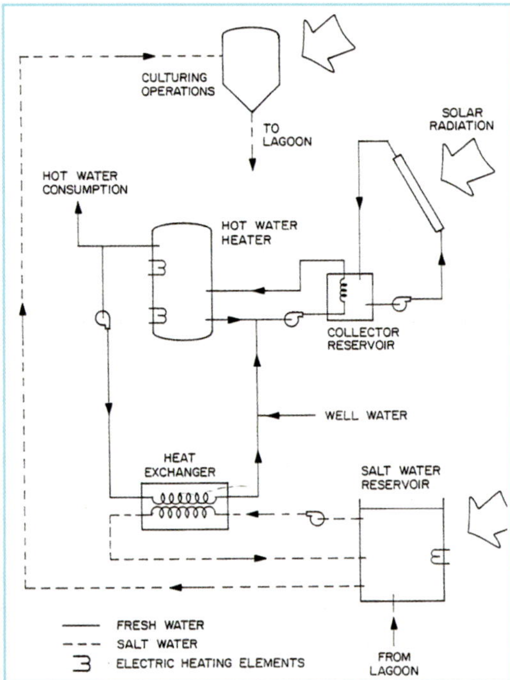

Figure 12.13 Schematic of a solar-assisted shellfish hatchery on Martha's Vineyard, Massachusetts. Courtesy of R. Karney.

collectors. Solar-heated water flows from the collectors to an electric hot water storage tank; both water from the electric hot water storage tank, and from the solar-heated collectors flows through a heat exchanger. Ambient seawater is pumped from the pond with a 2.5 HP swimming pool pump, and heated to the appropriate temperature as it passes through the heat exchanger; larvae are reared at 24°C. Heated seawater is thereafter pumped to the larval and post-set tanks. The combination of electric and solar-assisted energy source is sufficient for cultivating shellfish during the colder months. More specifically, it has the capacity to heat a static 500 L volume from 5°C to 22°C in approximately 8 hours; when seawater is used continuously, this system heats a 500 L volume an average of 2°C/hour above ambient, and up to 3–4°C/hour above ambient on sunny days.

Solar energy is also used for intermediate culture of bivalve seed in floating upweller systems (FLUPSY). A FLUPSY nursery system consisting of a floating platform, with shellfish seed suspended on 'upweller' screens in the marine environment, relies on the pumping of surrounding seawater moving a constant flow of naturally occurring phytoplankton and nutrient-rich water through the seed. The mesh size for screens varies according to seed growth. Pumped water flows through the screen to the surface. This results in a rapid and uniform growth of seed, and has become an increasingly popular method among oyster and clam growers, as it alleviates the demand on microalgal food and in this way reduces seed production costs. The pumping of seawater is conventionally achieved using a motorized paddle connected to shore power or to a generator.

More recently, solar energy has been considered as the source of power for pumping seawater through a FLUPSY (Fig. 12.14), relying on a shaft-driven axial flow pump. Construction costs are comparable to those for a conventional FLUPSY, and small-scale commercial operations are starting to install solar-powered FLUPSY, preferring to reduce the use of fossil fuel. The capacity to operate FLUPSY systems using solar energy increases the selection of sites available to this nursery system; given the independence from shore-based power and coastal properties, FLUPSY nurseries can be installed in more remote, and less controversial areas. This alleviates the multi-conflict challenges often faced by aquaculturists. One such off-grid, photovoltaic FLUPSY has been designed relying on 254 W solar panels charging a 12 V battery, which in turn operates the pumping system for seawater supply to the seed. In an evaluation of various upwelling systems, flow rates of solar-powered FLUPSY were higher than those of a conventional FLUPSY (300 L/min compared to 255 L/min). This not only indicates a comparable output for both systems, but suggests the potential for rearing a higher number of seed in a

Figure 12.14 Solar-powered FLUPSY. Courtesy of D. Leavitt.

given surface area. If this proves true, the solar FLUPSY may prove to be a benefit advantage to an aquaculture operation.

Chilling seawater

Chilling seawater is often necessary for broodstock conditioning. In small-scale operations, with flow rates less than 160 L/min, in-line air-cooled chillers with titanium evaporator coils can be installed prior to entry of seawater into a culture tank; a direct immersion titanium sensor is connected to the digital temperature controller.

For an operation with several tanks requiring chilled seawater, larger air-cooled chillers with insulated heat exchangers and commercial grade condensing units mounted on a stainless steel frame are suitable; these can be used outdoor, but need protection from direct weather. Temperature is regulated by a built-in digital temperature controller. Such chillers range from 2 HP to 20 HP, accommodating flows from 120 L/min to 500 L/min, and are useful for broodstock conditioning.

Large scale operations, or for operations chilling to a high differential not achieved using a single air-cooled chiller, a compressor and evaporator with the capacity to control water temperature for all tanks are required.

Adding to the sustainability of the operation, the heat produced by the cooling system can be captured by heat exchangers, which are refrigerant to water systems, connected to a condensing unit or refrigerant loop. This heat can then be used for larval and post-larval needs.

Water source heat pumps, referred to previously in this section, can also be used to cool water during the warmer months.

Algal production: culture process, seawater treatment, light, air, and commercial alternatives

The schematic layout of the hatchery complex illustrates four main areas for the algal culture facility (Fig. 12.7): seawater treatment units; dry laboratory area housing stock (or master) cultures and related equipment, and probiotic cultures if used; starter and intermediate algal culture area; and large algal culture area. Depending on the scale of the aquaculture operation and the processes used for algal production, additional indoor space allocation for other equipment, such as photobioreactors, and outdoor space – for further algal production supplementing intermediate nursery – are needed. Outdoor space requirements refer mainly to operations aiming to grow seed to larger size in ponds or tanks.

The algal culture sector is a vital part of the aquaculture operation, and a key component of molluscan hatcheries; it provides a reliable supply of high-quality unicellular microalgae on a daily basis, critical to the survival and growth of larval and post-larval stages. A failure in the algal culture production will have repercussions on the entire operation, and can be catastrophic to the hatchery. This is the sector of the operation, where extreme care cannot be over-emphasized; cleanliness and careful microbiological transfer techniques must be applied daily without fault. This will ensure the production of healthy monocultures of selected algal species, grown and harvested from large vessels, whose production is in turn reliant on a reservoir of small master or stock cultures. For a healthy algal system, it is crucial to maintain this reservoir free of contaminants and excessive bacteria. Contamination of algal cultures may occur via the seawater supply, air supply, and cross-contamination from nearby algal cultures.

During the design of an aquaculture operation, careful consideration of the algal sector layout and associated seawater treatment system will assist in minimizing the risks of contamination and ensure the production of high-quality algae.

One such consideration, is the separation of stock cultures and starter/intermediate cultures from the larger vessels (Fig. 12.7); the maintenance of stock and starter cultures in a climate-controlled room, isolates these from daily inoculating and harvesting of larger vessels that may resuilt in potential contamination.

A modular approach to the seawater treatment consists of filtration, sterilization, and temperature-control units dedicated to the algal culture sector; such a system provides an independence to the functioning of the algal cultures, and a safeguard from any failure of other non-algae related equipment (Fig. 12.7). In this case, incoming seawater is obtained from the coarsely filtered seawater line and treated within the algal culture sector. An alternative route (shown as a dashed line) is the use of pre-treated seawater from one larger system supplying both hatchery (larvae and spat) and algal culture sectors. The latter option does create an interdependence with hatchery seawater demands, and is not recommended.

An algal culture facility requires the capacity for:

- filtration (1 μm) of seawater for all vessels – from stock to large harvestable cultures
- sterilization for all vessels
- heating seawater (in northern temperate regions)
- light
- filtered aeration
- carbon dioxide (CO_2) input into airline
- freshwater for cleaning.

Each area of the algal culture sector has access to some or all of the components above, depending on the processes used. Not all hatcheries will have the capacity or need for each algal sector outlined below; small-scale operations can often grow starter to large cultures within the same space allocation. One such layout is described in Sarkis and Lovatelli (2007). Ideally, an algal culture facility will include the following.

1 **Dry laboratory** with climate control has a sink with freshwater access; a fume hood for inoculations; an incubator for storing stock cultures, or alternatively counter space with

a lighting system; separate areas for probiotic culture if needed with related equipment; a 25 mm PVC delivery pipe fitted with a flexible hose and regulated with a 25 mm one-way valve, supplying treated seawater on demand. Other equipment for the storing of nutrients, samples and/or commercial diets (refrigerator and freezer), and for the monitoring of algal cultures (such as microscope, cells, pipettes, and so on) are kept in this laboratory area. Please note: Heated seawater is not usually mandatory here, as seawater used for culture is often autoclaved or pasteurized.

2 **Starter and intermediate culture** areas receive heated and sterilized seawater similarly to the dry laboratory area; as this is used for filling vessels for batch culture, one pipe with a fitted hose supplying water on demand is sufficient. A 25 mm PVC airline runs along the wall supplying constant air to each vessel regulated by small valves. Air can be provided from a main compressor supplying the entire greenhouse, or from a smaller unit installed in this area; if the latter, a separate CO_2 supply must also be provided. Counters and shelves are installed along the wall for algal vessels, with light banks at each level for illumination. Climate control, and freshwater access with a sink of this area is also a must. Counter space for cleaning, preparing and storing of vessels is needed. This area should be considered a 'dry' area where all algal vessels are closed, and seawater supplied via a hose for inoculations, discarding old cultures externally; there is no running seawater flowing continuously in this section.

3 **Large algal cultures** receive heated and sterilized seawater. Procedures vary among aquaculturists, however, all vessels need access to treated seawater whether it is on demand or continuously. PVC pipes are fixed to the wall or ceiling for delivery of seawater; most often 25 mm diameter pipe is sufficient, but larger diameters may be required for large-scale operations. For continuous cultures, one or more header tanks (1000–5000 L) are used for ensuring constant 24 hour flow of treated seawater to culture vessels. A main aeration line (25 mm or above), runs alongside the seawater line, fitted with small valves regulating the flow to each vessel. As for starter and intermediate cultures, air supply can be from a main or dedicated compressor. Light banks are installed for each vessel. Freshwater access and hose for cleaning, a well-drained floor, and climate control are all required. Climate control can be difficult in this larger area; often an extraction fan assists in controlling heat output from the lighting in tropical areas. This area is also the site of algal harvesting on a daily basis; dependent on the delivery procedure, collector tanks to which algae is pumped for distribution to hatchery are installed strategically for easy transfer to the hatchery tanks.

The following sub-sections outline the types of filtration pertaining to algal culture, temperature-control systems, illumination requirements, and aeration supplied; these are discussed in the context of the algal culture procedures required to achieve target production.

Live microalgal culture

Molluscs filter feed on phytoplankton throughout their lives as larvae, juveniles, and adults. Several species of microalgae are used as food organisms in hatcheries; they vary in size, environmental requirements, growth rate, and nutritional value as a live food (Fig. 12.15). Their success as food species depends not only on their nutritional quality, but also on their tolerance to temperature, salinity, and light. Because of this, algal species will exhibit different growth

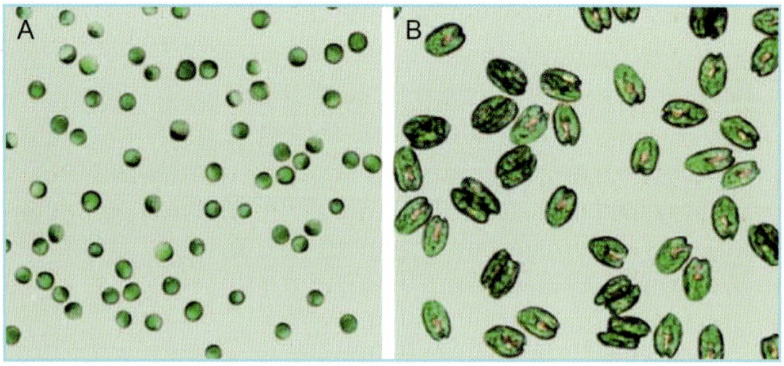

Figure 12.15 Standard species of unicellular live microalgae cultured for molluscan rearing: (A) *Isochrysis* sp., (B) *Tetraselmis* sp. (Helm et al., 2004).

performance depending on the hatchery site. Knowledge of the optimal growth responses of microalgae under local conditions is of great benefit, but a first selection can be based on the literature, as much has been reported on biochemical constituency of microalgae with respect to environmental factors. The first criterion in selecting a range of microalgal species is thus dependent on the environmental conditions of the hatchery site itself.

The second criterion is the matching of the algal biochemical constituents with the biochemical needs of the life stages of the target molluscan species; larval and juvenile growth and survival are dependent on the content and nature of biochemical constituents in algal food, as is gametogenic development in broodstock.

The variety of algal species grown serves different needs throughout the production cycle with respect to size and digestibility, culture characteristics, and nutritional value. There is a wealth of information in the literature relating the nutritional value of unicellular algae to specific molluscan species. It is recommended that live diets be composed of at least three algal species, which will provide all of the essential constituents suited to the requirements of the life stage. In total, this translates to the large-scale culture of four to six different algal species for the feeding of larvae, seed, and broodstock. Details on most commonly used algal species in molluscan production are given in Chapter 13.

Algal cultures are started under controlled environmental conditions; large-scale algal production ranges from single species culture (mono-specific) reared indoors under controlled environmental conditions to outdoor ponds or raceways, supplied either with ambient incoming seawater and additional nutrients promoting a bloom of the naturally occurring assemblage of microalgae, and/or supplemented by dedicated outdoor algal production tanks or bags.

In summary, given the multi-faceted activity of microalgal culture, the size, location, and engineering specifications implemented depend on the following.

- Site characteristics, where incoming water quality and temperature will dictate the filtration, sterilization, and temperature-control procedures.
- Food ration requirements of the target species, which will dictate the volume of algae produced for the early life stages, and to some extent the algal species selected for culture.
- The type of culture to be practiced; this especially relates to the algal culture of larger volumes (>100 L), which can be done as batch culture, semi-continuous, or continuous.
- The available illumination: in tropical areas outdoor algal ponds will thrive in existing light,

and larger algal culture vessels can also benefit from natural light; in more temperate regions, a total reliance on artificial light is needed.

- The production goals which will be key to sizing the algal culture facility.

These considerations are interdependent, and each should be treated as a part of the whole. For a better understanding of design and layout considerations, an overview of the algal culture process is given below.

Algal culture process: progressive batch culture to large harvestable cultures

Typically, microalgal culture facilities utilize incoming seawater similar in chemical composition to that used to culture the animals. Following treatment, seawater is enriched with a nutrient media – consisting primarily of nitrates, phosphates, essential trace elements, and vitamins, and for the culture of diatoms, additional silicates (Fig. 12.7). In some laboratories, synthetic seawater is used for small-scale, stock microalgae cultures, but it is prohibitively expensive for large-scale production in commercial hatcheries.

Progressive batch culture

The quantities of algal cells required for hatchery production of molluscan larvae are produced through a process called *progressive batch culture.* This entails the transferring of small volume cultures of concentrated algal inocula into larger volumes of treated, nutrient-enriched water. This stage is conducted in a 'dry' laboratory area, climate-controlled, and isolated from the remaining algal culture areas. In order to ensure a contaminant-free culture, sterile microbiological techniques are used during sub-culturing or start-up of solutions. These techniques rely on the flaming of all equipment prior to use, and the continuous working by the flame during addition of nutrients, transfer of cultures, and so on. Some operations install a fume hood in the dry laboratory for this process. Techniques are detailed in protocols outlined in Sarkis and Lovatelli (2007) and Chapter 13.

A guide on volumes required for inoculum throughout the progressive batch culture process is given in Box 12.1. An example of the timeline from stock culture to harvestable volumes is given in Fig. 12.16. Stock cultures can be maintained in test tubes or 125 mL flasks; starter cultures are

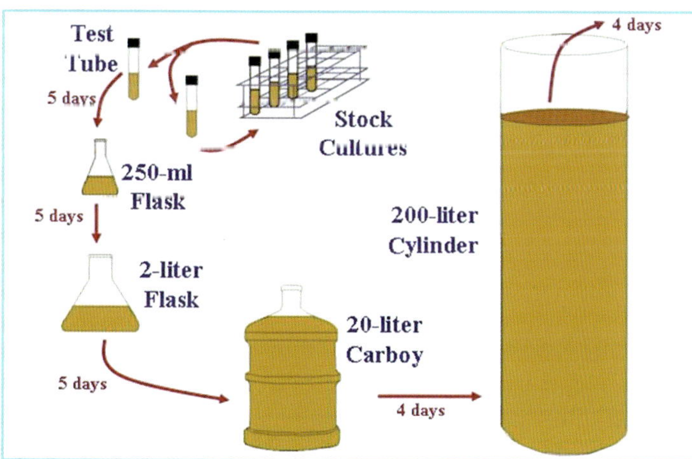

Figure 12.16 Example of progressive batch culture timeline and vessels used for the culture of a mono-specific microalgal species. Illustration by P. van Wyck, modified by R.L. Creswell.

Box 12.1 Progressive batch culture process.

Axenic stock or master cultures are usually maintained in test tubes or 125 mL Erlenmeyer flasks.

Starting with cells taken from an axenic stock culture, microalgae are cultured in an enriched medium through a series of culture vessels of increasing volume (Fig. 12.16). The algae grown in each culture vessel serves as the inoculum for the next larger vessel, until the quantity of cells required for feeding is reached. A typical series for large-scale production:

1 25 mL test tubes (10 mL stock culture) inoculates ...
2 500 mL flasks (250 mL starter culture) inoculates ...
3 2.8–4 L flasks (1000 mL culture) inoculates ...
4 20 L carboys (16 L culture) inoculates ...
5 250 L cylinders (180 L culture) inoculates ...
6 12,000 L tanks (10,000 L culture) inoculates...

Starter cultures which vary in volume, ranging from 500 mL to 1 L flask cultures, are used to inoculate intermediate cultures (2–25 L), which are subsequently used to inoculate even larger volume cultures for final production before harvest and feeding. Starter cultures are grown similarly to stock cultures, usually in 500 mL flasks with 250 mL of sterile medium; about 50 mL of the starter culture is transferred to similar volume flasks to maintain the line, while the remaining 200 mL are used to inoculate intermediate culture containers (typically from 4 L flasks to 20 L carboys). The vessel size inoculated subsequent to the 4 L flask is dependent on the volume of algae required and species production scale; for example, a 4 L flask culture can be used as a direct inoculum to start a 100–500 L vessel. In turn, a 500 L vessel can be used as an inoculum for a 5000 L tank.

grown from 250–500 mL flasks; intermediate cultures range from 2 L to 20 L flasks and carboys; large-scale harvestable cultures are grown from 100 to >5000 L vessels.

Dependent on budget and space available, stock cultures can be maintained in an incubator, where temperature and light are controlled. For a more basic facility, stock cultures are kept in an isolated section of the dry laboratory area, in order to prevent contamination from other flasks, and grown under similar light and temperature conditions as starter cultures. Flasks are closed loosely – a cotton/foam plug, or screw top – allowing some oxygen in the culture; similarly, if test tubes are used, the cap is loosely closed. They are not aerated, but manually shaken daily to facilitate gas exchange and mixing, and are maintained through monthly re-inoculations.

Starter cultures (250–500 mL flasks) are grown at 18–24±1°C at a distance of 15–20 cm from 40 W or 80 W cool fluorescent lamps, giving a level of illumination at the culture surface of 4000 to 5250 lux (Fig. 12.17). Depending on the resources available, various types of vessels are used by aquaculturists, beakers sealed with parafilm and aerated are used as starter culture vessels; others use Erlenmeyer flasks closed with cotton plugs, or rubber stoppers through which a glass rod connected to the airline is fitted (Fig. 12.17B). Starter and intermediate culture vessels may or may not require access to an airline, depending on the preference of the aquaculturist. If

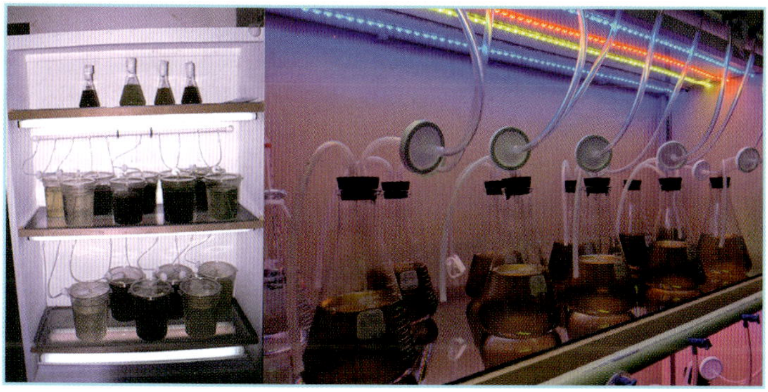

Figure 12.17 Starter cultures reared in Elenmeyer flasks and sealed beakers. Courtesy of Michael C. Voisin Oyster Hatchery.

aerated, a supply of air/carbon dioxide (CO_2) mixture is given to maintain a pH between 7.5 and 8.5, ensuring sufficient carbon for photosynthesis as the algal culture grows. Usually, 2–5% CO_2 is sufficient to maintain an adequate pH; this is tested onsite at the beginning of the hatchery season. An in-line filter (0.3–0.5 µm) on the delivery tubing is necessary to prevent contaminants that can be introduced through aeration. A dedicated air compressor fitted in the small algal area, or air supplied from a blower supplying the entire facility are used dependent on the design of the hatchery. At least two CO_2 bottles should be available, with one connected to the airline, for constant injection and mixing of CO_2.

Intermediate culture volumes, typically 4 L flasks to 20 L carboys, are used to inoculate larger vessels, typically 100–500 L translucent fiberglass cylinders or polyethylene bags or even larger fiberglass tanks (up to 5000 L) and raceways (Fig. 12.18). Similar to the starter cultures, sterile, nutrient-enriched seawater with an inoculum is aerated with a supplement of CO_2 carried in compressed air. Illumination for culture growth is provided by fluorescent lamps, usually mounted externally to the culture vessels. The number of lamps used is determined by the height and diameter of the culture vessels with the object of providing 15,000 to 25,000 lux measured at

Figure 12.18 Intermediate culture using 20 L carboys at Harbor Branch Oceanography Institute. Courtesy of R.L. Creswell.

Figure 12.19 Large-scale culture systems, batch, semi-continuous, and continuous: (A) cone-shaped 200 L cylinders; (B) round 5000 L tanks of mass algal culture; (C) preparation of steel-mesh frame for support; (D) polyethylene bags as algal vessels suspended with a steel-mesh frame; (E) algal bags without frame supported by stainless steel rails. Courtesy of (A) R. Karney; (B) UMCES Horn Point Oyster Hatchery; (C) Auburn University Shellfish Laboratory; (D) Bay Shellfish, Tampa, Florida; (E) Michael C. Voisin Oyster Hatchery.

the center of the empty culture container. Two 65 W or 80 W lamps are sufficient to illuminate 3 L glass flasks, which are about 18 cm diameter, whereas five lamps of the same light output are necessary for 20 L carboys.

Large cultures 4 to 8 days old from carboys (20 L) are used to inoculate 100–200 L translucent fiberglass cylinders or polyethylene culture bags (Figs. 12.18 and 12.19). Similarly, large-scale vessels are filled with filtered, UV irradiated and usually chlorinated/dechlorinated seawater, enriched, and inoculated.

Fiberglass cylinders (cone or flat-bottomed shape) are most often used for semi-continuous culture; where algae are harvested daily until one-quarter of the total culture is used, and thereafter re-boosted by the addition of nutrient-enriched water (Fig. 12.19A). Such a culture can be maintained for 1 month; at which time, it is discarded and started with a new intermediate culture inoculum. A solid frame – often epoxy-coated wooden custom-built – holds the cylinders, and top lids prevent the entry of air-borne contaminants. For harvest, one-way PVC ball valves are fitted at the bottom of the cylinder. It is a cost-effective system for small-scale operations, but requires too much labor to be considered for a large-scale facility.

Large (5000 L) round tanks are also used for batch and semi-continuous cultures in some commercial hatcheries; light is provided above the tank artificially or using solatubes to direct the sunlight (Fig. 12.19B).

For continuous culture of microalgae, heavy gauge polyethylene tubing is frequently used instead of fiberglass cylinders; this is purchased in a variety of widths, and by cutting a suitable length and heat-sealing one end a sterile, flexible culture container can be formed, either as a cylinder or as an oblong bag. Containers formed in this way can be strengthened by supporting them within a plastic-coated, steel-mesh frame (Fig. 12.19C, D); some algal species are reported to be light limited in such a system, with cell density remaining at a low level. Alternatively, the cylinders can be suspended, with or without lateral support mesh, if the diameter of the bag is less than 30 cm and the height less than 200 cm. Bags without frames are reported to enable increased algal densities up to threefold for certain species, as they are not light limited (Fig. 12.19D).

Continuous microalgal culture using polyethylene bags holds several advantages: (1) as a sealed container they are less likely to become contaminated than rigid containers with open tops or lids; (2) they do not require daily maintenance and cleaning; and (3) they have lower installation costs and utilize space more efficiently. Bags are the least expensive way of constructing large-scale culture vessels, and such containers can be used indoors with artificial illumination or outdoors taking advantage of natural light.

In most cases, these larger-volume cultures are housed in greenhouses and receive natural light, reducing electricity costs for adequate illumination using fluorescent bulbs. Under optimal environmental conditions, the culture will be harvested in 4 days following inoculum or can be used to inoculate large-scale cultures in tanks or outdoor ponds. A greenhouse system is not illustrated in Fig. 12.7; however, this large culture area shown in the diagram can be installed as a greenhouse, with a roof allowing natural light into this section of the facility. Alternatively, a separate building in proximity to the intermediate culture sector can be installed as a greenhouse.

Large-scale outdoor bag cultures are often positioned horizontally to maximize sunlight penetration. Such large-volume systems are often used to induce multi-species blooms that are best suited for feeding juvenile shellfish in nursery systems, or adult shellfish in broodstock systems (Fig. 12.20).

Photobioreactors

There are several newly designed microalgal production systems on the market collectively termed AAP or advanced algal production systems, also termed photobioreactors (PBR). It is proposed by several first trial users that the consistent density of harvested cells from the reactors will allow for continued automation of downstream hatchery processes not easily accomplished by the conventional procedure for algae production.

Photobioreactor systems can provide higher algae densities, and have more efficient space usage (a smaller footprint), and lower labor requirements. They operate in continuous or semi-continuous production, and are closed to the external environment, resulting in longer production cycles

Figure 12.20 Outdoor large-scale algal culture grown using polyethylene bags and natural sunlight. Courtesy of R.L. Creswell.

Figure 12.21 (A) Photobioreactor (1250 L and 2500 L). (B) Close up header tank. (C) Inoculation and harvesting pump. (D) Nutrient bottles with close up computerized control system. Courtesy of (A) S. Glover, Industrial Plankton Inc., (B)–(D) S. Sarkis.

with less contamination. In general, PBR can be classified into three categories: (1) tubular photobioreactors; (2) column or cylinder photobioreactors; and (3) flat-panel or plate photobioreactors. Many of these designs are still under design and evaluation, but some have been tested in commercial scale operations (Fig. 12.21).

The underlying principles of the bioreactor operation are similar to those for traditional algal cultures. Incoming seawater is filtered (to 0.5 µm) and sterilized (using UV); it is stored in a 200 L header tank, and further filtered to 0.1 µm using absolute filters and ultrafilters before being pumped to the reactor. An inoculum of algae is added using a positive displacement pump at the base of the reactor; a 20 L carboy is used to inoculate an initial 200 L volume. As density increases, sterilized seawater is added to ultimately sustain a 1250 L culture. Nutrients are added automatically though 0.1 µm filters, with flow regulated and settings entered and displayed on the computer panel. An inoculum of algae is added using the pumping system at the base of the reactor; the same pump is used for harvesting. The flow of incoming water can be regulated to either simulate a semi-continuous system, or continuous system. Lighting can be adjusted as density increases to prevent any photo-inhibition. The culture is aerated, ensuring mixing and exposure to light throughout. All tubing is autoclavable, and quickly disconnects to ensure that risk of contamination during inoculation, harvesting, and nutrient refill is minimal. Harvesting of the culture is achieved using the same pump as initially used for the inoculum, at the base of the bioreactor. Between 100 L and 1200 L of algae culture is harvested daily, depending on species and density, and new seawater and nutrients are automatically added to compensate for the harvest.

Such a system has demonstrated the production of high-quality algae, which can be used for feeding during the early life stages of molluscan species (Fig. 12.21). A range of standard phytoplankton species used in aquaculture operations was cultured in an experimental setting, namely flagellates (*T. Isochrysis galbana*, *Tetraselmis* sp., *Pavlova* sp.), and diatoms (*Thalassiosira pseudonana*, *T. weisflogii*, *Chaetoceros* sp., *Skeletonema* sp.). Peak densities achieved for these are reported as 20 million cells/mL, with maintenance of clean cultures for up to 3 months; algal cell density did decrease with time to 2 million cells/mL. In a real life situation, a commercial oyster hatchery tested such a bioreactor and reported an algal production of 300 L/day of each *Tetraselmis* sp. and *Isochrysis* sp. (*T-Iso*) microalgae which was fed directly to the land nursery on a continuous basis. Harvested algae from the reactors were also used as inoculum for open culture algae tanks. Other first trial users concur that the use of bioreactors appears to be a cost-effective means of producing high volumes of dense microalgae for direct feeding as well as inoculum of larger culture volumes. The full-scale automation allows one user to add two units without additional labor requirements, effectively increasing daily algae production volume by 600 L at 6 million cells/mL. One of these bioreactors requires a space of 2.2 m × 2.4 m, using a total surface area of 5.28 m^2; it is estimated to replace 27 m^2 of 500 L continuous culture bags, and thus substantially reduces the footprint required for algal production.

Not only is the footprint requirement improved, but labor is also significantly reduced compared to the traditional algal culture methods. Dependent on the PBR, various levels of labor saving automation are built into each step of the production cycle post inoculation. Incoming water flow, nutrient addition, temperature, and lighting can be automated. On the other hand, training staff is essential, and a good understanding of how the system operates, with careful set-up and monitoring, is needed. If a PBR is considered in the installation of a new hatchery, or as an addition to an existing operation, a cost–benefit analysis is required, considering not only costs and footprints, but also the skillset available for a reliable high-quality algal production.

Water filtration and sterilization

Pre-treatment of water, be it saltwater or freshwater, is the most important step in successful culture of microalgae. Culture water should be free of suspended solids, plankton (for example, protozoans, ciliates, and other algae species), bacteria, dissolved organic compounds (DOC), and dissolved metals and pesticides. Pre-treatment typically includes mechanical and chemical filtration, and sterilization by a variety of methods based on volume requirements and cost. Seawater to the algal culture section of the facility, used for small and large algal cultures reared under controlled conditions, is supplied from the main hatchery line (Fig. 12.1). The additional filtration required depends on the site-dependent characteristics of the incoming seawater quality; however, for all algal cultures grown under controlled conditions, sterilization follows filtration. Space allocation for filtration and sterilization is an important consideration during hatchery design, some equipment, such as an autoclave, are best located in isolated areas of the greenhouse, as the resulting heat during operation will impact the ambient temperature and smaller cultures. All filtration and sterilization equipment should be easily accessible for cleaning and repairs.

Mechanical filtration

Removal of suspended solids, plankton, and bacteria can be accomplished through mechanical filtration, and is usually used in concert with other forms of sterilization described below.

In most cases, incoming seawater to the algal culture facility has been previously filtered at least through coarse filtration systems and in most cases through 25 μm filtration. A mechanical filtration system dedicated to the algal culture facility, similar to that described earlier and using cartridge or disk filters results in 1 μm-filtered seawater. This level of filtration is sufficient for cultures greater than 500 mL (intermediate and large-scale), pre-sterilization. For small volume culture preparation (stock cultures), the seawater can be additionally filtered to remove bacteria using 0.22 μm or 0.45 μm membrane cartridge filters, pre-sterilization.

Chemical filtration

Algae are sensitive indicators of poor water quality. Dissolved inorganic and organic compounds (DIC, DOC), hydrogen sulfide (H_2S), metals, pesticides, and other contaminants can prevent or impede microalgal growth. There are several methods to filter such chemical components if needed: (1) activated carbon (charcoal) filtration is helpful in reducing DOC; (2) deionization is effective in removing metals; (3) reverse osmosis (RO) can also be used to treat small volume culture water; as a side note, this is also applied for nitrogen removal in aquaculture wastewater in areas where regulations are stringent.

In this case, additional space in the algal culture sector is required for the installation of this equipment; it is strongly recommended that an 'equipment' room is dedicated to this, in order not to affect the conditions under which algal cultures are grown.

Heat sterilization

Pre-filtered seawater can be sterilized by autoclaving at 1.06 kg/cm^2 for 20 minutes. Autoclaving is most suitable for small volume preparations (more commonly to 1 L volumes, and dependent on the autoclave up to 4 L maximum). For larger volumes, batch or continuous pasteurization at 65–70°C is commonly used as a heat sterilization method. Pasteurization for 8 to 10 hour exposure at 50°C is also reported to be effective, as is heat sterilization using a glass-lined water heater or 500–1000 W submersion heaters. Microwave sterilization is useful for small volumes of pre-filtered seawater (1–5 μm) for 8 to 10 minutes per 1–1.5 L using a 700 W unit. A 700 W microwave on high power will effectively kill microalgae in 5 minutes, bacteria in 8 minutes, and fungi in 10 minutes for volumes of 1.5 L of seawater.

As noted above, autoclaves or pasteurizers should be isolated from the algal cultures, as they will generate heat, require maintenance, all of which will affect nearby algal culture vessels. A seawater treatment area housing required filtration and temperature units is recommended (Fig. 12.7).

Chemical sterilization

Chlorination is the most common and simplest method of chemical sterilization for culture volumes 4 L and greater. This does not greatly affect the hatchery design, and usually requires counter space and a storage area for vessels in treatment; it can add to the surface area of the algal culture facility if large-scale algal production in 1000 L or greater volume vessels is planned, warranting sterilization. A 24 hour period is generally allocated to this sterilization process.

Protocols for chemical sterilization are widely available. In brief, pre-filtered seawater can be sterilized with sodium hypochlorite solution (regular household bleach), or granular swimming pool chlorine for larger volumes (such as raceways). Bleach is added in standard doses of 0.5 mL/L or 1 mL/L of total seawater volume, and sterilization occurs in 10 to 30 minutes; although standard practice is to chlorinate for at least 4 hours, and up to 24 hours. Following the sterilization period, residual chlorine is neutralized by adding sodium thiosulfate ($Na_2S_2O_3$). Sodium thiosulfate is sold in various forms, a stock solution made with standard doses added to the bleached seawater. For neutralization, in general, aquaculturists add an excess of the sodium thiosulfate stock solution. Common swimming pool chlorine test kits can be used to determine the presence of residual chlorine; this is especially useful in larger operations, where the follow through is conducted by more than one staff member.

Ultraviolet irradiation (UV) and ozone (O_3) sterilization

Sterilization can also be achieved by using UV or ozone; both are used following the removal of suspended particulates from seawater through mechanical filtration; practically, seawater is filtered to 1 μm or at least 5 μm before passing through a sterilization unit. Sterilization units are therefore installed closely after filtration units, and housed in the seawater treatment area (Fig. 12.7).

The more common of the two methods is UV, in large part because it does not impart any hazardous by-products, while ozone at high levels can produce chloramines, toxic to marine animals. Ozone is a strong oxidizing agent that is particularly effective in removing dissolved organics, pesticides, color, and nitrates. It is highly unstable and quickly reverts back to O_2, but it is also highly corrosive and requires handling with special materials. In-line ozone generators are the most common and usually have monitors/controls to provide an adequate level of ozone yet avoid residual build-up.

UV irradiation is provided by mercury vapor bulbs at a wavelength of 254 nm; however, the killing power of UV is affected by turbidity/coloration of the incoming water, distance from the source, and exposure time (flow rate). Minimum dosages vary widely for different species, 22,000 μwatt-sec/cm^2 for water born algae, 35,000 μwatt-sec/cm^2 for bacteria/viruses, 100,000 μwatt-sec/cm^2 for protozoans, and as much as 330,000 μwatt-sec/cm^2 for *Aspergillus niger* (mold). UV sterilizer vendors are helpful to provide product specifications and advise customers for their particular application.

Temperature control

Since most of the microalgal species preferred by culturists are tropical/subtropical, most strains grow best at temperatures ranging from 16–27°C (60– 80°F) – with an optimum temperature of 24°C (75°F). Cultures should be maintained at the lowest temperature that is consistent with good yield to avoid encouraging bacterial growth. When considering temperature characteristics for an enclosed culture room, one should consider: (1) the size of the room; (2) heat sources (such as lights and ballasts); and (3) the volume and temperature of air pumped into the culture vessels.

Depending on the hatchery site, incoming seawater used for algal cultures may need to be heated. Heated seawater can be supplied through an independent algal culture seawater treatment system, or be delivered from a main hatchery complex system, receiving the same seawater as the larval and spat sectors (Figs. 12.1 and 12.7). It is recommended, and especially if the algal culture area is further removed from the main seawater treatment area, that a second fine-filtration and sterilization unit be installed right before delivery to algal tanks. Delivering treated seawater over any distance can entrain contamination, but also loss of heat, especially in northern temperate regions.

For large algal cultures reared indoors using batch and semi-continuous methods, sterilized and heated seawater is required on demand; where large volumes of treated seawater are utilized at a given time during the preparation of culture vessels for inoculation. In these cases, heated seawater demand is intermittent. On the other hand, for continuous bag culture systems (500 L), a steady incoming seawater flow is required over a 24 hour period for the duration of the algal production; in this case, there is a non-stop demand for heated seawater system for each 500 L vessel, in addition to that required for intermediate and small cultures on an intermittent basis. The algal culture process used thus affects the design and construction of the algal sector.

Figure 12.22 Starter cultures grown using LED strip lights, with illumination enhanced by use of mirrors as top and bottom shelving. Courtesy of Michael C. Voisin Oyster Hatchery.

Illumination and light source

Light intensity, spectral characteristics, and photoperiod are the parameters to be considered in an illumination regime. Artificial light is usually preferred over sunlight, where the duration and intensity cannot be easily controlled and may result in overheating, insufficient irradiance, or photo-inhibition if the natural light is too intense for too long. Irradiation of 2500–5000 lux (or 250–500 footcandles) is optimal for microalgae photosynthesis, with 10,000 lux near maximum.

Artificial lighting

Artificial lighting is mostly used for indoor controlled algal culture production; a range of bulbs is used for illumination. Although, 30 W 'cool white' bulbs are commonly used, their spectral characteristics are not ideal for intensive microalgae production, while bulbs with red and blue wavelengths (Gro-Lux^{TM}) are reported by some algologists to support higher yields. LED strip lights available in blue and red wavelengths are another option (Fig. 12.22); they generate considerably less heat than standard bulbs, and require transformers to connect to standard 120 V current. Indoor facilities should have bulbs placed 15–25 cm from stock cultures, and if possible the ballasts, or transformers, should be outside the culture room to maintain temperature control by minimizing input of heat generated from these. This is often not possible, however, and in this case, the algal culture facility design should include a climate-controlled enclosed area for algal cultures ranging from 250 mL to 20 L. For the larger algal culture vessels section (100–500 L), an extractor fan can be installed to counteract the heat generated by the lamps. Light banks using similar bulbs are installed for each vessel. To illuminate larger cultures (1800 L or greater), metal halide lamps (750–1000 W) are typically used; these generate considerable heat, and should be placed at least 30 cm above the surface in open, well-ventilated greenhouses.

Artificial lighting can be controlled by a simple timer or light monitor. The lighting regime ranges from a minimum of 16 : 8 hours light/dark per day (minimum) to 24 hour illumination (maximum) for indoor cultures, dependent on the aquaculturist's preference. Although artificial lighting can be precisely controlled in terms of quality and quantity, it is costly, accounting for almost 95% of the cost to culture microalgae.

Natural light

Natural light is a key component for outdoor raceway algal cultures suitable for feeding seed, or broodstock. In some operations, it is also used for large-volume cultures in greenhouses. Different materials are used to allow light into a greenhouse; insulated glass blocks used for windows in subtropical greenhouses allow light in, while keeping heat from sun's rays out; for the roof, clear/translucent polyethylene or corrugated translucent fiberglass sheets are often used. Some operations use natural light in combination with fluorescent or metal halide bulbs to provide evening illumination (see solar-assisted hatchery description, the 'Renewable technology: geothermal and solar energy' section). When natural light is used, it is best to use the morning sun, with 40% shade cloth on the west side of the building and 60% shade cloth down the middle of a north-south oriented building, especially during the warmer months.

Aeration and mixing

Aeration provides three important functions in the culture of microalgae: (1) air supplemented with CO_2 is a source of carbon for photosynthesis; (2) CO_2 provides essential pH stabilization; and (3) aeration physically mixes the culture so that nutrients and cells are evenly distributed, reduces self-shading and/or photo-inhibition (a decrease in photosynthesis due to excess light), and avoids thermal stratification in outdoor culture systems.

Air diffusers (airstones) create small bubbles that maximize oxygen/CO_2 transfer, and they are often used for small volume cultures. Another common alternative is the delivery of air into flasks using pipettes or glass rods for cultures ranging from 250 mL to 20 L. In larger culture containers, the fine bubbles from air diffusers create spray and foam that promotes bacterial growth and should not be used; larger bubbles (no airstones) actually do a better job of mixing the culture with minimal foaming. A common method for fiberglass cylinder or bag culture is the supply of air from a main line, regulated by a small valve (8 mm) and delivered through a 7 mm I.D. Tygon tubing, connected directly to the bottom of the tank; for cylindrical fiberglass vessels, the air connection is prior to the harvest valve (see 'Algal culture process' section). This provides a stream of large air bubbles mixing nutrients and keeping microalgal cells in suspension.

Aeration throughout the algal culture facility is provided by air pumps of 2 HP in small-scale operations, or air blower systems of up to 6 HP (Fig. 12.23). In order to increase air output, a tandem set-up, where the first blower pushes air into the second blower, is an alternative to purchasing one large air blower. Air supply into the blower is typically filtered; an excess air pipe, generated by back pressure, is fitted on the delivery side of the blower. In addition, a 0.25 µm filter is fitted in-line prior to each algal vessel to reduce the risk of bacterial contamination (Fig. 12.17B).

Common alternatives for mixing larger culture systems include jet pumps, paddle wheels (in ponds), continuous recirculation, and airlift pumps.

Figure 12.23 Air blower supplying entire algal culture facility, with Hepa air filter on supply side of blower, and excess back pressure release pipe on delivery. Courtesy of R. Karney Michael C. Voisin Oyster Hatchery.

Commercial diets

Commercially available diets can be used as a partial substitute to live microalgae for the later stages, and in the case of certain species, as a complete substitute to live microalgae; they can serve as a back-up when there is a failure in the algal production system, or as a supplement in case of reduced algal supply. This alleviates the demand on live microalgal production in an operation but does entrain costs in purchasing commercial diets. Careful consideration is needed to assess the most cost-effective method in feeding all life stages of the target species, comparing costs and performance of live algal production and commercial diets on survival and growth of targeted life stages. If commercial diets are used, storage in a freezer and/or refrigerator will be required; depending on the scale, an additional unit to that installed in the dry laboratory will most likely be needed and best suitably located in the large culture area of the facility.

Several diets suitable for juveniles and broodstock of several molluscan species can be purchased commercially in liquid or paste form. Some of these are also used routinely for larval stages of certain bivalve species, namely for oysters.

Liquid microalgae concentrates can be purchased in 1–10 L containers, often as a mix of several marine microalgae (standard flagellate and diatom species); some are used routinely in commercial operations and yield suitable growth and survival of oyster larvae, and post-set of quahogs and scallops.

Instant algae are highly concentrated mono-specific cultures of microalgae, with non-viable but intact algal cells. They are sold in 1 L bags, with concentrations per liter equivalent to 1800 L of a 2 million cells/mL live microalgal density (for *Isochrysis* spp.).

Spray-dried technology is used for the production of a commercially available dry white powder; the basis of which is single-cell protein, derived from algae and yeast, nutritionally balanced with fats, carbohydrates, vitamins, and amino acids. Some formulas contain high levels of omega-3 fatty acid (DHA, docosahexanoic acid), beneficial to molluscs. This product is found to be a live microalgal substitute in bivalve broodstock food ration; the powder is blended with seawater and added directly into the feeding tank. Dosage needs to be assessed by the aquaculturist, dependent on molluscan species targeted.

Sizing an algal culture facility

The size of the algal culture facility is largely dependent on the target species and the production output planned. This will define the quantity and composition of microalgal diet. A good understanding of the nutritional requirements and food quality required for optimal growth and survival rate of the larvae is key to setting up the algal cultures.

Molluscan species differ markedly in their food ration requirements; for example, early life stages of oysters require a much higher food ration than scallops. The first consideration is the scale of larval production, and the food ration required for this. The second is the performance (growth and survival) of early life stages of target species; for example, oyster larvae have been successfully reared on a commercial diet, whereas scallop larvae require a 100% live microalgae diet to achieve competency for metamorphosis, but do demonstrate expected growth and survival rates on a mixed diet of live and commercial algae at the seed stage.

Following settlement, food ration, and composition changes, post-larval growth becomes quickly exponential, requiring an increasing food ration; and the biochemical changes for this life stage call for a change in food composition. Many molluscan species will grow well with a mixed diet of live microalgae and commercial diet, after 1 mm size. The ration of commercial food is progressively increased, and can eventually become the major component of the diet for seed greater than 3 mm, comprising 85–100% of the food provided. The longer the seed is maintained in the hatchery, the greater load is imposed on food supply.

Algal production also sustains broodstock, especially for operations with a conditioning program. Commercial diets can also be used in part for broodstock conditioning; again, the relative proportion of microalgae used in completing gametogenesis is species and site dependent.

It follows that the quantity of live microalgae required will differ based on (1) the nutritional demands of the target molluscan species cultured, (2) its performance when fed commercial diets, (3) the requirement for broodstock conditioning, and (4) the seed strategy for optimal transfer based on culture method for seed and size to be achieved prior to transfer.

Pertinent questions need to be considered when estimating the spatial requirements of an algal culture facility (Worksheet 12.2). The design and size of the algal culture facility and its associated components depend on the type of seawater treatment required, the logistics of delivering algae to all cultured stages (larvae to broodstock), maximal food production output for various target molluscan species life stages, the techniques selected for large-scale algal production (batch/continuous/PBR/outdoor), the use of commercial diets, and the cultivation of probiotic bacteria.

Continuing along the same scenario as Example 12.1, Example 12.2 provides a real life calculation for a target production of 5 million, 5 mm seed of the calico scallop, *Argopecten gibbus*. This example illustrates the requirements for culturing live microalgae under controlled conditions or for seed averaging 1.5 mm, enabling the sizing of the algal culture facility. It excludes the feeding requirements for seed >2 mm, as other factors come into play; such as (1) the importance of stocking density and cultured biomass in assessing food ration, (2) the strategy for post-set

>1.5 mm and how it dictates diet composition and food ration – namely (a) the introduction of commercial diet at this stage, and (b) the method of rearing selected from 1.5 mm on. (See the 'Hatchery: larval and post-set rearing systems' and 'Intermediate nursery' sections for further details.)

Broodstock facility

Conditioning tanks for ripe broodstock are maintained independently in a quiet area of the plant where animals will not be subject to frequent disturbance (Fig. 12.7). Most bivalve species respond to shadows and vibration by closing their shell valves; such disturbances affect their feeding, and, when ripe, may trigger spontaneous spawning.

Another reason for maintaining broodstock separately is that broodstock holding and conditioning systems are recognized to be a major source of *Vibrio* spp. in a hatchery operation, and considered a high-risk area, potentially contaminating larvae and spat cultures. Vibriosis caused by pathogenic species constitutes the main bottleneck in the bivalve production process during the early stages of development, leading to high mortality rates and the rapid loss of production batches. Keeping the broodstock separate from the hatchery/nursery sections will minimize bacterial contamination. Nonetheless, to facilitate operations during spawning, selecting broodstock and induction to spawn should be conducted in an allocated space in-between the broodstock and hatchery sector, such as the spawning table (Fig. 12.7). This also facilitates the transfer of fertilized eggs to larval tanks; the dry laboratory for counting and monitoring egg fertilization and larval development should be strategically located for ease of access from both the Hatchery and Nursery 1 sectors (Fig. 12.7).

Broodstock are fed with live microalgae produced in-house and/or commercial diets if kept for any length of time in the hatchery. They also obtain supplement nutrients from incoming seawater; for this reason, seawater pumped from the source only requires coarse filtration. It is sufficient to remove larger organisms, and yet allows for the diversity and abundance of natural food species to pass through, enriching the diet of adult molluscs. Depending upon the conditioning regime, which is species- and site-specific, incoming seawater may require heating, or chilling, or both; a separate seawater temperature control unit for the broodstock sector (Fig. 12.7). From there, incoming seawater is distributed to all broodstock tanks. For operations with a more complex conditioning protocol, lasting the duration of the larval production season, a system enabling more than one temperature regime simultaneously will be needed. This can be achieved using smaller temperature-control units supplying water to various parts of the broodstock sector. Similar heating and chilling units, as those described in the 'Temperature control: broodstock, larvae, seed, algae' section, are used and adapted for temperature control of broodstock.

This section focuses mainly on broodstock tank systems. Tank type and design for holding broodstock aim to maintain and/or obtain a sufficient number of reproductively mature adults until spawning is induced. There are a number of water quality related factors which play a role in broodstock health, impacting the metabolism of the targeted species and ultimately their reproductive output. They include: temperature, salinity, dissolved oxygen, dissolved carbon dioxide, and nitrogenous wastes. Ensuring adequate water quality becomes increasingly important the longer broodstock are held in the hatchery; for this reason, it is key to ensure the following when designing a broodstock tank system:

- constant flow of incoming seawater
- even distribution of incoming water and food throughout tank
- climate control and/or seawater temperature control, avoiding temperature fluctuations in tank
- adequate exchange of seawater within tank
- avoidance of feces build-up in tank leading to bacterial contamination
- space allocation within tank to avoid overcrowding (species dependent)
- sufficient tanks to maintain males and females separately.

Simple broodstock tank

Continuous-flow open systems are often used for broodstock, taking advantage of the natural phytoplankton production in the incoming seawater as a food source. This is sufficient for animals brought in ripe less than a week prior to spawning. If animals are brought in earlier during the latter stages of gametogenesis and require additional food to complete gamete maturation, a semi-recirculation system will ensure the retention of added food and increase the opportunity for consumption by the broodstock. This is easily achieved by using an airlift system, consisting of a bottom pipe drilled with holes allowing for water from the bottom to be pulled to the surface using an airline (Fig. 12.24). For tanks without airlifts, a simple airline can provide a back-up to maintain dissolved oxygen levels in case of pump failure overnight. Tanks ranging 100–600 L are commonly used in hatcheries for broodstock holding and conditioning. Broodstock tanks can be installed in a series, with one main pipe connected to the coarsely filtered seawater system, and individual valves regulating the flow in each tank. Incoming seawater is distributed at the surface at one end of the tank, and outgoing seawater flow is at the opposite end of the tank. Set-up is

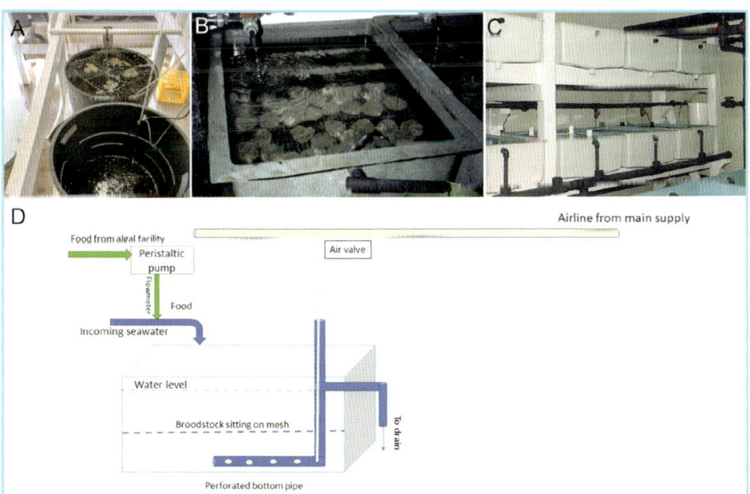

Figure 12.24 Broodstock tank design examples. (A) Simple static barrel for quahogs. (B) Flow through tank with suspended mesh supporting oysters. (C) Series of flow-through tanks for oyster culture. (D) Schematic of flow-through broodstock conditioning tank showing airlift. Courtesy of (A, B) R. Karney, (C) Helm et al. (2004) and (D) Richard C. Karney Solar Shellfish Hatchery.

such that outgoing flow removes water from the bottom, usually by a standpipe. A drain plug at the bottom of the tank is required for thorough cleaning.

Tanks made of fiberglass or fiber-reinforced plastic, usually square or rectangular are standard for broodstock. Insulated tanks with lids are preferable, which helps to minimize temperature fluctuations if tanks are located in a poorly climate-controlled area or outside. This also reduces the risk of contamination.

The number of tanks required depends on the extent of conditioning required; the more extensive the conditioning program, the greater the number of tanks needed, allowing for the regular introduction of new stock. In this way, a rotation of ripe stock is possible, ensuring a continuous supply of spawners. Examples of designs used are illustrated in Fig. 12.24.

For species naturally occurring on top of the substrate, broodstock are held in meshed trays or in nets suspended off the bottom in the tank, keeping the animals away from particulate organic matter accumulated over time. Fecal production by broodstock can be substantial, especially during conditioning, when a daily ration is given to ensure that energy ingested exceeds the maintenance of metabolic demands and losses for reproductive growth to occur. Suspending broodstock allows for feces to fall the bottom of the tanks. A rigorous cleaning protocol is required to maintain a healthy environment, when particulates are siphoned routinely, and tanks thoroughly cleaned.

For species naturally burrowing in the substrate, such as surfclams (*Spisula* spp. and *Mactromeris polynyma*) and softshell clams (*Mya arenaria*), precautions to prevent natural gaping when they are removed from the substrate are needed. These species depend on the pressure of being immersed in a substrate to close their valves; when removed from sediments, gaping ensues, leading to a deterioration of clam condition, and ultimately negatively impacting survival and reproduction. To simulate their natural environment and to preclude gaping, aquaculturists use rubber bands for conditioning, thus creating 'clam sandwiches' (Fig. 12.25). The sandwiches are constructed of two pieces of vinyl-coated lobster trap wire (45.7 cm × 45.7 cm) and two pieces of extruded plastic netting of similar size (6.4 mm mesh). The pieces of wire form the outer portion of the sandwich, while the mesh netting forms the inner portion of the sandwich. Twenty to 30 clams are placed on the mesh netting that sits on top of one piece of trap wire. The other piece

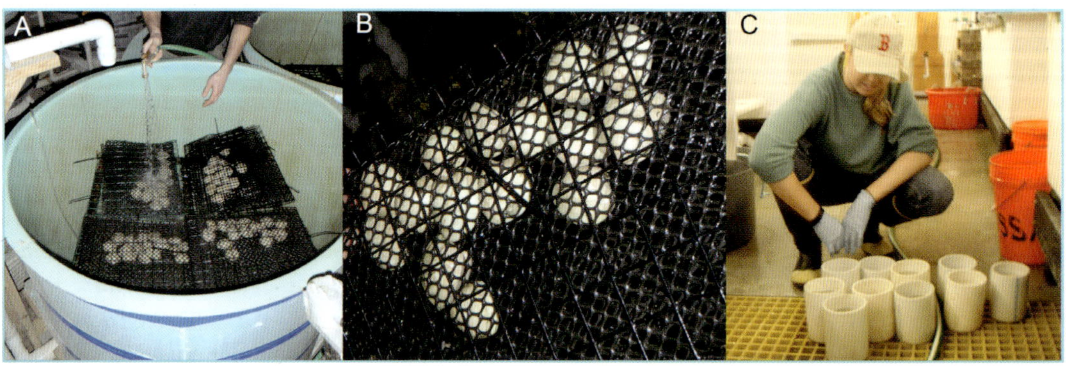

Figure 12.25 Clam sandwiches: (A) in holding tanks; and (B) close up view. (C) An alternative tube system used at Cat Cover Marine Laboratory (Massachusetts, USA). Courtesy of (A) and (B) B. Beal; (C) J. Buttner.

of mesh netting is then placed over the clams and the second piece of trap wire placed over the netting. The arrangement is as follows: (1) trap wire; (2) mesh netting; (3) clams; (4) mesh netting; and (5) trap wire. Finally, nylon cable ties are used to cinch the two sides together, placing enough pressure on the valves of the clams to simulate burial in sediments. Approximately six sandwiches, each with up to 30 clams, are used. Each sandwich can then be handled as a single unit. Softshell clams have been kept alive for a year in these sandwiches.

Other species, such as scallops, can be active swimmers, and require more space than fixed bivalves, such as mussels and oysters; scallops maintained in high densities will result in clamping each other, damaging the mantle, and negatively impacting conditioning, survival, and growth.

In general, the number of individuals held in a broodstock tank depends on live biomass per adult and the species-specific behavior and requirements for water quality and physical space.

For all conditioning tank systems, water flow and food ration (for conditioning) need to be adjusted for the density of animals (or total live biomass) per tank, where turnover of water needs to be high enough to maintain optimal oxygen levels, but not so high that added food is flushed out before being consumed. If conditions are poor in the tank system, gonads will resorb, and spawning will be unsuccessful. A total water exchange rate of at least 90 minutes minimizes loss of cultured algae, and gives broodstock the time to filter 60–80% of the food added.

Different species will differ in their requirement for water flow; scallops fare well in highly oxygenated water. A water flow of 4.8 L/minute (288 L/h), with a 100% water exchange every 90 minutes will successfully condition scallops held at a live biomass of 7 g/L. On the other hand, with a standard flow of 25 ml/minute/adult, a tank of 150 L stocked with 50 oysters of 75–100 g live weight (30 g/L) will require a flow of 1.25 L; this gives a total exchange rate of 120 minutes. Understanding the requirements in water flow for broodstock contributes to a realistic estimation of pumping demand during the hatchery design stage.

Sub-gravel filtration system

A sub-gravel (or sub-sand) filtration system, consisting of a substrate on the bottom of the tank, with an airlift system below the substrate for semi-recirculation of water, is ideal for species naturally recessing in substrate, such as certain scallop species and clams. This system does not require routine siphoning or cleaning throughout the conditioning period, as particulate matter produced is absorbed and filtered by the gravel. Animals can be kept throughout the hatchery season in such a tank with minimum maintenance.

Details of a sub-gravel filter are given in Sarkis and Lovatelli (2007). In brief, a sub-sand filtration system for a 380 L tank uses as a substrate locally sourced naturally occurring sand passed through a 1 mm screen to retain larger particles. The sand layer is supported 5 cm off the bottom of the tank, and sits on a acrylic sheet drilled at regular intervals to allow passing of water through sand to bottom of tank. Sand is retained by a 500 μm Nitex mesh covering the acrylic sheet, preventing sand particles from falling through. The semi-recirculation system is driven by an air lift, constructed with a 25 mm pipe placed on the bottom of the tank, and connected to a vertical pipe, into which an airline is inserted. Tank water is passed through the sand, filtered, and collected below the plexiglass sheet at the bottom of the tank; filtered seawater is pulled by the air and recirculates back into the tank through a T-junction at the top of the vertical pipe. The tank is continuously supplemented with new incoming seawater (coarsely filtered) supplied directly

Figure 12.26
Compartmentalized oyster broodstock table. Courtesy of Horn Point Oyster Hatchery.

from the pump house. Inflow of new seawater is regulated by a 25 mm one-way PVC ball valve at the end of the line. Seawater flows in at the top of the tank, and outflow is at the opposite end. Animals continuously receive new incoming seawater and recirculated seawater.

A slower water exchange in such a system (every 9 hours) enables temperature control and is successful for the maturation of a low live biomass (4 g/L). Monitoring flow rate is essential throughout the conditioning period; flowmeters at the incoming pipe are often installed in-line.

Compartmentalized broodstock system

When working with different populations of broodstock collected from different locations, constructing a compartmentalized system of shallower trays is an alternative to reducing the surface are required for broodstock tanks, dependent on scale of production. Compartmentalized broodstock tables allow for the production of selected strains of seed, and/or the genetic manipulation of broodstock, and is especially useful at the R&D scale (Fig. 12.26). The added advantage of compartmentalization is that it helps to control spontaneous spawning by minimizing the risk of triggering the whole population as would occur if all adults were within one tank. Incoming seawater and food ration are individually regulated for each compartment; outflow is at the bottom of each compartment, to ensure no cross-contamination between compartments. For these shallow trays, there is no need for air or airlift.

Feeding set-up for broodstock

Broodstock can be batch fed or fed using a continuous system with automated feeding manifolds.

Batch feeding is simple, and a 20–100 L bin fitted with a spigot at the bottom of the bin, and an airline to maintain algal feed in suspension, is sufficient for a smaller number of broodstock tanks. Food flow is regulated by the spigot or 8 mm valve to ensure an even distribution over a 24 hour period.

Food can also be dosed continuously into the water delivery line by a peristaltic pump (Fig.

12.24D). This generates a better mixing of food, although pipes will foul more rapidly. For a larger number of conditioning tanks, algae can be harvested automatically from the algal culture facility, stored in a dedicated reservoir tank, and pumped to each tank regulated by individual valves and a digital timer. The use of a reservoir tank allows for the addition of commercial diet to supplement the live diet if needed.

Spawning broodstock set-up

Spawning requires the isolation of ripe broodstock and collection of gametes for fertilization, in a dedicated area enabling exposure to the stimulus used for spawning induction – air exposure, thermal shock, or chemical stimulus – and a collecting vessel for eggs and sperm, for fertilization. Access to fine filtered and sterilized seawater for eggs and sperm, and temperature-controlled seawater, for broodstock, and for eggs is required.

Depending on the strategy for breeding, some hatcheries need to take great care in maintaining broodstock of different source locations separate. This will require a more complex spawning area, such as a compartmentalized table (see 'Compartmentalized broodstock system' section), or alternatively for a more basic approach, a series of non-toxic plastic trays. If broodstock is sourced from one site, or there is no genetic selective breeding by the aquaculturist, a simple spawning table can be used.

Basic spawning table

A saltwater table is easily constructed using plywood and sealed with epoxy resin. This can, in fact, have multiple uses during hatchery operations; it can be used for spawning a large number of ripe broodstock, but also as a working table during larval water exchange, and as a collector for seed prior to transfer out of the nursery. The saltwater table should have the capacity to receive fine filtered seawater as well as heated seawater as for larvae. A drain at one end can be fitted with a standpipe, and is adapted to collect eggs into a second reservoir during spawning. Seawater flow from ambient fine filtered, or heated fine filtered lines to the saltwater table is regulated by a 25 mm threaded ball valve, which in turn is fitted on the outflow end with a threaded hose barb. A clear flexible hose supplies water to the saltwater table on demand (Fig. 12.27).

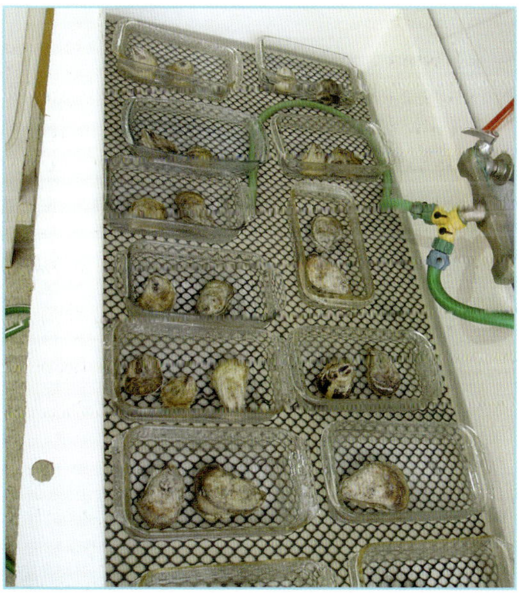

Figure 12.27 Basic spawning table with oysters spawning. Courtesy of Richard C. Karney Solar Shellfish Hatchery.

Sizing the broodstock sector

The number of broodstock maintained at any one time in a hatchery depends on the ease of access to the adult organisms the fecundity of the species cultured, the level and duration of conditioning, and scale of one larval production cycle. The difference in incoming water volume, tank capacity, space allocation, and algal requirements between a broodstock facility conducting extensive conditioning and one bringing a near-ripe stock shortly before spawning is substantial, as is that between using a hermaphroditic versus a gonochoristic species.

Some guidance on the thought process for assessing the space requirements for the broodstock sector is outlined in Worksheet 12.3.

Two scenarios in broodstock conditioning, demonstrating for each the tankage capacity needed, is provided in Example 12.3. This example is a follow-up to Examples 12.1 and 12.2, using the real life production of *Argopecten gibbus*, a hermaphroditic scallop species. Additional information is given if a gonochoristic species was used. From this simple example, the marked difference in tank requirements alone for different conditioning protocols and species dependent reproductive systems is evident.

Hatchery: larval and post-set rearing systems

Larval system

This section provides guidelines on facilities required for the development of eggs to metamorphosis.

All larval tanks need access to:

- main supply of UV sterilized seawater for embryogenesis
- main supply of fine filtered seawater for larval rearing
- temperature control if required
- aeration, with an in-line individual air filter to tanks.

A description of the main seawater supply lines is given in 'General principles in hatchery construction', and details of hatchery and nursery seawater treatment in 'Larval production: fine filtration, sterilization and temperature control'. Each larval tank is supplied with treated temperature-controlled seawater and air with seawater and air delivery lines installed above all tanks (Fig. 12.8A); nursery tanks are not shown here, but are fitted similarly, especially for those used for settlement and early post-set rearing (see Nursery 1, Fig. 12.7). Modifications to both seawater treatment and air flow are required to nursery tanks as animals grow; these are discussed further in the 'Nursery systems' section.

Aeration

In cases where air quality is a concern, air is filtered prior to entry into the tanks, as is seen for algal cultures (Fig. 12.23, 'Illumination and light source' section). The hatchery and nursery sectors are each supplied with a dedicated air compressor, or with a shared larger air blower. The former reflects a modular concept, where each sector has its own dedicated seawater and air supply, and reduces the interdependence of sectors and respective production outputs. A low aeration provides mixing of feed throughout the tank, keeping larvae suspended off the bottom where particulate organic matter and bacteria may accumulate, and maintaining adequate oxygen levels.

Air is provided through a 25–50 mm diameter PVC pipe, drilled with 8 mm holes, threaded for the fitting of labcock ball valves, or similar. For larger diameter tanks, more than one airline is required for adequate aeration, supplied by installing a 15 mm PVC manifold to airline, fitted in turn with labcock ball valves regulating the flow. Tygon tubing cut to reach the bottom of the tank is weighed down, aerating the tank from bottom to top, with a slow stream of bubbles. Airstones are not advised as larvae may get entrapped. Details are given in Sarkis and Lovatelli (2007).

Larval tanks: static and flow-through specifications

Tanks should be constructed from either new or non-recycled polyethylene or fiberglass. Toxic substances potentially harmful to larvae may leach from the surface of new plastics, and new vessels have to be soaked in seawater with weekly water changes for at least one month (two to four months are recommended in the literature) before use. Some report that steam curing of fiberglass tanks substantially reduces the 'soaking' period. Many tanks currently manufactured have ultraviolet inhibitors for outdoor use.

Traditionally, larval rearing is conducted in static tanks. Many hatcheries still use these, as they are a reliable method for rearing larvae. The disadvantages are the high labor demand and the necessity to control bacterial contamination. In the past decade, the shift has been toward the sole use of flow-through larval tanks, where a continuous open flow of treated seawater simulates the natural environment. The advantages are the capacity to increase the density of larvae reared within the given footprint of the larval facility, the reduced risk of bacterial contamination, and the lower physical labor demand.

Larval density during rearing can be increased in a flow-through system, as there is a high turnover of complete water exchange. For example, a 200 L tank with a standard threefold water exchange over 24 hours is equivalent to 600 L; hence, rearing larval density can be increased threefold in a given space.

Bacterial control is achieved by the continuous removal of wastes, as water flows continuously out of the tank; anti-microbial agents are therefore not required to mitigate contamination of the larval culture, as they are for static tanks; studies have shown comparable survival rates of bivalve larvae to metamorphosis between static and flow-through culture systems.

Both static and flow-through systems require the complete drainage of the tank for larval collection, referred to as tank 'take-down,' or 'draindown.' Tanks can be fitted if needed with a through-hull fitting glued into an existing drain hole or similar, at the bottom of the tank. On days of larval collection, a flexible hose/pipe is clamped on using a quick connect or threaded PVC adapters; outflow of water onto a meshed sieve or tray is regulated by a one-way ball valve. Flow is such that larvae are not subjected to high pressure, which will cause crushing for more sensitive species. Meshed sieves are kept in plastic non-toxic trays, off the bottom to avoid direct contact of larvae with the hard surface of the tray and subsequent damage to larval shells. The top of the sieve is above tray level, and monitoring of flow rate and water level is ongoing throughout entire tank take-down to prevent clogging of sieves and larval loss.

Further details are given below specific to static and flow-through larval tank system design.

Figure 12.28 Examples of larval tanks. (A) Small-scale conical and insulated semi-square tanks. (B) Flat-bottom tanks. (C) Large-scale round flat-bottom tanks. (D) Series of conical tanks for flow-through larval system with flow meters. Courtesy of (A) S. Sarkis; (B), (D) Michael C. Voisin Oyster Hatchery; (C) UMCES Horn Point Oyster Hatchery.

Static tanks

Hatcheries use rearing vessels ranging from 20 L to greater than 25,000 L as static tanks for larval rearing. In general, the larger the tank the better; it is preferable to minimize the surface area to water volume ratio, as surfaces tend to have higher numbers of bacteria. The size of the larval rearing vessel used depends on rearing larval density and production goal per cycle for the target species.

Tanks may be semi-square (square with rounded corners), or round with flat bottoms or coned bottoms (Fig. 12.28).

Flat-bottom tanks are placed directly on the floor, resting on timber footings at each corner or along the sides of the tank if needed. A gentle slope is required for complete drainage of the tank, by raising the back side using additional timber wedges. The drain valve is located on the front of the tank.

Conical tanks can be ordered with legs or a stand can be constructed in the hatchery. In both cases, consideration must be given to the clearance between the drain hole (in the center of the tank) and the floor for larval collection.

For embryo development, semi-square or round with flat-bottom, or slightly tapered bottom tanks are commonly used; the reason is that aeration is non-existent or very low during this phase in order to not disturb the delicate stage of embryogenesis, and, as eggs are immobile, they will tend to aggregate at the bottom of a steeper coned tank. This is a species-dependent process and can be circumvented by adding a low aeration (one bubble at a time) to the tanks for those species considered more robust. Please note that the mechanical disturbance created by aeration can lead to abnormal embryo development.

For larval rearing, tapered bottom tanks are advantageous, as particulate organic matter, algae, and larval waste will fall to the bottom of the cone; this facilitates the collecting of larvae, separates larval waste from the remainder of the healthy culture, and ensures a thorough draining of the tank.

For optimal temperature control, it is recommended that insulated tanks be used, especially if there is poor climate control in the larval facility. For example, tanks often used for fish transport, with double walls filled with foam, molded legs and lids prove excellent for embryo development and for scallop larval rearing until the pediveliger stage; such insulated tanks maintain rearing temperature relatively constant, a cost-effective attribute, especially in cases when larval water is heated.

Static tank culture is based on a minimum three-times a week tank take-down for a 100% water exchange, and monitoring of larvae. Aquaculturists use a range of variations on the percent water exchange protocol throughout larval life, dependent on species and seawater quality. Details on static tank set-up and larval collection are given in Sarkis and Lovatelli (2007).

Flow-through tanks

Flow-through larval rearing systems – also referred to as high-density rearing systems – generally improve the water quality conditions in a culture tank; this is achieved mainly by the removal of waste organic materials at a constant rate, and thus minimizes bacterial proliferation and the risk of high larval mortalities.

Design of the flow-through system is key to successful larval rearing. It must achieve a constancy in flow rate, reliability of continuous temperature control, appropriate surface area of outflow banjo to avoid larval impingement, and a reliable feeding system for constant input. The lack of routine water exchange as conducted for static tank culture, makes difficult the assessment of the flow-through larviculture performance. For this reason, it is recommended that protocols for a flow-through system be evaluated during preliminary trials through the frequent take-down of the entire culture. Once all procedures are in place, a flow-through tank may require as little as one take-down per cycle for the fractionation and culling of larvae, dependent on the species and the duration of the larval life.

A general flow-through tank design (Fig. 12.29) consists of the following.

- Incoming treated seawater supplied through the bottom of a conical tank.
- A header tank to ensure a constant flow in the tank, minimizing any surges occurring in the main seawater line.
- Gravity or pump-supplied seawater from the header tank, regulated by one-way ball valve.
- Air supply mixed with incoming seawater – airline tubing is drilled into incoming seawater pipe at the bottom of the cone, prior to entry into larval tank.
- Outflow of water at the top of the larval tank, passing through a fully submerged banjo filter; this prevents larvae from being discharged with the water flow.
- A range of banjos with different mesh size; meshed on both sides; diameter is dependent on water flow rate, in order to minimize suction pressure at the outflow. A discharge flow rate of less than 0.8 mL/min for every square cm of mesh area is sufficiently slow to prevent trapping of larvae against the banjo.
- Flow rate of larval tank adjusted to desired turnover of complete water exchange.
- Collection of larvae from the bottom, where a T-connection is fitted to the incoming seawater line, and regulated by a dedicated one-way ball valve.

A steeper cone is recommended for flow-through larval tanks with an angle in the range 30–60°. This depends in great part on desired tank volume, available space in proposed hatchery, and number of tanks required; the steeper the cone, the higher the tank has to be for a given

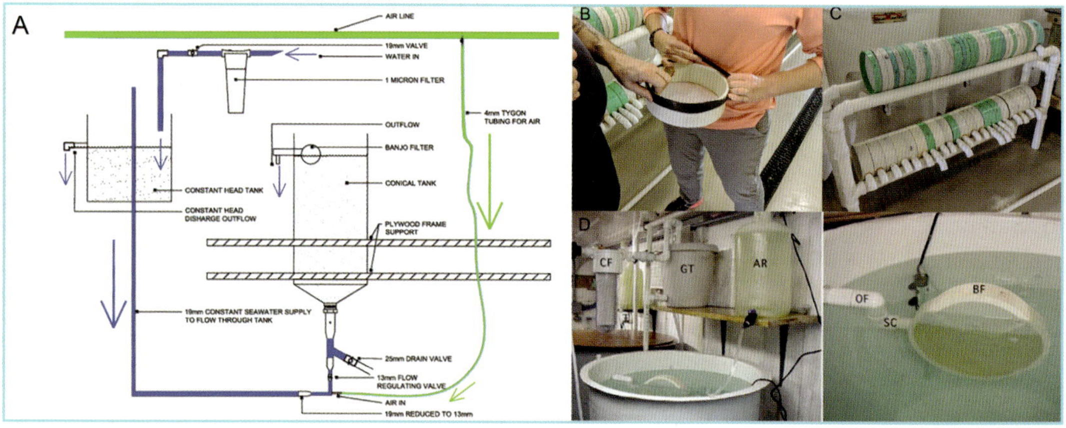

Figure 12.29 Flow-through (or high-density) larval rearing system. (A) Schematic of a general conical flow-through tank set-up. (B) One-way flow to make a banjo. (C) A series of banjos with varying mesh size. (D) Set-up of feed and incoming water in larval flow-through tank (left), with close up of banjo at outflow (right). Abbreviations: CF, cartridge filter; GT, header tank; AR, algal ration; OF, outflow; BJ, banjo filter; SC, swivel connection. Courtesy of (B), (C) Michael C. Voisin Oyster Hatchery; (D) S. Sarkis, Bermuda Biological Station Hatchery.

volume. Several hatcheries have flow-through tanks ranging from 200 L to 3000 L. Flow rate is adjusted to achieve the equivalent of two to four complete water exchanges over a 24-hour period. For example, for a 200 L tank, a continuous water flow of 0.480±0.16 L/minute, equates to 3.4 complete water exchanges over a 24-hour period.

A number of questions need to be answered for a realistic assessment of the space requirements for a larval hatchery, based on species-dependent culture techniques, expected survival rates to metamorphosis, the selection of larval culture system, and the scale of larval production (Worksheet 12.4). In doing this exercise, it becomes evident that using a flow-through system reduces the footprint of the larval sector, as the larval production output can be achieved by increasing larval density in a given volume.

Nursery systems

The nursery sector calls for a flexible seawater system, as it is the one sector in the hatchery which accommodates several critical stages of production: the settlement of mature larvae, rearing of post-larvae, and rapidly increasing food ration and water flow demand of young seed.

The nursery sector requires access to high-quality seawater treated similarly to larval tanks, but also needs the capacity to gradually modify temperature and filtration level, as animals begin to acclimate to natural environmental conditions (Fig. 12.7). Settlement conditions and initial post-larval rearing techniques require the same filtration and temperature control as for larvae. Older spat (>1 mm) benefit from a reduced filtration to 10 µm, allowing a supplement of naturally occurring plankton to the diet; in addition, if rearing temperature is higher than ambient, acclimation of spat or seed to ambient farm conditions is a must during the last phase of nursery rearing.

For these reasons, parallel and independent PVC lines are installed in the nursery to provide:

(1) a fine filtered and temperature-controlled seawater; (2) an ambient coarsely filtered seawater fitted with additional 10–20 μm filter system prior to the post-set culture vessels; and (3) an aeration system providing homogeneous water circulation for post-set; the latter is especially critical for circular tanks housing spat on substrate.

Careful consideration of all steps involved in setting larvae, growing post-larvae, and successfully culturing young seed in preparation for transfer out of the hatchery facility is recommended for a smooth nursery operation.

The first consideration relates to settlement and post-settlement rearing techniques; these are species dependent, and vary according to the timing and size of seed for transfer at sea to farm sites. The type of settlement system used defines in great part the footprint of the nursery sector, and is in turn, somewhat dependent upon the eventual spat cultivation system to be deployed. The selection of a nursery method and strategy is dependent on the species, but is also a balancing act between achieving the production target with optimal survival and a realistic labor demand in a given nursery footprint with the available staff. The longer seed are kept in the hatchery, the more labor intensive and space demanding it becomes, but the higher the survival rate when transferred to the farm.

There are several methods for setting competent larvae – circular tanks, raceways, and those more species-specific, such as fluidized bed bottles. Cost-effectiveness and species performance are the factors defining system selection.

Circular nursery tanks: preparation and plumbing requirements

Bivalve larvae can be set on cultch or artificial substrate submerged in circular, usually flat-bottom tanks. Seed can be reared until 5–8 mm in this way, but typically, seed on substrate is transferred to the intermediate nursery or on longlines when greater than 2 mm size. It is key that initial setting density be low enough to allow for the exponential increase in biomass as spat grow; if setting density is too high, the available surface area for growing seed will not be sufficient, leading to overcrowding and increased mortality. At this time, seed can be washed off the substrate and transferred to appropriate grow-out systems. This method maximizes space in the hatchery; however, survival is reduced compared to the raceway system. Please note that conical tanks are not often used for settlement, as competent larvae tend to gravitate to the bottom and clump.

Tank surface

A first precautionary measure by some aquaculturists is to paint the surface of setting tanks with liquefied wax (paraffin), in order to avoid settlement of larvae that can attach on the bottom of tanks.

Air

Aeration becomes increasingly important in a tank system as seed grow. Initially, air is provided in a similar manner to larval tanks, through a PVC manifold with a series of valves regulating flow to several Tygon tubes placed around the perimeter and center of the tank. This minimizes disturbance to competent larvae and allows them to set on the given substrate.

Once metamorphosed larvae are fixed on the substrate, homogeneous circulation of water and food is enhanced by using an airlift system, avoiding 'dead' zones and consequently post-set mortality.

Figure 12.30 Airlift system for bivalve spat set on substrate and reared in a semi-recirculated system. Courtesy of S. Sarkis, Bermuda Biological Station Hatchery.

In order to prepare for this change in air distribution, an airlift system is installed prior to the use of the tank for setting. A simple airlift system consists of four vertical pipes, into which an airline tube is inserted, and thus pulls water from the bottom using drilled transverse pipes. Vertical pipes are connected to four surface transverse pipes; water flows out through the surface pipes drilled at even intervals. In this way, recirculation of water is created, where water from the bottom of the tank is driven by airlift up the vertical pipe, and flows out through the holes of the transverse pipe (Fig. 12.30).

Seawater access

Setting tanks should be initially plumbed to the temperature-controlled filtered seawater line. In order also to have the capacity to reduce the filtration level (to 10–20 µm), in-line cartridge filters fitted to the ambient coarsely filtered seawater line are installed. The initial static setting system eventually becomes a semi-recirculating system with a constant flow of new seawater. An independent temperature-control system for the nursery is preferable, in order to acclimate seed gradually to natural conditions; however, if this is not possible, supply from ambient seawater and heated seawater from two distinct pipes can be mixed in order to achieve the target rearing temperature. Flow rates must be regulated by flowmeters to minimize the fluctuations incurred by the natural daily variations of ambient seawater. Construction of the nursery must allow for the change from static to flow-through, and to adjust the aeration system accordingly.

Preparing cultch

Setting tanks are filled with the selected cultch or substrate for settlement of larvae. Cultch or substrate used depends on the species; scallop species settle on various types of netting, such as black polyethylene mesh, mussels settle on ropes. The behavior of the older seed and the handling required for transfer to the farm and grow-out must be taken into account, as must be the end product targeted. The oyster example below illustrates this.

Oysters, which permanently cement to surfaces shortly after setting, must be provided with movable substrates to facilitate further culture and transfer out of the hatchery. If oysters are to be cultured as singles for market, the setting larvae are provided with microcultch, usually in the form of ground oyster or poultry shell sieved to a size of about 300 μm to facilitate the setting of only one larva per chip and ensure single seed oysters. If produced for stock enhancement, eyed oyster larvae are set on shells, and the land-based nursery stage is eliminated (see 'Feeding intermediate nursery systems' section for further details).

Initial static system

A static setting tank allows competent larvae to settle, and spat to attach to the cultch or substrate with minimal disturbance. High-quality incoming seawater (filtered and temperature controlled) is regulated using a one-way ball valve from the main supply line, passing through a second 1 μm filter fitted prior to each tank, as for larvae (Fig. 12.8A). Incoming water is distributed at the surface. Outgoing water is at the opposite end through a standpipe threaded into the bottom drain. It is critical to secure a standpipe in nursery tanks, to avoid any loss of water and exposure of substrate – upon which larvae are fixed – to air. Aeration, filtration, and feeding are much the same as for static larval tanks.

Post-set system

Airlines are inserted in the pre-installed airlift system, and a recirculation system is initiated – in which water is drawn from the bottom and recirculated to the surface. Dependent on species, water exchange of setting tanks may continue as for larvae until spat are older. Subsequently, tanks are placed on an open flow-through system, with a continuous airlift system uniformly distributing nutrients and thoroughly oxygenating the cultured water. Flow rate of incoming seawater is adjusted as seed grow to accommodate the increasing biomass.

Feeding system

Continuous feed is distributed to nursery tanks using a magnetic drive pump, installed to a food reservoir. The reservoir can be supplied from the algal culture facility (refer to 'Algal production' section on collection tank for harvested microalgae). A floating valve in the reservoir tank ensures that the pump is turned off when reservoir is close to empty; this prevents the burning out of the pump. From the food reservoir, a PVC feeding pipe is placed or fixed on the side of each nursery tank, running the perimeter of the nursery area, and returns excess food back into the reservoir. Food ration can be regulated to each tank through a series of 8 mm labcock ball valves or automated feeding manifolds. Commercial diets, if used, are also added to the reservoir, mixing both live microalgae and commercial food for the required food ration. The feed added is diluted with fine filtered seawater to ensure a 24 hour continuous supply (Fig. 12.31). Incoming seawater, feed, and air are illustrated for a few tanks; this demonstrates the set-up required for each nursery tank. Although not illustrated here, seawater pipes can be extended to Nursery 2, as needed; a second feeding tank is recommended in Nursery 2 as diet will change with seed size in both composition and quantity.

Downwelling and upwelling raceways

Certain bivalve species fare better when set on a downwelling system of sieves (or 'silos') in a raceway; this system can be subsequently modified to an upwelling system as spat grow. Seed

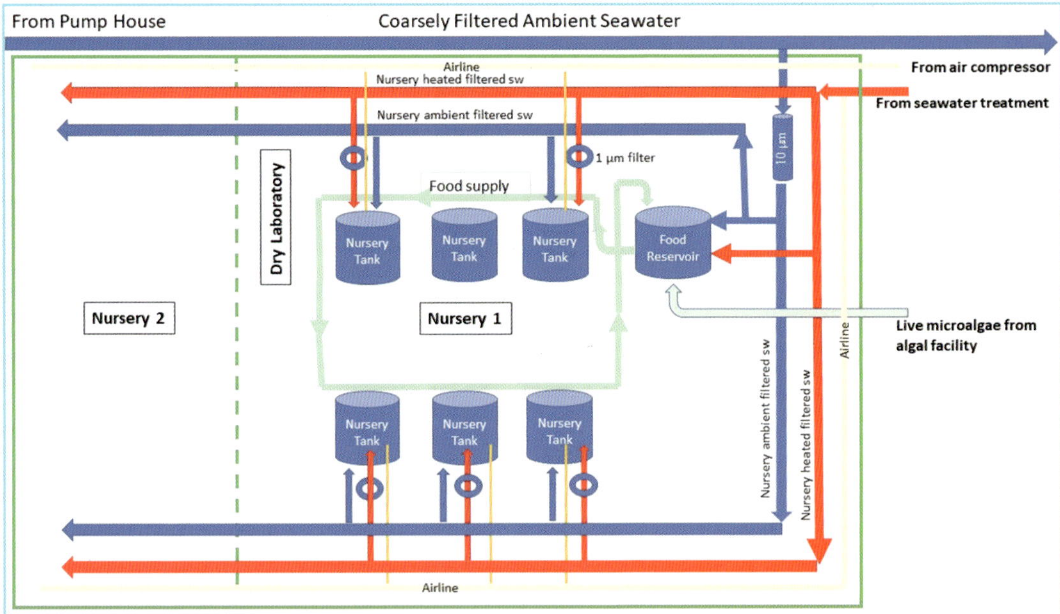

Figure 12.31 Schematic of nursery sector showing independent heated and ambient seawater supply, air supply, and food reservoir for distribution to nursery tanks.

is successfully reared in this way up to 10 mm size. The raceway system requires a large surface area, as stocking density is restricted by the surface area of each sieve. It is also a labor-intensive method, which requires the routine grading and thinning of seed as animals grow, in order to maintain a suitable biomass in the raceway system. On the other hand, a raceway system generally yields high survival rates, which is a major advantage in the culture of more delicate species.

The advantage of having a raceway system is its flexibility in use. Aside from their use as downwellers or upwellers, raceways can be used as a single large tank, or can be installed in a series of smaller tanks, increasing the tank capacity or compartmentalizing seed of various sizes or sources. Details on the installation of raceways are given in Sarkis and Lovatelli (2007).

Sieves

Sieves made of PVC pipe, buckets or other materials are suspended in the raceways. They are constructed with mesh size appropriate for setting larvae and older seed; interiors can be waxed to prevent larvae from attaching onto the sieves.

Downwelling system

Incoming seawater is supplied via a main PVC line, running the length of the raceway; holes are drilled at intervals and fitted with tubing. Each sieve receives its own seawater supply through a dedicated tubing. Incoming water flows down through the animals and mesh, and flows out of raceway standpipe.

This downwelling system is efficient for young spat, especially suitable to scallop species, and encourages settlement as competent larvae are essentially forced down toward the bottom of the sieve. Scallop species will attach to the mesh on the sieve, and eventually detach, making their handling and transfer to nets or bags relatively easy. Oysters can be provided with a substrate (cultch) at the bottom of the sieve onto which they will remain fixed until adult size.

Semi-recirculating system

Once post-larvae are set, the system is connected as a semi-recirculating system, where a constant inflow of treated seawater is provided, and partial semi-recirculation is achieved with the use of a sump/reservoir tank. This minimizes the waste of food and ensures constant temperature. Outflowing water from the raceway is returned to the sump tank, consisting of partially unconsumed food. This recovered water is mixed with new seawater, and recirculated back to the raceway. In this way, animals in the raceway receive new incoming seawater from the main supply line and recirculated water with food from the sump tank. Excess water flows out of the sump tank through its overflow pipe.

Upwelling system

As spat grow, fecal production increases, and an upwelling system is recommended for optimal growth and survival. With an upwelling system, seed is not trapped with excessive fecal or detritus material. Essentially, flow as described for the downwelling system is reversed. Seawater and food pass through the bottom of each sieve and flow out at the top of the sieve into a common drainage channel (Fig. 12.32). This is an open system for older seed, where food is distributed continuously, pumped from a reservoir tank at a rate optimizing consumption.

Upwelling silos using the same principle as described above are used for infaunal bivalve species, such as clams, which naturally depend upon immersion in a substrate to keep their shells closed. Without a substrate, these species expend added energy to contract their muscles and keep their valves closed. Providing a substrate under culture conditions simulates natural conditions and enhances growth and survival. Alternatively, packing clams densely together in upweller silos seems to duplicate the required pressure. When packed in this way, razor clams assume the natural vertical positions in which they would be found in nature, but instead of being surrounded by substrate, they are pressed up against each other. This is a successful method for culturing high densities of both quahogs and razor clams, provided that they are supplied with adequate flows of seawater to meet nutritional needs.

Figure 12.32 Close up view of a central drainage channel in an upwelling raceway. Courtesy of Murder Point Oyster Hatchery.

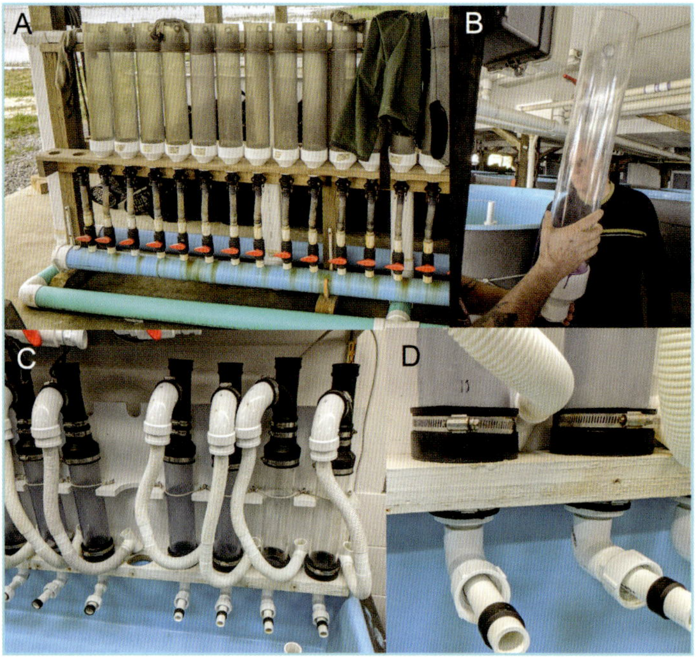

Figure 12.33 Fluidized bed bottle nursery. (A) Whole nursery with main drain. (B) Making one cylinder. (C & D) plumbing details for a bottle upweller nursery. Courtesy of (A) Murder Point Oyster Hatchery; (B) Auburn University Shellfish Laboratory; (C) and (D) Roger Williams University Shellfish Hatchery.

Fluidized bed bottle nursery system

Oyster and clam spat can be stocked at high density, layered one above the other. For this reason, 1 mm spat or greater can be reared in a relatively small footprint in the nursery. Plastic soft drink bottles of 1–3 L in volume are inverted and can be used as upwelling cylinders. Cylinders are installed in a series, with 10 cylinders or more in a series (Fig. 12.33).

Instead of a mesh screen containing the spat, a ball or large marble is placed inside to cover the opening of the neck. This serves as a non-return valve. The flow of water from the bottom keeps the juveniles suspended in the water column inside the cylinder, but if water pressure is lost the ball or marble seals off the neck so that no juveniles are lost. A flow rate of 25–40 mL/minute/g of spat is adequate to lift and circulate spat. Flow rate depends on spat size and weight, and on the diameter of the cylinder.

Discharge water from a series of upwelling bottles is passed over a screen to collect any juveniles that may accidentally float away.

This is an optimal rearing technique, especially for oyster species, as spat will form clusters as they grow, if flow is not sufficient. The upwelling flow also minimizes fecal deposits accumulating around the spat.

Other construction considerations

Space must be allocated in the nursery for grading as seed grow, and for a transfer system to the farm for grow-out. Both of these are labor intensive, requiring a full team working simultaneously, and the capacity to manipulate seed such that disturbances are minimal during transfers, packing, transport, and loading to dock or ponds. Space allocation for this phase of operation cannot be underestimated.

For grading and thinning, a homemade grading table, constructed with a series of trays of various mesh facilitates the fractionation of the seed. Space allocation for this is equivalent to a minimum of two saltwater tables (Fig. 12.27). In the grading area, additional portable saltwater tables or raceways can be used for subsequent transfer of seed. Additional tanks are recommended and can be used on a rotational basis during grading, or for the subsequent setting cycle.

When sizing a nursery, there are questions relevant to the setting and spat rearing techniques that need to be considered (Worksheet 12.5). The type of system used, the number of tanks, raceways, silos or specific systems such as the fluidized bed bottle system for oysters, are all dependent on the ultimate goal, production output, and seed strategy, that will dictate the footprint of the nursery.

Intermediate nursery

Transferring seed as soon as possible out of the hatchery is a strategy favored by most aquaculturists, as costs for seed production increase with seed growth. This is due in great part to the increased requirement for space, the increased demand on seawater volume and its associated prohibitive pumping costs, the exponential demand for food ration, and the labor necessary to the daily routine of maintaining larger seed. For all of these reasons from an economic perspective, it usually behooves the grower to move the seed into field nurseries as early as possible.

Some operations produce young seed in excess of the amount that can be cultured to market size. The strategy at this time is to transfer seed at a small size, eliminating nursery costs and labor, accepting a high mortality post-transfer, but resulting in sufficient seed of suitable size to meet the requirements of the farm. It results in an improved economic bottom line.

For other operations, natural bloom occurrences, and/or recruitment of predatory species impact seed survival post-transfer; in such cases, the timely transfer of seed is critical, and aquaculturists may have to hold seed in an intermediate nursery until environmental conditions are favorable. A similar situation may also arise for hatcheries manipulating broodstock, and thus extend the natural reproductive cycle with an early start to the larval culture cycles. This can lead to a first batch of seed ready for transfer before ambient temperatures or other environmental conditions are suitable to the cultured species' survival and growth. Design and size of an intermediate nursery depends on the target species, production output, and natural grow-out conditions. An intermediate nursery enables a more rapid turnover of larval cycles by freeing tankage volume in the nursery, if optimal transfer periods to grow-out farms are limited.

- **Water treatment**. Filtration to a certain degree is still required for seed 2 mm and up reared in raceways, tanks, and ponds; this avoids the entry and growth of undesirable organisms. Coarse filtration is considered sufficient for this sector of the facility (approximately 50 μm) (Fig. 12.1); a sand filter is an easy, cost-effective way to prevent undesirable organisms and larvae of predatory organisms to enter, yet allowing for natural phytoplankton species to pass through and supplement food ration. Complete water exchange is adjusted to achieve at least 80% of food consumption, and thus minimize loss of added food. This translates into the preference for a dedicated pump and supply line to the intermediate nursery sector.

- **Preparing outdoor area**. Raceway and tank systems are installed on cement pads. In most cases, some protection is required from sunlight, wind and rain. A range of shading material is used, such as sheets of corrugated plastic, HDPE rolls which can be cut to size, webbing, or vinyl shade sails or structures.
- **Access to protected waters**. FLUPSY systems differ in that they are used for growing seed already acclimated to natural environmental conditions, but need to attain a larger size prior to transfer to farm sites.
- **Siting of intermediate nursery**. Ideally, the intermediate nursery area should be located as close to the dock for transfer and transport of seed to the farm. This is discussed in greater detail in the 'Feeding intermediate nursery systems' section.
- **Feeding**. Although seed at this time relies in great part on nutrients in incoming seawater, in most cases, a supplement is required to optimize growth. Commercially available diets can be used to a certain extent, but more often, microalgal cultures are grown outdoors in large vessels, and pumped to the intermediate nursery rearing set-up. Space for culturing algae outdoors and supplying it to the seed must be taken into account.

Approaches used for the continued culture of seed 3-mm up to 10-mm, or size ready for transfer vary among species. As spat and seed grow, they develop forms and behaviors more closely linked to that of the adults they will become; this change in lifestyle requires the culturist to modify rearing methods to more closely simulate the natural environment of the target species. Some of the approaches used are:

- an additional upwelling raceway or silo system, shaded, with 25 µm filtered seawater, supplemented with commercially purchased diet and/or externally produced microalgae – this is usually constructed outdoors
- an external continuous circular tank system, shaded and with same water quality and food as for external raceway, with an aeration system
- ponds supplied with coarsely filtered seawater (50 µm), either fertilized for enhanced primary production, or supplied with live microalgae externally produced
- floating sand boxes for certain infaunal bivalve species
- FLUPSY system as discussed in the 'Renewable technology: geothermal and solar energy' section.

Outdoor upwelling raceways and tanks

The set-up for an outdoor raceway system or tank system is similar to that described in the 'Hatchery' section. The main difference in the upwelling system is that there is no need for a connection to an airline, as water is forced to move from the bottom of the raceway, through the sieve, and out the top.

A raceway system for growing seed greater than 3 mm requires a larger additional footprint in the hatchery complex. Such a system is more often used for scallop species; one commercial operation, functional at the time of writing, demonstrates the feasibility of routinely producing 5 million 5 mm scallops in a raceway system. Similar to indoor raceways, labor is intensive, requiring weekly grading and thinning to maintain the stocking biomass optimal; its cost-effectiveness should be carefully considered during the design stage.

Figure 12.34 Rectangular troughs for mussel rearing. Courtesy of John T. Hughes Hatchery and Research Station

Similarly, a system of circular tanks resembling that described for early post-set ('Nursery' section) can be used as an intermediate nursery. Round flat-bottom tanks from 1000–10,000 L, or rectangular troughs can sustain seed suspended in bags, trays, or mussel socks while awaiting transfer to the grow-out sites (Fig. 12.34). This will require a high flow of seawater to establish a current for distribution of food.

Floating sand boxes

Burrowing species, such as northern quahogs and other clams, are often reared in floating sand boxes, mimicking their natural environment. A protected area is required for anchoring of sandboxes, avoiding heavy wave action which may wash both sand and seed away.

Sandboxes provide a burrowing substrate environment and are constructed using a standard 1.2 m × 2.4 m sheet of plywood to which lengths of 5 cm × 10 cm boards are attached to create a box (Fig. 12.35). The box is attached to two floats with ropes so that it is suspended about 1 m below the surface of the water. The box is filled with several inches of filtered beach sand and can be planted with up to 100,000, 1 mm quahog seed. A netting cover further protects seed from predatory crawling crabs.

Ponds

Pump-fed rectangular or square ponds (100–400 m^2), have potential use in molluscan culture as an intermediate nursery for larger seed. Supplying the pond via pumping enables controlling seawater quality to a degree through coarse filtration and leads to a better management of the pond water quality. Ponds are operated on an open continuous seawater system, and once stocked with the target species are maintained running until the end of the production cycle. For this reason, care must be taken to design ponds in such a way that they will effectively remain an environment with good water quality throughout the duration of seed production. If designed

Figure 12.35 Floating sandbox for rearing quahog seed. (A) A schematic diagram for construction. (B) Sandboxes in a protected bay. (C) Close up view of a sandbox showing seed. Courtesy of R. Karney.

correctly, intense labor is limited to the start of the season, when ponds are emptied, cleaned, and prepared for seed.

For rearing bivalves in ponds, seed can be transferred to pearl or lantern nets/trays directly suspended in the pond with surface floats using a longline system similar to that of the farm. The depth of the pond limits stocking, and production output in ponds needs to be assessed during the design. The key to successful culture in ponds is the even distribution of water, both horizontally and vertically, for thorough circulation and maintenance of suitable water quality.

General recommendations and considerations in pond design and operation include:

- ensuring easy access to seed in the pond for monitoring and harvest
- limiting cross-contamination if more than one pond; parallel ponds with an independent water supply is preferred over a linear series of ponds
- establishing a standard width of all ponds dedicated to same use for ease of construction, and management – generally, ponds are twice as long as they are wide
- providing drainage canals to collect excess seawater, usually serving a row of ponds
- lining ponds with impervious material, such as plastic or rubber sheeting
- setting a turnover rate based on stocking density and sufficient to maintain optimal water characteristics – temperature, salinity, and oxygen
- routine monitoring of vertical water quality profile – namely dissolved oxygen and salinity.

Continuous-flow design in ponds

For clams, geoducks, or other burrowing bivalves, the pond can be designed to enable burrowing of post-set (Fig. 12.36). Care must be taken in constructing ponds where complete drainage is possible; pond bottoms should be gently sloping, for total drainage at the outflow end. If a leak-proof drain is not possible, the pond can be emptied using a trash pump (2 HP is sufficient in most cases) (Fig. 12.36). Outflow is through a stand-up pipe and effluent is directed to a main discharge pipe. Drainage channels and main drainage pipe are constructed in various materials; a PVC half-pipe of 100–150 mm in diameter, running along the outflow end, collects water for discharge from several ponds.

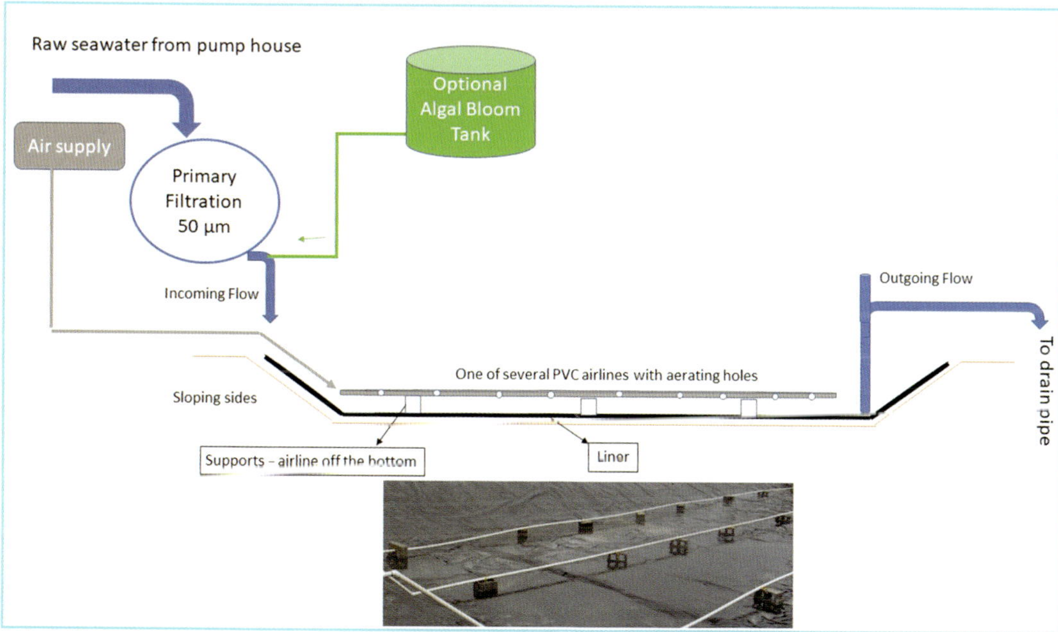

Figure 12.36 A schematic of a pond, applicable to the culture of non-burrowing bivalves.

A sand filter or other coarse filtration system can be installed at the incoming seawater line and supply several ponds with ambient seawater.

Aeration in ponds

Aeration in the pond is critical to maintaining food in suspension and providing sufficient oxygenation to live biomass, especially as stocking density increases. A brief summary is given here on oxygen level in ponds for a better understanding of the importance of aeration.

Dissolved oxygen (DO) is one of the most critical elements of water quality in pond culture. The DO in pond water is depleted by: (1) diffusion back into the atmosphere; (2) respiration by the target species and naturally occurring organisms in incoming coarsely filtered seawater; and (3) decomposition of organic material by microbial activity (or biochemical oxygen demand – BOD). On a daily basis, the amount of DO in a pond is highest during the day when photosynthesis is underway, and lowest at night when respiration and BOD depletes oxygen.

Stocking density and temperature affect the DO. The higher the stocking density, the greater the consumption of oxygen; and the higher the temperatures, the lower the solubility of oxygen, and the lower the DO.

In addition, DO will vary with pond depth as a result of: (1) increased surface temperatures; (2) decreased sunlight intensity with depth; (3) turbidity from suspended sediment or high algae concentrations close to the bottom. This will lead to concentration at the surface, and anoxic conditions at depth where BOD usually exceeds DO. The degree of stratification is related to the depth of the pond, with deeper ponds showing greater stratification.

In terms of seed production, low DO leads to poor water quality; this results in less energy for

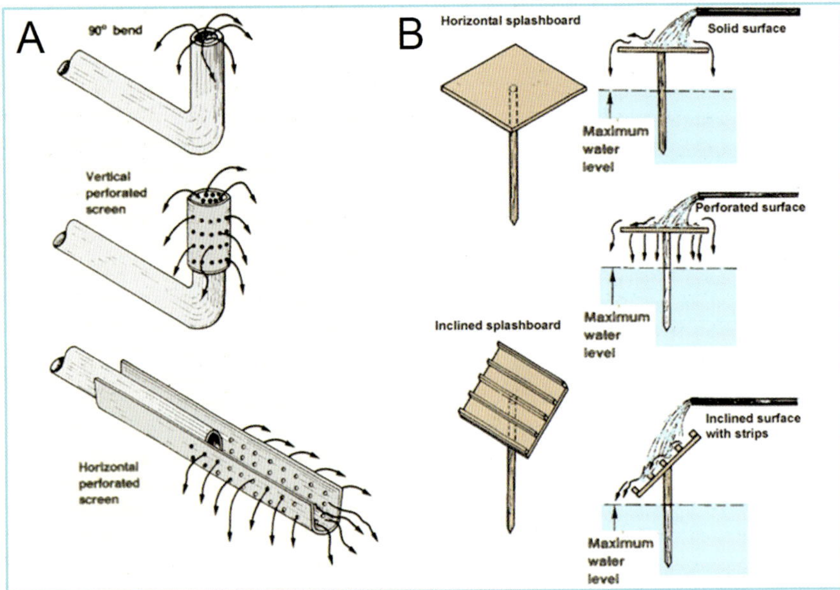

Figure 12.37 Simple aeration methods adapted to inlet. (A) Inlet perforation. (B) Horizontal and inclined splashboard (FAO, 2003).

growth, and yield in pond is negatively impacted. Monitoring DO in ponds is part of the daily routine, assessing a full vertical profile by sampling at several key locations – incoming, middle, and outgoing flow sites.

Pond design is critical to adequate aeration, which is in turn necessary to prevent stratification. In this way 'dead' areas leading to anoxia or bacterial proliferation are avoided, and food is evenly distributed. A proper aeration system depends on the physical characteristics of the pond (size, depth, shape), seasonal changes (for example, ice cover), and aeration type and efficiency. It is not the intent of this chapter to provide information on all pond and aeration systems, but to focus on those most likely to be useful for molluscan culture.

There are several ways of achieving aeration, ranging from a basic fitting at the incoming pipe to increase splashing (Fig. 12.37), to an air compressor connected to a system of PVC lines at the bottom of the pond, to vertical pump aerators which discourage ice formation in higher latitudes. The selected aeration system also has to accommodate the harvest methodology for the species produced. The use of an air compressor and system of PVC pipes is most applicable to bivalve seed production.

As for other sectors of the facility, air can be supplied from a main air compressor line supplying the facility, or a separate smaller compressor if ponds are further away from the hatchery complex. A system of 40 mm PVC lines, drilled at even intervals with 8 mm holes, is secured at the bottom of the pond, and lines are weighted down by concrete blocks or cement bags. Three parallel lines in a pond of 12 m width is sufficient. The release of small bubbles facilitates oxygen transfer (as surface area: volume ratio is greater than for larger bubbles). Suspension of bottom sediments is minimized through this airlift system as bubbles rise and expand, entraining water to the surface. It is a relatively easy approach to avoiding dead areas and distributing food evenly.

Feeding intermediate nursery systems: natural and controlled algal production

In most cases, it is necessary to supplement the naturally occurring plankton and organic nutrients with cultured algae. Large outdoor rectangular concrete tanks (up to 40,000 L), and circular tanks (12,000 L) are used for large volumes of algae; this microalgae supply is of lesser quality than that produced in the algal culture facility but is adequate for seed 3 mm and up. Agriculture-grade fertilizers can be used as nutrient medium. There are two approaches. The first relies on naturally occurring phytoplankton and encourages the bloom of the more dominant species; the second is based on the monoculture of standard algal species. The approach selected depends on site, phytoplankton composition, and benefits to the target species.

If a pond is used as an intermediate nursery system, the more basic approach of 'blooming' is achieved by directly fertilizing the pond. Accurately dosing the fertilizer, and close monitoring of the phytoplankton bloom in the pond is necessary to avoid an 'uncontrolled' bloom, which will negatively affect water quality. This system is best for larger seed, which is less vulnerable to bacteria and reduced water quality.

For a more controlled approach, and for supplementing circular and raceway systems, monospecific algal cultures from indoor production are used as inoculum for dedicated external algal vessels (see 'Algal production' section). For a 12,000 L tank, an inoculum of 1000 L with algal count averaging 8 million cells/mL is a minimum. The stronger the inoculum, the faster the culture will grow, and the higher the potential to reach densities high enough for supplementing the food ration.

The outdoor algal production vessels supplement the intermediate nursery by automated pumping of algae into the incoming seawater. Protection from temperature fluctuations and sunlight are recommended to ensure a more reliable production of algae. This type of system is vulnerable to poor batches and low-quality cultures due to exposure to weather and sunlight. Controlling water quality to a certain extent, and, if possible, using an in-line UV sterilizer leads to better algal production results. Microalgal species with faster growth rates are selected for this type of culture, thus minimizing the risk of crash in the algal culture due to exposure to unfavorable environmental conditions. One such algal species, *Skeletonema costatum*, is also recognized to be of benefit to bivalve seed.

An automated supply of algae from the bloom tank to the intermediate nursery system – or pond for example – uses a small pump (1.5–2 HP) and algae is supplied directly into incoming seawater line (Fig. 12.36). A float valve within the algal tank ensures that the pump stops, once algal level is low. The algal tank is ideally set-up as a semi-continuous culture, with UV sterilized seawater and nutrients added once algal culture is harvested to 25% of bloom tank. If this is not possible, two algal tanks may be necessary, and algae cultured on a rotational basis for a daily supplement to pond. The output of the outdoor algal production depends on:

- the maximum live biomass cultured in the pond
- the average algal cell density obtained
- the natural phytoplankton composition and abundance in the pond
- the target species.

Figure 12.38 Oyster spat set on shell. Courtesy of UMCES Horn Point Oyster Hatchery.

Remote setting

Some species, namely oysters and mussels, are suited to remote setting, or the direct transfer from the pediveliger stage to field nurseries. This greatly alleviates the costs involved in growing seed in the land-based hatchery, and in the transport of post-set from the hatchery to the farm. The remote setting technique is often used for stock enhancement purposes; and is also used for oysters, where eyed larvae are set on shells, for a 'spat-on-shell' end product (Fig. 12.38). In this case, it is often more economical to set remotely in tanks set-up on shore, close to the grow-out site. The remote setting site is often a long distance from the hatchery. Eyed larvae are transported using the 'coffee filter' method – in which larvae are concentrated into a bundle on a piece of mesh, kept cool, and wrapped in damp paper towel (or other similar absorbent material such as coffee filter, burlap, cheese cloth, or cotton). So prepared, larvae can be successfully shipped in an insulated cooler, remaining viable and ready for set over a period of days. Oysters remotely set in shore-side tanks become well fixed and ready for planting on beds or transfer to field nurseries for grow-out within one week of growth in the shore-side tanks.

The shore-side site will require a source of electricity, a submersible pump to provide for water changes, and air pumps. If done out of season, some means of heating the water may also be required. Commercially available diets (see 'Live microalgal culture' section) are convenient and the preferred food for remote-set systems.

Remote setting with some variations in method can also be applied to mussel larvae and spat of certain scallop species. Care must be taken with scallop species, as they will detach when taken out of the water, but do re-attach when submerged. Some aquaculturists transport scallop spat or young seed (0.5–2 mm) in bundles to the grow-out farm site, with the whole bundle transferred to a mesh spat bag upon which scallops will re-attach and grow. Spat bags can be laid horizontally into trays or lantern nets, or can be suspended directly from longlines.

Final considerations

Moving seed out of hatchery facility

The final hatchery steps of moving seed out of the land-based facility are not to be underestimated in terms of time, labor required, and costs incurred.

When transporting seed to dock or ponds, it is key to minimize the manipulation of seed and exposure to air, wind, or marked changes in temperature. Transport procedure for molluscan species differ, and details for this are beyond the scope of this chapter; however, the design of the hatchery needs to incorporate these requirements at the construction phase for a successful operation, as it may prove to be a substantial part of the nursery. For example, scallops are the most delicate bivalve, and can be transported moist for a limited time period. Ideally, they are kept immersed in clean seawater for the greater part of the handling and transport. Large-scale commercial operations achieve this by the use of an open water trough running externally the length of the nursery to the dock, moving scallop nets using a current. Scallops are directly loaded into holding tanks or wells on the boat. Other methods are also successful, where seed is packed moist still attached in substrate, or washed and collected in 'coffee filters' as briefly mentioned in the 'Intermediate nursery' section.

Operating efficiently is key to success of a nursery. The larger the seed becomes, the more labor and cost intensive the routine tasks become, and designing an efficient operating system is critical. For this reason, the location of the nursery and intermediate nursery on the facility needs to be carefully assessed, in consideration of the transfer and transport of seed to the farm. Siting ponds or other intermediate nursery facility adjacent to the dock – if a boat is used for transport to the farm – or road – if seed are transported by vehicle – is ideal. Between ponds and dock, an enclosed holding tank and saltwater table area, can be used to prepare seed for transport, and to receive seed for sorting from the farm, or market size adults for sale. The proximity and easy access to the loading area from the hatchery results in minimal disturbance, optimizes seed survival, and reduces economic and logistic costs in this stage of operation.

Processing sector

Depending on the target goal of the operation, processing the final product may be required to a certain extent. For several species, processing may occur in a location separate from the land-based hatchery complex. For some, an independent section of the facility may be sufficient for holding the final market size product, brought in from the farm and shipped for sale. The key points are that the requirements for processing need to be identified at the onset of designing a facility. Sufficient surface area for the processing section should be set aside and created as a stand-alone unit with no impact on the remaining hatchery/nursery operations. If animals ready for sale are collected from the farm by boat, the docking area and transport to the processing section also need to be separate from hatchery/nursery traffic.

General management practices

Diseases in hatcheries occurring during larval and post-set stages are minimized through best management practices. The diseases most often occurring in bivalve hatcheries are caused by bacteria living in the marine environment, whether or not bivalve larvae are present, but that opportunistically take advantage of the high densities of larvae in the hatchery. Because of this, opportunistic diseases are often called management diseases, the most common is vibriosis. It is now generally considered that most bacterial diseases of bivalve larvae are management diseases.

Poor management practices allow opportunistic harmful bacteria to gain entrance or proliferate at any number of sites in a hatchery. The overall approach to good management of these diseases is as follows.

1 Maintain pathogen-free algal stocks and expanded cultures.
2 Maintain an absence or low levels of pathogenic *Vibrio* spp. and other disease-causing microorganisms in the system (water column and surfaces) by proper water filtration, hygiene of system surfaces, and frequent water changes.
3 Isolate infected stocks and associated equipment at first sign of disease.
4 Discard infected stocks and sterilize equipment.
5 Identify source of contaminants, and modify and clean system.
6 Use probiotic cultures.

Additional considerations: expect the unexpected

Every site will differ, and the hatchery design must take into consideration site-specific issues. Not all issues can be anticipated, and for this reason, the hatchery must be sufficiently flexible in its design, with extra space to accommodate site-specific challenges that will most likely arise. For example, large fluctuations in salinity resulting from river runoff in some regions, may require buffering of larval and post-larval culture water through an additional tank and pumping system increasing the salinity to suitable culture requirements. Acidification of incoming seawater from climate change related causes cannot be ignored, and may have to be addressed by altering water chemistry, and adjusting incoming seawater treatment systems to the hatchery, nursery, and greenhouse areas. Similarly, a change in broodstock selection or breeding will require additional space for genetic manipulation.

Designing and constructing a hatchery facility requires the input of aquaculturists and biologists, and also that of engineers, water supply and treatment technicians, water chemists, and oceanographers to address the many considerations in achieving a land-based hatchery facility most suitable to the site and target species. Creating and operating a shellfish culture facility is a continuously challenging task, but also provides a constant source of stimulating opportunities for creative solutions. The satisfaction the culturist derives from successfully operating the life-support systems of a shellfish hatchery provides personal rewards that often surpass the economic income derived from the final saleable product. The advances in the technology are a testament to the creative and collaborative efforts of many over the years and no doubt the excitement of growing molluscs will continue to spawn new culturists and further innovation.

References cited and further reading

Andersen, R. (ed.) (2005) *Algal Culturing Techniques*. Elsevier Academic Press, London.

Creswell, R.L. and McNevin, A. (2009) Best management practices for bivalve molluscan aquaculture. In: Tucker, C.S. and Hargreaves J.A. (eds.) *Environmental Best Practices for Aquaculture*. United States Aquaculture Society, Sorrento, LA, pp. 427–486.

Dolan, E., Murphy, N., and O'Hehir, M. (2013) Factors influencing optimal micro-screen drum filter selection for recirculating aquaculture systems. *Aquacultural Engineering*, 56: 42–50.

Dubert, J., Barja, J.L., and Romalde, J.L. (2017) New insights into pathogenic Vibrios affecting bivalves in hatcheries: present and future. *Frontiers in Microbiology*, 8: 762.

FAO (2003) Simple methods for aquaculture. Méthodes simples d'aquaculture. Métodos sencillos para la acuicultura. FAO Training Series [CD-ROM]. Rome

Guillard, R.R.L. (1975) Culture of phytoplankton for feeding marine invertebrates. In: Smith, W.L., and M.H. Chanley (eds.) *Culture of Marine Invertebrate Animals*. Plenum Press, New York, pp 29–60.

Helm, M.M., Bourne, N., and Lovatelli, A. (2004) Hatchery culture of bivalves; a practical manual. FAO Fisheries Technical Paper. No. 471. FAO, Rome.

Huguenin, J.E. and Colt, J. (1989) *Design and Operating Guide for Aquaculture Seawater Systems*. Developments in Aquaculture and Fisheries Science, Volume 20. Elsevier Science, Amsterdam.

Kamermans, P., Galley, T., Boudry, P., Fuentes., J., McCombie, H., Batista, F., Blanco, A., Dominguez, L., Cornette, F., Pincot, L., and Beaumont, A. (2013) Blue mussel hatchery technology in Europe. In: Allan, G., and Burnell, G. (eds.) *Advances in Aquaculture Hatchery Technology*. Woodhead Publishing, Sawston, pp. 339–373.

Magnesen, T., Jacobsen, A., and Moepi, M.H. (2013) Effect of different filter methods on seawater quality at a marine scallop hatchery. *Journal of Aquaculture Research and Development*, 4: 168.

Neu, A.K., Mansson, M., Gram, L., and Prol-Garcia, M.J. (2014) Toxicity of bioactive and probiotic marine bacteria and their secondary metabolites in *Artemia* sp., and *Caenorhabditis elegans* eukaryotic model organisms. *Applied and Environmental Microbiology*, 80: 146–153.

Prado, S., Romalde, J.L., and Baria, J.L. (2010) Review of probiotics for use in bivalve hatcheries. *Veterinary Microbiology*, 145(3–4): 187–197.

Sarkis, S. and Lovatelli, A. (2007) Installation and operation of a modular bivalve hatchery. FAO Fisheries Technical Paper. No. 492. FAO, Rome.

Worksheet 12.1 Sizing a hatchery based on the reproductive capacity of the target species and the survival rate in the early life stages. *This is intended as a guide only.*

The final number of seed targeted defines three main key elements of the land-based hatchery: the pumping capacity, the number of broodstock required and the tankage capacity for larval and spat culture. Seawater demand can be satisfied with as little as 200 L/min using a single 2 HP pump, to 1500 L/min using more than one 15 HP pump. Highest seawater demands are often for nursery rearing and algal cultures.

The first set of general questions to answer include those relevant to the market strategy of the proposed operation.

1. What is the final product of the hatchery operation? Market size, seed, or both? If the operation anticipates selling market size animals, processing and packing will require additional space.
2. What is the optimal size for transferring seed to the farm? Or, what is the maximum size of seed to be reared in the hatchery facility?
3. How is setting of larvae conducted? Are tanks or a raceway system used?
4. Is intermediate nursery required for seed, and is the system land-based (ponds requiring further pumping capacity) or water-based (FLUPSY with independent pumping system)?
5. Once spat are settled, how long do they remain in setting system?
6. What water flow and what volume of spat rearing vessels are required until target size is achieved?

The second set of essential information relates to biological components, the growth and survival rate between life stages will inform the tankage volume required (see Appendix I for definitions of life stages):

- broodstock fecundity per female
- percentage embryo development and number of D-larvae
- rearing density of D-larvae
- percent survival to pediveliger (setting) and number of larvae to set
- larval growth rate and duration of larval life
- setting density
- post-larval growth rate
- percentage survival to 1 mm spat and no. of spat
- spat stocking density
- wet weight of spat with size for assessing live biomass and monitoring stocking density
- percentage survival to 2 mm seed and number of seed
- percentage survival to 4 mm seed and number of seed (if within production target goals).

The expected duration of larval and post-larval life to time of transfer out of hatchery dictates the number of larval and post-larval tanks required per cycle, and/or number of larval cycles anticipated in one hatchery season.

The number of tanks for broodstock, larvae, post-larvae, raceways if used, and pump size are determined once this information is collected.

Example 12.1 Sizing a hatchery for pilot scale production of calico scallops, *Argopecten gibbus. Please note: this example is to be used as a guide only, not for exact replication.*

Target Production: 5 million seed (4 mm)

Fecundity: 2.5 million/female

Given the fecundity and known survival rates between stages, a total of 60 females need to be spawned per cycle, resulting in 90 million D-larvae, 31.5 million pediveligers to set, 6.3 million settled spat, and 4.9 million 4mm seed.

To culture 90 million D-larvae at a density ranging from 3–5 larvae/mL, a total larval tankage capacity of 30,000 L is required. Larval rearing densities and yields are species dependent; oysters and clams (*Crassostrea* and *Tapes*) are reared at higher densities: 10–15 larvae/mL.

To set 32 million pediveligers at a density of 1–2 larvae/mL, a total of 30,000 L is required, if set in tank systems. In general, at least the same amount of tankage is required for settlement of pediveligers as for initial larvae. Smaller scallop production can make use of raceway systems, where settling larvae are set in sieves suspended in a downwelling raceway. This demands space, with initial setting densities of 60,000 per 25 cm^2, and never exceeding 50 mL/min/g wet weight of spat. Setting density is also species-dependent: ranging from as low as 20,000 per 25 cm^2 for oysters to 100,000 per 25 cm^2 for bay scallops (*Argopecten irradians*). As a general rule of thumb, older post set (>1 mm) should not cover more than 50% of the setting surface area.

Since hatchery production is continuous, and simultaneous larval and spat production occurs, a minimum of 60,000 L tankage capacity is required in this example.

In order to spawn 60 females, the availability of at least double and preferably triple the number of available broodstock is recommended. For *A. gibbus* broodstock, with adults of average shell height of 55 mm, one rectangular 300 L tank is sufficient to hold 30 scallops. A minimum of two 300 L tanks are required for broodstock in this case. Adult size and behavior dictates broodstock tank set-up and size; for larger scallop species (Japanese or Lion's paw scallops), one 500 L tank is needed for 30 scallops.

The volume of seawater required at peak times for each section of the hatchery is calculated for accurate pump sizing. For this *A. gibbus* example, two 2 HP pumps with a 200 l/min pumping capacity is sufficient for supplying seawater to broodstock, algae, larvae, and spat (up to 4 mm).

Worksheet 12.2 Sizing the algal culture facility. This is intended as a guide only.

Considering seawater treatment and logistics of delivery for daily feeding.

1 How and where will water be treated for algal cultures? Does it come from the main supply as pre-filtered, and further sterilized in the greenhouse, or is all treatment conducted in the greenhouse? Sufficient space needs to be allocated for water treatment.
2 How will air be supplied to the algal cultures? Will this be tied into the hatchery or be a separate air compressor system dedicated to algal culture facility; additional space is required for the latter.

3 How will algae be delivered to the larval and post-set tanks? In a small-scale operation, algae can be harvested and distributed manually. For larger operations, a collector and delivery tank with supply lines need to be integrated in the large culture area.

Estimating the size and layout of each area (laboratory, intermediate, large, and outdoor).

1 Will broodstock need feeding, is there a conditioning program? Will broodstock be fed with live microalgae, a combination of commercial diet and live microalgae, or commercial diet only?
2 Daily microalgal food ration for larvae at end of larval life, assuming all larval tanks full.
3 Estimated average algal cell density when harvested (cells/mL).
4 This will give an estimate of algal volume (L) to be harvested on a daily basis for larvae feeding.
5 Daily microalgal food ration for post-set until commercial diet can be supplemented (if used); assuming nursery post-set tanks are all full.
6 Based on average algal cell density produced, algal volume to be harvested on a daily basis for post-set larvae is estimated.
7 Information from points 3 and 5 provide an estimate of total volume of algae to be harvested daily. This will be divided by the number of microalgae composing the daily food ration. Keep in mind that food composition will differ as larvae and post-larvae grow with their changing biochemical requirements.
8 An estimate of the algal volume which can be harvested daily from any one vessel is needed. For 500 L continuous culture bags, approximately 100 L/day can be relied upon.
9 It is recommended that at least three replicate vessels be cultured simultaneously for the same species; this allows alternating harvest among bags, allowing algae to grow in-between harvests; it is also a risk-management approach in the case of an algal crash in one of the bags.
10 Keep in mind that new cultures are started weekly to ensure continuity in production; this means that additional vessels for stock cultures are required simultaneously with cultures harvested for the hatchery.
11 What is the time estimated for newly inoculated large culture vessels to be ready for harvest? This can range anywhere between 4 days to 10 days, and is largely dependent on the efficiency of the operator and the system.
12 Based on the volume of high-quality algae required (or algae grown under controlled conditions), the number of smaller intermediate vessels – which are used to inoculate the larger vessels – can be estimated. As a guide: 1 × 4 L culture at density of 10–15 million cells/mL can be used as an inoculum for a 500 L vessel.
13 For intermediate nursery feeding (for example in ponds), where a dedicated algal tank located outdoors is used as a live microalgae supply, volume of required algal inoculums need to be taken into account (this is discussed further in the 'Intermediate nursery' section). Surface area available outdoors must be considered during the sizing of the greenhouse.
14 Take into account the following space requirements: independent section for the autoclave; independent section for stock cultures; independent lab for probiotic preparation and culture.
15 How many broodstock (males and females) will be kept at any one time in the hatchery requiring feeding? By estimating the total live biomass for broodstock, an indication of live microalgae required on a daily basis can be obtained.
16 Algae for broodstock can be cultured by expanding the continuous 500 L bag culture system, or can be cultured in a separate semi-continuous or batch system using larger 1000–5000 L vessels.

Example 12.2 Sizing the algal culture facility for pilot scale production of calico scallops, *Argopecten gibbus*. This is intended as a guide only.

Referring to Example 12.1
Total number of D-larvae: 90 million with a total tank capacity of 30,000 L.

Food ration is calculated based on volume of larval culture. Food ration ranges from an initial 6,000 cells/mL to pre-metamorphosis 21,000 cells/mL. Assuming an average algal cell count of 8 million cells/mL, maximum algal volume required (at the end of larval life or pre-metamorphosis), is as follows:

Algal volume required to feed daily (L) = 21,000/8,000,000 × 30,000 L = 78L

Maximum algal requirements for larvae at this cell concentration is 78 L.

Assuming (as in Example 12.1), 32 million pediveligers to set in 30,000 L. Food ration ranges first week post-set from 21,000 cells/mL to 60,000 cells/mL at 1.5 mm. Assuming (as in Example 1), 32 million pediveligers to set in 30,000 L, maximum algal volume required for the first 6 weeks of post-set, is as follows:

Algal volume required to feed daily (L) = 60,000/8,000,000 × 30,000 L = 225 L

Maximum algal requirements for 1.5 mm post-set (six weeks old) is 225 L.

This means that the greenhouse needs to provide 225 L of live microalgae on a daily basis for 1.5 mm scallop spat. Food ration consists of at least three microalgal species, given as a 1 : 1 : 1 ratio, or 75–100 L harvested on a daily basis for each (of three) microalgal species.

For large scale culture sizing

If using continuous culture with 500 L vessels. Assumption:

- 100 L of algae is harvested per day per vessel.
- One species cultured in three replicate bags at any one time, or total of nine bags for three species.
- New cultures, or back up cultures are required, resulting in a further nine bags.

Total: 18 bags for post-set requirement (to 1.5 mm), for culture of three algal species.

Practically, at least four to five species of algae are cultured at any one time, and number of 500 L bags required would increase.

Estimated capacity for large vessel area: 30 × 500 L vessels.

For small and intermediate culture sizing

Assumptions:

- 1 × 4 L inoculum per 500 L vessel.
- Inoculation of three new bags per species every two weeks.

Total: 15 × 4 L cultures are required every two weeks, or 7–8 per week.

Estimated capacity for small and intermediate area: 10 × 4 L flasks, and 20 × 500 mL flasks.

Worksheet 12.3 Sizing a broodstock facility. This is intended as a guide only.

1. Are animals conditioned or brought in from the field a few days before spawning?
2. What is the goal of the conditioning period? More specifically, does broodstock only need ripening (last 3–4 weeks of gametogenic cycle); or does it require development for most of the gametogenic cycle (species dependent and may extend several months).
3. How many animals are used for one spawn? At least twice the number need to be available for selection of most ripe. In case of gonochoristic species, consider the number of males required for fertilizing available females.
4. Are animals brought in from several sources (populations or locations) which need to be kept separate?
5. Is a breeding program selecting broodstock source planned?
6. Is there a turnover of animals used for broodstock – where spent animals are replaced by new ones brought from the field – or are new broodstock added routinely to those conditioned? The latter requires additional tank space.
7. What temperature control is required? Where will water be treated for broodstock? Chillers or heaters for larger-scale operations require a dedicated area; for smaller-scale hatcheries, space by broodstock tanks for in-line installations is sufficient.
8. What is the average size of broodstock? What are the species spatial requirements? What are its water quality requirements? This dictates the volume of a broodstock tank, aeration, and holding method for animals.
9. How is food delivered to broodstock tanks? Batch feeding – in this case, no additional space is required; continuous feeding – a broodstock feeding tank with pump receiving algae from greenhouse and delivering to tanks needs to be set up. Its size is dependent on number of broodstock fed.
10. How is spawning conducted? How many spawning tables are required?
11. What is the duration of oocyte release, and will animals be allowed to spawn completely in isolated tanks if spawning continues overnight? This requires additional 'spawning tanks' for broodstock.
12. How will eggs be collected – in buckets or in a dedicated tank fitted to the spawning table?

Example 12.3 Sizing a broodstock facility. Please note: this example is to be used as a guide only, not for exact replication.

Referring to Example 12.1.

Candidate species: *Argopecten gibbus*
Reproductive system: Hermaphroditic
Total no. of females required per spawn: 60
Production output: 5 million seed

Conditioning scenario 1

- Relying on natural reproductive cycle.
- Minimal conditioning.
- Collecting animals from farm one week prior to spawning induction.

Number of animals collected: 90 ripe adults are brought in for the selection of 60 best females.

Broodstock tank characteristics: One 300–400 L, with incoming coarsely filtered seawater at a flow rate of 4.8 L/min (for an open flow system), and a constant temperature 1–1.5°C below peak spawning.

Temperature control: Two aquarium heaters, maintaining constant temperature; dependent on gametogenic state and natural seawater fluctuations, water temperature in broodstock tank can be increased by 0.5°C every day.

It is assumed that scallops will be returned to the farm after spawning, and a new batch brought in for a subsequent spawn. Only one tank is required for hermaphroditic species; two tanks are required if a gonochoristic species, to separate males from females. For a gonochoristic species, a minimum of 20 males should be brought in. One 100 L bin fitted with a spigot valve is required for feeding the first few days until 24–48 hours prior to spawning induction.

Conditioning scenario 2

- Extensive conditioning for the greater part of the gametogenic cycle.
- Conditioning goal is to ensure that a sufficient number of animals complete gametogenesis and develop fully ripe gonads with viable eggs, and to mitigate for any mortality during conditioning. At the end of the conditioning period, optimal broodstock is selected for spawning. A program staggering conditioning periods with at least two batches is recommended.

Total number of scallops collected from farm: 120 females (and 30 males if gonochoristic sp.) per conditioning cycle.

Total no. of tanks: Four tanks holding females for two conditioning cycles, (and four tanks for males if gonochoristic). Additional tanks for scallops post-spawn are required, in order to keep induced or spent scallops separate from those which require further conditioning. This means a minimum of 8 tanks for broodstock.

Worksheet 12.4 Sizing a larval hatchery. This is intended as a guide only.

The use of a static tank will require a larger surface area and a larger tank than a flow-through tank to yield the same larval production. If possible, flow-through systems are recommended.

1 Is species amenable to flow-through culture techniques?
2 If flow-through tanks are used, an estimate of the flow rate per tank is needed. This is species-dependent.
3 What is the optimal larval density initially and pre-metamorphosis? Generally, density is reduced as larvae grow (see Example 12.1). Optimal larval densities for static and flow-through systems are required.
4 What is the expected survival rate to pediveliger stage? In other words, how many tanks will be required pre-metamorphosis.
5 Will the hatchery consist of both static and flow-through larval tanks?

Answers to questions 1–5 above will indicate the size and number of static or flow-through tanks which are needed.

Feeding for flow-through tanks is continuous and delivery of live microalgae needs to be planned, and incorporated in the design of the hatchery; this can be achieved through:

1 the use of a continuous feed supply tank, receiving algae pumped from the algal facility, to a reservoir supplying flow-through larval system with a 24 hour food ration
2 direct connection to a photobioreactor
3 batch-fed, where algae is supplied once a day manually to a dedicated feed container; a drip-feed from feed tank distributes algae over a few hours, rather than in one batch. This method is not efficient for larger flow-through tanks, or for medium and large-scale hatcheries, but very cost-effective for small-scale operations.

Worksheet 12.5 Sizing a nursery. This is intended as a guide only.

The number of spat that can be held in a nursery system depends on type of tanks used and on the live biomass, and more specifically on their size/weight, which are species dependent.

Questions to be considered for sizing a nursery are as follows.

1 What setting system is used?
2 What is the maximum size at which seed will be kept in the hatchery in the setting tanks?
3 What is the setting density?
4 What is the expected survival to post-settlement or to size at which survival will stabilize (1 mm for several bivalve species)?
5 What is the expected shell growth and estimated wet weight? This will provide an indication of live biomass, stocking density, and optimal tank size and flow rate required.
6 How will spat/seed be collected for transfer to the intermediate nursery or to the farm? If seed is transferred using setting substrate, a system for keeping seed and substrate submerged during transfer is needed. If seed is collected from substrate for transfer into meshed bags, nets or trays, one or more grading tables, with saltwater tables for re-distribution into new bags, will be needed.
7 A dry area for quantifying seed production (scale; dissecting scope) is advantageous.

Appendix I. Definition of early life stages for bivalve species, adapted from FAO (Food and Agriculture Organization, United Nations).

Early life stage	Description
Embryo	Organism in early developmental stage; in bivalves, prior to larval stage.
Veliger	Larval stage characterized by presence of a velum (or ciliated locomotory organ of larva).
Pediveliger	Larval stage of bivalves, prior to settlement; larva still has swimming ciliated organ (velum), and foot needed for settlement.
Spat	Newly settled or attached bivalve (also termed post-larval or juvenile in bivalves). In this chapter, defined as settled larvae to seed stage. Also referred to as post-set juveniles.
Post-set juveniles	Any stage following post-set; it includes spat, seed or young juvenile for non-bivalve species.
Trochophore	First free-swimming planktonic stage of a molluscan larva or bivalve embryo.
D-larvae	Early veliger larval stage of bivalves, also known as straight-hinge larva.
Post-set	Stage following metamorphosis of larvae.
Seed	Young bivalve with shape of adult stage with no definition of size. In this chapter, defined at >2 mm shell height. Also referred to as post-set juveniles.

Appendix II. Nominal NPT pipe size to nominal metric size conversion chart.

Nominal pipe size (inches/US)	Nominal pipe size (metric)
1/8"	6 mm
3/16"	7 mm
1/4"	8 mm
3/8"	10 mm
1/2"	15 mm
5/8"	18 mm
3/4"	20 mm
1"	25 mm
1 ¼"	32 mm
1 ½"	40 mm
2"	50 mm
3"	80 mm
4"	100 mm
6"	150 mm

CHAPTER 13

Algal culture for shellfish aquaculture

Steve L. Morton

Introduction

The molluscan aquaculture industry is sustained with hatchery seed production, a process of producing seed under completely controlled conditions, including broodstock selection and conditioning, spawning and gamete collection, fertilization, swimming larval culture, metamorphosis, and post-metamorphosis rearing. Hatchery seed production provides opportunities to increase seed quantity and quality and allows for broodstock breeding and genetic manipulations (see Chapters 12 and 14). Additionally, shellfish hatcheries can overcome the limitation of natural shellfish spawning season to produce seed with extended spawning periods.

Microalgal culture is an essential aspect for hatchery seed production, because unicellular microalgae are the food providing the proper nutrients for all stages of cultured filter-feeding bivalves. Generally, the microalgal species used in shellfish hatcheries include species from different phyla, depending on their size and cell structure, nutrient value, and ease of culture. Algal culture typically accounts for about 40% of hatchery expenses until bivalve seed reach 5 mm in shell length.

Microalgal species – which one to choose

There are potentially thousands of microalgal species that can be cultured for shellfish aquaculture, and there are a number of species that have been cultured successfully, are used routinely, and are available through different culture collections (Table 13.1). These microalgae used as feed vary in size, environmental requirements, growth rate, and nutritional value (Table 13.2). When selecting a species for culture, it is important to take all of these parameters into consideration. Most hatcheries grow a variety of species that serve different needs throughout the shellfish life cycle with respect to size, digestibility, culture characteristics, and nutritional value.

Table 13.1 Culture collections that contain common species used by the aquaculture industry. Starter cultures are available for purchase from many of these collections.

Culture collection	Website
NOAA/NMFS Milford Culture Collection	https://www.nefsc.noaa.gov/nefsc/Milford/
National Center for Marine Algae and Microbiota (NCMA)	https://ncma.bigelow.org/
University of Texas Culture Collection (UTEX)	https://utex.org
American Type Culture Collection	https://atcc.org
Culture Collection for Algae and Protozoa (CCAP)	http://ccap.ac.uk
Australian National Algae Culture Collection (CSIRO)	http://csiro.au

Table 13.2 Common aquaculture culture strains and growth requirements. Temperature range and cell size from the National Center for Marine algae and Microbiota.

Name	Strain	CCMP number	Temperature (°C)	Size (µm)
Isochrysis galbana	ISO	CCMP1323	3–28	2–6
Dunaliella tertiolecta	DUN	CCMP1320	11–28	6–9
Tisochrysis lutea	T-ISO	CCMP1324	11–36	4–8
Tisochrysis lutea	CISO	CCMP463	17–30	4–8
Nanochloropsis oculata	NANO	CCMP525	11–25	1–2
Tetraselmis sp.	WTET	CCMP908	17–26	8–18
Thalassiosira weissflogii	THAL7	CCMP1051	11–30	8–20
Chaetoceros calcitrans	CCAL	CCMP1315	17–26	3–7
Chaetoceros muelleri	CHGRA	CCMP1316	15–30	4–10
Skeletonema marinoi	SKEL	CCMP1332	11–30	6–14
Thalassiosira pseudonana	3H	CCMP1335	4–25	4–6
Skeletonema marinoi	SZN	CCMP2092	17–22	4–6

Culture population dynamics

Microalgal cultures proceed through a typical growth pattern of lag phase, log phase, stationary phase, and senescence phase. Initially, a newly established culture is required to become acclimated to new growth medium and will not grow for a period of time, the lag phase. This phase varies in length depending upon the size of the initial inoculum, algal species, and environmental condition. Once acclimated, the cells will grow exponentially (log growth phase). Cultures are usually harvested for use in shellfish feeding at this stage. The maximum culture density attained is dependent upon the algal species, water temperature, water salinity, light intensity, nutrients, aeration, and culture vessel. When the rate of cell division declines and there is no further increase in cell density, the culture reaches the stationary phase. This decreased growth is the result of nutrient limitation, self-shading, and changes in the culture medium, such as increased pH and accumulation of metabolic waste products. As a general rule, stationary phase cultures should not be used in shellfish aquaculture as the cultures are generally not as nutritious and have a larger bacterial concentration, including pathogens, such as *Vibrio*. The stationary phase is followed by a senescent phase in which the density of the culture decreases.

Estimation of cell number

Estimating algal density is an inherent part of any algal production system. Biomass estimates are used to determine when to transfer an inoculum through a serial dilution to larger volume culture and to determine harvest volumes of culture in production. A number of different techniques are available to estimate cell number in algal cultures. The first is manually counting cells using a microscope. This method employs a counting chamber of known volume and the number of cells is counted giving a number of cells per milliliter. Counting chambers used include the Palmer-Maloney counting chamber and the Sedwick-Rafter counting chamber. The Palmer-Maloney counting chamber has a total volume of 0.1 mL, while the Sedwick-Rafter chamber has a total volume of 1 mL. Automatic counting instruments are available. The most commonly used in the past was the Coulter Counter, first developed to count blood cells, which can be used to count algal cells. An algal cell passes through a small aperture and a slight electrical current passes between two electrodes. Each time a cell passes between the electrodes, the current is impeded and the cell is counted. The disadvantage of the Coulter Counter is the fact that it will count cells and other particles or precipitates in the culture. There are numerous other options for counting cells, including flow cytometry, but they are expensive and requires specialized training to operate accurately. Another option that has bridged several technologies is an instrument known as the FlowCam (Fluid Imaging Technology). This instrument combines the functionality of a flow cytometer and a microscope, giving the user pictures of each cell or particle that passes the detector, giving a greater precision of the cell count. A recent and relatively inexpensive option for hatcheries is Luna-II Automated Cell Counter. Like the FlowCam, the Luna-II couples automated cell counter with microscopy.

General culture procedures

Pretreatment of the water is one of the most important steps in successful microalgae culture. Water should be free of suspended solids, bacteria, high amounts of dissolved organic compounds, dissolved metals, and pesticides. Water treatment includes mechanical and chemical filtration, sterilization/disinfection followed by nutrient enrichment. A reliable, consistent source of water is imperative for long-term consistent microalgal production.

Mechanical filtration removes suspended solids, bacteria, plankton, and other particles unwanted when culturing microalgae. The type of filtration required is dependent on the source of water; however, a mechanical filter usually consists of a series of filters that will remove increasingly smaller particles. Chemical filtration removes dissolved inorganic and organic compounds from source water, which can reduce microalgal growth. Activated charcoal is an efficient but expensive method of removing these contaminates.

After filtration and before nutrient enrichment, water needs to be either sterilized or disinfected to eliminate contaminating bacteria and virus. Two different methods can be utilized to sterilize water, heat sterilization (autoclaving, pasteurization, and microwaving) and chemical sterilization. Heat sterilization is useful for small amounts of water, but is unrealistic for the larger amounts of water required for mass culture of microalgae for shellfish aquaculture. Chemical sterilization using chlorination is a simple method of eliminating microbial organisms in large volumes of water. For disinfection, two methods can be employed, ultraviolet (UV) irradiation and ozone. UV irradiation uses low-pressure mercury bulbs usually at 254 nm light wavelength. The effectiveness of UV is dependent upon the exposure time of the water. Ozone is a strong oxidizing agent and is extremely effective in removing dissolved organic materials. Ozone itself is highly unstable and quickly reverts to oxygen and is highly corrosive.

Table 13.3 Artificial seawater base AK (modified from Keller et al., 1987).

Compound	Concentration (mol/L)	Weight/L
NaCl	4.1×10^{-1}	24.0 g
$MgCl_2$ $6H_2O$	8.0×10^{-3}	0.6 g
$MgSO_4$ $7H_2O$	1.48×10^{-2}	3.0 g
$CaCL_2$ $2H_2O$	2.7×10^{-3}	0.4 g
NaBr	1×10^{-5}	1.03 mg
H_3BO_3	9.7×10^{-5}	27.0 mg
$SrCL_2$ $6H_2O$	1×10^{-4}	6.0 mg
NaF	1×10^{-5}	0.42 mg
KI	2.0×10^{-7}	33 μg
$NaHCO_3$	2.38×10^{-3}	0.2 g

Source of seawater

An important factor in consistent success in algal culture is the source of consistent, high-quality seawater. If natural seawater is used, minimal variation in salinity, nutrient concentration, pollutants, and other factors is necessary. In most cases, artificial or synthetic seawater is required for mass culture of algae. Recipes for artificial seawaters, sometimes referred to as defined media can be grouped in two types. The first imitates natural seawater in ionic ratios, while the second type differs from natural seawater in their ionic ratios. An often used artificial seawater recipe is shown in Table 13.3.

Many manufacturers exist for commercially available synthetic seawaters. These include, but are not limited to, Marinemix, Seachem Reef Salt, Kent SeaSalt, Coralife Salt, and Instant Ocean. It should be noted that some of these formulations have added nutrients.

Nutrient enrichment

"Numerous enriched and synthetic media have been formulated which together with generally trivial modifications, almost equal the number of investigators" (McLachlan, 1973, p. 26). Among the vast number of recipes in the scientific literature the Guillard f/2 formula stands out (Guillard, 1975). This formulation has been the backbone of algal culture and shellfish research since its inception (Table 13.4). The original medium used ferric sequestrene, which can be substituted with sodium ethylenediaminetetraaceticacid (EDTA) and iron chloride. To simplify routine media preparation, the recipe is subdivided into working stock solutions. Adding small volumes of liquid stock solutions is easier and quicker than weighing individual dry chemicals. It is better to add these stock solutions directly to seawater, either natural or artificial, and mix thoroughly. Nutrient enrichment should occur after sterilization to eliminate undesired

Table 13.4 f/2 Medium (Guillard, 1975).

Final medium: To 950 mL of filtered or sterile seawater add:

Compound	Stock solution (g/L of H_2O)	Quantity used	Concentration in final medium (M)
$NaNO_3$	75	1 mL	8.82×10^{-4}
NaH_2PO_4 H_2O	5	1 mL	3.62×10^{-5}
$NaSiO_3$ 9 H_2O	30	1 mL	1.06×10^{-4}
Trace metal solution	See following recipe	1 mL	
f/2 vitamin solution	See following recipe	0.5 mL	

f/2 trace metal stock solution: Into 900 mL of H_2O dissolve EDTA and other components. Bring to the final volume of 1 L with H_2O.

Component	Primary stock solution (g/L of H_2O)	Quantity used	Concentration in final medium (M)
Na_2EDTA		4.36 g	1.17×10^{-5}
$FeCl_3$ $6H_2O$		3.15 g	1.17×10^{-5}
$MnCL_2$ $4H_2O$	180	1 mL	9.10×10^{-7}
$ZnSO_4$ $7H_2O$	22.0	1 mL	7.65×10^{-8}
$CoCl_2$ $6H_2O$	10	1 mL	4.20×10^{-8}
$CuSO_4$ $5H_2O$	9.8	1 mL	3.93×10^{-8}
$NaMoO_4$ $2H_2O$	6.3	1 mL	2.60×10^{-8}

f/2 vitamin stock solution: Into 950 mL of H_2O dissolve the thiamine and add 1 mL of the primary stocks. Bring final volume to 1 L with H_2O.

Component	Primary stock solution (g/L of H_2O)	Quantity used	Concentration in final medium (M)
Thiamine HCL		200 mg	2.96×10^{-7}
Biotin	1	1 mL	2.05×10^{-9}
Vitamin B_{12}	1	1 mL	3.69×10^{-10}

precipitations which sometimes occur with autoclaving complete media. If a diatom is not the cultured alga, silica can be eliminated from the final media. Complete Guillard f/2 formulas are commercially available from a number of venders.

Maintaining starter cultures

Stocks of unialgal cultures should be kept in specialized maintenance media, usually f/2 under controlled conditions, including temperature and light intensity. Stock cultures serve as inoculum for the large-volume production used in shellfish feeding. These cultures are kept in small containers such as 25 mL test tubes or 125 mL flasks. Every effort should be made to make sure these cultures are free of contaminating microorganisms.

Figure 13.1 Large-scale production of algae using calwall (left) or bag (right) system.

Batch culture and scale-up

To produce the number of algal cells required for feeding larvae and juvenile shellfish, a process of batch culture is utilized. Starting with cells from the stock culture, algae are cultured in enriched medium through a series of culture vessels of increasing volume. For best results, transfer of cultures should occur when the culture reaches mid- to late-log growth phase. For most cultures, this is the stage where the cells are most healthy and growing fastest with the lowest chance of contamination. The stock culture is used for the inoculum for a 250 mL culture flask which inoculates a 2–4 L flask which in turn then inoculates a 20 L carboy. The 20 L carboy is used to inoculate the final large-scale culture vessel. The final large culture vessels range from calwalls to sterile bags. Both systems have been used for mass culture of microalgae successfully (Fig. 13.1) in controlled environments. Generally, outdoor ponds are not used for the culture of microalgae for shellfish.

Photobioreactors are closed systems that theoretically provide higher cell densities than traditional batch cultures using calwalls or bags. These technologies, such as tubular and plate photobioreactors, have been used extensively by the biofuel industry for large-scale production of algae and are commercially available. Generally, this approach is not useful for the shellfish industry; however, a temperature- and light-controlled system produced by Industrial Plankton (Fig. 13.2) has been utilized to produce algae in large concentrations for the shellfish aquaculture industry. The main advantage of a bioreactor over both the calwall or bag systems is the control of all environmental conditions, such as temperature and light, without a specialized culture room. Other options include bioreactors such as those manufactured by Belach Biotecknik, and GMI (Solida Biotech Bioreactors).

Conclusion

Since algae are the essential food for all stages of cultured filter-feeding bivalves, algal culture is a lynch pin of a successful hatchery operation. A consistent source of healthy, quality algae will

Figure 13.2 Large-scale production of algae utilizing a photobioractor.

have an impact on every facet of the shellfish operation. The principles and methods described represent only a few of the many approaches to grow microalgae. The reader is advised to modify these guidelines as needed to suit specific hatchery conditions and shellfish species.

References cited and further reading

Andersen R. (ed.) (2005) *Algal Culture Techniques*. Elsevier Press, Amsterdam.

Guillard, R.R.L. (1975) Culture of phytoplankton for feeding marine invertebrates. In: Smith, W.L. and Chanley, M.H. (eds.) *Culture of Marine Invertebrate Animals*. Plenum Press, New York, pp. 29–60.

Guillard, R.R.L and Morton, S.L. (2003) Culture techniques. In: Hallegraeff, G.M., Anderson, D.M., and Cembella, A.D. (eds.) *Manual of Harmful Algae*. UNESCO, Paris, pp. 77–98.

Keller, M.D., Selvin, R.C., Claus W., and Guillard, R.R.L. (1987). Media for the culture of oceanic ultraphytoplankton. *Journal of Phycology*, 23: 633–638.

McLachlan, J. (1973) Growth media – marine. In: Stein, J. (ed.) *Handbook of Phycological Methods: Culture Methods and Growth Measurements*. Cambridge University Press, Cambridge, pp. 25–51.

Stottup, J.G. and McEvoy. L.A. (2003) *Live Feeds in Marine Aquaculture*. Blackwell, Oxford.

CHAPTER 14

Genetics in shellfish culture

Ximing Guo

Introduction

The long history of agriculture is basically a history of plant and animal domestication. Agriculture would not be possible without domestication and genetic improvement of crops and livestock. Shellfish aquaculture is relatively new and has relied on wild seed for much of its short history. Wild seed are unreliable and lack the characteristics desired for aquaculture, such as adaptation to high-density culture conditions, fast growth, disease resistance, and uniformity. The use of wild seed precludes genetic improvement, which has been a major contributor to increases in agriculture production since 1960. Further and more rapid improvements of agricultural crops and animals are being made through advanced genetics and genomics. Shellfish species are cultured in open waters and face more environmental stress than terrestrial animals. Climate changes, such as the warming and acidification of world oceans, present new challenges to shellfish and shellfish culture. More and more diseases and unexplained mortalities are affecting cultured molluscs. As cultured molluscs are either wild or have a short history of domestication, the need for genetic improvement is great. It is inconceivable that shellfish aquaculture can be sustained without serious efforts in genetic improvement.

The development of hatchery production has made genetic improvement of molluscan shellfish possible. Currently, over 30 species of molluscan shellfish are produced in hatcheries and subjected to some degrees of genetic improvement or domestication (Table 14.1). They include many species of oysters, scallops, mussels, clams, and abalone. Once the life cycle is closed, hatchery production will inevitably lead to genetic changes with or without intentional selection. Genetic change is made every time broodstock shellfish are picked to produce the next generation. Without a well-managed breeding program in place, hatcheries may unknowingly inbreed their stocks and suffer from the negative consequences. Therefore, it is critical that all shellfish hatcheries establish sound breeding protocols and proactively manage the genetic health of their stocks.

Genetics is the study of genes, genetic variations, and heredity of characters in living organisms. Extensive studies have been conducted on the genetics of molluscan shellfish including simple Mendelian inheritance of allozymes, inheritance of shell color and sex, population

Table 14.1 Hatchery production and genetic improvement of cultured molluscan shellfish.

Species	Location	Breeding and improvement
Argopecten irradians	China, USA	Growth, survival, shell color, hybridization
Argopecten purpuratus	Chile	Reduced escape response
Argopecten ventricosus	Mexico	Shell width, body weight
Chlamys farreri	China	Growth, shell color
Chlamys nobilis	China	Shell and meat color, growth
Crassostrea angulata	China, EU	Triploids in commercial production, tetraploids
Crassostrea belcheri	Southeast Asia	Induced triploidy
Crassostrea gigas	Worldwide	Growth, disease resistance, triploids, tetraploids, color
Crassostrea iredalei	Southeast Asia	Induced triploidy
Crassostrea virginica	USA, Canada	Disease resistance, growth, triploids, tetraploids
Cyclina sinensis	China	Growth, intraspecific hybridization
Haliotis discus hannai	China, Japan, Korea	Survival, growth, intraspecific hybridization
Haliotis diversicolor	China	Growth, survival
Haliotis iris	New Zealand	Growth
Haliotis rubra & laevigata	Australia	Growth, behavior, hybridization
Haliotis rufescens	USA, Chile	Heritability of growth and disease resistance
Hyriopsis cumingii	China	Growth, color, pearl production, hybrids
Mercenaria mercenaria	USA, China	Body size, shell color pattern
Meretrix meretrix	China	Growth, shell color, intraspecific hybrids
Mytilus chilensis	Chile	Live weight, shell length
Mytilus edulis	EU	Resistance to spring mortality, growth
Mytilus galloprovincialis	Australia	Growth
Panopea abrupta	USA, Canada	Population genetics and induced triploidy
Patinopecten yessoensis	China, Japan	Growth, color, introduced to China
Perna calaniculus	New Zealand	Shell length, meat weight
Pinctada fucata martensii	China, Japan	Shell growth and color, survival
Ruditapes philippinarum	China, EU, USA	Growth, shell color patterns
Saccostrea glomerata	Australia	Resistance to QX & *Bonamia roughleyi*
Scapharca broughtonii	China	Growth, intraspecific hybridization
Sinonovacula constricta	China	Growth, survival, intraspecific hybridization
Tegillarca granosa	China	Growth

genetics, estimation of heritability, development of genetic markers and maps, identification of quantitative trait loci (QTL) and functional genes (Guo et al., 2008b; Hollenbeck and Johnston, 2018; Newkirk, 1980). Genomes and transcriptomes of many molluscan shellfish have been sequenced, yielding rich information about genes and molecular mechanisms underlying molluscan development, function, immunity and physiology. While these studies have improved our understanding of molluscan genetics and biology, most of them have not led to actual genetic improvements for shellfish aquaculture. Academic studies on molluscan genetics are not the focus of this chapter. Rather this chapter focuses on practical molluscan breeding and related studies with the goal of providing some guidance to shellfish breeders and hatchery managers.

Selective breeding

Traditional selective breeding is an effective approach to genetic improvement and has been used for the improvement of many aquaculture species. Over 100 breeding programs have been established for aquaculture species, most of which are for finfish. Selective breeding for salmonids, tilapia, carps, and catfish is particularly effective and successful. For example, four generations of selection in Coho salmon produced over 60% improvement in body weight (Hershberger et al., 1990). Five generations of selection produced a 113% improvement in growth rate in Atlantic salmon (Thodesen et al., 1999). Three generations of selection produced a 21–29% improvement in body weight in catfish (Rezk et al., 2003). Selective breeding in rainbow trout produced strong resistance to infectious pancreatic necrosis, a significant viral disease (Okamoto et al., 1993).

While selective breeding has been tested for many molluscan species, mature or sustained breeding programs that have contributed to shellfish production are few. One of the longest molluscan breeding programs is the eastern oyster breeding program at Rutgers University. The program began in 1960, soon after the outbreak of MSX (multinucleated sphere X caused by protozoan *Haplosporidium nelsoni*) disease in Delaware Bay, and has continued to this day. The program used individual or mass selection to improve MSX resistance by selecting survivors from field exposure, as MSX cannot be artificially transmitted in the laboratory (Ford and Haskin, 1987). Significant improvement in MSX resistance was achieved after five generations of selection, with selected lines surviving 8.8 times better than the susceptible control (Fig. 14.1). Several MSX-resistant lines derived from different geographic populations were developed over a period of 30 years. Dermo disease (caused by protozoan *Perkinsus marinus*) appeared in Delaware Bay in 1990 and caused significant mortality in the MSX-resistant lines. To restore genetic variability and breed for Dermo resistance, two synthetic strains, NEH and DBH, were produced by crossing MSX-resistant lines originated from the Northeast (Long Island Sound) and Delaware Bay, respectively (Allen, 1993). Crossing the inbred MSX lines restored genetic variability and led to improvements in larval survival. In 1998, fast growth was added to the breeding program by selecting the largest 20% of oysters that survived 3 years of field exposure to MSX and Dermo. Shell shape or appearance was also considered, and oysters with a flat left shell and a curled hinge area were rejected. New genetic materials were introduced, and sublines were created and managed through progressive rotational crossbreeding (see below). In several evaluations, Rutgers NEH strain demonstrated strong resistance to MSX and moderate resistance to Dermo (Guo et al., 2003; Proestou et al., 2016). Although the NEH strain has not been exposed to *Roseovarius* oyster disease (ROD, caused by bacterium *Roseovarius crassostreae*), it has shown resistance to ROD probably due to its hardiness or non-specific resistance to bacterial infections (Gomez-Leon et al., 2008). The Rutgers NEH strain is developed in a highly stressful intertidal environment at Cape Shore, New Jersey, where oysters experience 3–4 hours of air exposure twice a day, air temperature may reach 50°C in summer, and MSX, Dermo and *Vibrio* spp. (infecting larvae) are endemic. Besides disease resistance, hardiness may be an important contributor to the improved performance of the NEH strain. Because of its strong MSX resistance and hardiness, the NEH strain are widely used by oyster farmers in NJ and Northeastern US.

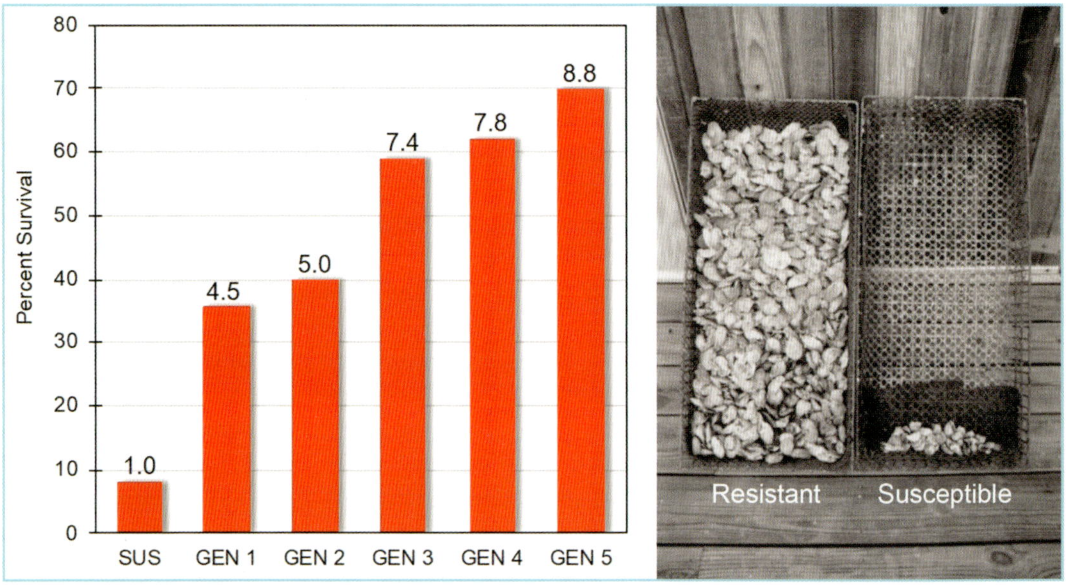

Figure 14.1 Improvement of eastern oysters in MSX resistance after five generations of mass selection. Left: progressive improvement in percent survival under MSX standardized with the susceptible control (SUS) (Guo et al., 2008a). The number above each bar represents the improvement in survival compared to SUS (i.e., the first generation had 4.5 times as many survivors as the average SUS group). Right: survived oysters in susceptible control and MSX-resistant lines that started with the same number. Courtesy of S.E. Ford.

Similarly, mass selection at Virginia Institute of Marine Science (VIMS) has produced strains that showed moderate dual resistance to MSX and Dermo (Calvo et al., 2003). Recently, family-based selection has been applied to the breeding program at VIMS. Disease-resistant strains of VIMS are widely used by oyster farmers in Chesapeake Bay. In 1986, University of Maine started breeding eastern oysters for fast growth and survival under ROD by testing several strains including that from Rutgers and Frank M. Flower and Sons Inc (Flowers). The UMFS (University of Maine Flowers Select) line was developed from the Flowers strain. UMFS oysters exhibited ROD resistance with cumulative mortality of only 11% compared to 96% in wild controls, during a ROD outbreak in the Damariscotta River, Maine (Barber et al., 1998). The Flowers Company has been breeding oysters as part of their commercial production, targeting mainly fast growth. Selection for ROD resistance at Flowers was brief, but surprisingly successful. Significant improvement in survival of ROD, by up to 7 to 35 times, was achieved after one to three generations of selection (Farley et al., 1998). Recent data, however, indicate that the Flowers strain may have lost its ROD resistance (Gomez-Leon et al., 2008). Several other commercial hatcheries in Maine, Massachusetts, and New York have also been breeding their own oysters as part of their hatchery production. While their breeding history and performance are not known to the public, these commercial stocks have contributed to oyster farming and represent valuable genetic materials for future breeding. An eastern oyster breeding program also existed in Connecticut, and new programs are being developed in North Carolina, Canada, and the Gulf of Mexico. Nearly all eastern oyster hatcheries in the US now use stocks that have gone through some degree of selective breeding, and the development of disease-resistant strains and triploids have contributed significantly to eastern oyster culture.

In the Pacific oyster, families resistant to summer mortalities were developed after three generations of selection (Hershberger et al., 1984). Resistance to the Ostreid herpesvirus 1 μVar (OsHV-1) was obtained after four generations of mass selection (Degremont et al., 2015), and OsHV-1 resistant strains have been developed in New Zealand and Australia. Pacific oysters with superior growth have been developed through family-based (de Melo et al., 2016) and mass selection (Zhang et al., 2019). In the European flat oyster, resistance to bonamiasis caused by *Bonamia ostrea* was observed after three generations of mass selection (Naciri-Graven et al., 1998) and 4–10 generations of natural selection (Elston et al., 1987). In the Sydney rock oyster, resistance to winter mortality caused by *Bonamia roughleyi* and QX (Queensland unknown) disease caused by *Marteilia sydneyi* was obtained after three to four generations of mass selection (Dove et al., 2013; Nell and Perkins, 2006). Many other studies have also demonstrated significant improvements in growth and survival in other molluscs through selective breeding. Disease resistance, the most important trait for aquaculture production, is often easy to acquire. The fact that strong resistance to certain diseases, such as MSX, ROD and OsHV-1, are obtained within a few generations suggests that resistance to these diseases has high heritability and may be under the control of a few major-effect genes. Conversely, progress in breeding Dermo resistance has been slow, indicating that Dermo resistance may have low heritability or under the control of many small-effect genes.

Molluscs have colorful shells. Certain shell colors have aesthetic values and are useful tools for marketing and managing selected strains. Shell color is usually controlled by one or a few genes following simple Mendelian inheritance, making them relatively easy to select for. Often, wild populations are dominant by a common color type, and other colors occur at a low frequency, but can be easily purified by selection. Shell color is an important trait in molluscan shellfish cultured in China where red and golden colors are desired. Red strains of Chinese and Japanese scallops, red and purple strains of bay scallops, golden Pacific oysters, and golden Portuguese oysters, golden variety of noble scallops, red *Meretrix* clam, and 'zebra' Manila clams have been developed through selective breeding and used for commercial production in China. For hard clams cultured in the US, *notata* shell color pattern has been selected for and become a recognized marker for hatchery produced clams.

Inbreeding and inbreeding depression

Selective breeding relies on genetic variation, or more importantly, additive genetic variation. The fact that selective breeding is highly effective in producing genetic improvement in molluscs indicates high genetic variation. Genome sequencing revealed high levels of genetic variation in most molluscs including oysters, mussels, scallops, and clams. Genetic studies in most molluscs have also revealed moderate to high heritability, a measure of additive genetic variation, for most production traits. High heritability means that genetic improvement can be easily obtained through selective breeding.

Selective breeding inevitably leads to loss of genetic variation and results in inbreeding. Once the life cycle of an organism is closed in the hatchery, inbreeding will ensue with or without intentional selection. Severe inbreeding creates two problems for selective breeding. First, once all loci are fixed or all genetic variation is lost due to inbreeding, no further genetic improvement can be made by selection. Rapid inbreeding will limit response to selection over time. Second, severe inbreeding leads to decreases in fitness, a phenomenon known as inbreeding depression. Highly inbred lines of molluscs are often difficult to sustain due to reduced fertility and survival.

Inbreeding depression is well-documented in cultured shellfish species affecting growth, development, survival, and fecundity. In the Pacific oyster, significant inbreeding depression in yield, individual growth rate, and survival was observed in families with an inbreeding coefficient (F) of 0.203 (Evans et al., 2004). In the Catarina scallop *Argopecten ventricosis* (=*circularis*), selfing ($F = 0.5$) produced a 12.4% depression in larval growth and a 43.9% depression in larval survival (Ibarra et al., 1995). Inbreeding depression in growth and survival was reported in a selected line of pearl oyster with white shells (Wada and Komaru, 1994). Significant inbreeding depression in metamorphosis success has been documented in the Pacific abalone (Deng et al., 2005).

Inbreeding depression is likely caused by deleterious recessive genes. As molluscan genomes are highly heterozygous or polymorphic, they may contain high levels of deleterious recessive genes or genetic load (Launey and Hedgecock, 2001). The high polymorphism and genetic load make molluscs particularly sensitive to inbreeding. Positive correlation between heterozygosity and growth rate is commonly observed in marine bivalves, underscoring the importance of maintaining genetic diversity in selective breeding. Slow or controlled inbreeding coupled with purifying selection may purge deleterious recessive alleles over time and reduce the negative effects of inbreeding.

Managing inbreeding

Family-based selection

Controlling or managing inbreeding is one of the most important tasks for a breeding program. It is particularly important for highly heterozygous molluscs with high genetic load that are sensitive to rapid inbreeding. Inbreeding can be managed through family-based selection where pedigree information can be used to avoid mating between close relatives. Family-based selection is also advantageous for traits that have low heritability or cannot be measured on the individuals being selected such as adductor muscle size in scallops. Family-based selection, however, requires the production and evaluation of large numbers (often hundreds) of families, which are expensive and challenging for most research programs and commercial hatcheries. Some molluscs cannot be easily tagged and pooled for field evaluation. Culturing families in separate tanks and bags creates family-specific environments that may confound genetic and environmental effects on phenotype.

Pooling families to remove confounding environmental effects is crucial for family-based selection. If animals cannot be reliably labeled, genetic markers such as microsatellites and single-nucleotide polymorphisms (SNP) may be used to assign pooled individuals into families. Individuals from mass spawns can be genotyped and assigned to their parents, which restores family structure and enables family-based selection. Genetic markers can also be used in mass selection to avoid and monitor inbreeding. Genotyping with genetic markers is expensive; the cost may be reduced by developing microsatellite multiplexes, SNP arrays, and newer genotyping techniques such as genotyping-by-sequencing. Genetic markers and marker-assisted breeding are valuable tools for genetic improvement of molluscan shellfish.

Equalized factorial cross

Mass or individual selection, which is simple, effective and widely used for shellfish breeding, is prone to inbreeding and must be carefully managed. Inbreeding in mass selection can be controlled in several ways including: (1) using large number of breeders; (2) balancing sex ratio and parental contribution; (3) maintaining multiple lines; and (4) crossbreeding between lines.

In a randomly mated population, the rate of inbreeding (ΔF) is determined by the effective population size (N_e), so that $\Delta F = 1/2N_e$. Effective population size equals census population size under equal sex ratio and parental contribution. When sex ratio deviates from 1 : 1, N_e is determined by $4N_fN_m/(N_f + N_m)$, and $\Delta F = 1/8N_f + 1/8\ N_m$, where N_f is the number of females and N_m is the number of males. Under unbalanced sex ratio, the rate of inbreeding is mostly determined by the underrepresented sex. Using a small number of males, which is a common tendency in shellfish hatcheries, can lead to reduced N_e and rapid inbreeding. Therefore, it is important to use the same number of males and females, even though one or few males can fertilize all eggs. The use of 10 females and 10 males produces 2.5% inbreeding per generation if all 20 parents contribute equally to the next generation, which is an unrealistic assumption. The use of at least 25 females and 25 males, which theoretically produces 1% of inbreeding per generation, is recommended for most molluscan species.

The rate of inbreeding predicted by N_e assumes random mating, equal contribution of gametes and no selection, which are often violated in shellfish hatcheries. Selection occurs in all breeding programs by design, which creates unequal contribution to the next generation and accelerates inbreeding. For example, the largest 10% of oysters may come from one or a few families, and selecting them can lead to rapid inbreeding. Fecundity is exceptionally high and variable in molluscs. Adult Pacific oysters may produce 10 to 150 million eggs per female, and eastern oysters may produce 1 to 80 million eggs per female. To maintain genetic diversity, it is important to ensure random mating and equal contribution of gametes from all parents. In species such as oysters where artificial insemination is possible, mating is preferably done using equalized factorial crosses where eggs from all females are equalized, pooled and then divided into equal aliquots, with each aliquot being fertilized by one male (Fig. 14.2). This way all males have an equal chance to mate with all equally contributing females. Equalized factorial crossing is not possible for species where controlled fertilization is difficult or impossible. For hermaphroditic bay scallops, production hatcheries usually use thousands of scallops for mass spawning, where inbreeding may be quite low.

The use of large numbers of breeders and equal parental contributions are important for maintaining genetic diversity in mass selection. Even when large numbers of breeders with equal contributions are used, intensive selection can lead to rapid inbreeding. It is crucial that inbreeding is closely monitored with genetic markers and by looking for signs of inbreeding depression, such as reduced fecundity, larval survival, and growth. When inbreeding depression is apparent, corrective measures should be taken.

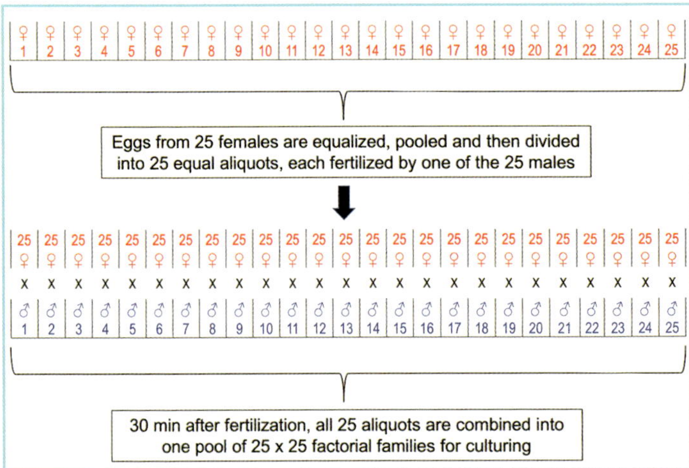

Figure 14.2 Equalized factorial crossing used for oyster breeding for maximizing genetic combinations and balancing parental contribution. The use of 25 females and 25 males produces 625 factorial crosses or families.

Crossbreeding

Inbreeding can be corrected by introducing new genetic material and crossbreeding between inbred lines. Crossbreeding between inbred or pure lines can restore genetic diversity, produce heterosis, and enhance uniformity of culture stocks. Heterosis has been observed between inbred lines of oysters and scallops. Crossbreeding is widely used in crop improvement. In crop breeding, pure lines are developed through in-line crossing and selection over many generations, and subsequently tested for general and specific combining abilities. Hybrids are produced by crossing selected pure lines in different schemes including single (A×B), three-way ((A×B)×C) and double ((A×B)×(C×D) crosses. Single crossing has been used for molluscan breeding and production. Commercial hatcheries often cross their own stocks with selected lines from other sources. Three-way and double crosses can combine genetic materials or characters from multiple lines as well as restore vigor and fecundity of F1 hybrids/broodstock, although they require more lines and an additional generation to produce F2 hybrids for production. Double crossing of four lines of open-pollinated crops are expensive and not widely practiced. Cross fertilization in dioecious molluscs is easy, and double-cross hybrids may be useful for breeding molluscan species, where time to sexual maturation is short and the F1 hybrids can be easily maintained. Overall, in molluscs, the number of pure lines is few, and there has been little effort in developing and systematically testing multiple pure lines for combining abilities. The development (with selection) and evaluation of large numbers of molluscan lines are essential for producing the best performing hybrids through single, three-way and double crossbreeding.

Progressive rotational crossbreeding

Rotational crossing of pure lines is commonly used for genetic improvement of livestock, but its application in shellfish breeding has been limited. Pure or inbred lines are easy to produce but difficult to maintain in molluscs, as molluscs are mostly wild, highly polymorphic, and sensitive to inbreeding. Highly inbred lines often fail, due to low fecundity and poor survival, within a

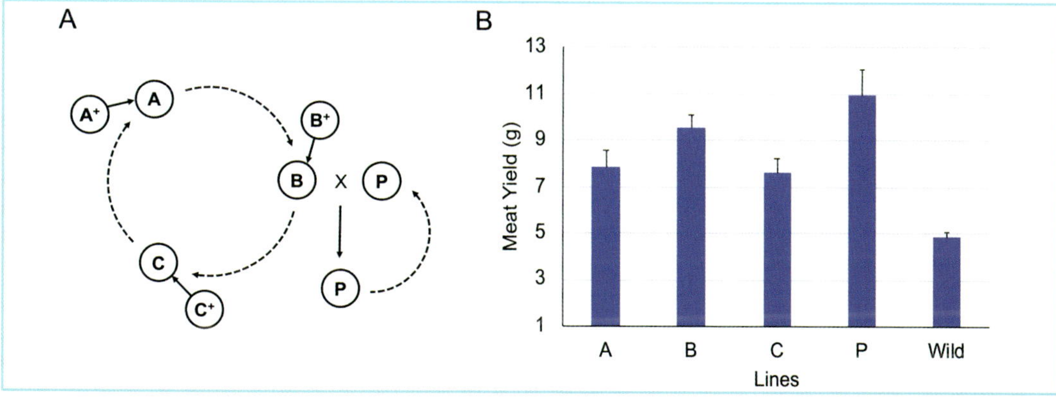

Figure 14.3 Genetic improvement of eastern oysters by progressive rotational crossbreeding (PRC). (A) PRC with three core lines (A, B, and C) derived from different sets of parents from the wild population, progressively improved with strong selection and crossed in rotation with the production line (P). The three core lines are independent and never crossed with each other, each being supported by a bridging line (A+, B+ or C+) that is produced by progressive outcrossing, in-crossing and backcrossing between the core line and the source population. (B) Improvement in meat yield of the production (P) line compared with the three core lines (A, B, and C) and a wild control in the eastern oyster.

few generations of closed breeding. Thus, rotational crossing of pure or highly inbred lines may not be practical in molluscan breeding.

A modified scheme of progressive rotational crossbreeding (PRC) is used for eastern oyster breeding at Rutgers University. Since PRC does not rely on pure or highly inbred lines, new lines are established from wild populations, and progressively improved through strong mass selection and crossed in rotation with a production line to maintain genetic diversity and heterosis (Fig. 14.3A). A minimum of three independent core lines are established using different sets of individuals from the wild population. The independent core lines are subjected to strong mass selection and progressively improved. They are kept separate, never crossed with each other, but crossed in rotation with a production line, which is heterozygous and used for production. With three core lines, inbreeding in the production line will stabilize at 14% after six generations of rotational crossing and with four core lines, inbreeding will stabilize at 7% after four generations (Table 14.2). It means that 86% and 93% of the heterozygosity are persistently retained in the production line after six and four generations of rotational crossing with three and four core lines, respectively. Genetic diversity and heterosis are maintained in the production line, regardless inbreeding levels in the core lines.

To prevent the collapse of the core lines from inbreeding, a bridging line is produced by crossing each core line with the source population and used to bring in new genetic material (Fig. 14.3A). The F1 hybrids of the bridging line are in-crossed to expose both favorable and deleterious recessive alleles in F2 hybrids to purifying selection. After selection, F2 hybrids are backcrossed to the core line, and the backcross is in-crossed (BC × BC) again to expose deleterious recessive alleles to selection. Alternate backcrossing and in-crossing are continued with strong selection. When inbreeding depression is apparent in the core lines and the performance of the bridging and core lines are comparable, the bridging lines are crossed into their perspective core

Table 14.2 Genetic composition and inbreeding coefficient of the production (P) line when mated with three or four core lines in progressive rotational crossbreeding.

Generation	Cross	A fraction	B fraction	C fraction	D fraction	Inbreeding
With 3 lines: A/B/C						
1	P1=A×B	0.500	0.500	0.000		0.000
2	P2=P1×C	0.250	0.250	0.500		0.000
3	P3=P2×A	0.625	0.125	0.250		0.250
4	P4=P3×B	0.313	0.563	0.125		0.125
5	P5=P4×C	0.156	0.281	0.563		0.125
6	P6=P5×A	0.578	0.141	0.281		0.156
7	P7=P6×B	0.289	0.570	0.141		0.141
8	P8=P7×C	0.145	0.285	0.570		0.141
9	P9=P8×A	0.572	0.143	0.285		0.145
10	P10=P9×B	0.286	0.571	0.143		0.143
11	P11=P10×C	0.143	0.286	0.571		0.143
12	P12=P11×A	0.572	0.143	0.286		0.143
With 4 lines: A/B/C/D						
1	P1=A×B	0.500	0.500	0.000	0.000	0.000
2	P2=P1×C	0.250	0.250	0.500	0.000	0.000
3	P3=P2×D	0.125	0.125	0.250	0.500	0.000
4	P4=P3×A	0.563	0.063	0.125	0.250	0.125
5	P5=P4×B	0.281	0.531	0.063	0.125	0.063
6	P6=P5×C	0.141	0.266	0.531	0.063	0.063
7	P7=P6×D	0.070	0.133	0.266	0.531	0.063
8	P8=P7×A	0.535	0.066	0.133	0.266	0.070
9	P9=P8×B	0.268	0.533	0.066	0.133	0.066
10	P10=P9×C	0.134	0.267	0.533	0.066	0.066
11	P11=P10×D	0.067	0.133	0.267	0.533	0.066
12	P12=P11×A	0.533	0.067	0.133	0.267	0.067

lines to increase genetic diversity. In PRC, selection within core lines utilizes additive genetic variation, while rotational crossing restores heterozygosity and creates heterosis in the production line by capturing non-additive genetic variation.

Progressive rotational crossbreeding has led to improvements of eastern oysters where the production line shows improved yield than the core lines and unselected controls (Fig. 14.3B). The PRC scheme requires only eight to ten lines and can be easily implemented in research and commercial hatcheries.

Hybridization

Hybridization is an effective tool for genetic improvement. Hybridization may provide two benefits: heterosis and the combination of desirable traits from different stocks or species. As discussed above, intraspecific hybridization or crossbreeding between inbred lines, which restores genetic diversity and produces heterosis and uniformity, is extremely valuable and widely used for genetic improvement. Intraspecific hybridization between non-inbred, cultured or wild populations may also produce genetic improvements and are commonly used in shellfish breeding (Table 14.1). In the Pacific abalone, hybridization between Chinese and Japanese populations produced significant improvement in survival and growth. The use of the intraspecific hybrids virtually saved the abalone aquaculture industry from devastating disease and morality in China.

Although less common, interspecific hybridization can also result in genetic improvements. Interspecific hybridization has contributed significantly to finfish aquaculture. Many successful hybrids have been developed and used for finfish aquaculture including hybrid tilapia, hybrid carps, hybrid catfish, and hybrid striped bass. Interspecific hybrids have been produced and tested in many cultured molluscs, but few have provided genetic improvements and contributed to aquaculture production. Interspecific hybrids between the weathervane scallop (*Patinopecten caurinus*) and the Japanese scallop (*P. yessoensis*) showed fast growth and strong resistance to disease or mortalities during grow-out. The hybrids have been cultured in British Columbia, Canada. Hybrids between scallops *Argopecten irradians* and *A. purpuratus* demonstrated heterosis in growth. While large-scale production of the hybrids is difficult due to hermaphroditism in the two *Argopecten* scallops, the hybrids are used for genetic improvement of bay scallops.

Interspecific hybridization between abalone species is exceptionally fruitful. Hybrids between the green-lip (*Haliotis laevigata*) and black-lip (*H. rubra*) abalone are cultured in Australia. The hybrids grow faster and do not aggregate or escape in culture, a trait that is desired by farmers. Hybrids between *H. discus hannai* and *H. fulgens* exhibited heterosis in growth, survival, and high temperature tolerance. The hybrids are used for commercial production in China. Hybrids between *H. rufescens* and three other sister species *H. sorenseni*, *H. fulgens* and *H. corrugata* showed improvements in growth and survival (Leighton and Lewis, 1982).

The success of interspecific hybridization depends on the divergence of the two species involved. Hybridization may occur naturally between some closely related species such as *Mytilus* mussels and between oysters *Crassostrea gigas* and *C. angulata*. Between distantly related species, pre- and post-gametic barriers may prevent hybridization by affecting fertilization, early development, growth, and gametogenesis. Hybrids between the eastern and Pacific oysters, two species that are separated by about 83 million years, are not viable. Fertilization and development to D-stage are apparently normal, but hybrid larvae cannot survive beyond 120 µm. Hybrids between distant molluscan species may be less fit, sterile, or partially sterile.

Ploidy manipulation

Triploids

Molluscan shellfish can be improved by ploidy manipulation or more specifically through the production of triploids. Triploids are organisms with three sets of chromosomes. Because of the extra set of chromosomes, triploid molluscs exhibit several characteristics that are desired for aquaculture including sterility, fast growth, and improved meat quality. Consequently, triploids have become an important part of the oyster aquaculture industry. Triploids do not contain any foreign genes and are not considered as genetically engineered (GE) organisms. Many agricultural crops such as rice, wheat, potato, banana, grapes, strawberry, and seedless melons are natural or induced polyploids.

Triploids have been produced and studied in dozens of molluscan species. In most species studied so far, triploids grew significantly faster than diploids (Guo et al., 2009). Triploid eastern oysters grew 32–50% faster and survived 26–38% better than diploids in Delaware Bay. The best performing group was the hybrid triploids between diploid females from Flowers strain and tetraploid males from Rutgers NEH strain, which in body weight yielded 108% more than the diploid hybrids and 61% more than pure NEH triploids (Guo et al., 2008a). The hybrid triploids performed better than pure NEH triploids probably due to increased heterozygosity and heterosis. In the Chesapeake Bay, triploid eastern oysters had lower mortality than diploids, greater shell height, whole weight, and yield, and a higher proportion of market size oysters than diploids (Degremont et al., 2012).

Several hypotheses have been proposed attributing the superior growth of triploid molluscs to sterility, increased heterozygosity, and/or increased cell size. All three factors may contribute to the superior growth of triploids. Sterility or partial sterility results in reduced gonadal development and increased energy for somatic growth in triploids. Sterility may contribute to fast growth of triploids after sexual maturation and reproduction, but it cannot explain why triploids grow faster than diploids during larval and juvenile stages. Triploid and tetraploid larvae are significantly larger than diploids, which may be explained by increased cell size. Large polyploid cells may lead to large organs and organisms if the cell number is not reduced under mosaic development in molluscs. Increased heterozygosity is also an important contributor to the increased growth of triploids. A positive correlation between heterozygosity and growth has been observed in triploid oysters (Fig. 14.4). The fact that hybrid triploids grow faster than pure line triploids supports the heterozygosity hypothesis. Triploid oysters and scallops have significantly enlarged adductor muscle compared with diploids, which support the cell size hypothesis. Adductor muscle is the primary product of scallop aquaculture. Enlarged adductor muscle would make triploids ideal stocks for scallop culture, although commercial scale production of triploid scallops has not been possible so far.

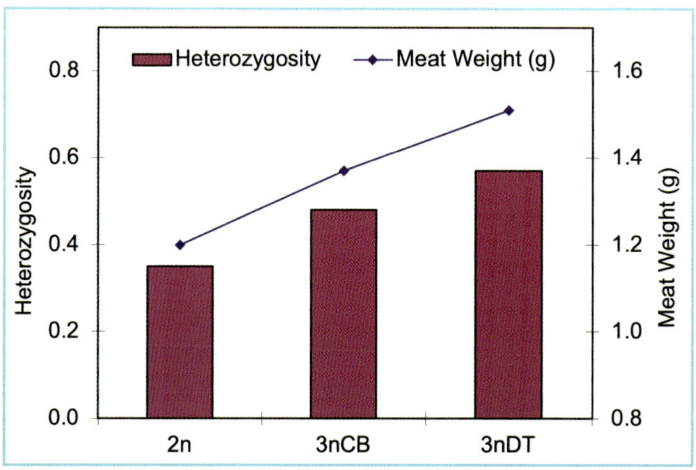

Figure 14.4 Correlation between heterozygosity and meat weight (three replicates combined) of normal diploid (2n), triploids produced by inhibiting meiosis II with CB (3nCB) and triploids produced from diploid x tetraploid mating (3nDT) in the Pacific oyster (Wang et al., 2002).

Another benefit of triploid oysters is improved meat quality during the spawning season. Most bivalve molluscs allocate a large portion of their body weight to gonad production. Sexual maturation and spawning in diploids often result in poor meat quality. The issue is especially important for oysters that are consumed raw, where mature or spawned out oysters cannot be marketed during summer months. Because triploids have limited gonadal development, their meat quality is not affected by sexual maturation and spawning. Triploid oysters provide a high-quality product that can be marketed year round. Improved summer meat quality is the primary reason why triploid Pacific oysters are widely cultured.

Triploid molluscs may have improved health or survival during reproductive seasons because they do not produce much gonad and deplete their energy reserve, although abnormal or unexplained mortality of triploid oysters have also been observed in some areas. Whether triploids are more resistant to specific pathogens is uncertain. Triploid eastern oysters are equally susceptible to Dermo (Barber and Mann, 1991), but less susceptible to MSX than diploids (Matthiessen and Davis, 1992). Fast growth in triploid eastern oysters shortens times to market and reduces their exposure and losses to Dermo and MSX diseases, which cause heavy mortality around 2 years of age as oysters reach market size. Triploid eastern oysters had lower ROD-caused mortality than diploids (Guo et al., 2008a). Triploid Sydney rock oysters survive two times better than normal diploids under infection by the parasite, *Mikrocytos roughleyi* (Hand et al., 1998). Triploid Pacific oysters showed about the same level of resistance to OsHV-1 and summer mortality as diploids (Degremont et al., 2015). Triploid eastern oysters cultured in Chesapeake Bays and the Gulf of Mexico experienced high and unexplained mortalities. Such mortalities have not been observed in Delaware Bay and the northeastern US.

The use of sterile triploids in shellfish culture benefits the environment. Selected stocks are genetically different from the wild population and may cause genetic pollution through escape and interbreeding. Although triploids are not completely sterile in some molluscs such as oysters, the chance for triploids to reproduce is greatly reduced. As more selectively bred hatchery stocks are used in aquaculture, the use of sterile triploids may help to reduce potential impact of selected stocks on wild populations. This is particularly important for non-native shellfish species that are introduced for aquaculture. Many molluscan shellfish are introduced and cultured as non-native species across the globe including the Pacific oyster, Portuguese oyster, European flat oyster, Kumamoto oyster, bay scallop, Japanese scallop, sea scallop, Peruvian scallop, Manila clam, hard clam, and abalones. While non-native species are important for shellfish culture, their introductions carry certain risks that must be carefully considered and managed.

Tetraploids

Triploid shellfish can be produced by inhibiting the release of polar body 2 with heat shock, cytochalasin B (CB), or 6-dimthylaminopurine (6DMAP). The two chemicals are toxic, and the success of polar body inhibition is highly variable. Inconsistency of polar body inhibition is a major problem for commercial production of induced triploids. The best way to produce triploids is through diploid × tetraploid mating, which is 100% effective and does not involve any treatment. The mated triploids have higher levels of heterozygosity than chemically induced triploids, and heterozygosity is positively correlated with fast growth in triploids (Fig. 14.4).

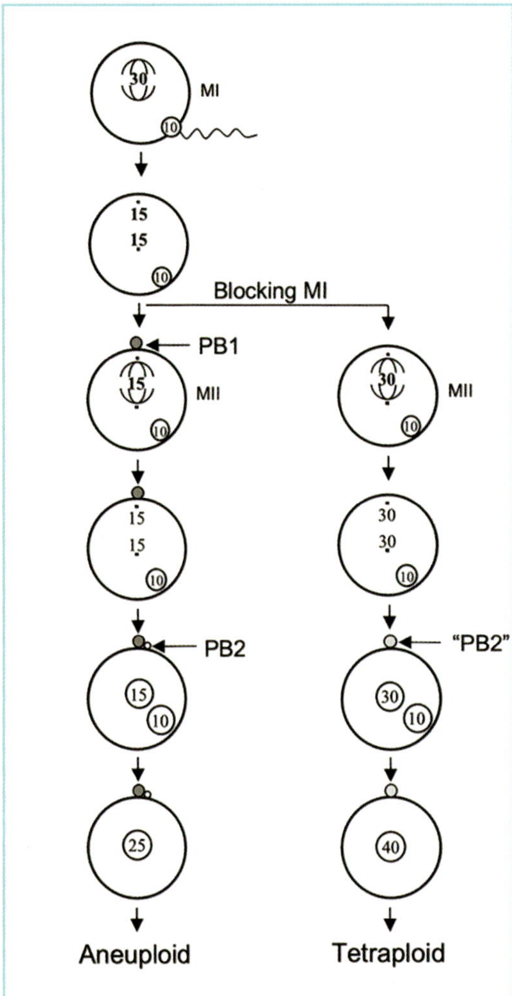

Figure 14.5 Production of viable tetraploid by inhibiting polar body I in eggs from triploids fertilized with normal haploid sperm (Guo, 1991).

Viable tetraploids were first developed in the Pacific oyster in 1993 by blocking the release of polar body I in eggs from triploid females fertilized by normal haploid sperm (Fig. 14.5). While triploid oysters are functionally sterile and have greatly reduced gonadal development, some triploids produce small numbers of eggs and sperm, which without genetic intervention are aneuploid and not viable. Inhibiting the first polar body in eggs from triploids restores triploidy, which become tetraploids when combined with the one set of chromosomes from normal haploid sperm (Fig. 14.5).

The successful production of tetraploids using eggs from triploids may be attributed to the large eggs from triploids. Tetraploid molluscs can be induced with eggs from diploids by polar body I inhibition, gynogenesis, mitosis I inhibition, and cell fusion, but the viability of tetraploids produced from eggs of diploids is extremely low, possibly because the limited cytoplasm in the eggs cannot support the large tetraploid nucleus. Eggs from triploids are about 50% larger than eggs from diploids, providing more cytoplasm for the large tetraploid nucleus. Another possibility is that the surviving triploids have a genome that has already adapted to polyploid conditions, that is, the genetic makeup of the triploid female gives higher survival of tetraploids. Nevertheless, small numbers of tetraploids have been produced with eggs from diploids in several species, but the low survival of tetraploids has made establishing breeding populations of tetraploids difficult. For tetraploid induction with eggs from triploids, it is important to obtain high quality eggs with synchronized development. Treatments (CB, 6DMAP or heat shock) should be optimized to precisely target the first polar body without overly impacting other developmental processes.

Tetraploid oysters are fertile and produce diploid gametes. Diploid × tetraploid crosses produced 100% triploids, which grew significantly faster than diploids and induced triploids (Guo et al., 1996). In practice, mated triploids are produced by crossing diploid females and tetraploid males. Hatcheries obtain tetraploid males or sperm from breeding centers for crossing with females from their own stock. The hybrid triploids often exhibit heterosis in growth and inherit characteristics from both diploid and tetraploid parents. Tetraploid oysters take longer to condition and have lower fecundity than diploids. Sperm of tetraploid oysters can be collected 'dry' (without the addition of seawater) and kept at 4–6°C for 2 to 3 days. They can be shipped overnight to remote hatcheries for triploid production. Sperm of tetraploids have a large head and swim slowly compared with normal haploid sperm. It is important that enough sperm is used and given sufficient time (30 minutes) for fertilization. It is critical that tetraploidy of the males is individually confirmed by flow cytometry, and eggs from diploids are not contaminated by haploid sperm from diploid males or hermaphrodites. Triploid larvae are larger and may set at a larger size than diploids.

The development of tetraploid oysters has led to rapid commercialization of mated triploids. Triploid Pacific oysters have become an important part of oyster farming in many countries including the US, EU, Australia, China, Chile, and Korea. Mated triploids now account for 30–70% of Pacific oysters cultured in the US, France, Australia, and China. Tetraploid eastern oysters were developed in 2001 and quickly commercialized for aquaculture production in the US, now accounting for 30–40% of hatchery production. Tetraploid Portuguese oysters have been developed and used for commercial production in China. Tetraploid Suminoe (*Crassostrea ariakensis*) oysters have been developed but are not used for commercial production due to the lack of an aquaculture industry. Small numbers of tetraploids have been produced by inhibiting polar body I in normally fertilized eggs in several molluscs including blue mussel, pearl oyster, tropical oysters, Chinese scallop, bay scallop, Catarina scallop, Manila clam, and dwarf surfclam, but no breeding populations of tetraploids have been successfully established for commercial production.

Improvement of triploids and tetraploids

As triploids and tetraploids have become a significant part of the oyster farming industry, their improvement is important. Triploid oysters generally show improved performance, although their performance may vary depending on the environment and quality of the tetraploid stocks used. Triploid oysters may suffer unexplained mortality in certain environments as in the case of triploid eastern oysters cultured in Chesapeake Bay and the Gulf of Mexico. Triploids are mostly sterile and cannot be directly improved. They can be improved by improving the performance of their diploid and tetraploid parents.

The genome of triploids and tetraploids are unstable. Triploid oysters can lose chromosomes and revert to diploids and 2n/3n mosaics, although the frequency of reversion is low or typically 3–10%. The genome of tetraploid oysters is seriously unstable. In the eastern oyster, tetraploids may quickly revert to aneuploids (4n– and 3n+), triploids and mosaics that contain tetraploid, triploid and/or aneuploid cells (Fig. 14.6). The rate of reversion varies among cohorts and environments. In some cohorts, over 50% of tetraploids may loss chromosomes and become 3n/4n mosaics in two to three years (Table 14.3). The reversion primarily occurs in somatic cells particularly in hemocytes, mantle, and gill. Germline cells are mostly unaffected, and most 3n/4n mosaics may be used for triploid production. There is no clear evidence that reversion in somatic cells affect the growth and survival of tetraploids. The viability of mated tetraploids is still lower than triploids and diploids. As the tetraploid father contributes two of the three sets of chromosomes of triploids, it has more influence on the performance of the triploid progeny than the diploid mother. Therefore, genetic improvement of tetraploids is critical for improving triploid performance.

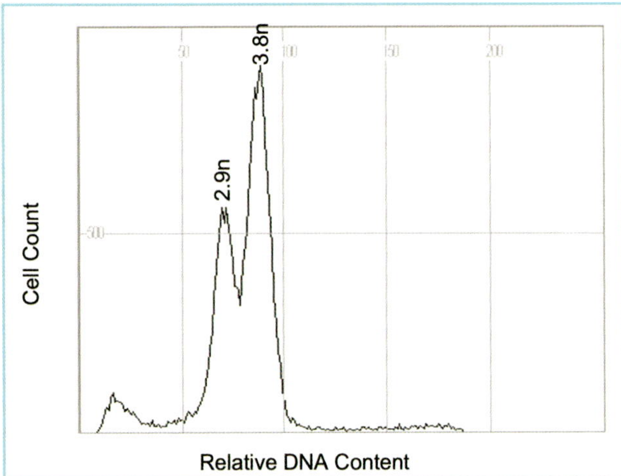

Figure 14.6 Flow cytometry analysis of chromosome loss in gill of tetraploid eastern oysters showing two populations of cells at 3.8n and 2.9n.

Tetraploid and triploid eastern oysters can be improved through systematic interploidy breeding (SIB). In SIB, the first-generation or *de novo* tetraploids are produced with the best performing triploids and diploids (largest 10–20% of survivors) (Fig. 14.5). The best performing tetraploids are selected to produce mated tetraploids by 4N × 4N crosses. Another set of the best performing tetraploids are mated with the best performing diploids to produce mated triploids (Fig. 14.7A). Like the breeding of diploids, triploids and tetraploids are subjected to strong selection, and only the best performing individuals are selected as parents for the next generation. The best performing triploid females (largest 10–20% of survivors) are selected again to mate with the best diploid males from a different core line to produce new or *de novo* tetraploids. Concurrently, mated tetraploids are produced by 4N × 4N crosses and subjected to strong mass selection (largest 10–20% of survivors). After every three generations of selection, *de novo* and mated tetraploids are crossed to produce the next generation of tetraploids, completing one round of SIB (Fig. 14.7A). Each round of SIB involves crossing with diploids from two different lines, bringing two sets of chromosomes into *de novo* tetraploids, which in turn brings one set of chromosomes into the tetraploid line by crossing with mated tetraploids.

Systematic interplay breeding has significantly improved the performance of triploid and tetraploid eastern oysters. The first generation of induced tetraploids were small and experienced heavy unexplained mortality. The survival and growth of tetraploids were greatly improved after SIB. Triploids and tetraploids produced in 2016 showed significant improvement in growth or

Table 14.3 Ploidy composition of two tetraploid eastern oyster populations over time.

Group/age	N	4n	4n/3n	4n/3n/2n	3.2–3.4n	3n	3n/2n	2n
4N06A								
6 months	50	100.0	0.0	0.0	0.0	0.0	0.0	0.0
24 months	47	29.8	46.8	2.1	2.1	12.8	4.3	2.1
4N06B								
6 months	50	100.0	0.0	0.0	0.0	0.0	0.0	0.0
24 months	50	56.0	28.0	0.0	6.0	6.0	2.0	2.0

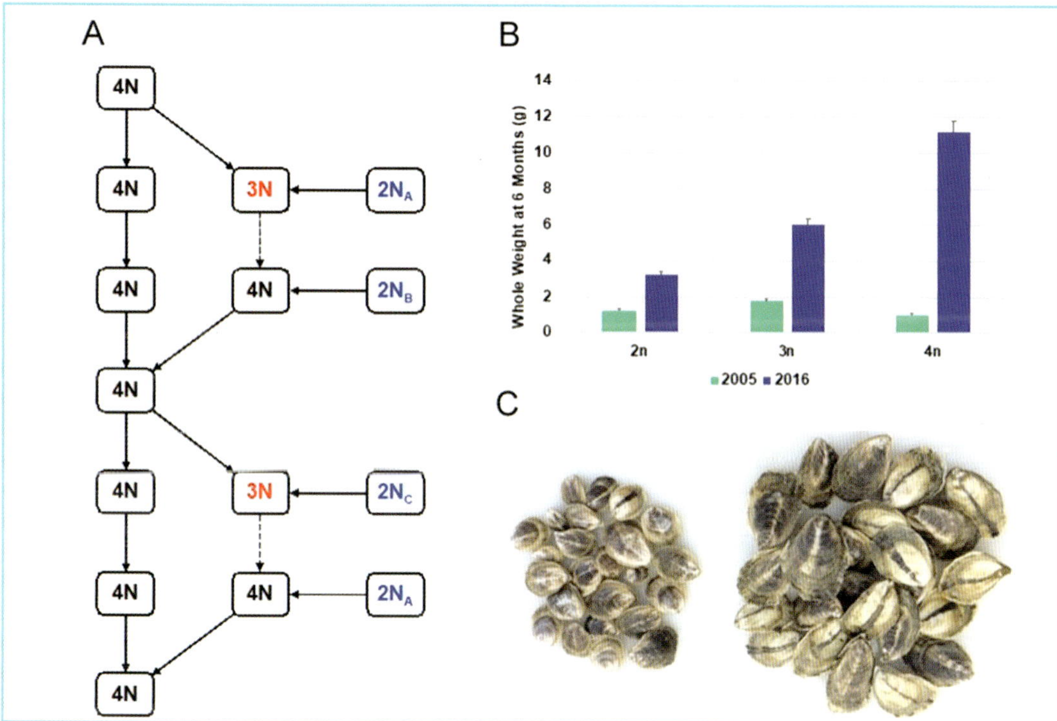

Figure 14.7 Systematic interploidy breeding (SIB) of diploid (2N), triploid (3N) and tetraploid (4N) eastern oysters. (A) Schematic presentation of SIB integrating selection and genetic improvement from 2N, 3N, and 4N phases. Dashed lines represent the production of *de novo* tetraploids by inhibiting polar body I in eggs of triploids fertilized by sperm from diploids. Each SIB cycle brings two sets of chromosomes from different diploid lines into *de novo* tetraploids through a triploid phase, where genes supporting superior triploids are enriched and brought into tetraploids. (B) Body weight of diploid, triploid and tetraploid eastern oysters before (2005) and after (2016) SIB. (C) Diploid (left) and tetraploid (right) oysters of the same age.

body size. At 6 months of age, the 2005 year-class tetraploids were about the same size as diploids, while the 2016 year-class tetraploids after SIB were 250% heavier than diploids (Fig. 14.7B, C). The 2005 triploids were 50% heavier than diploids, while the 2016 triploids were 88% heavier than diploids. The 2016 triploids were 233% heavier than the 2005 triploids. By selecting the best performing diploids, triploids, and tetraploids, SIB brings selection and genetic gains from diploid, triploid and tetraploid phases together in new tetraploids, which in turn produce better performing triploids. By introducing new genetic materials from two diploid lines every three generations, SIB maintains genetic diversity in the tetraploids and sustains long-term breeding.

Genomic selection and gene editing

Recent advances in molluscan genetics and genomics have opened the door to more sophisticated approaches to genetic improvement. Genetic markers and maps have been developed for all major cultured species. They have been used to identify quantitative trait loci (QTL) related to production traits. QTL for growth and disease resistance have been identified in oysters providing insights into the genetics of these traits and candidate markers for selection.

RNA sequencing has been extensively used to study molluscan development, reproduction and transcriptomic responses to infections and environmental stress. Transcriptomic studies have greatly advanced our understanding of the molecular mechanisms of growth, immune and stress responses, shell formation, and sex determination. Large numbers of genes related to production traits have been identified, and variations in these genes provide candidate markers for marker-assisted selection. Marker-assisted selection is effective for traits controlled by few major-effect genes, such as disease resistance. It is relatively ineffective for polygenic traits that are controlled by many small-effect genes.

Whole genome sequences have been obtained for over 40 molluscan species. The availability of whole genome sequences and high-throughput genotyping platforms has paved the way for genomic selection. Genomic selection requires high-density genetic markers covering the entire genome. Genomic selection does not require prior knowledge of the function of genes and markers involved. It uses all markers and marker effects on the target trait in the training population to predict the genomic estimated breeding values (GEBV) for individuals of the breeding population based on their genotypes. It removes the need for phenotyping, which can be expensive or problematic, for traits such as disease resistance. Genomic selection has been shown to be more efficient than marker-assisted or phenotype-based selection. Genomic selection is being tested in several molluscan species including the Pacific and eastern oysters. High-throughput genotyping of SNP markers is conducted with SNP arrays or genotyping-by-sequencing approaches. As genotyping cost declines, genomic selection may become a powerful tool for molluscan breeding.

High-throughput genotyping is usually performed by professional service providers. Commercial hatcheries can potentially biopsy individuals and send samples out for genotyping and genomic selection. Genomic selection requires pedigree information, which is often obtained from family-based breeding programs. Genomic selection may also be applied to mass selection programs with pedigree information reconstructed from genotypes.

Transgenics is widely used for genetic improvement of agricultural crops. Transgenic fish have been produced in several species, but only transgenic salmon has been approved for human consumption in the US. While producing transgenic molluscs is technically possible, it has not been used in shellfish aquaculture because of concerns for GE organisms. Molluscan shellfish are broadcasting spawners mostly cultured in open waters, making the containment of transgenics challenging. Transgenic shellfish should not be used for aquaculture without community support and careful review by regulatory agencies.

Gene editing is a new technology that permits the modification of genes already in the species. It does not involve the introduction of new genes from another species. Gene editing creates changes in genes that also occur naturally. Thus, gene editing may pose less of a concern and faces less regulation than transgenics. Still, community acceptance should be considered, and careful risk assessment should be conducted before gene-edited shellfish are used for aquaculture production.

Conclusion

Genetic improvement of cultured stocks is essential for the sustainable development of shellfish culture. Genetic improvement has contributed to shellfish culture through the development of disease-resistant and fast-growing varieties, triploids and hybrids, although progress to date is insufficient in meeting the need of the shellfish culture industry. As cultured molluscs are either wild or at early stages of domestication, intensified efforts are needed to establish more breeding programs and to develop large numbers of selected or domesticated lines. Closing the life cycle in hatchery production inevitably leads to genetic changes and inbreeding. It is important for hatcheries to implement sound breeding protocols and proactively manage the genetic health of their stocks.

Molluscan shellfish are highly polymorphic and rich in deleterious recessive genes, which makes them particularly sensitive to inbreeding. Therefore, maintaining genetic diversity is critical in molluscan breeding, and crossbreeding among selected strains, inbred lines, and divergent populations is expected to generate heterosis and be a powerful breeding strategy. Crossbreeding in molluscs does not necessarily require pure or highly inbred lines, which are difficult to maintain. Progressive rotational crossbreeding can be used where independent core lines are established from wild populations, progressively improved through strong mass selection, kept separate but crossed in rotation with a production line that maintains genetic diversity and produces heterosis. Crossbreeding is a simple and effective strategy for shellfish breeding and when maintaining multiple lines is not possible in commercial hatcheries, crossbreeding of the local stock with selected stains from a different source can be fruitful. Interspecific hybridization may produce genetic improvement between some species but hybrids between distantly related species may be inviable, less fit, or sterile. Triploids and tetraploids have become an important part of the oyster culture industry and can be effectively improved through SIB where genetic gains from the diploid, triploid and tetraploid phases are brought together. Even for triploid production, crossbreeding between diploids and tetraploids from different strains produces heterosis and further improvements. Genomic selection, which can be implemented in commercial hatcheries in collaboration with geneticists and genotyping providers, may greatly improve selection efficiency and benefit shellfish culture.

References cited and further reading

Allen, Jr., S.K. (1993) Development of high survival resistant lines in oysters using MSX-resistant strains. *Journal of Shellfish Research*, 12(1): 103.

Barber, B.J. and Mann, R. (1991) Sterile triploid *Crassostrea virginica* (Gmelin, 1791) grow faster than diploids but are equally susceptible to *Perkinsus marinus*. *Journal of Shellfish Research*, 10(2): 445–450.

Barber, B.J., Davis, C.V., and Crosby, M.A. (1998) Cultured oysters, *Crassostrea virginica*, genetically selected for fast growth in the Damariscotta River, Maine, are resistant to mortality caused by Juvenile Oyster Disease (JOD). *Journal of Shellfish Research*, 17(4): 1171–1176.

Calvo, L.M.R., Calvo, G.W., and Burreson, E.M. (2003). Dual disease resistance in a selectively bred eastern oyster, *Crassostrea virginica*, strain tested in Chesapeake Bay. *Aquaculture*, 220: 69–87.

De Melo, C.M.R., Durland, E., and Langdon, C. (2016) Improvements in desirable traits of the Pacific oyster, *Crassostrea gigas*, as a result of five generations of selection on the West Coast, USA. *Aquaculture*, 460: 105–115.

Degremont, L., Garcia, C., Frank-Lawale, A., and Allen, Jr., S.K. (2012) Triploid oysters in the Chesapeake Bay: comparison of diploid and triploid *Crassostrea virginica*. *Journal of Shellfish Research*, 31(1): 21–31.

Degremont, L., Nourry, M., and Maurouard, E. (2015) Mass selection for survival and resistance to OsHV-1 infection in *Crassostrea gigas* spat in field conditions: response to selection after four generations. *Aquaculture*, 446: 111–121.

Deng, Y., Liu, X., Zhang, G., and Guo, X. (2005) Inbreeding depression and maternal effects on early performance of Pacific abalone. *North American Journal of Aquaculture*, 67: 231–236.

Dove, M.C., Nell, J.A., and O'Connor, W.A. (2013) Evaluation of the progeny of the fourth-generation Sydney rock oyster *Saccostrea glomerata* (Gould, 1850) breeding lines for resistance to QX disease (*Marteilia sydneyi*) and winter mortality (*Bonamia roughleyi*). *Aquaculture Research*, 44: 1791–1800.

Elston, R.A., Kent, M.L., and Wilkinson, M.T. (1987) Resistance of *Ostrea edulis* to *Bonamia ostreae* infection. *Aquaculture*, 64: 237–242.

Evans, F., Matson, S., Brake, J., and Langdon, C. (2004) The effects of inbreeding on performance traits of adult Pacific oyster (*Crassostrea gigas*). *Aquaculture*, 230: 89–98.

Farley, C., Lewis, E., Relyea, D., Zahtila, J., and Rivara, G. (1998) Resistance studies for juvenile oyster disease (JOD) 1997: some early insights into genetic aspects. *Journal of Shellfish Research*, 17: 352–353.

Ford, S.E. and Haskin, H.H. (1987) Infection and mortality patterns in strains of oysters *Crassostrea virginica* selected for resistance to the parasite *Haplosporidium nelsoni* (MSX). *Journal of Parasitology*, 73: 268–376.

Gomez-Leon, J., Villamil, L., Salger, S.A., Sallum, R.H., Remacha-Trivino, A., Leavitt, D.F., and Gomez-Chiarri, M. (2008) Survival of eastern oysters *Crassostrea virginica* from three lines following experimental challenge with bacterial pathogens. *Diseases of Aquatic Organisms*, 79: 95–105.

Guo, X. (1991) Studies on tetraploid induction in the Pacific oyster, *Crassostrea gigas* (Thunberg). PhD dissertation, University of Washington.

Guo, X., Debrosse, G., and Allen, S. (1996) All-triploid Pacific oysters (*Crassostrea gigas* Thunberg) produced by mating tetraploids and diploids. *Aquaculture*, 142: 149–161.

Guo, X., Ford, S., Debrosse, G., Smolowitz, R., and Sunila, I. (2003) Breeding and evaluation of eastern oyster strains selected for MSX, Dermo and JOD resistance. *Journal of Shellfish Research*, 22(1): 353–354.

Guo, X., Wang, Y., Debrosse, G.A., Bushek, D., and Ford, S. (2008a) Building a superior oyster for aquaculture. *Jersey Shoreline*, 25: 7–9.

Guo, X., Wang, Y., Wang, L., and Lee, J.-H. (2008b) Oysters. In: Kocher, T.D. and Kole, C (eds.) *Genome Mapping and Genomics in Fishes and Aquatic Animals*. Springer, Berlin, pp. 163–175.

Guo, X., Wang, Y., Xu, Z., and Yang, H. (2009) Chromosome set manipulation in shellfish. In: Burnell, G. and Allen, G. (eds.) *New Technologies in Aquaculture: Improving Production Efficiency, Quality and Environmental Management*. Woodhead Publishing, New York, pp. 165–194.

Hand, R.E., Nell, J.A., Smith, I.R., and Maguire, G.B. (1998) Studies on triploid oysters in Australia. XI. Survival of diploid and triploid Sydney rock oysters *Saccostrea commercialis* (Iredale and Roughley) through outbreaks of winter mortality caused by *Mikrocytos roughleyi* infection. *Journal of Shellfish Research*, 17, 1129–1135.

Hershberger, W.K., Perdue, J.A., and Beattie, J.H. (1984) Genetic selection and systematic breeding in Pacific oyster culture. *Aquaculture*, 39: 237–245.

Hershberger, W.K., Myers, J.M., Iwamoto, R.N., Mcauley, W.C., and Saxton, A.M. (1990) Genetic changes in the growth of coho salmon (*Oncorhynchus kisutch*) in marine net-pens, produced by ten years of selection. *Aquaculture*, 85, 187–197.

Hollenbeck, C.M. and Johnston, I.A. (2018) Genomic tools and selective breeding in molluscs. *Frontiers in Genetics*, 9: 253.

Ibarra, A.M., Cruz, P., and Romero, B.A. (1995) Effects of inbreeding on growth and survival of self-fertilized catarina scallop larvae, *Argopecten circularis. Aquaculture*, 134: 37–47.

Launey, S. and Hedgecock, D. (2001) High genetic load in the pacific oyster *Crassostrea gigas. Genetics*, 159: 255–265.

Leighton, D.L., and Lewis, C.A. (1982) Experimental hybridization in abalones. *International Journal of Invertebrate Reproduction,* 5: 273–282.

Matthiessen, G.C., and Davis, J.P. (1992) Observations on growth rate and resistance to MSX (*Haplosporidium nelsoni*) among diploid and triploid eastern oysters [*Crassostrea virginica* (Gmelin, 1791)] in New England. *Journal of Shellfish Research*, 11(2): 449–454.

Naciri-Graven, Y., Martin, A.G., Baud, J.P., Renault, T., and Gerard, A. (1998) Selecting the flat oyster *Ostrea edulis* (L.) for survival when infected with the parasite *Bonamia ostreae. Journal of Experimental Marine Biology and Ecology*, 224(1): 91–107.

Nell, J.A. and Perkins, B. (2006) Evaluation of the progeny of third generation Sydney rock oyster *Saccostrea glomerata* (Gould, 1850) breeding lines for resistance to QX disease *Martelia sydneyi* and winter mortality *Bonamia roughleyi. Aquaculture Research*, 37(7): 693–700.

Newkirk, G.F. (1980) Review of the genetics and the potential for selective breeding of commercially important bivalves. *Aquaculture*, 19: 209–228.

Okamoto, N., Tayama, T., Kawanobe, M., Fujiki, N., Yasuda, Y., and Sano, T. (1993) Resistance of a rainbow-trout strain to infectious pancreatic necrosis. *Aquaculture*, 117: 71–76.

Proestou, D.A., Vinyard, B.T., Corbett, R.J., Piesz, J., Allen, Jr., S.K., Small, J.M., Li, C., Liu, M., Debrosse, G., and Guo, X. (2016) Performance of selectively bred lines of eastern oyster, *Crassostrea virginica*, across eastern US estuaries. *Aquaculture*, 464: 17–27.

Rezk, M.A., Smitherman, R.O., Williams, J.C., Nichols, A., Kucuktas, H., and Dunham, R.A. (2003) Response to three generations of selection for increased body weight in channel catfish, *Ictalurus punctatus*, grown in earthen ponds. *Aquaculture*, 228: 69–79.

Thodesen, J., Grisdale-Helland, B., Helland, S.J., and Gjerde, B. (1999) Feed intake, growth and feed utilization of offspring from wild and selected Atlantic salmon (*Salmo salar*). *Aquaculture*, 180: 237–246.

Wada, K.T. and Komaru, A. (1994) Effect of selection for shell coloration on growth rate and mortality in the Japanese pearl oyster, *Pinctada fucata martensii. Aquaculture*, 125: 59–65.

Wang, Z., Guo, X., Allen, Jr., S.K., and Wang, R. (2002) Heterozygosity and body size in triploid Pacific oysters, *Crassostrea gigas* Thunberg, produced from meiosis II inhibition and tetraploids. *Aquaculture*, 204: 337–348.

Zhang, J., Li, Q., Xu, C., and Han, Z. (2019) Response to selection for growth in three selected strains of the Pacific oyster *Crassostrea gigas. Aquaculture*, 503: 34–39.

CHAPTER 15

Shellfish diseases

Roxanna Smolowitz

Knowledge of the diseases most commonly affecting cultured invertebrates helps to provide a basic understanding of how to prevent a disease or mitigate its effects, hopefully resulting in a healthy population of animals and a profitable crop. This chapter addresses significant diseases affecting invertebrates of commercial interest. It is not an in-depth study of diseases, but provides basic information. Further reading is recommended for more advanced understanding. The chapter is divided into sections based on phyla and classes of invertebrates, then proceeds to describe important diseases for each type.

Important general concepts concerning disease

Disease is the result of the interaction of the host (in this case invertebrates) with the environment in which the host lives and the disease-causing agent (virus, bacteria, fungus, parasite, or other agents). Infection does not always result in disease and some animals may either rid themselves of the agent or remain infected but not develop disease. Moving an animal may allow an infectious agent to encounter another strain of the same species of host resulting in the introduction of a disease in the new setting. Alternately, some agents may not cause disease in one species of animal but will cause disease in another species. If infected animals are moved to a new area with a susceptible species, a new 'disease' may appear in the unfortunate receiving community of animals. This is one of the most important reasons for preventing movement of invertebrates (and vertebrates) and must always be considered, and this applies to moving shellfish.

While there are many options for the treatment and prevention of infection and disease in vertebrates and even terrestrial invertebrates, there are essentially no treatment methods available for invertebrates cultured in open waters. Vaccines do not work on invertebrates (the white blood cell that responds to vaccines is not present in invertebrates), and use of antibiotics or other water borne chemicals is not allowed in most countries. Note that none of the diseases discussed in this chapter infect humans (unless indicated otherwise).

The term 'disease tolerant' has been used, especially in communities of bivalve molluscs. The term disease tolerant refers to animals that are infected and have a disease, but at a low enough level that it does not impair the functioning of the animal. The term disease resistant is used interchangeably with disease tolerant. But, disease resistance refers to animals that either do not become infected or become infected but do not develop disease. Either group may, or may not, rid themselves of the infectious agent over time. In this chapter, for brevity and because it often is not known whether the animals are resistant or tolerant, the term disease resistant will be used to cover both concepts. Disease resistant animals are developed by selecting survivors of a disease outbreak and using them as broodstock for the next generation. This method hopefully increases the occurrence of genes in the population that may confer resistance in the next generation. In some cases, significant resistance is developed in only one or two generations, but in most cases it takes sequential generations of survivors used as broodstock to develop resistant strains of the invertebrate (see Chapter 14). Additionally, many hatcheries may outbreed with local stock to enhance survival and growth (which may be depressed during the development of resistant broodstock). Unfortunately, this outbreeding may decrease the 'resistance' previously developed.

Parasite life cycles

Parasites come in all forms and vary in their ability to cause disease. Some parasites are opportunistic, that is they can survive in the environment without infecting a host (in this case an aquatic invertebrate). Many parasites are obligate parasites. To survive, they must find a host for at least part of their life cycle. Most obligate parasites can be divided into two broad groups with either direct or indirect life cycles.

An example of a direct life cycle is the infection of eastern oysters by the parasite *Perkinsus marinus* (Dermo), in which infectious forms leave the diseased eastern oyster host, move through the water column and then directly infect a naïve (healthy) eastern oyster. Parasites with indirect life cycles require two or more hosts. The trematode, *Proctoeces maculates*, found in blue mussels is one that requires two hosts. The adult form of the flat worm is found in the intestine of fish (the definite or final host). Eggs enter the water column with the fish feces, are filtered and infect an intermediate host (in this case a blue mussel). Intermediate forms of the trematode proliferate in the intermediate host by cloning thus providing a way for one egg/parasite to become a thousand parasites in an intermediate host. These intermediate forms can infect the fish when it eats the mussel and proceed to form new adult worms in the fish intestine.

World Organization for Animal Health (OIE/WOAH)

OIE/WOAH is an international organization that monitors the occurrence of disease agents and the diseases they cause in all types of animals around the world (http://www.oie.int/animal-health-in-the-world/oie-listed-diseases-2018/). Most countries are voluntary members of this organization. The advantage of being a member is that information concerning infection/disease is available to all other members, thus helping countries to prevent entry of infected animals into their country. The disadvantage is that the occurrence of agents on the OIE list and diseases caused by them must be reported to OIE/WOAH by each member country when the agent is detected, thus potentially reducing the ability to sell animals/animal products to another country. The long-term benefits of belonging greatly outweigh the short-term loss of revenue.

Phylum Mollusca, class Bivalvia, superfamily Ostreoidea (oysters)

Agent: Perkinsus *spp.*

There are several species of dinoflagellate-like *Perkinsus* that infect oysters. The two most important are *Perkinsus marinus* and *P. olseni*. Both of these disease agents are on the OIE notification list.

Commonly named Dermo, *Perkinsus marinus* infects *Crassostrea virginica* (the eastern oyster) along the eastern and southern coasts of the US. It has helped to deplete populations of *C. virginica* along the coastline since 1940 when Ray first identified the organisms as the cause of significant morbidity and mortality (sickness and death). Since then the agent and associated disease has spread north along the eastern coastline of the US. Based on 2018 data, it appears that almost all populations of eastern oysters are affected along the mid-eastern, southeastern, and Gulf coasts of the US. Many eastern oyster populations are affected along the more northeastern US coastlines and the occurrence there may continue to increase as global climate changes continue.

Gross findings

Infected oysters have no distinctive abnormalities (lesions) visible upon shucking. As the disease progresses, the oyster becomes thin and watery (an appearance that can be due to other diseases as well as normal physiological cycles). The animal eventually dies of the disease.

Microscopic findings

The agent infects the oyster primarily through the stomach, but also the gill and mantle epithelium. Hemocytes (the blood cells of oysters), engulf the parasite. Normally this process would result in death of the parasite, but the parasite produces an inhibiting substance that prevents the hemocyte from destroying it and instead the organism proliferates within the hemocyte (in the cytoplasm). Two distinctive forms can easily be identified in the oyster tissues; the signet ring (meront) and rosette (schizont) (Fig. 15.1). Other intermediate forms are present also. When the oyster dies, the parasite undergoes transformation into flagellated zoospores (which become motile) and enters the water column where they can infect adjacent oysters.

Method of transfer from one animal to another

This parasite has a direct life cycle. Infectious organisms move directly from the infected oyster to the next oyster by flagellated, thus motile, zoospores produced in dying oyster tissue. But, all forms of the parasite are considered infectious when released from the degenerating oyster body into the water column. Infectious forms can also be released through the feces of living oysters providing another method for transfer of the agent from infected animals (see general references; Bushek et al., 2002).

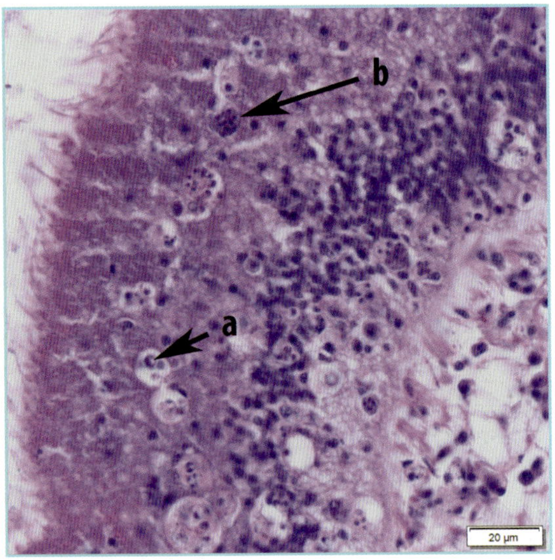

Figure 15.1 An eastern oyster infected with *Perkinsus marinus*. (a) A signet ring form and (b) a rosette (meront) containing developing merozoites (hematoxylin and eosin stained 6 µm section of paraffin embedded tissue).

Promoting environmental conditions

Optimal growth of *Perkinsus marinus* occurs in oysters held in waters with temperatures greater than 18°C (65°F) and salinities of 15 to 30. Unfortunately, while *P. marinus* does not grow in oysters held in salinities of 3, it does not die in the oysters held at that salinity (Smolowitz, 2013). Growth of the parasite in the oysters is affected but not inhibited between 3°C and 15°C.

Diagnostic methods

The Ray/Makin fluid thioglycollate method (RFTM) (see general references in Howard et al., 2004) has been used to identify infections by invertebrate pathologists for several decades. This method involves culturing pieces of the oyster for 5–7 days in the fluid media. During that time another form of the parasite, a hypnospore, is produced. The tissue can be stained with a dilute solution of iodine, which stains the capsule of the hypnospore black. Microscopic examination of the stained tissue can identify the tissues as positive or negative for the infectious organism and can provide an estimation of the degree of infection (number of stained hypospores) (Fig. 15.2). Results are usually reported as degrees of severity from 0 to 5 (with 5 being severely infected). This method is sensitive, but not specific since other *Perkinsus* species also form stainable hypnospores.

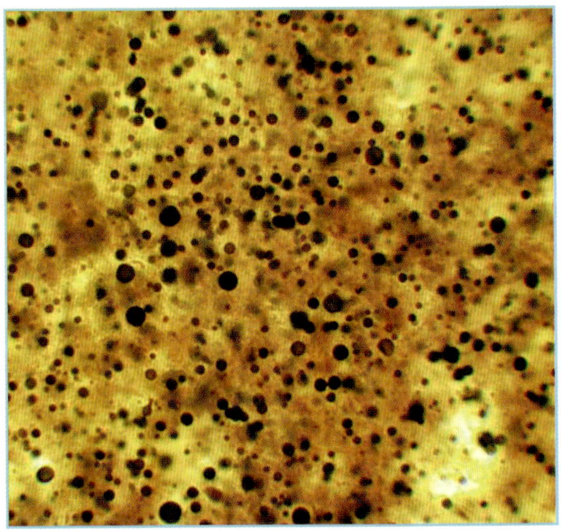

Figure 15.2 An iodine stained sample of oyster tissue cultured in Ray's thioglycollate media. Note the numerous black stained hypnospores.

Molecular methods using polymerase chain reaction (PCR) and quantitative PCR (qPCR) are used in laboratory settings. These methods are sensitive and specific. The qPCR methods can provide quantification information similar to that developed for RFTM.

Histological examination of tissues is specific but is usually not as sensitive as other tests especially in early or mild infections (few distinctive forms can be identified in the tissues).

Management/treatment methods

In the northeastern US, the highest mortality usually occurs in the fall of the year (after a summer of proliferation), but more variation in time of mortality is noted in the southern US waters.

Several universities have developed oysters that appear resistant to *Perkinsus marinus.* Seed (juvenile bivalves) produced from resistant broodstock do not develop disease. In areas with high potential for *P. marinus* infection, it may be advisable to culture resistant seed.

Other methods include health exams of seed, decreased density in cages, increased water flow, monitoring of populations for disease, spawning and raising seed at low salinities.

Agent: **Perkinsus olseni**

See below for a discussion of *P. olseni.*

Agent: **Haplosporidium** *spp.*

General information

Commonly named multinucleated sphere unknown (MSX), *Haplosporidium nelsoni* is a protozoan organism that infects eastern oysters. It was first identified and described as a cause of morbidity and mortality in the 1950s by F. Haskins in New Jersey, US. It has played a role in depleting populations of wild and cultured eastern oysters along the mid and northeastern coastlines of the US. Recently, MSX was identified in cultured eastern oysters on the east coast of Canada (Stephenson et al., 2005). This infectious agent was considered the most important cause of eastern oyster disease before the occurrence and spread of *P. marinus.* Interestingly, work by Burreson et al. (2000) has shown that MSX was probably introduced into populations of eastern oysters by the transfer of infected (but not diseased) Pacific oysters, *Crassostrea gigas,* to the east coast.

Gross findings

Infected oysters have no distinctive abnormalities (lesions) visible upon shucking. The body of moribund animals becomes thin and watery (an appearance that can be due to other diseases as well as normal physiological cycles).

Microscopic findings

Two distinctive forms of the infectious protozoan agent are identified histologically, the multinucleated cell (a plasmodium) (Fig. 15.3) and a sporangium containing spores (see general references; Bushek et al., 2002). The infectious agent (a spore) released into the water column enters into the mantle cavity where it infects through the gill epithelium. Once it has penetrated the open vascular system (blood), it grows and proliferates in the vascular channels (vessels). The

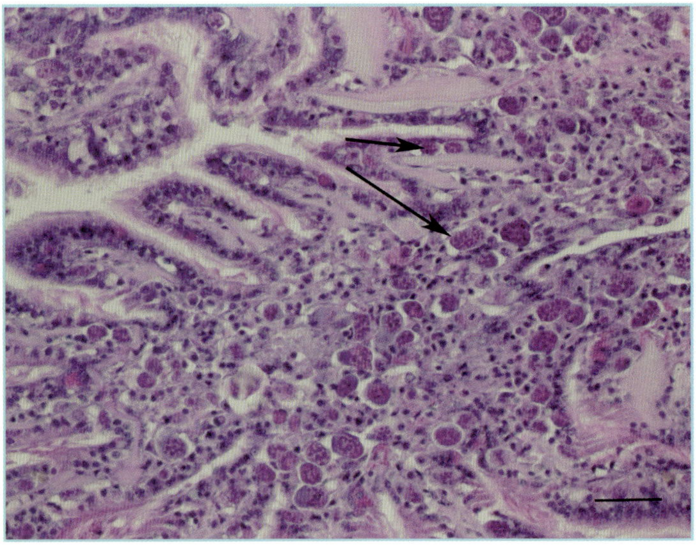

Figure 15.3 Gill tissue from a MSX-infected eastern oyster showing many MSX organisms. The arrows point to one of the MSX organisms (hematoxylin and eosin stained 6 μm section of paraffin embedded tissue, bar = 20 μm).

large, round, multinucleated cell can become abundant in the vascular channels. The hemocytes (blood cells) react to the parasite in the initial stages of infection, but as the vascular system fills with the MSX plasmodia, inflammation (and the number of hemocytes) decreases. Rarely, in some juvenile eastern oysters the second form of the parasite (a sporangium containing maturing spores), is identified in the digestive gland epithelial cells.

Method of transfer from one animal to another

The MSX parasite is considered to be indirectly infective. The origin of the infectious spore that invades the gill is unknown. Spores produced within the juvenile eastern oysters are not infectious for other eastern oysters. Importantly, the other host species has never been identified in spite of the efforts of several researchers. It is known that oyster drills can transfer the infectious form, but that the drill is not the other host in the parasitic life cycle.

Interestingly, this infectious agent and resulting disease can be identified in only a few individuals in many cultured populations in sequential years without causing noticeable morbidity and mortality in that population. In some years, however, numerous oysters in the population are infected resulting in high morbidity and mortality. It is probable that the year to year variation is a result of the numbers of the other host species in the area, and the severity of infection in the other host (both factors resulting in increased forms of the infectious agents in the water column that can then infect oysters).

Promoting environmental conditions

The protozoan grows best in oysters held in waters of greater than 18°C (65°F) and in salinities between 15 to 28. MSX organisms do not survive in oysters held at low salinities. The highest mortality occurs in the late summer and early fall in the northeast US. MSX disease of eastern oysters is less common in other southeastern areas and no mortality patterns have been identified there in the southeastern US.

Diagnostic methods

Microscopic examination of tissues is the most common method of diagnosis, but the multinucleated sphere (plasmodium) is sometimes similar in appearance to the plasmodium noted in *Haplosporidium costale* infections so histological diagnosis may not be specific. Molecular methods can be used to differentiate and, in the case of qPCR, quantify the infections. This tests are both specific and sensitive.

Management/treatment methods

Development of resistant strains of eastern oysters has been accomplished by several universities and are used as broodstock in many hatcheries. Most are resistant to both Dermo and MSX. Other methods include health examinations of seed, monitoring populations for disease, and spawning and raising seed at low salinities.

Agent: Haplosporidium costale

General information

The commonly termed seaside organism (SSO), *Haplosporidium costale*, is a protozoan and a close relative of *H. nelsoni*. It was first identified in the Delaware Bay in 1962 where it was observed to cause high mortality in eastern oysters. Since that time, it has been identified in high salinity coastal areas of the US from Maryland to Maine; however, unlike Dermo and MSX, SSO is not considered a major cause of mortality in eastern oyster populations.

Gross findings

Some sources report a 'grainy' appearance to the tissues in oysters heavily infected by this parasite; however, most infected oysters have no distinctive abnormalities (lesions) visible upon shucking. The body of moribund animals becomes thin and watery (an appearance that can be due to other diseases as well as normal physiological cycles)

Microscopic findings

As in MSX, there are two distinct forms of the infectious protozoan agent in the tissues (see general references; Bushek et al., 2002). A multinucleated plasmodia (Fig. 15.4) is the common form identified histologically during most of the year. It is similar in appearance to the plasmodium identified in MSX, but is usually slightly smaller and slightly blue in hematoxylin and eosin stained histological sections. Unfortunately, histological differentiation between SSO and MSX may not be possible. Other clues help in the diagnosis. The SSO infectious spore appears to invade through the gastric epithelium and in early infections the plasmodia may be found in the surrounding connective tissue. Plasmodia are rarely found in the tissue in the late summer to fall time period, and when they do occur, are usually accompanied by numerous hemocytes (inflammation). In the spring from March to the end of May/early June, there is an increase in the number of plasmodia and a decrease in hemocytes and inflammation. If the animal contains abundant plasmodia it becomes markedly dehabilitated during this period. At the end of May/ beginning of June the plasmodia undergo synchronous sporulation. The plasmodia develop into sporangia containing numerous spores, identified histologically. If the oyster is heavily infected, it will die during the sporulation event. If not, the hemocytes appear to destroy the sporangia

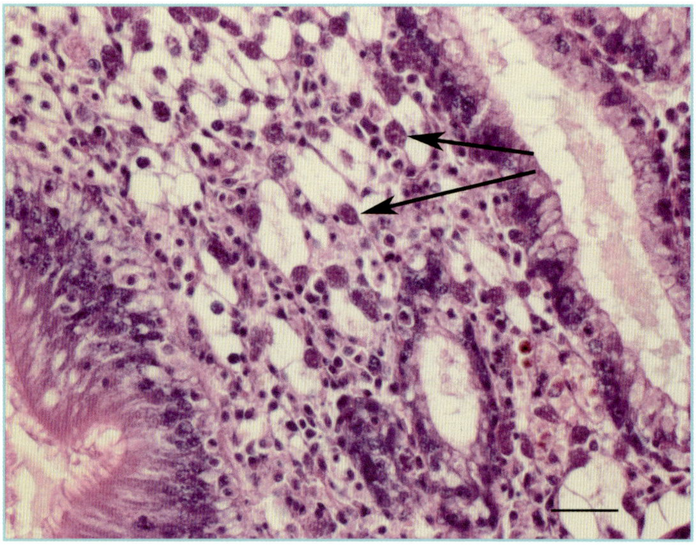

Figure 15.4 Connective tissue adjacent to gastric and digestive gland ductal epithelium of an eastern oyster. The arrows point to two of the several seaside organisms (SSO) in the tissue (hematoxylin and eosin stained 6 μm section of paraffin embedded tissue, bar = 20 μm).

and spores successfully, and the oyster, although temporarily set back in growth, will survive and continue to grow. SSO plasmodia cannot usually be found in the oyster tissues after sporulation until mid- to late summer.

Method of transfer from one animal to another

The SSO organism has an indirect life cycle. The origin of the infectious spore that invades the stomach is unknown. Spores produced by the oysters during synchronous sporulation are not infectious to other oysters. As with MSX, the other host in the indirect life cycle has never been identified in spite of the efforts of researchers. Similar to MSX, the SSO agent can be identified in only a few individuals in many cultured eastern oyster populations in sequential years without noticeable morbidity and mortality in the population; however, in some years, it can infect numerous animals resulting in moderate morbidity and mortality. Morbidity and mortality are usually never as severe as that caused by MSX and occur in the spring, not the late summer/fall as is noted in MSX. It is probable that the year to year variation is a result of the numbers of the other hosts in the area and the severity of infection in the other host (both factors resulting in increased infectious agents in the water column).

Promoting environmental conditions

Warming spring temperatures appear to promote growth of SSO in the oyster tissues. The agent is not found in oysters held in salinities of less than 25.

Diagnostic methods

Microscopic examination of tissues is the most common method of diagnosis, and as discussed previously, histological diagnosis may not be specific. Sporulation events occur only in SSO-infected oysters. Molecular methods are both specific and sensitive and, in the case of qPCR, can quantify the infections.

Management/treatment methods

The highest mortality occurs in oysters in the May to June time period, just before and during sporulation. Management methods include health exams, monitoring populations for disease, and spawning and raising seed at low salinities.

Agent: Roseovarius crassostreae *(juvenile oyster disease, JOD;* Roseovarius *oyster disease, ROD)*

General information

Classified as an alphaproteobacteria, *Roseovarius crassostreae* is a disease first detected in juvenile eastern oysters in New York and Maine in the late 1980s; however, it took several years to identify the bacterial agent (Maloy et al., 2007). Since the initial detection, the disease has been found in cultured eastern oysters in other disparate locations in Rhode Island and Maine as well as New York, and continues to be a problem at certain aquaculture locations. The disease has not been identified in wild eastern oysters.

Gross findings

Juvenile oysters less than 25 mm in shell height are the most severely affected. Over a course of 2 to 3 weeks, an infected animal shows cupping of the right valve and shorting of the left valve with mantle retraction. When shucked, the body of the animal is thin. The hallmark of infection is the easily identified buildup of a yellow-brown, moist, flaky deposit, termed conchiolin, on the inner surface of the shells (Fig. 15.5), especially the cupped valve. Mortality is high and can reach 90% of the population 6 weeks after initial infection. Juvenile oysters larger than 25 mm can become diseased, but do not exhibit the severe changes and mortality seen in the smaller oysters.

Microscopic findings

Histological examination shows severe inflammation in the mantle especially affecting the shell epithelium, the portion of the mantle that produces the nacre (inner surface) of the shell. Lesions progress from mild inflammation and sometime hyperplasia (increased cells) of the epithelium to destruction of the epithelium and underlying mantle with buildup of necrotic/dead cells and poorly formed nacre between the mantle and the shell resulting in the characteristic conchiolin deposit. Other tissues of the animal may exhibit increased hemocytic inflammation throughout.

Figure 15.5 A photograph of contrasting two shucked normal eastern oysters (N) and two shucked oysters severely infected with *Roseovarius crassostreae* (D).

Method of transfer from one animal to another

Little is known about this bacteria, it is not known where it originated or where it is found in the environment. It appears to be an opportunistic agent since locations that have had outbreaks of ROD tend to have repeated incidences of the disease over several years. The bacteria appear to reside within sediment or other animals in those locations. Research is needed.

Promoting environmental conditions

The disease occurs when water temperatures reach 22°C (69°F) and salinities are greater than 25.

Diagnostic methods

This disease can be diagnosed by observing the external changes in conformation of the shell of the juvenile oyster and finding the characteristic conchiolin deposit upon shucking the oyster. Diagnosis can be confirmed by histological examination and by culturing the bacteria, coupled with molecular identification of the cultured bacteria.

Management/treatment methods

Major methods of disease prevention have been developed. Early spawning is used to allow juvenile oysters to reach a size greater than 25 mm before the spring/summer water temperatures reach the critical 22°C. Oysters that survive the disease pass down marked resistance to the first generation of offspring. Several universities have developed oysters with combined resistance to *P. marinus*, MSX, and ROD.

Agent: Bonamia *spp.*

Both *Bonamia ostreae* and *B. exitiosa* are very small protozoans. While they are the two most important species of *Bonamia*, other infectious *Bonamia* species are found in bivalves throughout the world. Both *B. ostreae* and *B. exitiosa* are on the OIE list of reportable disease agents.

Bonamia ostreae

General information

This protozoan infects the European oyster, *Ostrea edulis*, the Suminoe oyster, *Crassostrea ariakensis*, the Pacific oyster, *C. gigas*, and some other oyster species. Disease caused by *B. ostreae* (microcell disease) was identified in the 1960s in California but ultrastructural and molecular identification occurred in the early 1980s. Data strongly support distribution of this parasite to other locations (Maine, Washington, and Europe) and is due to movement of infected *O. edulis* from California in the late 1970s. Significant morbidity and mortality of *O. edulis* in many locations in Europe are caused by *B. ostreae* (see general references in Bower, 2010).

Gross findings

Upon shucking, the oyster tissue may present a yellow discoloration or may exhibit ulcers of the gills and mantle; however, tissues of most animals only appear thin and watery.

Microscopic findings

There is little consensus of how this small, 2–5 μm protozoan enters the body, but it is taken up by and proliferates within the hemocytes. Upon lysis of the hemocytes, several new protozoans are released to infect new hemocytes eventually causing marked depletion of hemocytes and slow degeneration of the body tissues.

Method of transfer from one animal to another

The protozoan is directly infective from one oyster to the next.

Promoting environmental conditions

Highest prevalence of infection and resulting disease occurs during the springtime in most locations in Europe, but the agent is detected year-round in infected populations.

Diagnostic methods

Gross findings can be helpful. Cytological examination of stained hemolymph from imprints of the heart on microscopic slides provide a quick diagnostic method. Histological examination is considered moderately sensitive. Molecular detection methods are very useful.

Management/treatment methods

Translocation of infected oysters should be prevented. Other methods include reducing oyster density, raft culturing (off bottom culturing) and marketing early before the disease has debilitated the oysters.

Bonamia exitiosa

General information

Otherwise known as 'hemocyte disease' of fished oysters, *Bonamia exitiosa* is primarily found in New Zealand in the dredge oyster, *Ostrea chilensis*. Recent work, however, shows it is in many locations around the world and in several oyster species (Hill-Spanik et al., 2015). In New Zealand it caused heavy mortalities between 1974 and 1992. Since closing the fishery there, the dredge oysters have repopulated, although the oysters are still infected by the parasite. The consensus is that reduction of environmental stress (extreme temperature, high salinity, and starvation that occurred due to disruption of the sediment during that time period) was the cause of the outbreak in New Zealand. It was hypothesized that development of resistance in the wild population has helped increase the resistance of *O. chilensis* to the infectious agent. This may indicate that *B. exitiosa* can infect other oyster species and not cause disease.

Gross findings

No gross lesions have been reported.

Microscopic findings

Microscopically, findings are similar to *B. ostreae* but *B. exitiosa* organisms are slightly bigger (4–4.5 μm).

Method of transfer from one animal to another

This organism is directly infective.

Promoting environmental conditions

Destruction of the benthic habitat by heavy dredging has been hypothesized as a disease promoter. In the laboratory, temperature extremes, high salinity, and starvation have promoted the disease.

Diagnostic methods

The same methods are used as for *B. ostreae*.

Management/treatment methods

Highest prevalence was noted in April (autumn in New Zealand). Using lighter dredges, decreasing density of oysters, and prohibiting translocation helped to control infection and disease.

Agent: Oyster herpes virus

General information

A specific type of herpes virus found only in oysters (OsHV) has been detected in several locations in the world since the late 1990s (Mineur et al., 2015) including the west coast of the US (especially California), Spain, New Zealand, Japan, the UK, France, Italy and many other locations. It is primarily a disease of *Crassostrea gigas*, the Pacific oyster. The Pacific oyster is not indigenous to many of these locations but is being used extensively in aquaculture. OsHV has been identified in other species of oysters, but is not usually considered a cause of morbidity and mortality in other species. There are two basic groupings of pathogenic OsHV: OsHV-1 and OsHV-1 µVar. Both variants cause morbidity and mortality, but OsHV-1 µVar is considered more pathogenic (better able to cause disease). OsHV-1 µVar, especially, can cause acute severe morbidity and mortality primarily in juvenile oysters but can cause mortality in any age animal.

Gross findings

Animals show no visible lesions except poor growth and relatively quick death.

Microscopic findings

Histological examination of infected cultured Pacific oysters in Europe (primarily infected with the OsHV-1 µVar) show sparse hemocytic inflammation and intranuclear inclusions (Cowdry type A) in various epithelial and connective tissue cells; however, OsHV-1 – infected Pacific oysters from California do not show inclusions, although they do show hemocytic inflammation and tissue destruction.

Method of transfer from one animal to another

The virus is directly infective. Adults may be infected with the virus, but not show disease. Even juveniles and larvae have been identified as infected, but not diseased, in some studies.

Promoting environmental conditions

Temperatures above 21°C (70°F) appear to promote the development of disease, in juveniles especially, but in older Pacific oysters also.

Diagnostic methods

Histological examination of oyster tissues can be diagnostic if inclusion bodies are noted in the cells. Molecular identification is very important in identifying the variant of OsHV and in determining the abundance of the virus in animals in the hatchery and grow-out areas.

Management/treatment methods

Controlling translocation of animals, especially seed, prevents spread. Also important is preventing the mixture of potentially infected adults with larvae or juveniles, reducing oyster density and culturing oysters in cooler waters. Development of resistant Pacific oysters is underway.

Agent: Marteilia refringens

General information

The protozoan parasite, *Marteilia refringens,* of the European oyster (*Ostrea edulis*) is on the OIE list of reportable infectious disease agents (see general references in Bower, 2010). It continues to cause extensive morbidity and mortality in cultured European oysters along the western coast of Europe from the UK to Spain and has been identified in cultured European oysters from Greece and Morocco as well. The protozoan infects the digestive gland inhibiting digestion and absorption of nutrients, resulting in the slow death of the infected animal. The disease causes morbidity and mortality in the Calico scallop (*Argopecten gibbus*) in the coastal waters of Florida, US.

Gross findings

The digestive gland, normally dark brown/black, is reported to be discolored (pale), the animals are thin and watery with mantle retraction. Reduced growth is usually described before the disease becomes apparent.

Microscopic findings

Histological examination of the digestive gland epithelium shows various stages of proliferation of the parasite in the epithelial cells resulting in cell enlargement (hypertrophy), inability to function in digestion and, finally, gland-wide sporulation that results in digestive gland epithelial necrosis. Refringent granules are formed in the epithelial cells as part of the proliferative process.

Method of transfer from one animal to another

Indirect infection is considered the most likely method. A copepod (*Acartia grani*) has been identified as a probable intermediate host for *O. edulis* in French shallow water ponds.

Promoting environmental conditions

Water temperatures greater than 17°C (63°F) have been associated with sporulation and transmission. Salinities above 35 appear to inhibit development of the parasite in the digestive gland epithelium.

Diagnostic methods

Microscopic examination of squashes of digestive gland or stained imprints of the digestive gland can identify animals with generalized infections. Histological examination is more sensitive, as is the use of molecular methods.

Management/treatment methods

European oysters from infected areas should not be translocated to uninfected areas. Delaying deployment of seed until after July and August (a time of higher temperatures) will reduce exposure of the seed to the infectious agent. Culturing in deeper, colder waters is recommended, as is reducing the population density.

Agent: Polychaete worms (mud blisters)

General information

While included here with oyster diseases, because it occasionally infects the oyster tissues, polychaete worms affect almost all bivalve species. The worms produce borrows in the valves of the shells dragging in mud and sand. The bivalve continuously produces new nacre on the inner surface of the shell to prevent the worm from breaking through into the soft tissues of the body.

Gross findings

Upon shucking, brown/black tunnels ('blisters') are noted in the shell (Fig. 15.6). Very rarely in oysters (but not in other bivalves in this author's experience) an abscess develops because of a break in the nacre covering the mud blister, resulting in development of an abscess from exposure of the oyster tissues to the debris and bacteria present in the mud blister. In most instances, the abscess is adjacent to the adductor muscle and the oyster can successfully 'wall off' the inflammatory process, thus preventing further tissue necrosis and death. Occasionally the abscess wall is composed of a thin, but hard, outer surface.

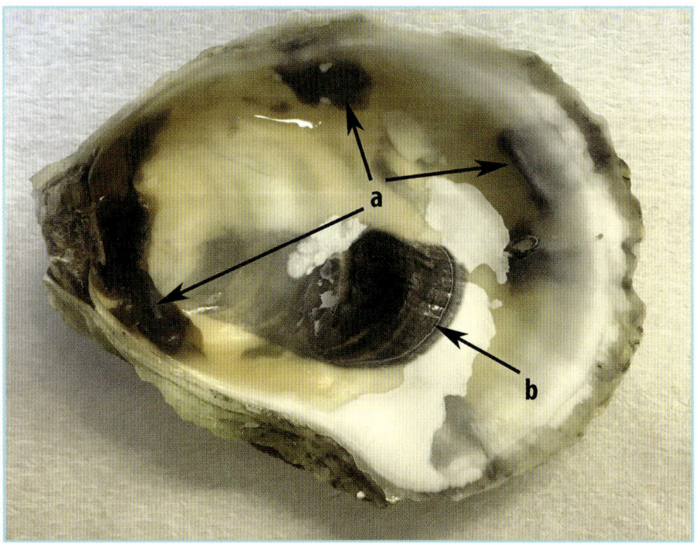

Figure 15.6 The inner surface of an eastern oyster shows numerous (a) mud blisters and (b) a black muscle scar.

Microscopic findings

Hemocytic abscesses are identified in oysters with ruptured inner shell surfaces.

Method of transfer from one animal to another

Polychaete worms are directly infective.

Promoting environmental conditions

Abundant polychaete worm infestations may result in increased numbers of mud blisters and rarely tissue abscesses in oysters.

Diagnostic methods

Gross observation of mud blisters and an abscess upon shucking the bivalve.

Management/treatment methods

Numerous methods have been tried to reduce the polychaete worm infestation of the shells. While not considered a threat to the health of the oyster, infestations do reduce the commercial worth of the oysters.

Phylum Mollusca, class Bivalvia, order Venerioda (clams and cockles)

Agent: Perkinsus olseni *(and* P. atlanticus*)*

General information

The dinoflagellate-like protozoan *Perkinsus olseni*, and other very similar species such as *P. atlanticus* (but not *P. marinus*), infect several species of clams, cockles, and a few oyster species. Only *P. olseni* is on the OIE notification list.

Infection does not always lead to morbidity and mortality, but in some species and locations, such as in the cultured grooved carpet shell clam, *Ruditapes decussatus*, in Portugal and the Manila clam, *Venerupis philippinarium*, cultured in China and Korea, the infection has produced devastating disease (see general references in Bower, 2010). In general, the infections appear to draw upon the bivalve energy resources that would otherwise go into growth and reproduction. Only under certain environmental conditions can the disease cause mortality.

Gross findings

White firm nodules are noted in the gill, or in the body of the animal. Infected animals may have reduced growth.

Microscopic findings

Infection incites a localized hemocytic inflammatory reaction forming granulomas (encapsulated nodules) of hemocytes that attempt to wall off the parasite from the surrounding tissues (producing the white nodules seen grossly). Parasitic forms infect hemocytes and are similar histologically to those noted in *Perkinsus marinus* although the parasite forms are much larger in *P. olseni* and *P. atlanticus*.

Method of transfer from one animal to another

Perkinsus olseni is directly infective.

Promoting environmental conditions

There is no consensus on the seasonality of infection and morbidity, or on promoting temperatures and salinities. Some investigators noted that temperatures above 15°C (59°F) promoted disease.

Diagnostic methods

The Ray/Makin fluid thioglycollate method (RFTM) can be used to identify *P. olseni* and other *Perkinsus* species as well as *P. marinus*; however, because the method does identify so many species, it is not specific. Histology examination can be used, but is not as sensitive as RFTM and is also not specific. Molecular methods have been developed for *P. olseni* and other species, but there is overlap in specificity using this method.

Management/treatment methods

Management methods include health exams of seed before translocation, decreasing oyster density in cages, increasing water flow and monitoring populations for disease. Prohibiting importation of bivalve broodstock from other countries is important and recommended by the OIE.

Agent: Vibrio tapetis *(brown ring disease [BRD] of Manila clams)*

General information

The bacteria *Vibrio tapetis* was identified as a cause of morbidity and mortality in cultured Manila clams, *Venerupis philippinarum*, in 1987. It has caused significant disease in that clam species along the west coast of France and on the coast of Korea. Clams show progressive debilitation before death. While the bacteria can infect other clam species (*Venerupis aurea*, *Thylacodes decussatus,* and *Dosinia exoleta*), disease has not been detected in those species.

Gross findings

A brown, moist, flakey to granular conchiolin accumulates at the palladial line and can extend to the shell edge in juvenile and adult clams. Clam growth is usually stunted.

Microscopic findings

Inflammation and necrosis of the mantle tissue overlying the circumferential ring of abnormal conchiolin is seen histologically. Abundant bacteria are associated with the conchiolin and degenerating mantle tissue in later stages of disease.

Method of transfer from one animal to another

This is a directly infective disease.

Promoting environmental conditions

Water temperatures of 14°C (39°F) or less promote the bacterial disease. Temperatures above 21°C (70°F) can result in recovery from the disease. Salinities lower than 20 may promote disease development.

Diagnostic methods

Infected animals show the brown circumferential conchiolin ring in mid to late disease stages. Culturing of the bacteria from the lesions, combined with molecular identification of the bacteria, is the most common method of positive identification. Note: brown ring-like disease occurs worldwide in some bivalves, but usually on an individual animal basis. To date, BRD has only occurred as disease outbreaks in Manila clams.

Management/treatment methods

Decreasing planting density and the use of uninfected seed have been suggested as management methods.

Agent: Quahog parasite unknown (QPX)

General information

Quahog parasite unknown, a protozoan, is a thraustochrid (phylum Labyrinthulomycota) and is closely related to the organisms that cause eel grass disease. It was first identified in 1963 in the wild northern quahogs (hard clam) (*Mercenaria mercenaria*) fished from the Gulf of St. Lawrence, New Brunswick, Canada. In 1995, it was identified as the cause of major morbidity and mortality in cultured quahogs in Wellfleet and Provincetown, Massachusetts, US. In following years, it was identified in cultured clams in other areas of Massachusetts, Virginia, New York, and Rhode Island. The most severe morbidity and mortality appear to occur in cultured quahogs in Massachusetts.

Gross findings

Sick animals often are debilitated, slightly gaping, and can be found lying on the surface of the sediment. In Massachusetts, one to several small to large tan nodules or swellings are noted at the junction of the siphon and the mantle or in areas of the membranous mantle (Fig. 15.7), a location where pseudofeces is usually stored in the clam before periodic elimination.

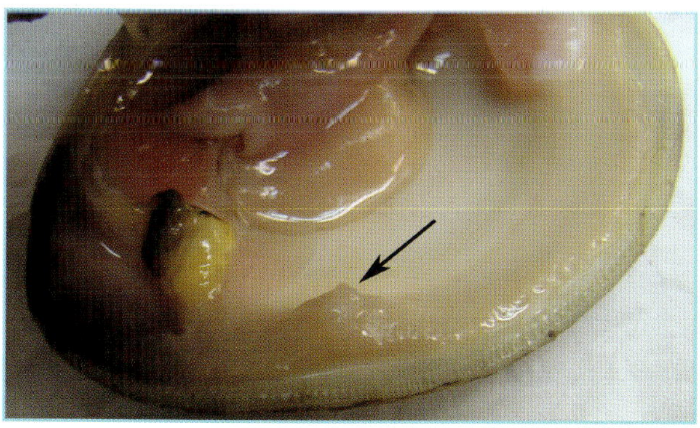

Figure 15.7 An arrow points to a QPX nodule in the mantle adjacent to the siphon of the northern clam (hard clam, *Mercenaria mercenaria*).

Microscopic findings

Nodules and swellings noted above are composed of varying numbers of QPX organisms, the mucus they produce, and hemocytes (the inflammatory reaction of the clam). In the spring and fall, nodules and swellings are predominately composed of QPX organisms and mucus. In the summer, the nodules and swellings are usually composed of lesser numbers of QPX organisms and amounts of mucus and instead show a marked hemocytic inflammatory response.

Method of transfer from one animal to another

QPX is an opportunistic parasite. QPX can survive in the environment without being associated with a host. In some circumstances, such as the yearly introduction of naïve clam seed for culture, or when there is a heavy set of wild clams in the area, the pathogenic potential of the organism may increase, resulting in a more infectious organism than is normally encountered, thus resulting in abundant disease. Interestingly, it was determined that seed from broodstock acclimated over several generations to the warmer waters of the southeastern US, but cultured in the northeast, were more likely to be infected, develop disease and die, than seed from broodstock acclimated to cooler waters of the northeast. This may indicate an inability of southern seed innate immune system to function at cooler temperatures.

Promoting environmental conditions

While the QPX organism grows best in culture at 24°C (76°F) and salinity of 32, paradoxically, it causes the highest morbidity and mortality in the spring in most areas. This mortality pattern reflects the inability of the quahog to eliminate the infectious organisms from the tissues at lower temperatures and is most likely due to a sluggish innate immune system in the clam that occurs in the spring and fall, and not the ability of QPX to grow well at that time.

Diagnostic methods

Identification of nodules along the edge of the mantle and adjacent to the siphon when shucking the clam indicates a probable QPX infection. Histological examination is accurate and specific. Quantitative PCR can also be used but may not be as accurate as histology since tissues selected for molecular exam may not have parasites present.

Management/treatment methods

Monitor populations for QPX. If possible, remove infected clams from the aquaculture site. Use locally produced seed. Decrease density of clams in the affected sites.

Agent: Neoplastic cell

General information

The disease has been called by many names including: disseminated neoplasia, disseminated sarcoma, hemocytic leukemia, and hemocytic neoplasia. This condition is considered to be a contagious tumorous disease that causes morbidity and mortality in both wild and cultured stocks of several species of bivalves (including oysters and mussels). It is especially a problem in the soft shell clam, *Mya arenaria*, the northeast Atlantic common cockle, *Cerastoderma edule*, and the Baltic clam, *Macoma balthica* (Metzger et al., 2016).

Gross findings
The animals show poor condition and white or tan color of the hemolymph in heavily neoplastic animals (Smolowitz, personal observation).

Microscopic findings
The infectious cell is a large round neoplastic cell with large round to kidney shaped nuclei and prominent nucleoli. The neoplastic cells proliferate and fill the vascular spaces of the bivalve causing blockage of vessels, loss of normal hemocytes, and probable lack of nutrient flow to the tissues.

Method of transfer from one animal to another
This disease is directly infective. It is one of only three types of contagious neoplastic diseases known in all animals throughout the world. In bivalves, the neoplastic cells are thought to originate from infected animals, travel through the water column, and invade the naïve animal (probably through the gills and mantle). The disease is most common in juvenile bivalves that are undergoing their first major reproductive event. For example, it is found in softshell clams of 30–40 mm in shell height and cockles (*Cerastoderma edule*) that are 18–23 mm in shell height. The disease is generally not identified in younger or older bivalves.

Promoting environmental conditions
The timing of highest morbidity and mortality varies between species and studies, but many researchers have identified a biannual occurrence of disease in the spring and fall of the year. Studies of neoplastic cells originating from diseased *M. arenaria* showed survival of the cells in salinities of 2 to 36 and at temperatures between 20°C and 40°C (68°F and 104°F).

Diagnostic methods
Microscopic examination of hemolymph withdrawn from the animal, with or without stains, and histological examination of tissues are commonly used.

Management/treatment methods
Prevent movement of infected seed. Do not move animals from infected areas to uninfected areas.

Phylum Mollusca, class Bivalvia, superfamily Unionacea (mussels)

Agent: Neoplastic cell

General Information
Disseminated neoplasia has been identified in several species of bivalve, but can cause high morbidity and mortality in the Pacific bay mussel, *Mytilus trossulus*, found in the northwestern coast of the US and Canada.

Gross and microscopic findings
Findings and method of transfer are similar to that previously discussed for clams and cockles.

Promoting environmental conditions

Unknown.

Diagnostic and management/treatment methods

See those described previously.

Agent: Marteilia refringens *and* M. maurini

General information

These protozoans have been identified in the blue mussel, *Mytilus edulis*, the Mediterranean mussel, *M. galloprovincialis*, and a few other mussel species along the Atlantic coast of Europe, the Persian Gulf, and Gulf of Thermaikos in northern Greece. While generally not thought to cause disease in most wild mussel populations, disease has been associated with high morbidity and mortality in cultured mussels, or when naïve mussels were moved from an area without infection, to an area in which the *Marteilia* sp. parasite was endemic. It is probable that 'tolerance' to infection develops in mussels exposed to the parasite over generations and when moving animals without tolerance into an endemic area, disease results from infection (http://www.dfo-mpo.gc.ca/science/aah-saa/diseases-maladies/mrmaurmu-eng.html).

Other characteristics are as noted previously for oysters and scallops.

Management/treatment methods

Do not move naïve mussels into infected areas.

Agent: Steinhausia mytilovum *and* Steinhausia *sp. (mussel egg disease)*

General information

A microsporidal protozoan parasite, *Steinhausia* infects maturing eggs in the gonadal tubules of *Mytilus edulis*, *M. galloprovincialis,* and *M. guyanensis* (mangrove mussel) and causes egg degeneration and necrosis. It also has been reported in other bivalves, generally at a lesser prevalence and severity. The inflammatory response can be significant at times. Reports suggest a resultant decrease in mussel fecundity and condition index occurs in infected individuals.

Gross findings

Heavy infections with marked inflammation may cause discoloration and swellings in the mussel mantles (where gonadal tubules are found).

Microscopic findings

The number of infected eggs varies from few to many. Hemocytic inflammation is associated with moderate to heavy infections resulting in degeneration of the eggs.

Method of transfer from one animal to another

Infection is thought to be direct and result occurs by consumption of infected eggs.

Promoting environmental conditions

Lower temperatures may result in higher prevalence and more severe infections.

Diagnostic methods

Histological examination is used to diagnose infections and disease.

Management/treatment methods

None.

*Agent: trematodes (*Prosorhynchus *sp. and* Proctoeces maculates*)*

General information

Many species of trematodes infect various species of bivalves, but most do not cause disease or significant mortality in the bivalve host; however, moderate to heavy infection of two trematodes, which use mussels as an intermediate host, do cause reduced mussel reproduction and mortality. Living in the gonadal tubules, *Prosorhynchus* sp. can cause parasitic castration and poor growth of the mussel. Living in the vascular system of the blue mussel, *M. edulis*, *Proctoeces maculates* causes castration by decreasing available nutrients for gonadal growth/ maturation (see general references in Bower, 2010). Both infections result in decreased condition of the mussel.

Infections caused by *P. maculates* can cause mortality without an environmental perturbation, but severe infections can result in catastrophic mortality events in mussels also (such as those that occur on a very hot day when infected mussels are exposed at low tide). Both of these trematode infections can reduce the number of wild animals in the environment. Work on *Prosorhynchus* sp. shows that infected fish inhabiting the area around cultured mussels will increase the number of infected mussels and the severity of infection over time.

Gross findings

In moderate to severe infections the mantle takes on a lumpy appearance with yellow to white bumps.

Microscopic findings

Hemocytic inflammation is usually minimal around the parasites when residing in the gonadal tubules; however, if *Prosorhynchus* sp. breaks out of the gonadal tubule into the surrounding vascular system, it incites a dramatic inflammatory response. In contract, *P. maculates* circulates throughout all parts of the vascular system of the mussel without inciting much inflammation until the end of the period of proliferation within the animal (concurrent with death of the mussel) at which time inflammation becomes common. Interestingly, if the blue mussel is lightly infected, one of the forms of *P. maculates* that can develop in the vascular system, a metacercaria, can migrate to the heart and develop into an adult trematode that produces eggs. This causes a severe pericarditis (another reason for death of the mussel). The occurrence is not common.

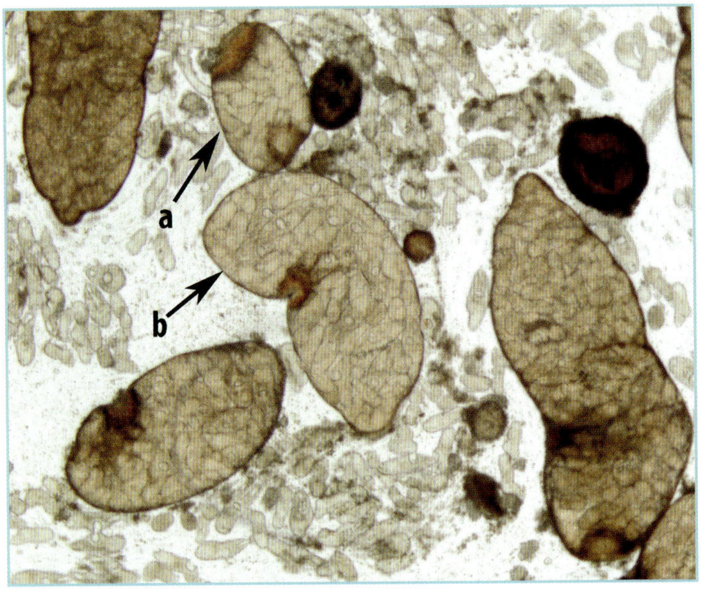

Figure 15.8 (a) A mother sporocyst and (b) daughter sporocysts as well as several small, developing sporocyts and cercaria released from a ruptured large daughter sporocyte are noted in this squashed tissue biopsy from the mantle of a *Proctoeces maculatus* infected blue mussel.

Method of transfer from one animal to another

This is an indirectly infective parasite. Feces, containing parasite eggs, from an infected fish are filtered and eaten by the mussel. Fish eat infected mussels or can be exposed to forms of the parasite in the water column that have exited from the mussel.

Promoting environmental conditions

Association with infected fish will promote mussel infection. Summer water temperatures increase the likelihood of proliferation of the parasite in the mussel tissues.

Diagnostic methods

A small biopsy of the mantle can be examined microscopically for parasite forms (Fig. 15.8). Histopathological examination is commonly used.

Management/treatment methods

Reducing fish in mussel aquaculture areas is advisable. Using uninfected mussel seed is important.

Phylum Mollusca, class Bivalvia, superfamily Pectinacea (scallops)

Agent: Mycobacteria *sp.*

General information

Sea scallops (*Placopecten magellanicus*) fished from the northwest Atlantic are affected by *Mycobacterium* sp. No mortality has been specifically identified, but examination suggests that mortality is probably a common sequelae since the adductor muscle function is affected. Additionally, no animals with healed lesions have been found to date. Adductor muscle meats

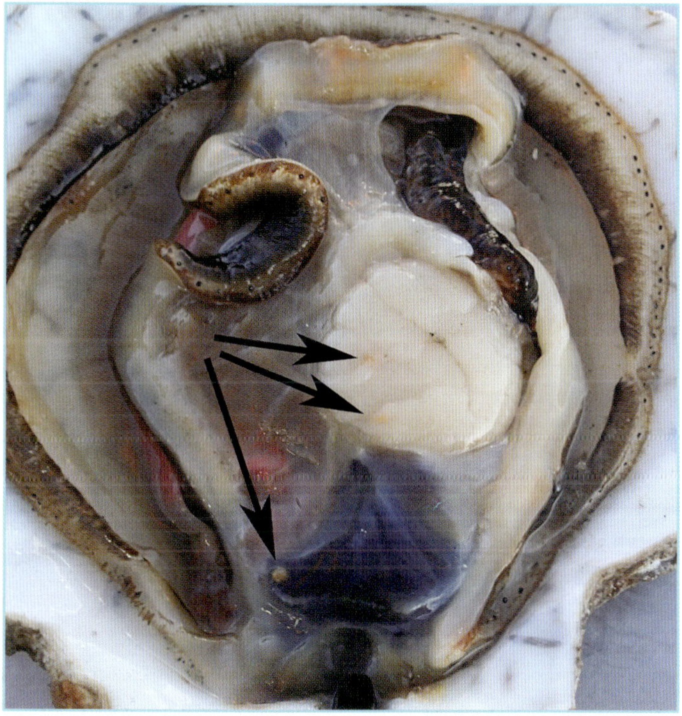

Figure 15.9 Mycobacterial infections in the adductor muscle and other tissues of *Placopecten magellanicus* (Atlantic sea scallop). Arrows point to infected foci.

are not saleable. Unlike other diseases of bivalves, the infectious agent found in scallop meats may be infectious for humans.

Gross findings

Upon shucking, pink/orange nodules of various sizes (usually 2 cm in diameter or less) (Fig. 15.9) are found in the adductor muscle and, especially in heavily diseased animals, throughout other organs.

Microscopic findings

The nodules are granulomas (encapsulations) composed of the hemocytic reaction to abundant proliferating bacteria.

Method of transfer from one animal to another

Unknown.

Promoting environmental conditions

Unknown.

Diagnostic methods

A special stain (acid fast) that identifies the bacteria in a histological section is used to differentiate *Mycobacterium* sp. from other bacteria. Molecular methods can identify the bacteria as *Mycobacteria* sp.

Management/treatment methods

None.

Agent: Perkinsus qugwadi *(previously named SPX)*

General information

The dinoflagellate-like protozoan *Perkinsus qugwadi*, infects the yesso scallop, *Mizuhopecten* (*Patinopecten*) *yessoensis*, along the coast of British Columbia, Canada (see further reading section; Chapter 10 in Shumway and Parsons, 2016 and general references therein). Very high prevalence and mortality have been noted with infection in both juveniles and 2 year old scallops. A genetically different *Perkinsus* sp. organism was reported to cause mortality in yesso scallops moved from the east coast of Russia to Japan, but no details have been provided for that disease occurrence.

Gross findings

Upon shucking, white/cream pustules up to 5 mm in diameter can be identified in the gonads and other tissues.

Microscopic findings

Findings are similar to those described for *Perkinsus olseni*; however, *P. qugwadi* produces zoospores in the tissues of the live scallops.

Method of transfer from one animal to another

This organism is directly infective.

Promoting environmental conditions

Unknown.

Diagnostic methods

The Ray/Makin fluid thioglycollate method (RFTM) can be used to identify the parasite but, as mentioned before, this method is not specific for *Perkinsus qugwadi*. Histological diagnosis is commonly used and molecular identification methods are available.

Management/treatment methods

Monitor animals and do not move infected animals.

Agent: Polychaete worms

General information

As noted previously, polychaete worms affect almost all bivalve species. Unfortunately, scallop species tend to suffer much more from polychaete shell infestations than other bivalves. The worms often burrow into the ventral valve (the valve lying on the surface of the sediment). Burrows in this position interrupt the function of opening and closing of the shell by the adductor muscle and thus affect movement of water through the animal. Infestations by various polychaete species in several species of scallops have been observed to cause reduced growth, and increased mortality. Additionally, the burrows, once vacated by the polychaete, can become a home for other organisms, even in the living scallop and can expand the original burrow.

Gross findings

Mud blisters are visible on the inner surface of the shell valves.

Microscopic findings

Poor growth of the scallop and possible poor condition of the scallop tissues may be noted.

Method of transfer from one animal to another

Polychaete worms directly infest the shell.

Promoting environmental conditions

Unknown, but scallops in some areas are more heavily affected than in other areas.

Diagnostic methods

Mud blisters can be seen upon observation of the shells when shucking.

Management/treatment methods

Numerous methods have been tried to reduce the polychaete worm infestation of the shells. No one method appears to 'cure' the bivalve of the infestation.

Agent: Merocystis kathae

General information

The protozoan, *Merocystis kathae*, infects Iceland scallops (*Chlamys islandica*), queen scallops, (*Aequipecten opercularis*), king scallops (*Pecten maximus*), and Atlantic sea scallops, (*Placopecten magellanicus*). Scallops are an intermediate host in this parasite life cycle (Kristmundsson and Freeman, 2018). The primary (definite/final) host is the common whelk, *Buccinum undatum*. Whelks predate scallops and, in doing so, acquire an infectious stage of the parasite from the scallop tissues. Whelk infections primarily occur in the kidney. Iceland scallops, queen scallops and, to a lesser extent, Atlantic sea scallops have all undergone periodic mortality events due to infections with this protozoan parasite.

Gross findings

The most obvious finding noted is a brown/grey discoloration of the scallop muscle in heavily infected animals. In the Iceland scallop, which appears to be most severely affected of the scallop species, lack of gonadal development is noted. Other tissues can be heavily infected and in poor condition.

Microscopic findings

Forms of the parasite termed sporoblasts and sporozoites are found in the intestinal tract, the muscles and other tissues of the Icelandic scallop, causing destruction of tissues and the grossly visible muscle discoloration.

Method of transfer from one animal to another

The parasite has an indirect life cycle. The infectious form of the parasite must be acquired by the scallop from the whelk. It is likely that the infectious forms are excreted into the water column with the whelk urine. The spores are filtered from the water column and invade the scallop through the digestive system.

Promoting environmental conditions

The parasite produces infective spores in the late winter and spring. Studies on Iceland scallops demonstrated the time of highest prevalence and severity of the disease in that species occurred then.

Diagnostic methods

Discolored adductor muscles (especially grey to brown coloration of the muscle) provides evidence of infection. Histological examination of tissues and molecular identification of the infectious agent are used.

Management/treatment methods

It has been suggested that fishing whelks in infected areas appears to reduce the disease in Icelandic scallops.

Phylum Mollusca, class Gastropoda, subclass Prosobranchia, family Haliotoidea (abalones)

Agent: abalone herpes-like virus (AbHV)

General information

Cultured and wild, juvenile and adult abalone, especially blacklip, *Haliotis rubra*, and greenlip, *H. laevigata*, in Tasmania and *H. diversicolor supertexta* in Taiwan become infected with this Herpes-like virus. High morbidity and mortality usually results from infection. AbHV is an OIE reportable disease.

Gross findings

The abalone foot edges curl and the animal cannot right itself. Mouth parts (radula) prolapse out of the mouth.

Microscopic findings

Nerve cells of ganglia are necrotic and show severe hemocytic inflammation (viral ganglioneuritis).

Method of transfer from one animal to another

This virus is directly infective and can move through the water column between animals.

Promoting environmental conditions

None known.

Diagnostic methods

Histological and transmission electron microscopic examination of diseased tissues and molecular methods are used to identify the disease and the viral agent.

Management/treatment methods

Do not move infected animals to an uninfected area. Experiments have not shown development of disease resistance in survivors.

Agent: Candidatus xenohaliotis californiensis *(rickettsiales-like prokaryote [RLP] withering syndrome)*

General information

The rickettsial-like bacteria, *Candidatus xenohaliotis californiensis*, can cause disease in abalone. The disease cause by RLP was first described along the coast of Central and Southern California in 2000 in both black abalone (*Haliotis cracherodii*), which is more severely affected, and red abalone (*H. rufescens*). Unfortunately, translocation of abalone from California is suspected to have resulted in spread of the infectious agent to several locations around the world. The rickettsia-like organism infects the digestive gland resulting in poor ability to digest and absorb food by the abalone. Experimental work has shown that it can take up to 245 days for the animal to starve and die. A similar rickettsial-like bacteria causing disease was identified in the small abalone (*H. diversicolor supertexta*) in Taiwan in 2008. Rickettsial infections are an OIE reportable disease in abalone.

Gross findings

The condition of the abalone slowly declines over time and the foot muscle atrophies. The animal becomes unable to grasp the substratum with its foot.

Microscopic findings

The digestive gland epithelium is hypertropic (enlarged) and contains abundant RLP organisms in the cytoplasm. The foot muscles especially show thinning due to starvation.

Method of transfer from one animal to another

The organism is spread directly from one animal to another.

Promoting environmental conditions

Warm waters (average of 18.5°C [65°F]) have been shown to promote RLP disease in California in red abalone, but not green. Disease did not occur in infected animals in Taiwan abalone until the water temperatures reached 30°C (86° F).

Diagnostic methods

Histological examination and molecular methods are used to diagnose the disease.

Management/treatment methods

Treatment with an antibiotic (oxytetracycline) kills the bacteria in the abalone. However, the drug is not approved for use in abalone in the US.

Agent: Vibrio *sp., a gram negative, rod shaped bacterium*

General information

Infection by several species of *Vibrio* bacteria, such as *V. carchariae*, resulting in disease and often high mortality, has been identified in cultured and wild abalone in several locations. Population of European abalone (*Haliotis tuberculta*) along the coastline of France have suffered high losses since 1997 due to *V. carchariae*. This species has also affected populations of the Japanese abalone (*Sulculus diversicolor supratexta*). Mortality caused by *Vibrio fluvialis* in abalone (*H. discus hannai*) in China was reported.

Gross findings

Occasionally, white pustules are noted on the foot but otherwise the animals are moribund with no external signs of disease.

Microscopic findings

Necrosis of tissue and edema (buildup of fluid) around vessels can be noted. Bacteria may be observed in the tissues.

Method of transfer from one animal to another

The bacteria are directly infective.

Promoting environmental conditions

Adult *Haliotis tuberculta* that are reproductively active were more likely to experience disease and mortality than juveniles. Temperatures of 17°C (63°F) appear to promote infection and disease. This is considered an emerging disease resulting from the warming waters along the coast of France and potentially other areas.

Diagnostic methods

Culture of bacteria and molecular identification are used.

Management/treatment methods

None are currently known.

Agent: Haplosporidium montiforti *and* Haplosporidium *sp.*

General information

Disease and mortality caused by a *Haplosporidium* sp., a protozoan parasite, has been documented in cultured abalone in New Zealand in paua, *Haliotis iris*, and Spain in the European abalone, *H. tuberculata*. The outbreak in Spain was attributed to juveniles that had been infected before being translocated from Ireland and France to Spain. It is proposed that the higher sustained temperatures of the Spanish waters promoted the disease.

Gross findings

The animals were lethargic, easily detached from the substratum and lacked a righting reflex.

Microscopic findings

Generalized necrosis of tissues and filling of the vascular system with parasites was identified histologically.

Diagnostic methods

Histological and molecular methods of examination of tissues are used.

Management/treatment methods

None.

Phylum Mollusca, class Cephalopoda, subclass Coleoidea (cuttlefish, octopus, squid)

Agent: Dicyemennea *sp.*

General information

Dicyemids are worm-like organisms found in the renal sac of many bottom dwelling octopi and cuttlefish. The dicyemids tend to be host specific (one species of dicyemid per species of host). They are a highly unusual organism, so unusual that they share a small phylum (Mesozoa) with another highly unusual, but unrelated, organism because neither can be related to any other organism in the world. Worm-like dicyemids proliferate within the lumen of renal sac (similar to a urinary bladder) (Furuya et al., 2004). While they usually are considered non-pathogenic (do not damage the cephalopod), in high numbers, in this author's experience, they were associated with inflammation and necrosis of the renal sac.

Gross findings

Potentially, cloudy urine may be identified.

Microscopic findings

Few to abundant worms can be seen histologically in the renal sac of the cephalopod and is sometimes associated with hemocytic inflammation and epithelial necrosis of the renal sac. Microscopic examination of a urine sample will demonstrate dicyemids in the urine.

Method of transfer from one animal to another

Larvae are released with urine from the urinary pore. The larvae move through the water column and directly infect the urinary sac of other cephalopods.

Promoting environmental conditions

None identified, but only species of bottom dwelling cephalopods are infected.

Diagnostic methods

Examination of fluid from the urinary sac and histological examination of tissues is used.

Management/treatment methods

None are currently identified.

Agent: Aggregatta spp.

General information

Six species of emeriid apicomplexan (coccidia-like) protozoan parasites, *Aggregatta* spp., have been identified in fourteen species of octopi and cuttlefish from around the world, including the commonly cultured octopus, *Octopus vulgaris*, and one species of cultured cuttlefish, *Sepia officinalis*. The parasite produces various forms in the connective tissues of the gastrointestinal tract and can infect all or parts of the tract but especially the cecum and intestine. Some reports have shown that infection is a limiting factor in growth of *O. vulgaris* in aquaculture due to poor digestion and absorption of food.

Gross findings

Part or all of the digestive tract contains white nodules (sometimes coalescing), visible on the serosa (outer surface), that often thicken the walls of the digestive tract (Fig. 15.10).

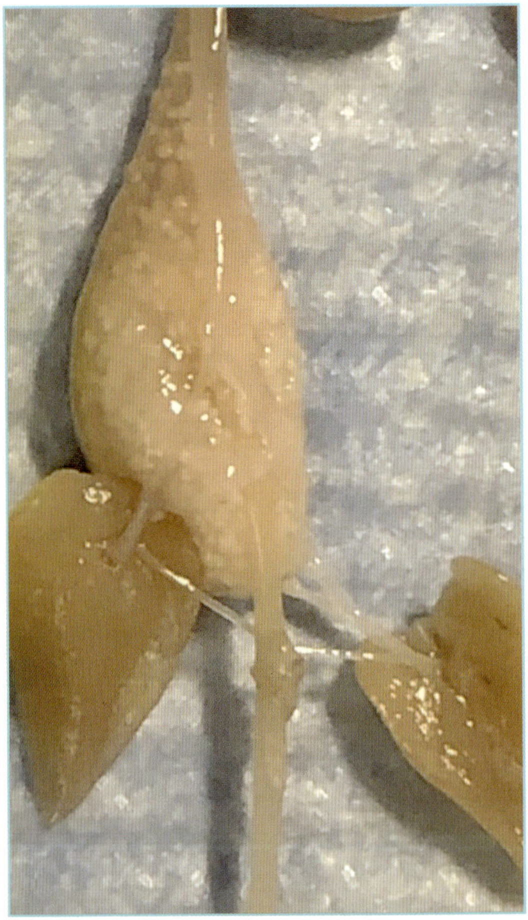

Figure 15.10 Numerous white nodules, resulting from *Aggregatta* sp. infection, are noted in the crop of an octopus (*Octopus bimaculatus*). Courtesy of Lisa Abbo, Marine Biological Laboratory, Woods Hole, MA.

Microscopic findings

Connective tissues of the cephalopod digestive tract underlying the epithelium are expanded by various forms of the parasite. Hemocytic inflammation is mild and localized in the intestine. In some severe cases, parasites can be found in the gills and muscles.

Method of transfer from one animal to another

These parasites have an indirect life cycle. Spores produced in the cuttlefish and octopi (the definite hosts) are consumed by an intermediate host (usually a crustacean) in which they proliferate. Numerous infective oocysts are produced in the crustacean, which then infect the cephalopod when the crustacean is consumed.

Promoting environmental conditions

Studies showed higher prevalence was noted in wild octopi found at deeper, cooler depths in high salinity waters. But, studies noted that such environments might result in higher predation of the intermediate host (bottom dwelling crustaceans) because of the lack of other food types, this negating direct effects of the environment. Aquaculture of octopi in floating cages in Spain has been severely hampered by this parasite.

Diagnostic methods

Necropsy and histological examination is used for identification. Molecular methods can be used for *Aggregata* speciation.

Management/treatment methods

Feed parasite-free food.

Agent: Trauma

General information

The epithelium of cuttlefish, octopi, and squid is only one cell thick with no overlying cuticle or shell. The retention of an unprotected epithelium probably occurred evolutionarily along with the occurrence of numerous interesting reflective and color producing cells in the dermis underlying the epithelium. A thin epithelium would allow the colors and patterns produced by the dermal cells to be better seen by other cephalopods and other animals; however, this thin epithelium is extremely vulnerable to trauma from many different causes and ulcers (holes through the epithelium into the underlying connective tissues) can be easily produced. In culture, three main types of traumatic lesions are noted (Smolowitz, personal observation). First, most cephalopods move in a direction opposite to the direction that they see (like backing the car up without looking in the rear view mirror), thus setting up a scenario for severely ulcerating the mantle apex and potentially cracking the cuttlebone if they back into a solid surface. Second, many of the animals, especially the mature males, are very aggressive and will fight head to head leaving deep sucker ulcers on the combatants (Fig. 15.11). Third, the use of plastic pipes or other materials with sharp/unfiled edges can result in ulcerative skin trauma. Unfortunately, cephalopods appear to have poor ability to heal ulcers or cuttlebones and secondary bacterial infections are common.

Figure 15.11 A formalin fixed cuttlefish (*Sepia officinalis*) showing (a) traumatic lesions resulting from suckers and (b) mantle apex ulceration (butt burn).

Gross findings

Mantle apex ulcers into the underlying dermis, sometimes into the pocket containing the cuttlebone of the cuttlefish or pen of the squid provides an opportunity for secondary infections (usually bacterial) and osmotic imbalance and/or broken cuttlebones (which affect the animal's ability to swim properly). Ulceration from sucker trauma can be severe.

Microscopic findings

Necrotic tissues usually show a poor inflammatory response.

Method of transfer from one animal to another

Fighting or flight from a threat can result in trauma to the epithelium.

Promoting environmental conditions

Poor culture conditions (sharp edges) and housing adult males and sometimes adult females with each other can promote trauma.

Diagnostic methods

Observation is needed to determine the cause of the trauma.

Management/treatment methods

Appropriate husbandry methods are used to prevent fighting and provide a safe, startle free environment.

Conclusions

How to detect and identify disease problems

Observance of the cultured organism is the most important activity that the culturist can do to prevent disease. Identifying when deviation occurs from the animals' normal behaviors, growth patterns, and appearance will result in early detection of a problem. If problems are identified, the culturist should contact a professional to help determine the cause. Extension agents, and similar governmental representatives (usually employed in a fisheries department), are often the first point of contact for a culturist. Invertebrate disease experts are present in laboratories throughout the world and often work closely with the extension agents and other government representatives to provide diagnostic abilities. (Note: the culturists should get to know these experts before a problem arises!) Both the extension agents and government representative can provide initial review of the problem, and will have knowledge of others in the areas with similar problems. Importantly, they can help direct the submission of animals to appropriate laboratories and sometimes can provide funds to pay for the laboratory diagnostics.

Translocation of animals and disease agents

As discussed throughout the chapter, translocation of animals from one water body to another (and sometimes even within the same water body) is to be avoided, especially as the animals become adults and are more likely to be infected. Seed purchased from any source or collected from the wild should be examined by diagnostic laboratories before movement occurs. Receiving states and countries often require health examination of seed by a state approved laboratory, although the requirements for type of infectious agent and timing of tests can vary greatly between states and countries. All of the experts noted above can help with disease determinations.

Remember, diseases are as common in molluscs as they are in all other animals. Importantly, preventing and responding quickly to disease outbreaks can often make the difference between a profitable or unprofitable culturing venture.

References cited and further reading

Bower, S.M. (2010) Synopsis of infectious diseases and parasites of commercially exploited shellfish. Last revised March 2010. Available at: http://www.dfo-mpo.gc.ca/science/aah-saa/diseases-maladies/index-eng.html.

Burreson, E.M, Stokes, N.A., and Friedman, C.S. (2000) Increased virulence in an introduced pathogen: *Haplosporidium nelsoni* (MSX) in the eastern oyster *Crassostrea virginica*. *Journal of Aquatic Animal Health*, 12: 1–8.

Bushek, D., Ford, S.E., and Chintala, M.M. (2002) Comparison of in vitro-cultured and wild type *Perkinsus marinus*. III. Fecal elimination and its role in transmission. *Diseases of Aquatic Organisms*, 51: 217–225.

Furuya, H., Ota, M., Kimura, R., and Tsuneki, K. (2004) Renal organs of cephalopods: a habitat for Dicyemids and Chromidinids. *Journal of Morphology*, 262: 629–643.

Gestal, C., Guerra, A., and Pascual, S. (2007). *Aggregata octopiana* (Protista: Apicomplexa): a dangerous pathogen during commercial *Octopus vulgaris* ongrowing. *ICES Journal of Marine Science*, 64: 1743–1748.

Hill-Spanik, K.M., McDowell, J.RR., Stokes, N.A., Reece, K.S., Burreson, E.M., and Carnegie, R.B. (2015) Phylogeographic perspective on the distribution and dispersal of a marine pathogen, the oyster parasite *Bonamia exitiosa*. *Marine Ecology Progress Series*, 536. 65–76.

Howard, D.W., Lewis, E.J., Keller, B.J., and Smith, C.S. (2004) Histological techniques for marine bivalve mollusks and crustaceans. NOAA Technical Memorandum NOS NCCOS No. 5.

Kristmundsson, A. and Freeman, M.A. (2018) Harmless sea snail parasite causes mass mortalities in numerous commercial scallop populations in the northern hemisphere. *Scientific Reports*, 8: 7865, DOI:10.1038/s41598–018–26158–1.

Maloy, A.P., Ford, S.E., Karney, R.C., and Boettcher, K.J. (2007) *Roseovarius crassostreae*, the eitiological agent of juvenile oyster disease (not to be known as Roseovarius oyster disease) in *Crassostrea virginica*. *Aquaculture*, 269: 71–83.

Metzger, M.J., Villalba, A., Carballal, M.J., Iglesias, D., Sherry, J., Reinisch, C., Muttray, A.F., Baldwin, S.A., and Goff, S.P. (2016) Widespread transmission of independent cancer lineages within multiple bivalve species. *Nature*, 534: 705–709.

Mineur, F., Provan, J. and Arnott, G. (2015) Phylogeographical analyses of shellfish viruses: inferring a geographical origin or ostreid herpesviruses OsHV-1 (Malacoherpesviridae). *Marine Biology*, 162: 181–192.

Shumway, S.E. and Parsons, G.J. (eds.) (2016) *Scallops: Biology, Ecology, Aquaculture and Fisheries*. Developments in Aquaculture and Fisheries Science, 40. Elsevier Scientific, Amsterdam.

Smolowitz, R. (2013) A review of current state of knowledge concerning *Perkinsus marinus* effects on *Crassostrea virginica* (Gmelin) (the eastern oyster). *Veterinary Pathology*, 50: 404–411.

Stephenson, M.F., Maillet, M., Veniot, V., Gagne, N., and Arsenault, J. (2005) Dynamics of MSX and SSO in *Crassostrea virginica*, American oysters in Atlantic Canada. *Journal of Shellfish Research*, 24(2): 677.

CHAPTER 16

Control of biofouling

Stephan Bullard, Sandra E. Shumway, and Alex Walsh

Biofouling has been a scourge in marine-related endeavors for centuries and is a ubiquitous problem in shellfish aquaculture. Biofouling organisms attach to natural and man-made surfaces and cover cultured shellfish and gear. Once attached, fouling organisms are extremely difficult to remove and require considerable amounts of time and effort to control. Unfortunately, aquaculture farms themselves promote the growth of biofouling in that they provide the physical structure needed for fouling attachment and cause increased nutrient levels that promote the growth of fouling macroalgal and increased phytoplankton production which is fed upon by biofouling invertebrates. In the US, over 90% of surveyed shellfish growers report that biofouling is a problem at their facility; ~30% say it is a big and constant problem (Adams et al., 2011).

Biofouling organisms can physically damage cultured shellfish, compete with shellfish for food and space, reduce water flow, induce low oxygen situations, interfere with shellfish feeding, increase the weight and drag of cultured organisms and gear, and lead to the spread of undesirable organisms and invasive species. Biofouling can also produce positive effects, such as when shellfish gain an associational defense from fouling organisms. Negative impacts, however, are far more common. Biofouling composition and abundance vary geographically and temporally and are affected by type of gear and species of shellfish being fouled. Growers can sometimes prevent fouling by deploying gear at sites with fewer fouling organisms or at times of the year when fouling organisms are not settling. Fouling can also be prevented by applying antifouling coatings to gear or by using biological control agents that consume fouling organisms. No universally effective prevention method is currently available. Once biofouling is attached, it can be removed manually (for example, by scraping or power washing), or killed by exposure to air or noxious materials (freshwater, brine, chemicals). These techniques can be highly effective, but are often time consuming, expensive, and do not always remove all fouling organisms. To manage biofouling, growers should become familiar with the biofouling assemblage at their sites and determine which mitigation efforts, if any, provide the most economical solution for their needs. Some measure of fouling prevention will probably be helpful at the majority of locations. Fouling removal may or may not be beneficial depending upon the type and degree of fouling present and the impact the fouling has on the production and marketability of product.

Figure 16.1 Examples of biofouling on shellfish gear.

Costs

Biofouling imposes significant economic losses to shellfish aquaculture worldwide. Losses come from expenses associated with fouling prevention and control, increased operating costs from the additional labor and specialized equipment needed to deal with fouled material, and reduced harvest and marketability of product. Costs associated with fouling vary geographically and between culture species and operations. In the US, efforts to control biofouling account for 14.7% of shellfish growers operating costs (Adams et al., 2011). In other areas, for example, pearl farms in Australia, mussels in New Zealand, oysters in Europe and Canada, costs frequently account for 20–30% of total operating costs. At any given site, the cost of controlling biofouling may or may not be economically beneficial. Cost–benefit analyses can be conducted to determine whether mitigation strategies are worth considering. If treatments are warranted, starting treatments early in the season when fouling organisms are small may be more cost effective than waiting until later in the season when fouling organisms have grown and become better established.

Types of fouling organisms

Biofouling occurs when organisms attach to hard surfaces in the marine environment. Algae can cause significant fouling, but sessile invertebrates are often more abundant and more problematic in aquaculture settings. Annelids, ascidians, barnacles, bivalves, bryozoans, hydroids, and sponges all contribute to fouling assemblages. These taxa possess dispersive larval stages but are sessile, or have very limited mobility, as adults. They are typically filter feeders and have very simple habitat requirements. Most only need a surface on which to attach and plankton to feed upon in order to flourish. Fouling intensity and species composition varies geographically and temporally. The specific composition of fouling organisms at each site will be unique. Species that are of minor importance at one location, can be major pests at others. This is especially true for invasive species. Some species are only present during certain seasons, while others are continually present but vary in abundance over the course of the year. The highest levels of fouling typically occur during the warmer months, although in many areas the problem exists year-round. The suite of biofouling taxa is similar worldwide and thus, most fouling mitigation techniques are effective in a wide variety of systems.

Effects of biofouling on cultured species

The impacts of biofouling on cultured shellfish can be direct (for example, shell damage) or indirect (for example, causing reduced growth due to competition for food). Both types can be significant. The exact effects that fouling will have at each site will vary depending on the species being cultured and the type of fouling present.

Biofouling can damage the shells of cultured shellfish or decrease the appearance of shellfish and reduce their aesthetic appeal. Some of the most damaging fouling organisms are those that burrow into the shells of cultured shellfish, such as boring sponges (*Cliona* spp.) and Spionid polychaetes (for example, *Polydora*, *Dipolydora*, and *Boccardia*). Boring species are not traditional biofouling organisms in that they excavate cavities into shells rather than attach to the surface of shells. They are still, however, generally considered part of the biofouling problem. The activities of boring sponges and polychaetes are highly destructive to shellfish and result in weaker and more fragile shells than non-burrowed shells. This makes affected shellfish more susceptible to predators and parasites and can lead to shell damage during shipping. The presence of boring species also forces shellfish to expend energy on shell repair instead of on other metabolic functions.

This can lead to reduced growth and decreased tissue condition. Some fouling organisms, like calcareous worms and barnacles cement hard structures onto shellfish shells. These structures may not actually harm the shellfish or damage the shell, but they can greatly lower the aesthetic appearance of shellfish. This is a particular problem for the high-end shellfish market where consumers want attractive, clean shells, and fouling material and associated shell damage are undesirable. The unsightliness caused by boring organisms and external fouling can reduce demand for product, lower market prices, or lead to product rejection.

Biofouling organisms can cause decreased growth of cultured shellfish. A variety of factors contribute to lower growth rates. One of the most common is when biofouling organisms compete with cultured shellfish for food or space. The majority of fouling organisms are filter feeders that consume similar particles as cultured shellfish. Fouling organisms can thus engage in interference competition with shellfish and scrub food out of the water before it reaches shellfish. Reduced shellfish growth can be caused by fouling of the cage material or fouling of shellfish shells. In general, younger and smaller shellfish may be more at risk from fouling than older and larger shellfish. Not all shellfish are affected by biofouling, however, and shellfish growth may not be impacted in some species or in some systems.

In addition to competing directly with shellfish for food, fouling organisms can clog or reduce the aperture size of nets and cages and reduce water flow to cultured shellfish. This can lead to decreased food availability and decreased waste removal and result in reduced shellfish growth. Heavy fouling can restrict water flow to such an extent that shellfish are smothered.

Fouling organisms can also cause shellfish mortality by physically preventing normal shellfish behavior (such as opening to feed). For example, heavy sets of barnacles can cause mortality when shellfish are prevented from opening and closing properly.

Increased weight of gear

Biofouling adds considerable weight to aquaculture gear and organisms. This is especially true for biofouling organisms that retain large amounts of water within their bodies, such as solitary ascidians. In Canada, heavy ascidian fouling (mainly by the solitary species *Ciona intestinalis*) adds significant weight to mussel lines ($\geq$10 kg per meter of line). The increased weight caused by biofouling dramatically increases the effort needed to work with aquaculture gear and can force growers to change the way they handle their gear. For example, before ascidian fouling was a problem in Atlantic Canada, mussel lines were pulled by hand. Once ascidians became abundant, specialized equipment was needed to haul the lines. Increased weight caused by biofouling can also lead to crop sloughing – when cultured organisms are pulled off of gear by the weight of biofouling organisms.

Introduction of unwanted and invasive species

Transportation of fouled gear and shellfish can introduce biofouling species to new areas. When moved to new locations, biofouling organisms can spread from aquaculture material to the surrounding substratum, or can release larvae into the water column that then colonize surrounding surfaces. This can lead to the introduction of species to new areas and aid the spread of invasive species. Ascidians are readily introduced in this way, but any benthic organism, including species like *Polydora*, can be spread through aquaculture transfers. It is generally a bad idea to introduce species into a new location, but it is particularly dangerous when invasive species reach new sites. Invasive species can be extremely damaging to aquaculture and the environment, and once established, they are nearly impossible to remove.

When biofouling organisms are moved to new locations on shellfish and gear, the biofoulers can also transport other noxious organisms along with them. For example, harmful algae can survive gut passage through the digestive tracks of ascidians. Ascidian gut clearance rates vary, but intact and viable harmful algal cells may be released for 48 hours, or more, after they have been ingested by ascidians. Thus, ascidians attached to shellfish can consume harmful algae in one area, be moved to a new location when the shellfish are moved, introduce harmful algae to the new location, and potentially cause a bloom.

Environmental impacts

Biofouling can alter the environment near aquaculture sites. For example, biofouling can greatly increase the levels of biodeposition around aquaculture facilities, especially in regions where the biofouling is very heavy. Increased biodeposition can affect nutrient cycling and lead to problems such as oxygen depletion and anoxia. Effects vary among sites, however, and in some cases nutrient depletion may not be a problem. Increased biodeposition may also alter benthic community structure near aquaculture facilities.

Positive effects on shellfish

While the impacts of biofouling on shellfish aquaculture are generally negative, biofouling can have positive effects on cultured shellfish in a few situations. Biofouling can increase food availability to shellfish by increasing primary production near aquaculture facilities. Biofouling organisms can also reduce predation on cultured shellfish by providing them with chemical or tactile camouflage, or by granting them associational defenses.

Factors that influence fouling intensity

Numerous physical and biological factors affect the level of fouling that occurs at any given location. Factors such as depth, surface coloration, surface orientation, salinity, and the physical distance from other sites can all influence fouling loads. Some of these factors can be manipulated to mitigate fouling, while others are difficult to control.

Fouling intensity often varies with depth, with higher fouling generally occurring near the surface. A wide variety of factors contribute to the depth distribution of settling invertebrate larvae and algal propagules. Larvae often position themselves in the water column in response to cues from light, gravity, chemicals signals, temperature, and food availability. Overall, the larvae of fouling organisms tend to congregate near the surface. Each species responds to its own set of settling cues, however, and higher fouling can occur at deeper depths for some taxa or at some sites. Depth can also affect the rate at which biofouling organisms grow, often due to differences in food availability. As with larval settlement, faster biofouling growth rates typically occur at shallower depths. Placing aquaculture gear at deeper depths can reduce overall fouling loads, but can lower shellfish growth rates due to reduced food availability.

Depth can also affect the species composition of fouling organisms, especially on fixed structures (as opposed to floating structures). There is often a distinct zonation of biofouling organisms with depth. If the fouling zonation for a given aquaculture site is known, gear can be deployed at depths with fewer, or less detrimental, fouling organisms.

Surface coloration, such as the coloration of aquaculture gear and structures, can affect fouling intensity. Darker colors generally attract more invertebrate fouling than lighter colors, and color may be the principle factor that affects settlement choice; however, the relative brightness of a surface, rather than its actual color, may be the stimulus that attracts or repels larvae. Some taxa, especially algal species, may settle at higher levels on light colored surfaces.

The orientation of material can also affect the type and intensity of fouling. Invertebrate fouling is often heavier on horizontal, downward facing surfaces. Algae is more abundant on vertical, upward facing surfaces.

Fouling organisms are readily attracted to aquaculture gear, and often foul gear at higher levels than surrounding natural surfaces. Differences in fouling levels are frequently noted on wood, concrete, and fiberglass surfaces. Additional testing is needed, however, to determine which aquaculture materials are more or less resistant to fouling.

Salinity greatly affects the growth and survival of biofouling organisms. Almost all biofouling species are marine in origin, so less fouling generally occurs in lower salinity areas. In estuaries, deploying aquaculture material further away from oceanic water can usually reduce fouling loads.

Distance from source populations can affect fouling levels, especially at newly established sites. Algae and marine invertebrates use dispersive propagules (that is, larvae and spores) to colonize new locations. Some invertebrates produce long lived, planktotrophic larvae that remain in the plankton for weeks to years before they settle into benthic communities. Others produce short-lived, lecithotrophic larvae that remain in the plankton for only a few minutes to a few days. Therefore, species with long-lived larvae possess a 'pool' of larvae in the water column that allows them to quickly colonize distant sites. In contrast, species with short-lived larvae tend to advance from the borders of their geographic range or use chance events to reach new locations. Newly established shellfish aquaculture facilities located far from source populations will likely be relatively free of lecithotrophic species during early operations. As species with short-lived larvae arrive, perhaps on contaminated gear or culture stock, they will begin to self-recruit and will likely become abundant.

Shellfish biology and physiology can affect biofouling levels. For example, the amount of biofouling can vary on the two valves of the same individual shellfish. Often, the upper valve will have more fouling than the lower valve, and lower valves may be nearly clean. In other cases, such as when boring sponges colonize oysters, lower valves may be more heavily affected than upper valves. Shellfish age can also affect fouling levels, with older individuals having more fouling than younger ones.

Control of biofouling

Biofouling control consists of preventing fouling from occurring and removing fouling once it has become established. Prevention strategies include deploying gear spatially or temporally to avoid fouling organisms, using antifouling coatings to keep fouling organisms from attaching, and using biological controls to consume fouling organisms (Table 16.1). Removal strategies include mechanically removing fouling, exposing fouled surfaces to the air, and dipping fouled material into noxious solutions (freshwater, brine, chemicals, high temperature baths). Growers typically use a combination of techniques to manage fouling at their facility. At present, the techniques primarily used are defensive husbandry (growing shellfish in areas with lower levels of fouling) and physical removal of biofouling. Different efforts are often combined to provide antifouling strategies for individual sites (Railkin, 2004; Watson et al., 2009; Dürr and Watson, 2010; Fitridge et al., 2012, and references therein).

Preventing fouling

Selecting sites with lower fouling levels, or deploying gear at times of the year with less fouling, can greatly reduce fouling loads. In particular, keeping gear out of the water when problematic species are settling can be very effective. Unfortunately, this technique can be difficult to employ. To be successful, growers must know the fouling regime for their site. These data are not always available, and because species often settle at different times in different areas, it can be hard to extrapolate these data from other locations. Even when fouling regimes are well known, most sites have diverse fouling communities composed of many different species, all with their own settlement profiles. Thus, even in the best situations, growers may be able to avoid the settlement of some species, but not all.

Antifouling coatings have not been widely used in shellfish aquaculture. Many existing coatings leach toxic substances or have detrimental effects on cultured organisms. Attempts to develop nontoxic antifouling coatings have so far been unsuccessful. Coating candidates have had poor performance against fouling, been expensive, or been toxic. Some of the more promising nontoxic coatings for aquaculture are fouling-release coatings based upon silicone elastomers and polydimethyl siloxanes. These materials make it very difficult for larvae to attach to treated surfaces. Trials of newer silicone release coatings in aquaculture settings have yielded positive results. For example, one commercially available silicone release coating significantly reduced fouling on lantern nets; the reduction was so substantial that a spring gear change was not needed on treated nets. It should also be noted that the efficacy of a variety of novel antifouling coatings, such as food grade oil (shortening), have also been tested with positive results. Additional development of these, and other coatings, may yield an inexpensive, environmentally friendly antifouling coating that can be used in shellfish aquaculture.

Grazer- and predator-based biological controls can reduce biofouling. The snail *Littorina* has been used to control algal growth on oysters. A variety of invertebrate predators have been used as biological controls in shellfish aquaculture, including crabs, shrimp, and sea urchins. Sea urchins are particularly attractive biological controls. They are sometimes co-cultured with

Table 16.1 General techniques for preventing and removing biofouling. Modified from Watson et al. (2009).

	Pros	Cons
Preventing biofouling		
Selecting sites with few fouling organisms (e.g., low salinity areas, deeper sites, areas with limited established fouling communities)	Can reduce the pool of settling larvae and lower fouling loads	Not always possible
Selecting material and placement of gear and structures with lower fouling potential (e.g., light colored surfaces, vertical orientation, different construction materials)	Can naturally reduce the willingness of settling larvae to attach to surfaces	Not always possible (especially adjustment of orientation); little information is available on the fouling properties of different construction materials
Avoiding seasons or times of the year when fouling organisms are reproducing	Can reduce fouling loads; if focused on specific problem species, can avoid/reduce fouling by particular species	Must know the settlement patterns of local fouling organisms; staggered reproductive periods for different species can reduce effectiveness
Antifouling coatings	Reduces overall fouling loads, sometimes dramatically	Historical formulations could harm shellfish, be toxic to the environment, or be expensive; newer formulations may avoid these issues
Biological controls	Once released, can consume fouling species with little additional intervention	Obtaining enough biological control organisms can be difficult; not always effective
Streams of air bubbles	Reduces overall fouling loads, often dramatically	Hard to employ on a large scale; bubbles must flow continuously
Removing biofouling		
Manual removal (e.g., scraping, power washing)	Very effective at removing fouling	Labor intensive; costly; can harm shellfish; can generate large amounts of waste material
Air drying	Very effective at removing fouling from gear	Cannot normally be used with shellfish – if tried can lead to shellfish mortality; may not control all fouling organisms; does not remove calcareous shells or tubes
Sprays and dips (freshwater, brine, hot water, acetic acid, lime)	Very effective at removing fouling in some situations	Often causes shellfish mortality; long exposure times (days) may be needed; can be costly; does not remove calcareous shells or tubes

shellfish so that they not only control fouling, but can also serve as an additional cash crop. Biological control of fouling can increase the growth and survival of cultured shellfish. Although potentially helpful, especially if used in conjunction with other fouling control techniques, the effectiveness of biological controls varies. There are differences in effectiveness among locations, challenges in obtaining enough predators, and differences in the composition of fouling communities. Additionally, some fouling species, especially invasive species, are not susceptible to native predators.

A novel form of fouling prevention is to use streams of air bubbles to continually strike a target surface. After 4 weeks of exposure, surfaces exposed to continuous aeration had only 4% the mass of fouling as control surfaces. Bubbles appear to inhibit larval settlement, and must be applied continuously to prevent fouling. While antifouling aeration is interesting in concept, scaling up bubble generating apparatus for use at an aquaculture facility would likely be challenging.

Removing fouling

Manual removal (for example, scraping, power washing, and so on) is perhaps the most common way to remove biofouling. Both gear and shellfish can be cleaned manually. All manual removal techniques are labor intensive, sometimes costly, can harm cultured shellfish, and can produce large quantities of waste material. If done in the field, some biofouling organisms, especially colonial species that readily fragment, can survive and reattach to nearby surfaces. This can lead to the spread of fouling organisms within an aquaculture facility. Additionally, some mechanical removal techniques can result in the reduced growth of cultured shellfish or in shell damage. The timing of cleaning will vary by site, depending on the types of fouling organisms present and the time that they recruit. Beginning cleaning early in the season when fouling organisms are still small may be the most effective way to control fouling.

Air drying is another common method used to control biofouling. Different air-drying techniques are available. In some cases, floating cages are periodically flipped to expose alternate sides to the air. This can dramatically reduce fouling on cage material, but does not affect fouling on the cultured shellfish. If the shellfish themselves are exposed to air, high levels of shellfish mortality can occur. As useful as air-drying can be, it is not necessarily a cure-all for biofouling. Many fouling organisms, including shellfish and solitary ascidians, are very resistant to air drying and may not be killed by exposure to air.

Biofouling can be removed with sprays and dips of noxious substances (freshwater, brine, hot water, acetic acid, lime, and so on). Some of these treatments are very effective in some settings and against some species of fouling organisms. Despite their potential value, there are some serious drawbacks to using sprays and dips. Most significantly, cultured organisms can be harmed by the treatments. Additionally, not all biofouling species are susceptible to sprays and dips, and very long exposure times (1–2 days) or repeated exposures may be needed for treatments to be effective. It can be expensive to use noxious substances in bulk, substances can contaminate the environment, and non-target species can be harmed. Finally, even if sprays and dips kill the biofouling organisms, the shells and tubes of barnacles and calcareous worms are left behind.

Conclusions

Biofouling will continue to plague aquaculture efforts. While physical means of removal are available, they will remain labor intensive and, in some cases, detrimental to gear and cultured animals. The future success of biofouling control on aquaculture gear will likely center on the development of nontoxic, affordable coatings. Effective coatings will prevent biofouling and reduce the need for manual fouling removal.

References cited and further reading

Adams, C.M., Shumway, S.E., Whitlatch, R.B., and Getchis, T. (2011) Biofouling in marine molluscan shellfish aquaculture: a survey assessing the business and economic implications of mitigation. *Journal of the World Aquaculture Society*, 42: 242–252.

Bannister, J., Sievers, M., Bush, F., and Bloecher, N. (2019) Biofouling in marine aquaculture: a review of recent research and developments. *Biofouling*, 23: 1–8.

Dürr, S. and Watson, D.I. (2010) Biofouling and antifouling in aquaculture. In: Dürr, S., and Thomason, J.C. (eds.) *Biofouling*, Wiley-Blackwell, Oxford, pp 267–287.

Fitridge, I., Dempster, T., Guenther, J., and de Nys, R. (2012) The impact and control of biofouling in marine aquaculture: a review. *Biofouling*, 28: 649–669.

Lacoste, E. and Gaertner-Mazouni, N. (2015) Biofouling impact on production and ecosystem functioning: a review for bivalve aquaculture. *Reviews in Aquaculture*, 7: 187–196.

Railkin, A.I. (2004) Marine biofouling: colonization processes and defenses. CRC Press, Boca Raton, FL.

Sievers, M., Dempster, T., Keough, M.J., and Fitridge, I. (2019) Methods to prevent and treat biofouling in shellfish aquaculture. *Aquaculture*, 505: 263–70.

Watson, D.I., Shumway, S.E., and Whitlatch, R.B. (2009) Biofouling and the shellfish industry. In: Shumway, S.E. and Rodrick, G.E. (eds.) *Shellfish Safety and Quality*. CRC Press, Boca Raton, FL, pp. 317–336.

CHAPTER 17

Shellfish farming: regulations, spatial planning, best practices, and certification

Daniel P. Cheney and Tessa Getchis

Introduction

There are many critical aspects of shellfish farming that affect production methods and product quality; however, there are also important elements of a successful shellfish farm which are governed by consumer and environmental regulations and policies. These are in place in an effort to minimize conflicts between aquaculture and other users of the waters and surrounding uplands, to ensure safe navigation, and to protect ecosystem and human health. In addition, tools are available to aid prospective producers with the site selection process. The adoption of best management practices and the use of sustainability certification programs may allow producers to access new markets and perhaps, increase profitability.

Regulations

Shellfish culture practices are coming under greater environmental and public scrutiny as consumers become more sensitive to food production methods. Also, domestic and international resource agencies are directing increasing attention toward protecting marine ecosystems. Resource agencies, particularly those engaged in water quality enforcement and regulation, have traditionally viewed the presence of a commercial shellfish industry as an indicator of a healthy estuarine environment (see also Chapter 11). This assessment has, however, within the last few years, been offset by regulatory and management actions calling for a more critical evaluation of the effects of shellfish farming activities. In the US, domestic regulatory policies of the US Army Corps of Engineers and other federal agencies, as well as state and local regulations and guidelines are increasingly playing a role in the siting of shellfish farms as well as cultivation and harvest practices. For example, endangered species listings for West Coast salmon stocks have raised concerns over aquaculture activities particularly as they affect issues of 'essential fish habitat' (EFH), salmon habitat and prey resources. Under the 'taking' provision of the Endangered Species Act, shellfish harvest and other culture activities can be banned or severely restricted during periods of peak use of estuarine areas used by threatened or endangered juvenile salmonids. Similar restrictions are already in place and imposed for aquaculture gear and activities that pose threats to protected species such as marine mammals, turtles, and birds.

The regulatory processes for shellfish aquaculture vary widely by region, state, and country, and involve a range of permissions to use the site, install structures such as bags, rafts, and floats, and harvest and sell the cultivated products. International environmental management standards, such as ISO 14000 and the United Nations Food and Agriculture Organization (FAO) Code of Conduct (Article 9, Aquaculture Development) and seafood production and processing requirements/standards are being enacted in many importing countries. Depending on the proposed location, type, and size of farm shellfish producers must obtain multiple national/federal, state and local permits before they can begin operations. It can typically take 1 to 3 years or longer from the time a farmer applies for these permits before receiving authorization to grow any shellfish on the farm. In addition, harvest, processing, product handling, and transport methods are typically prescribed and overseen by government entities and certification groups in collaboration with grower organizations and individual growers to ensure compliance with approved environmental and production practices, and to protect public health.

Public participation is an important part of the process and many resources agencies use public comment periods and public hearings as means to identify issues and potential conflicts with proposed aquaculture activities. These comments may shape the agencies' decisions with regard to a project.

Spatial planning and site selection tools

A number of countries have adopted spatial planning as a means to review shellfish aquaculture applications in the context of other human uses of the coastal zone, and to better assess the risk potential for interactions between cultivation activities and protected species and habitats (Aguilar-Manjarrez et al., 2017). In fact, in some places resource managers encourage or require applicants to use spatial planning tools, such as shellfish aquaculture map viewers, in the preparation of applications for aquaculture activity. These GIS-based viewers allow users to overlay various types of environmental and human use data to inform the site selection process. In turn, resource managers use these tools in their evaluation of applications to determine the suitability of a particular site. Municipal managers may also use shellfish aquaculture map viewers to help community members visualize how aquaculture activity would fit in local coastal waters.

Best management practices

The economic viability of a shellfish farm can be markedly enhanced if it takes an active role in preparing for, and putting into action, best management practices (BMP) and other changes that will be expected of the industry by consumers, and local, federal and international organizations. Producers who adhere to such BMP and use that as part of their marketing approach, or who gain sustainability certification through a third-party organization may gain access to new markets and increase profitability.

What are BMP? Most simply, they are a set of guiding principles defining specific actions, and a suite of managerial, operational, and structural performance measures. They are implemented by individual producers, shellfish farming organizations, and trade groups to address farm production, processing, marketing and sales, employee relationships, and on-farm environmental management. Government organizations, and public or private groups or individuals engaged in food safety and environmental protection may also create BMP, with and without input from shellfish producers, and sometimes with a regulatory intent.

Not limited to just shellfish farms, BMP are incorporated into the Code of Conduct for Responsible Fisheries by the United Nations Food and Agricultural Organization (UNFAO), as well as certification programs, such as the Food Alliance and the Best Aquaculture Practices (BAP) shellfish certifications, which are often requested by major seafood retailers and distributors. Currently, BMP are developed for many aspects of shellfish aquaculture in all growing regions in the US and elsewhere. In addition, port and harbor facilities, many marinas, and boat repair facilities also require their users to adhere to a written set of BMP.

Typically, BMP are intended to smooth the path for successful shellfish farming and facilitate effective business planning and regulatory review. Ideally, BMP are 'common sense' measures, with well-known outcomes accepted by the farming community. They can be targeted to specific farming practices and tailored to a localized regulatory environment. The best BMP are user-friendly, help producers improve productivity against a baseline, and give them the flexibility to develop their own ways to improve performance. These practices should be voluntary, simple, and low tech, and offer practical and affordable approaches without sacrificing productivity. For example, a reasonable suggested practice to minimize farming gear losses is for farmers to keep adequate records on total gear numbers, placement, and maintenance routines for each farming site. The farmer has a baseline of the gear purchased, the gear deployed on each site, and by making routine observations will have better control over the condition and the fate of the gear.

There are a number of excellent sources of information on general and shellfish species-specific applications of BMP. The following are excellent examples.

- *Environmental Best Management Practices for Aquaculture* (Tucker and Hargreaves, 2008) emphasizes farm level technical solutions. The approach to this subject was based on what the editors "perceived to be an imbalance in the literature dealing with aquaculture and the environment" where the discussions of environmental impacts are not balanced by guidance providing solutions to those problems. BMP are described as 'works in progress,' subject to change and further improvement. Practices are explored that may be implemented by large-scale or small-scale producers to improve environmental performance, with corresponding improvements in the overall sustainability of aquaculture. However, the text does not address social impacts of aquaculture, food safety, or policy guidance for larger-scale aquaculture development. Chapter 11 focuses on shellfish aquaculture BMP, and other chapters touch on relevant subjects. This is an excellent resource to gain an understanding of the broad nature of BMP with regards to shellfish farming practices.
- *Shellfish Aquaculture and the Environment* (Shumway, 2011) is a compilation and discussion of issues surrounding environmental sustainability of shellfish aquaculture. The 17 chapters in this book offer readers a glimpse of the current status (as of 2011) of this subject on a wide

range of topics. The following chapters are most applicable: Chapter 2. Shellfish aquaculture and the environment: an industry perspective; Chapter 3. Molluscan shellfish aquaculture and best management practices; and Chapter 15. Balancing economic development and conservation. These focus on the usage of BMP and interactions between shellfish farmers, regulators and other individuals who must respond to public scrutiny over shellfish aquaculture and the environment.

- *Northeastern U.S. Aquaculture Management Guide* (Getchis, 2014) describes aquacultural production hazards that may result in a loss of farmed product, closure of harvest areas, increased human health risk, and other factors which may adversely impact profitability, trade, and public perception. Includes science-based information about major production hazards facing farmers, including: predators, diseases, parasites, organisms that have the potential to cause aquatic animal illness and human illness (for example, toxic algae), biofouling, spread of invasive species, and other operational and environmental hazards. Chapter 1 outlines the basic types of risks associated with aquaculture ventures and provides short lists of BMP or 'tools and strategies' to manage those risks. Although this publication focuses on aquaculture in the Northeastern US, it offers invaluable guidance for anyone considering entrance to this business.

BMP and certification programs: regional and international examples

The following sections direct the reader to a broad range of selected local to international information sources for application of BMP, or more generally 'Environmental Codes of Practice' or Environmental Policy and Codes of Practice (ECOP) on shellfish farms. Most can be found online in the listed sources, and full-text documents are often available in university libraries or obtained via inter-library loan. They include several examples of farm or project specific BMP and application of BMP in a regulatory setting, and several farm and processor programs certifying those BMP. There is a considerable amount of information to digest, and it is suggested the focus should initially be on BMP documents developed for a specific state or region. This should be followed up with assistance from established and more experienced growers and processors, and extension agents. All BMP are 'living documents' that is, expect them to be modified whenever a significant change in management or operations is made in order to address a particular problem or regulation, or to improve farm efficiency. Examples of BMP applicable to proposed or newly established small shellfish farms are provided in Table 17.1 of the appendix. Finally, information is included on farm and processor certification programs that may be required or requested by some customers and/or viewed as a way to add value and consumer appeal to shellfish products.

US, regional, and states

NOAA Fisheries and Aquaculture

The NOAA Fisheries and Aquaculture shellfish BMP/standards/certification programs

This website provides links to documents containing both industry-driven and government mandated BMP. Listed are a selection of BMP documents, most of which are discussed below. This aquaculture webpage is also a good source of information concerning current aquaculture programs within the agency, along with links to other critical areas within and outside NOAA.

NOAA aquaculture webpage:

https://www.fisheries.noaa.gov/aquaculture/best-management-practices-standards-and-farming-manuals

https://www.nmfs.noaa.gov/aquaculture/index.htm

Interstate Shellfish Sanitation Conference (ISSC), The Nature Conservancy (TNC), and the NOAA Restoration Center – Shellfish Restoration Committee
Best Management Practices for Shellfish Restoration – 2011

This expansive report is an excellent source of information for growers interested in small-scale shellfish culture or 'shellfish gardens' or to explore methods to increase public awareness of the ecosystem benefits of shellfish culture. It is intended to guide future shellfish restoration projects that incorporate educational components designed to protect public health. It was built around workshops at regional ISSC and other professional shellfish management meetings held in the US. A broad list of BMP is recommended to: protect public health while restoring the environment; define goals and objectives of restoration projects; expand communication and education; expand community-wide restoration and; when practical, use noncommercial species in restoration efforts.

ISSC *shellfish* restoration manual:

http://www.issc.org/Data/Sites/1/media/publications/final%20draft%20bmps-01–23–12.pdf

West Coast US

Pacific Coast Shellfish Growers Association (PCSGA) – https://www.pcsga.org
Environmental Policy and Codes of Practice (ECOP) – 2011/2019

The PCSGA Codes of Practice (ECOP) and Environmental Policy were initially developed in 2001 by farmers in Alaska, Washington, Oregon, California, and Hawaii following a year-long series of regional meetings, and were revised in 2011 and 2019. The general principles of the ECOP are set out in a comprehensive environmental policy encompassing: environmental stewardship and responsible management; environmental excellence; regulatory compliance; waste management; and sharing resources. The 2011 ECOP revision is a 150+ page document with detailed information on farming, hatchery and processing methods, a large section on regulation and compliance, a farm plan checklist and a sample environmental farm plan. Unfortunately, the 2011 ECOP has already become somewhat dated due to significant and rapid developments in product types, and local, state, and federal regulations and policies. Changes in production methods, such as the recent expansion and installation of off-bottom oyster 'flip-bag' and floating culture in Washington State and elsewhere has necessitated the development of a set of culture-relevant BMP. PCSGA is conducting an extensive update will revise the ECOP so it better conforms to existing farming and regulatory practices. This updated document is expected to be available by late 2019.

Current PCSGA Environmental Policy:

https://pcsga.org/wprs/wp-content/uploads/2013/04/EPS.pdf

East Coast US

East Coast Shellfish Growers Association (ECSGA) – https://ecsga.org
Best Management Practices for the East Coast Shellfish Aquaculture Industry – 2010
Industry driven, this document is the final product of 22 workshops with input from hundreds of stakeholders from the majority of east coast shellfish growing regions. It includes a well-organized description of shellfish culture methods for oysters and clams, and three major sections: a code of conduct; a detailed and an inclusive BMP section with a discussion of issues and suggested practical BMP statements for specific farm operations. Also, provided as a bonus, is a shellfish farm template with suggestions for farm-level BMP elements and a sample farm management plan that demonstrates how this template can be used to put together farm-specific BMP. This template is available on the ECSGA website as an interactive Microsoft Excel file in macro format (.xlsm).

ECSGA BMP guidance and manual:

https://ecsga.org/wp-content/uploads/2018/01/BMPmanual.pdf

Individual states

California

California Fish and Game Commission / California Department of Fish and Wildlife (CDFW)
California Coastal Commission coastal development permit
Shellfish Aquaculture Best Management Practices
Stakeholders Discussions, July 17, 2017 and May 29, 2018
BMP are under development in California through discussions between shellfish farmers, regulatory staff, and others. During 2017 and 2018 progress was made on possible regulatory processes to incorporate farm-specific BMP into lease agreements. A review of options was considered to incorporate a BMP requirement either: (1) into the terms of each individual lease, or (2) through new regulations requiring the submittal of BMP covering specific categories. The focus was on outcomes coming from specific operational needs of the growers, the farm locations, production

methods and innovations, and the costs and benefits of those approaches. Current information and guidance regarding BMP status can be obtained from CDFW or growers in the state.

BMP discussion summaries (these access pdf files of the original documents):

https://nrm.dfg.ca.gov/filehandler.ashx?documentid=147222

https://nrm.dfg.ca.gov/FileHandler.ashx?DocumentID=160171

See also the PCSGA Environmental Policy and Codes of Practice (ECOPs).

Florida

Florida Department of Agriculture and Consumer Services, Division of Aquaculture
Aquaculture Best Management Practices Rule – 2016

This is regulatory document, with an extensive list of BMP specific to Florida aquaculture and required by Florida law for all aquaculture including hatcheries, producers leasing submerged lands from the State of Florida, and processors. They focus on largely prescriptive and some outcome-specific clam, mussel, scallop, and oyster BMP in the following areas: resource protection, genetic protection, health management, upland facility operations, submerged lands grow-out, mechanical harvesting, and public health protection. The BMP are intended to eliminate complex, duplicative and confusing environmental local, state and federal permitting and licensing requirements. This is an optimistic goal, but the BMP should assist in clarifying the required steps for permitting and regulatory approvals.

Florida BMP guide:

https://www.freshfromflorida.com/content/download/64045/1520653/bmp_rule_and_manual_final.pdf

Hawaii

Center for Tropical and Subtropical Aquaculture (CTSA)
Best Management Practices for Hawaiian Aquaculture – 2001

This manual is intended for use by potential aquaculture farmers, experienced commercial producers and extension personnel. While it included all existing and proposed aquaculture species in Hawaii, there is little direction specific to bivalve shellfish aquaculture. The primary goal of the BMP are to assist shrimp and finfish farmers in managing their facilities more efficiently and more profitably, while complying with effluent discharge regulations. Current information on the status of shellfish aquaculture and BMP in Hawaii is available from the Center for Tropical and Subtropical Aquaculture (CTSA) and the University of Hawaii Sea Grant program.

CTSA BMP manual:

http://www.ctsa.org/files/publications/CTSA_1486316728532840802601.pdf

Maine

Maine Aquaculture Association (MAA)/The Maine Aquaculture Innovation Center – https://www.maineaquaculture.org/
Recommended Code of Practice for Aquaculture in Maine – 2006
This document was prepared by the Maine Aquaculture Association (MAA) and covers both the shellfish and finfish aquaculture industry. The MAA Code of Practice is based on the UNFAO Code of Conduct for Responsible Fisheries. In addition, more than 40 other codes of practice, best management practice programs, environmental policies and certification programs were reviewed. A number of components from these documents are integrated into the MAA Code of Practice. This code requires MAA members to analyze their actions and determine what aspects of their operations may involve some risk and to take actions to control or manage these risks.

Members are expected to adhere to a code of practice based on the Guiding Principles for Responsible Aquaculture in Maine. This forms the basis for a crisp set of largely outcome-oriented BMP which offer guidance, and informal advice on the following topics: (1) conserving water resources; (2) proactive animal health management waste management; (3) grow out and husbandry practices designed to optimize use of aquatic resources; (4) animal welfare; (5) marine transportation; (6) land transportation; (7) community leadership; and (8) employee development. This is an excellent and practical tool for anyone engaged in shellfish farming activities.

> Maine Guiding Principles for Responsible Aquaculture:
> https://maineaquaculture.com/maa-code-of-conduct/

A full copy of the MAA Code of Practice may be obtained from the Maine Aquaculture Association.

Maryland

Maryland Aquaculture Coordinating Council
Best Management Practices a Manual for Maryland Aquaculture – 2007
Maryland produced a manual that was mandated by the state legislature and developed with industry input. It is presently a voluntary system, potentially moving toward regulation. The BMP were formed from existing state and federal laws and regulations, as well as recommended voluntary measures. Their purpose was to provide producers with a base of knowledge regarding expectations in the development of their businesses. In all, they comprise a roadmap for those entering the aquaculture industry to follow as they grow businesses in the state. Detailed and comprehensive guidance is provided with a focus on regulatory compliance, environmental stewardship, animal and public health, and community relations. Sections in the manual were prepared by teams of industry, government and academic representatives. They addressed a broad range of BMP topics, as follows: water resources and management, production systems, shellfish culture, non-native species, aquatic animal health, and shipping, transport, and sale. This is an excellent source of information on a wide range of BMP for all types of aquaculture, with a focus on applicable regulations, policies, and practices within the state of Maryland.

Maryland BMP manual:

http://www.nda.agric.za/doadev/sidemenu/fisheries/03_areasofwork/aquaculture/aquadocumentation/bestmanagementpracticeguide/bmp_marylandaquaculture.pdf

Massachusetts

Southeastern Massachusetts Aquaculture Center
Best Management Practices for the Shellfish Culture Industry in Southeastern Massachusetts – 2009

"There is no single best management practice for all shellfish crops at all sites. Therefore, the best practice for any individual grower will depend on site-specific circumstances, economic opportunities and environmental considerations."

Shellfish aquaculture BMP are a set of voluntary procedures developed by the Massachusetts shellfish aquaculture industry in collaboration with the Southeastern Massachusetts Aquaculture Center (SEMAC). The intent of the BMP are to provide information on 'normal industry practices' to help growers to farm profitably and in harmony with their neighbors and the surrounding environment. The approach is somewhat different than described in other shellfish BMP manuals in that input was provided by a broad group of stakeholders, including government regulators and environmental organizations. Publication of this manual is arranged in a loose-leaf format with the intention that sections will be replaced as better technology and/or practices become available. It is a large document (100+ pages) focused on five categories of standard practices: site selection and access, materials, operations and maintenance, improvement of shellfish survival and productivity, disease prevention and management, and maintenance of environmental quality. Recommended BMP are linked to each of the five categories and detailed advice is provided for specific types of activities and regulatory authorities. This is an interesting and comprehensive approach to the identification of shellfish farming BMP.

SE Massachusetts BMP manual:

http://www.mass.gov/eea/docs/agr/programs/bmp/shellfish-bmp.pdf

New Jersey

Rutgers Cooperative Extension
Recommended Management Practices for Aquatic Farms – 2004

This two-part manual combines a review of 'agricultural management practices' (AMP) with a detailed aquatic organism health management plan. The recommended AMP are intended for use by all holders of Aquatic Farmer Licenses in New Jersey. The farmer is also required to obtain any and all permits required by the appropriate state regulatory agencies (that is, New Jersey Department of Environmental Protection). The manual notes adherence to AMP does not exempt an aquatic farmer from other applicable agency permits. The overall approach in Part 1 of the manual is on AMP for land-based facilities, and there is little direction for the aquatic farmer in marine waters. The most relevant sections briefly cover these areas: federal permits; construction; shipment, transportation and sale; water resources; non-native and restricted non-native species; prohibited species; health management; mortality removal/disposal; chemical and drug handling; security and biosecurity; and marine bivalve facilities. The Aquatic Organism Health Management Plan and Appendix A of the plan includes useful information related to shellfish diseases and parasites (eastern oysters, hard clams, softshell clams, blue mussels, and bay scallops).

Main webpage:

http://www.state.nj.us/seafood/

AMP and aquatic animal health manual:

http://www.state.nj.us/seafood/aquacultureamp.pdf

Virginia

Virginia Institute of Marine Science
Best Management Practices for the Virginia Shellfish Culture Industry – 2008

BMP in this manual highlight operational parameters or 'normal industry practices' that novice culturists may want to implement. They are described as "intending to minimize the impact on the environment, while serving to maintain or increase crop production and quality." and addressing societal issues through 'good neighbor' practices. These are voluntary guidelines for aquatic farmers in Virginia, with four major management areas: social, operational, biological, and food safety and quality. There is no regulatory guidance, other than a BMP stating "Shellfish growers should obtain all necessary state, federal, and local permits for their business" (Issue 1–2).

Virginia Institute of Marine Science (VIMS) main webpage:

http://www.vims.edu/about/

BMP manual:

http://web.vims.edu/adv/aqua/MRR%202008_10.pdf?svr=www

Washington

Washington State Department of Health, Office of Shellfish and Water Protection (WDOH)
Best Management Practices during *Vibrio* Season – 2015

These are examples of BMP measures to minimize the likelihood of *Vibrio* contamination of farmed shellfish in Washington State. They were prepared through lengthy discussions between growers, processors, and public health specialists and represent a good example of cooperative and proactive development of BMP. The referenced webpage contains links to reports and data related to *Vibrio* control. Similar BMP recommendations are also available for British Columbia shellfish growers (see the British Columbia Shellfish Growers Association [BCSGA] code of practice, below).

WDOH *Vibrio* control plan webpage:

https://www.doh.wa.gov/communityandenvironment/shellfish/commercialshellfish/vibriocontrolplan

See also the PCSGA Environmental Policy and Codes of Practice (ECOP).

International

United Nations Food and Agriculture Organization (UNFAO)

Code of Conduct for Responsible Fisheries: Section 9 – Aquaculture Development
The Code, unanimously adopted on October 31, 1995 by the FAO Conference, provides a framework for national and international efforts to ensure sustainable exploitation of aquatic living resources in harmony with the environment.

http://www.fao.org/docrep/005/v9878e/v9878e00.HTM#9

Australia

South Australia Environment Protection Authority (EPA)
Code of Practice for the Environmental Management of the South Australian Oyster Farming Industry – 2017
The Environment Protection Authority (EPA) is a mandatory referral agency for Primary Industries and Regions South Australia (PIRSA) license applications. The majority of farmers in South Australia focus on the culture of Pacific oysters, *Crassostrea gigas*, with some farmers also culturing commercial quantities of native flat oysters, *Ostrea angasi*. The scope of PIRSA legislation applies only to licensed oyster farming sites and hatcheries.

The BMP "aim to be outcome based and not prescriptive and therefore should enable farmers to continue their own individual methods of farming oysters while meeting the overall environmental objectives of the code. Operators are encouraged to use their experience, knowledge and ingenuity within their own industry to develop tools to avoid pollution" (page 5).

The focus is on water and air quality, and noise with particular reference to EPA requirements. Farm activities and potential impacts are briefly described, and there is limited technical guidance and information. BMP are described as actions that 'Must' be followed, such as "Avoid practices that disturb or impact seagrass and other sensitive habitats;" and 'Should' be followed, such as "Regularly monitor/record seagrass presence on sites to quantify changes over time." There are good descriptions of waste management and reduction measures, and a complete glossary.

Main webpage:

http://www.epa.sa.gov.au/

Code of Practice webpage:

http://www.epa.sa.gov.au/files/13050_oyster_code_2017.pdf

Aquaculture Council of Western Australia (ACWA)
Mussel & Oyster Environmental Code of Practice – 2013

Assesses farm practice risk, suggests measures to reduce that risk, and explains applicable regulation/policy. "The focus is on best practice through a documented environmental management system (EMS) and a process where the business periodically reviews and evaluates its EMS to identify opportunities for improvement and their implementation. An aquaculture business can demonstrate conformity with this Code by undertaking an Internal Audit using the ACWA Audit Checklist which is available on the ACWA website" (page 5). In general, emphasis is on measures which meet specific state and national regulations.

ACWA webpage:

https://www.aquaculturecouncilwa.com/

Code of Practice webpage (2013):

https://www.aquaculturecouncilwa.com/wp-content/uploads/2019/05/ACWA_Mussel_Environmental_Code_of_Practice.pdf

Canada

British Columbia Shellfish Growers Association (BCSGA)
Environmental Management System Code of Practice – 2013

The BCSGA Environmental Management Codes of Practices (EMCP) were developed in consultation with industry stakeholders, both levels of government, NGO, and other marine user groups. The codes outline the commitment of the British Columbia shellfish growers to maintain and improve the operation of their industry by taking the lead as environmental stewards of the marine environment.

This substantial (75 page) document offers useful and generally practical guidance for growers to help them run a more efficient and environmentally sound business in compliance with regulatory and licensing requirements. It is a good overview of shellfish aquaculture practices and species cultured in British Columbia, a detailed list of codes of practices for shellfish aquaculture activities, and a review of applicable legislation and regulations. Each list is clearly outlined, for example, for harvesting with: a definition "the time when shellfish are collected from their grow outs for processing and distribution" (page 39); objectives "minimize environmental impacts of routine harvesting operations" (page 39); and guidelines, such as, "harvesting will only occur within the grower's tenure area except for the transportation of the product to shore" (page 39). BMP recommendations for minimizing *Vibrio* in British Columbia shellfish are also available (see below).

Main BCSGA webpage:

http://bcsga.ca

Code of Practice webpage (2013):

http://bcsga.ca/wp-content/uploads/2016/09/Envrionmental-Management-Code-of-Practice_BCSGA.pdf

Recommended *Vibrio* control practices (2016):

http://bcsga.ca/wp-content/uploads/2016/09/Vp-Control-Program_BCSGA-Recommended-Practices_v1.0-may-2016.pdf

Ireland

Bord Iascaigh Mhara (BIM)
ECOPACT Environmental Code of Practice for Irish Aquaculture Companies – 2003
This manual describes the environmental challenges and responsibilities faced by the Irish finfish and shellfish aquaculture industry and outlines state of the art management and mitigation measures that the Irish aquaculture industry employs. Each individual issue is given a defined environmental objective, together with environmental management options, that the Irish aquaculture industry can adopt to meet each objective. Also included are substantial appendices containing lists and descriptions of important bird species, protected wildlife species, and the Federation of European Aquaculture Producers (FEAP) Code of Conduct for European Aquaculture. The manual lists potential beneficial consequences to the farmer adhering to the agreed Code such as: "positive press coverage, good public image, increase in market share, smoother expansion and development, improved reputation, better relations with other stakeholders, and ability to counter false claims with facts" (page 11).

Main webpage for BIM:

http://www.bim.ie/

ECOPACT

http://www.bim.ie/media/bim/content/publications/ECOPACT.pdf

New Zealand

New Zealand GreenshellTM Mussel and Pacific Oyster Sustainable Management Frameworks
The A+ Sustainable Management Frameworks (SMF) for the Greenshell™ mussel, *Perna canaliculus*, and Pacific oyster, *Crassostrea gigas*, were developed as a result of a holistic programe to standardize and advance environmental management across the New Zealand aquaculture industries. These SMF supersede 2007 codes of practices.

The purpose of the SMF are to promote the sustainable management of New Zealand aquaculture by providing guidance for best environmental and social practice for the aquaculture industry and applies to all activities relating to the farming practices. The SMF are administered by Aquaculture New Zealand (AQNZ) on behalf of the New Zealand mussel, and oyster industries.

The SMF provides a process for continuous improvement of environmental performance and supports the industry to achieve best practice environmental and social outcomes with key principals, objectives, and targets to measure performance. For each section an overview of the process is provided followed by a summary of the key environmental issues for that activity and recommended operational best practices. Individual operators complete an annual checklist to record implementation of the objectives and targets for the SMF and measure the key indicators. Third party assessments are carried out annually for a proportion of operators to provide industry with the verification required to demonstrate its commitment to the sustainable management of its industry.

The A+ Sustainable Management Frameworks for both mussels and oysters are available from the Aplus website. Also available on the Aquaculture New Zealand website is a wealth of information about the past, present and future of New Zealand shellfish aquaculture.

Aplus website:

https://www.aplus.org.nz/
Aquaculture New Zealand

(AQNZ) website:

https://www.aquaculture.org.nz/

Additional information is available at:

https://www.fisheries.govt.nz/growing-and-harvesting/aquaculture/

Certifications

A shellfish certification allows a farm or processing facility to sell, label, and represent their products as meeting standards for production, harvest, processing, safety, and seafood quality. There are a number of certifying organizations providing a 'brand' to qualified growers and processors. They are similar to the USDA Organic Certification, in that all are based on a set of written standards, require periodic third-party certification, and are fee based. All are voluntary, with the exception of a Shellfish Harvest Site Certification or Shellfish Sanitation Certificate, which is required for commercial shellfish production and sale. Currently, some customers accept only shellfish certified products meeting either their standards or the standards of the certifying organization. This provides access to those markets and may permit producers to receive a premium price for their products. The following are examples of shellfish certifiers.

Aquaculture Stewardship Council (ASC)

The World Wide Fund for Nature (known as World Wildlife Fund [WWF] in the US) paved the way for the Aquaculture Stewardship Council (ASC) by initiating and coordinating aquaculture dialogues between growers, processors, regulators, and environmental organizations in 2004. These resulted in standards for seaweed, bivalves, crustaceans, and fish. The ASC applies a third-party certification system, meaning that certification of farms and product suppliers are done by independent certifiers. Certified farms are subject to an annual surveillance audit including a risk analysis, focusing on the farm improvement plans. Bivalve shellfish farms in the United Kingdom, France, Mexico, Chile, China, South Korea, and Japan are currently ASC certified.

An overview of the ASC process and grower/processor requirements are available in their general description and Bivalve Standards, listed below.

ASC general description:

https://www.asc-aqua.org/about-us/about-the-asc/

ASC Bivalve Standards (2020):

https://www.asc-aqua.org/what-we-do/our-standards/farm-standards/the-bivalve-standard/

Best Aquaculture Practices certification

Best Aquaculture Practices (BAP) certification is administered by the Global Aquaculture Alliance (GAA). BAP standards and guidelines are BMP applicable to the culture of all species of bivalves, such as clams (cockles, geoduck), oysters, scallops, mussels, and grazing gastropods (whelks, abalone). The BAP standards are achievable, science-based, and industry-reviewed global performance standards. They are designed to allow program applicants to perform self-assessments of the environmental and social impacts, and food safety controls of their facilities, and to lead to third-party certification of compliance. To obtain BAP certification, applicants must be audited by an independent, BAP-approved certification body. Pacific Seafood is an example of a shellfish growing and processing company recently certified in the US. Please refer to the BAP webpage and mollusk standards below.

BAP main webpage:

https://bapcertification.org/About

BAP mollusc farm standards (2016):

https://bapcertification.org/Downloadables/pdf/standards/PI%20-%20Standard%20-%20Mollusk%20Farms%20-%20Issue%201.0%20-%2001-May-2016.pdf

BAP certification for Pacific seafood:

https://bapcertification.org/blog/pacific-seafood-four-star-bap-oysters/

Food Alliance

Sustainability Standard for Farmed Shellfish Operations (oysters, clams, geoducks, and mussels)

The Food Alliance (FA) Sustainability Standards for Farmed Shellfish Operations addresses operational activities related to: soil and water conservation; fish and wildlife conservation; integrated pest, disease, and weed management; healthy, humane care for shellstock; safe and fair working conditions; and shared resource management. FA has developed a sustainability evaluation tool allowing growers to self-assess current sustainability performance – and set sustainability objectives – either as a prelude to certification or as a best management practice. This voluntary fee-based program gives growers a credible way to distinguish their sustainably managed operation and products in the marketplace, to customers and consumers. Taylor Shellfish Farms, Washington and Hog Island Oyster Co., California are examples of FA certified firms.

Introduction to the FA sustainability standard for farmed shellfish operations (2013):

http://fadocsstyle.wpengine.com/wp-content/uploads/2016/02/shellfishstandard-handout-2013.pdf

Food Alliance webpage:

http://foodalliance.org/shellfish-farms/

NSF International

Previously known as the National Sanitation Foundation, NSF International provides certification, auditing, consulting, training, farm services, and laboratory testing services to agricultural businesses, farmers, and growers worldwide. Areas of expertise include food safety, consumer assurance, environmental protection, animal health and welfare, biosecurity, traceability, sustainability, worker welfare and good agricultural practices (GAP). Also coordinates with the Best Aquaculture Practices (BAP) Certification of the Global Aquaculture Alliance (GAA) and is an accredited third-party certifier, and food safety, and seafood auditor.

Description of NSF certification:

http://www.nsf.org/services/by-industry/food-safety-quality/global-food-safety-certification/bap-certification

Seafood Watch

The Monterey Bay Aquarium Seafood Watch program evaluates the ecological sustainability of wild-caught and farmed seafood commonly found in the US marketplace. Seafood Watch defines sustainable seafood as originating from sources, whether wild-caught or farmed, which can maintain or increase production in the long-term without jeopardizing the structure or function of affected ecosystems. A simple scoring system is used to rate seafood as 'Best choice,' 'Good alternative,' or 'Avoid' with an overall seafood recommendation based on additional guidelines prepared in evaluations of each species. The ratings are widely distributed, and can in certain markets, impact consumer preferences. While Seafood Watch is not intended for certification of individual farms, the recommendations can have an effect on consumer preferences. All aquacultured shellfish fall within the best choice and good alternative scores.

Seafood Watch webpage:

www.seafoodwatch.org

Example of a Seafood Watch shellfish analysis (2015):

http://www.seachoice.org/wp-content/uploads/2014/03/MBA_SeafoodWatch_FarmedGeoduckReport.pdf

Whole Foods Market

Whole Foods Market is an important customer for premium raw and value-added shellfish products. In addition to prohibiting pesticide use and requiring water quality monitoring and protection of the coastal environment, Whole Foods has developed a set of detailed standards for farmed oysters, clams, mussels, and scallops requiring evaluations of the health of sediments on the seafloor to help protect bottom-dwelling animals and the ecosystem beneath the farms. The standards offer specific guidance for sampling and analyses of benthic communities and sediments, and identification and monitoring of potential human health risks. Whole Foods Market also supports a 'Responsibly Farmed' logo to help customers recognize that farmed shellfish are third-party verified to meet quality standards for farmed seafood.

General information on standards (2015):

http://www.wholefoodsmarket.com/farm-raised-seafood

Standards for farmed shellfish (2015):

http://assets.wholefoodsmarket.com/www/missions-values/seafood-sustainability/WFM_Quality_Standards-for-Farmed_Bivalve_Molluscs.Version%202.0_Jan.%201.2015.pdf

References cited and further reading

Aguilar-Manjarrez, J., Soto, D., and Brummett, R. (2017) Aquaculture zoning, site selection and area management under the ecosystem approach to aquaculture: a handbook. Report ACS18071. Rome, FAO, and World Bank Group, Washington, DC. 395 pp.

Creswell, R.L. and McNevin, A.A. (2008) Better management practices for bivalve molluscan aquaculture. In: Tucker, C.S. and Hargreaves, J.A. (eds.) *Environmental Best Management Practices for Aquaculture*. Wiley-Blackwell, Chichester, pp. 427–486.

Getchis, T.L. (ed.) (2014) Northeastern U.S. Aquaculture Management Guide: A manual for the identification and management of aquaculture production hazards. Connecticut Sea Grant and UConn Extension, University of Connecticut.

Shumway, S.E. (ed.) (2011) *Shellfish Aquaculture and the Environment*. Wiley-Blackwell, Chichester.

Tucker, C.S. and Hargreaves, J.A. (eds.) (2008) *Environmental Best Management Practices for Aquaculture*. Wiley-Blackwell, Oxford.

Appendix

Table 17.1 Examples of BMP for a small proposed shellfish farm on leased public lands in a lightly populated coastal embayment. The farm proposes raising oysters on the bottom and in suspended 'flip-bags,' and using both machine and hand harvest methods. The farm intends to construct a small dock and a storage/work structure with road access. Several private residences and small shellfish farms are in the immediate vicinity. This will be a full-time and year-round operation with three to five employees.

Issue	Objective	Example for this farm
Permits and regulatory compliance	Obtain necessary approvals for site leasing and construction, and shellfish harvest/water quality certification	Review permit and licensing requirements with local, state, and federal agencies. Work with extension agents, established growers and others familiar with the process. If required, submit an Army Corps of Engineers application complying with Nationwide Permit (NWP) No. 48 and other relevant NWPs.
Public engagement	Present farm development and operation plans	Engage in a dialog with your neighbors and others working and living in your farm area. Consider conducting an open house to educate the community about your proposed products and operations.
Access and property rights	Be a good neighbor	Recognize the needs of your neighbors and promote methods to avoid conflicts.
Aesthetics	Be aware of public aesthetic concerns	Ensure shapes, forms, profiles, materials and/or colors are consistent with the character of the adjacent environment.
Noise	Minimize noise impacts	Have a plan in place to control noise levels during regular farming activities.
Upland facility operations	Minimize environmental effects	Be sensitive to character of local community and shape business operations to blend with local community culture and avoid adverse impacts to receiving waters and uplands.
Vehicle operations	Minimize potential impacts on farm lands and adjacent property owners	Except during product movement and transfer, plan to restrict route selection and parking to the upper intertidal zone.
Vessels and marine equipment	Minimize possible negative impacts of marine equipment operations	Propose to follow 'clean marina' practices for maintenance of vessels and other in- and over-water equipment. Receive training in the operation and maintenance of all marine equipment.
Farm maintenance	Minimize environmental, navigation, and aesthetic issues of culture systems	Consider using identifying marks on any item that could become dislodged. Plan to routinely sweep the property and adjacent areas for old gear.
Site density	Maintain the productive capacity of shellfish growing waters	Have a plan in place to monitor shellfish growth to assess for changes as production is increased, and respond accordingly with appropriate husbandry practices. Consider the following for flip-bags arrays: orientation of the arrays with respect to currents and normal wave directions; the overall size of the arrays; and separation of the arrays from other farms (bottom or off-bottom).
Biofouling control	Minimize the amount and impact of biofouling discarded at the farm site	Propose cleaning gear on shore and disposing of fouling materials in an appropriate manner.

Issue	Objective	Example for this farm
Fish and wildlife interactions and predator control	Minimize and avoid impact on fish and wildlife	Ensure the farm will have limited impacts on fish and wildlife and their habitats, and predator control practices, if used, are targeted at specific organisms.
Disease management	Disease prevention	Have a capability to monitor and communicate mortality events to appropriate authorities and neighboring growers.
Protection of human health	Apply controls to prevent diseases associated with shellfish consumption	Work with public health personnel during certification of the farm and production/processing facilities.
Employee development and training	Be fully involved in proactive personnel management and training	Ensure that local, state and federal health and safety requirements are fulfilled and personnel are properly trained and equipped for each activity they undertake.

Note: Primary sources: ECSGA BMP manual, https://ecsga.org/wp-content/uploads/2018/01/BMPmanual.pdf; PCSGA environmental policy, http://pcsga.org/wprs/wp-content/uploads/2013/04/EPS.pdf; Maine BMP manual, http://www.maineaquaculture.com/code of practice v1.pdf.

CHAPTER 18

Environmental and social certification of aquaculture

Peter Cook

Introduction

At least three billion people on the planet depend on seafood for a significant portion of the protein in their diets. While the ocean is still a significant source of food, the demand for fish and shellfish is increasing each year and, unfortunately, the oceans are already unable to sustain current consumption demands – much less future ones. Many fish and shellfish populations are already exploited to their limits and, with demand for seafood predicted to rise 8% during the next decade, alternative sources will be required. It is estimated that about 30% of wild fisheries are already over-fished or significantly depleted, and the future of harvest fisheries is further complicated by issues such as illegal fishing, mislabeling, and the destruction of marine habitat. To accommodate the growing demand for more seafood, the world must increasingly turn to aquaculture as an alternative.

Aquaculture is the fastest-growing food production system on the planet, now being practiced in at least 190 countries. Aquaculture takes place in both freshwater and marine water and includes the production of a wide range of species, from tilapia and trout in freshwater, to salmon and oysters in seawater. Current estimates by the Food and Agriculture Organization of the United Nations (FAO, 2018) suggest that aquaculture now supplies more than one-half of all seafood produced for human consumption, and is expected to reach 109 million metric tons by 2030, representing a growth of almost 40% over 2016. More than 87% of the increase in aquaculture production in 2030 will originate from Asian countries.

Unfortunately, however, despite its importance as an alternative to wild fish, aquaculture creates its own sets of new problems. Ecological damage, water pollution, antibiotic overuse, threats to biosecurity, and excessive use of freshwater and energy, are all issues that could result from poorly managed aquaculture operations. To produce more seafood without increasing harm to the environment or to future human populations, it will be necessary not only to find new and innovative ways to operate farms, but also to find ways to recognize and reward farms that do the right thing. This chapter discusses whether internationally recognized certification schemes could help to achieve this.

There is certainly no shortage of sustainability standards around the world. As a response to the perceived environmental and social problems that aquaculture could potentially produce, a range of different standards have been developed. Sustainability certification is a market-based system and credible aquaculture certification schemes normally consist of three main components: (1) standards; (2) accreditation; and (3) certification.

Although the unit of certification varies between standards, most include the following steps.

1 Standards for farm-level ecological and social interactions are designed and set by a standard-holding body.
2 Independent auditors are appointed to check that farms comply with the set standards.
3 Entities that are judged to comply with the standards are permitted to use labels and logos owned by the standard-holding body.
4 Some type of 'chain of custody' system is instituted to ensure that certified products from the farm can be identified by end-user customers.

One problem, however, is that nowadays there are so many different standards that could be applied to aquaculture operations, many farmers, members of the seafood supply chain, non-governmental organizations (NGO) and consumers, find the entire sustainability landscape very confusing. This has led to confusion among producers, retailers, and consumers over how to recognize a credible seafood certification scheme, or how to purchase seafood that has been responsibly sourced. The following is a list of some of the better-known standards, and some NGO involved with aquaculture sustainability. Although not intended to be exhaustive, the list gives some indication of the range of alternatives available to consumers.

Seafood certification schemes

FAO Aquaculture Guidelines

The FAO created *Technical Guidelines on Aquaculture Certification* (FAO, 2011) in the belief that the application of certification in aquaculture is a potential market-based tool for minimizing potential negative impacts and increasing societal and consumer benefits and confidence in the process of aquaculture production and marketing (see Chapter 18). The guidelines provide guidance for the development, organization, and implementation of credible aquaculture certification schemes.

Aquaculture Stewardship Council

The Aquaculture Stewardship Council (ASC) is an independent, international, non-profit organization and labelling organization, that establishes standards and protocols for farmed seafood. The standards are designed to encourage responsible aquaculture. The Netherlands-based Sustainable Trade Initiative (IDH) joined forces with WWF to create the ASC in 2010. Its vision is a world where aquaculture plays a major role in supplying food and social benefits for mankind while minimizing negative impacts on the environment. Its mission is to transform aquaculture toward environmental sustainability and social responsibility using efficient market mechanisms that create value across the supply chain.

Global Aquaculture Alliance

The Global Aquaculture Alliance (GAA) is an international non-profit trade association, dedicated to advancing responsible aquaculture. Established in 1997, GAA works with industry, the NGO community, governments, academia, and the investment community to address the challenges and opportunities facing aquaculture. Its mission is to promote responsible aquaculture practices through education, advocacy and demonstration. It produces a bimonthly *Global Aquaculture Advocate* magazine and runs an annual Global Outlook for Aquaculture Leadership (GOAL) conference. GAA has been running a comprehensive third party aquaculture certification program (BAP) since 2002.

GlobalGAP

GlobalGAP is a farm assurance program, translating consumer requirements into Good Agricultural Practice. It is a global organization with the objective of creating safe, sustainable agriculture worldwide. It sets voluntary standards for the certification of agricultural (including aquaculture) products around the globe. EurepGAP is a common standard for farm management practice, created in the late 1990s by several European supermarket chains and their major suppliers. Many European customers for agricultural products now demand evidence of EurepGAP certification as a prerequisite for doing business.

Seafood Watch

Founded in 1990 as a project of the Monterey Bay Aquarium, Seafood Watch is a well-known sustainable seafood advisory list, which has influenced many similar programs around the world. Although not a standard-holder, it is best known for developing science-based seafood recommendations that consumers, chefs, and business professionals use to inform their seafood purchasing decisions.

Friend of the Sea

Friend of the Sea is a non-profit NGO whose mission is the conservation of the marine habitat. It encompasses a system for the certification and promotion of seafood from sustainable fisheries and sustainable aquaculture. It uses the same logo to certify both wild and farmed seafood. It was founded in 2006, and certified products include many species that are traded internationally as well as fishmeal, fish feed, and omega-3 fish oil. Friend of the Sea certification requirements follow the FAO – Guidelines for the eco-labelling of fish and fishery products.

Global Sustainable Seafood Initiative

The Global Sustainable Seafood Initiative (GSSI), although not itself a standard-holding body, represents an attempt to establish a collective, non-competitive approach to provide clarity on seafood certification and ensure consumer confidence in certified seafood. The GSSI approach is to use a series of reference documents to produce a bench-marking system. At the heart of the process are documents such as the FAO Code of Conduct for Responsible Fisheries (CCRF), the FAO Guidelines for Eco-labelling of Fish and Fishery Products from Marine/Inland Capture Fisheries, and the FAO Technical Guidelines for Aquaculture Certification (FAO Guidelines).

The GSSI Global Benchmark Tool includes GSSI Essential Components that are based on the CCRF and the FAO Guidelines and which seafood certification schemes must meet to be recognized by GSSI. These are defined by GSSI Expert Working Groups and are grounded in ISO normative standards and International Social and Environmental Accreditation and Labeling (ISEAL) codes. The overall objective of the GSSI Global Benchmark Tool is to work toward a collective objective to minimize the overall environmental impact of catching, growing, and delivering seafood.

Sustainable Fisheries Partnership

Sustainable Fisheries Partnership (SFP) is an organization that is attempting to reshape the world of corporate responsibility in the seafood industry through the creation of powerful information tools and a methodology that enables companies to directly engage with suppliers of natural resources. The SFP vision is one of healthy marine and aquatic ecosystems; secure seafood supplies; and a thriving, responsible seafood economy.

National and International Regulations

An extensive national and international legal framework already exists for various aspects of aquaculture and its value chain, covering such issues as aquatic animal disease control, food safety and conservation of biodiversity. Legislation is particularly strong for processing, export and import of aquatic products. National authorities are often empowered to verify compliance with mandatory national and international legislation. Many operators have suggested that strong aquaculture regulations in their region negates the requirement for additional certification schemes. The problem, however, is that not all countries have either the knowledge, experience or manpower to initiate and police aquaculture regulations effectively, and this provides the opportunity for independent, voluntary certification to become a means to demonstrate that a particular aquaculture system is managed responsibly. Where national aquaculture regulations already exist, the FAO (2011) Aquaculture Guidelines state that "third-party certification schemes implemented in application of these guidelines should not replace related certification schemes or official certificates issued by States" (Clause 175, p. 37).

Credibility of certification schemes

Given the range of certification alternatives, it is important to determine what constitutes a credible seafood certification system. According to the FAO (2011) aquaculture guidelines, credible certification schemes should preferably include at least the following aspects: (1) environmental integrity; (2) socio-economic factors; (3) animal health and welfare; and (4) food safety. The guidelines acknowledge that sustainable development of aquaculture depends on three factors, social, economic and environmental sustainability, all of which have to be addressed. In addition, to ensure credibility, the FAO guidelines cover:

- the standard setting processes required to develop and review certification standards
- the accreditation systems needed to provide formal recognition to a qualified body to carry out certification assessments
- the qualifications and practices of the certification bodies required to verify compliance with certification standards.

The guidelines provide information on the institutional and organizational arrangements for aquaculture certification, and include governance requirements that are designed to ensure that conflicts of interest do not occur.

One of the most important indicators of a credible certification scheme is compliance with international requirements set down by the International Social and Environmental Accreditation and Labeling (ISEAL) Alliance. ISEAL is a global association for credible sustainability standards. Members of ISEAL are sustainability standards-holders that meet 'Codes of Good Practice' and promote measurable change through open, rigorous and accessible certification systems. Members are supported by international accreditation bodies, which are required to meet accepted international best practice (see also Chapter 17).

To ensure objectivity and to avoid conflicts of interest, the actual process of farm assessment should be carried out through a third-party process. Third-party programs offer the highest level of assurance and mean that the outcome is unbiased. The difference between the various possible levels of assurance, are outlined below.

- **First-party** – An organization, product, or service meets standards it has set for itself.
- **Second-party** – An organization, product, or service meets standards established by peers, for example by an industry association.
- **Third-party ('certification')** – An independent assessment shows that the organization, product, or service meets standards that have been established by impartial experts, often in consultation with stakeholders. A certificate is issued to prove that the standard has been met.

To be credible, seafood certification schemes need to ensure the use of a third-party certification process. Farm assessments to determine compliance with standards are normally carried out by a conformity assessment body (CAB). Another indicator of the credibility of a scheme is the way in which the approved CAB itself is accredited and monitored. The more credible schemes

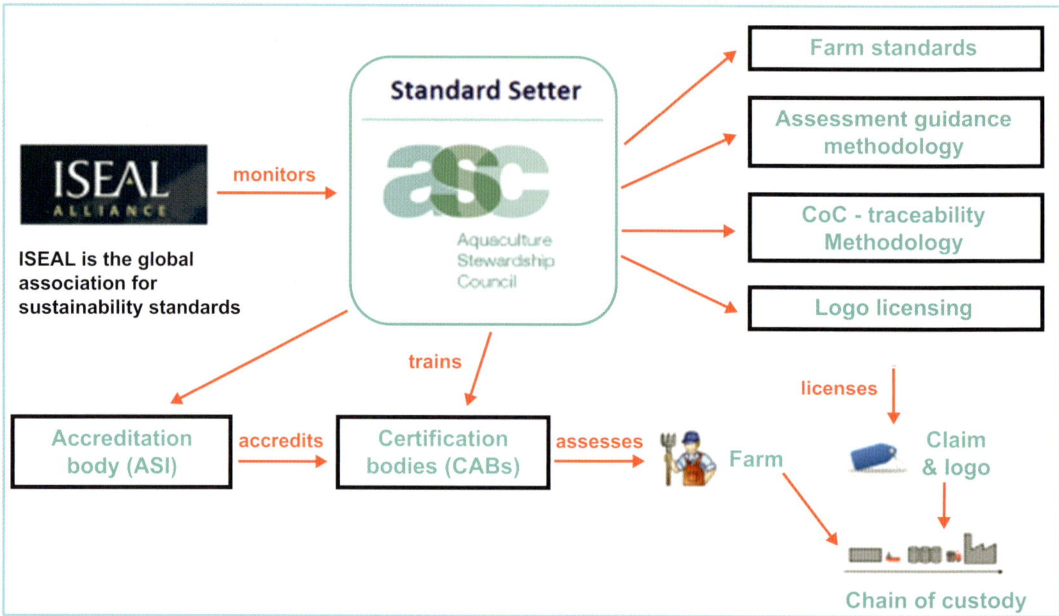

Figure 18.1 Accreditation and certification process used by the Aquaculture Stewardship Council.

usually use ASI (Assurance Services International, formerly Accreditation Services International) as a quality assurance process for their approved by a CAB.

The ASI is an international assurance partner for leading voluntary sustainability standards and initiatives, being involved both in the initial accreditation of a CAB and in monitoring CAB activities to ensure that they adhere to high international standards. Completely independent of all standard-holders, ASI helps to ensure the third-party nature of certification processes. If a particular certification scheme does not insist that all of its approved CAB are accredited and monitored by ASI, or at least by an internationally acclaimed alternative, then the credibility of that scheme would be in question.

As an example of how the accreditation and certification process works in practice, the scheme used by the ASC is illustrated in Fig. 18.1.

In addition to setting the various standards, the standard-holding body may publish guidelines for auditors on how to apply their standards (assessment guidance methodology). In addition, the standard-holder also selects a system for ensuring that the products certified on the farm are the same products that eventually get to the end-users (chain of custody [CoC] – traceability methodology). In the case of the ASC, the CoC methodology used is an established process that is administered through the Marine Stewardship Council. Perhaps the most important attribute of the ASC certification process is transparency – where all phases of the process are publicly available on the ASC website.

Another important aspect of the certification process is the use of a specific logo that a producer, processor, or marketer can use to demonstrate to end-users that the product they are purchasing has been approved through a specific certification process. This logo is often displayed on the final product packaging that a customer would see, for example, on a supermarket product.

The license to use the logo is normally held by the standard-holding body and in most cases that body will charge for its use.

The cost for a farm to go through the certification process and the cost for a farm, processor, or marketer to use a specific logo, varies between the different standard-holding bodies. Again, as an example, the costs associated with ASC certification will be summarized.

For a farm to become certified, the main cost is the hire of an accredited company (CAB) to undertake the farm-based certification survey. The cost of this depends upon the size and complexity of the farm, and on its location and the associated travel implications for the auditors. Because these costs are extremely variable, no attempt will be made, in this chapter, to estimate such costs. In many cases, more than one auditor may be required to perform the on-site audit. If the CAB finds that the farm satisfies the requirements of the standard, the auditor will request ASC to issue a compliance certificate. The ASC does not charge for this certificate. Processors or marketers can request that certified products from that farm are permitted to display the ASC logo, and this is where the ASC derives its income.

There are two different types of costs payable by organizations that use the ASC logo, annual fees and royalty fees. Organizations wanting to use the logo must first apply for a license to be legally covered to do so. The costs involved in using the logo vary depending upon the sector in which the organization operates. The following sectors do not have to pay either of these fees: certified farms and accredited assessment bodies; non-commercial organizations without food-service operations; the media; and accredited educational institutions (that is, universities and schools). Most supply chain, retail, and foodservice companies who handle ASC products pay only the annual fees while supply chain, retail, and foodservice companies who are the point in the supply chain where the ASC logo is put onto products, have to pay both the annual and royalty fees.

The annual fee depends upon the value of ASC certified seafood that is sold. For annual sales from US$0–200,000 the fee is US$250, for US$200,000–500,000 it is US$1000, and for sales above US$500,000 the annual fee is US$2000.

If a company uses the ASC logo on consumer facing (retail) products, it will need to pay royalty fees in addition to the annual fee. Royalty fees are calculated at 0.5% of the wholesale price of seafood that is sold. Within the supply chain, royalties are only collected once, for a product. The following example illustrates how fees are calculated.

A company that expects to sell US$100,000 of ASC logoed products would pay an annual fee of US$250, at the beginning of the year. Then, at the end of each charging period, they would calculate their total sales. If, for example, they sold US$110,000 of ASC logoed products in the year they would pay US$550 in royalty fees (0.5% of US$110,000) and so the total cost for this organization to be licensed to use the logo for a year would be US$800 (US$250 + US$550).

It should be emphasized that the above example is for the ASC only and that other certificate-holding bodies have their own, unique charging structures.

If we accept that to produce more seafood without increasing harm to the environment, we need to recognize and reward farms that do the right thing, we then need to consider whether internationally recognized certification schemes have contributed to this objective. In an attempt to address this question, ISEAL commissioned a report in 2018 that aimed to understand the effectiveness of sustainability standards and certification tools in driving the adoption of more sustainable practices (Petrokofsky and Jennings, 2018). The report concluded that there was a strong suggestion that the technical support that the certification process offered to farmers may have been critical in supporting them to adopt more sustainable practices. There was some evidence that certification and standards can contribute to the adoption of improved practices, but it must be noted, however, that most research was on the coffee and forestry industries, and it is not known, therefore, how relevant the findings would be for other sectors, such as aquaculture.

In conclusion, it is obvious from the above, that international aquaculture certification is a complex and potentially confusing process. If the overall objective, however, is to demonstrate that products have been produced, processed and marketed in a responsible manner, then the most important thing is to check the credibility of the scheme being used. Potential users of aquaculture certification are advised to research the various options carefully, before committing to a particular scheme.

References cited and further reading

FAO (2011) *Technical Guidelines on Aquaculture Certification*. FAO, Rome.

FAO (2018) *The State of World Fisheries and Aquaculture 2018 – Meeting the Sustainable Development Goals*. FAO, Rome.

ISEAL (2018) ISEAL code of good practice for setting social and environmental standards. Available at: https://www.isealalliance.org/get-involved/resources/iseal-codes-good-practice.

Petrokofsky, G. and Jennings, S. (2018) The effectiveness of standards in driving adoption of sustainability practices: a state of knowledge review. Available at: https://www.isealalliance.org/get-involved/resources/report-effectiveness-standards-driving-adoption-sustainability-practices.

CHAPTER 19

Marketing

Carole R. Engle

Introduction

Molluscan shellfish have served as an important food source for people living in coastal areas from the beginnings of human existence. Shellfish are a good source of protein, omega-3 fatty acids, vitamins, and minerals, and, hence, it is no surprise that they have been an important part of human diets and nutrition from the time that humans learned to eat them. In many areas, shellfish harvest and consumption became an important part of cultural traditions, both as food on important holidays and shells as tokens of appreciation. For example, in Santiago de Compostela, Spain, the final portion of the pilgrimage on the Way of Saint James includes impressions of scallop shells in the sidewalks; pilgrims used scallop shells as a cup for water and for begging for food. Native Americans in the northeastern USA used quahog shells to make their sacred white and purple wampum beads. Shells have been collected for centuries for aesthetic purposes (Duncan and Ghys, 2019). Royal shell collections have been maintained in several European countries, an international trade has developed in shells, and online shell retailers now cater to collectors and artists. Shells have also been used for industrial purposes for their calcium content in both agriculture and building, to produce lime for mortar and plaster. Given the importance of shellfish as a food source, a cultural symbol, an ingredient in agriculture and building supplies, and as works of art, it is no surprise that shellfish gathering evolved from simple gathering into commercial wild harvest and then into farming.

This chapter begins with a description of the global market for molluscan shellfish and then describes important considerations related to effective shellfish marketing. Various aspects of shellfish products as these relate to consumer preferences and demand will be described as well as guidance on how to set prices for various products. Supply and value chains are described as are the general types of regulations that affect shellfish marketing. The chapter concludes with discussion on how to develop, implement, and evaluate a marketing plan and strategy.

Global market for molluscan shellfish

The global market for a product like molluscan shellfish includes both production and sales as well as all the steps that must occur for product to reach its end consumers (such as shucking, packaging, shipping, and advertising). The wide range of shellfish species, varieties, and product forms produced and sold worldwide result in complex markets, challenges, and opportunities.

The total supply of shellfish worldwide includes both wild harvested and farmed shellfish and has grown rapidly. Farmed and wild harvested product compete in the marketplace, but the overall mix has shifted to predominantly farmed production over the years. In 2017, the total volume of farmed shellfish was nearly seven times greater than that harvested from the wild (FAO, n.d.). Farmed shellfish production has grown rapidly both in terms of volume and value over time (Fig. 19.1) and began to exceed that from wild harvest in 1983. In fact, the volume of wild harvested shellfish peaked in 1984 and has exhibited a declining trend since then, whereas farmed production of shellfish continued to grow at an annual average rate of 6.6% per year over that same time period.

Major categories of farmed shellfish (by volume in 2017) include clams and cockles (36%), followed by oysters (36%), mussels (14%), scallops (13%), and abalone (1%) (Fig. 19.2). Oysters historically had been the largest category of farmed shellfish; however, growth worldwide in the

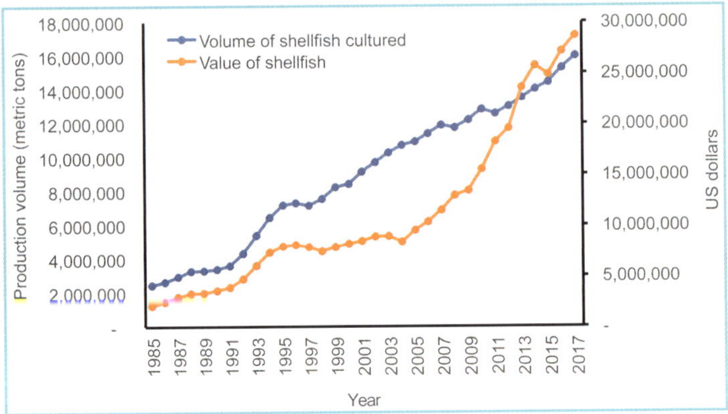

Figure 19.1 Growth in molluscan shellfish aquaculture, 1984–2017. Data from: FishStatJ database (FAO, n.d.).

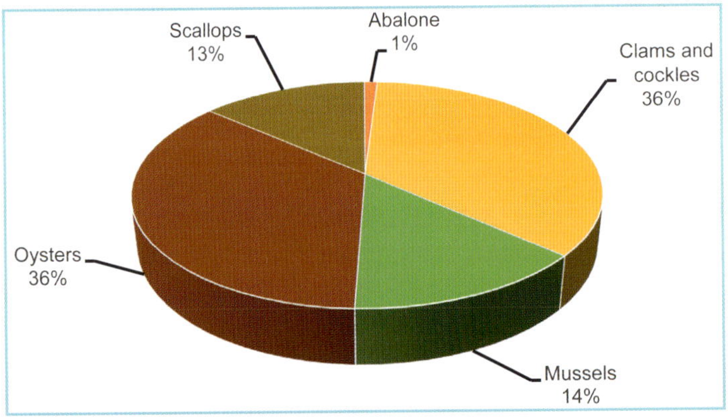

Figure 19.2 Relative proportions (by volume, MT) of types of shellfish produced, 2017. Data from: FishStatJ database (FAO, n.d.).

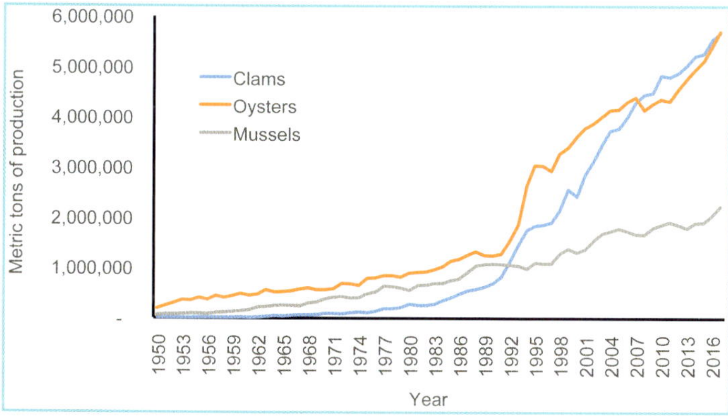

Figure 19.3 Growth of clam, oyster, and mussel categories, 1950–2017, by volume. Data from: FishStatJ database (FAO, n.d.).

volume of both oysters and clams began to accelerate in the early 1990s, with the volume of farmed clams supplied to the market exceeding that of oysters in the mid-2000s (Fig. 19.3).

Shellfish are farmed worldwide, with 211 countries reporting farmed production in 2017 (FAO, n.d.). China was the world's largest producer of shellfish overall as well as of all major categories of shellfish (oysters, clams, mussels, scallops, and abalone) in 2017 (Table 19.1); however, a number of other countries have played important roles over the years in the development of production techniques for various types of shellfish. Malaysia led the world in production of clams and cockles from 1962 to 1974; France led the world in oyster production from 1950 to 1956. France, Spain, the Netherlands, and Sweden have long histories of mussel culture in Europe that date back to the 13th century (Bardach et al., 1972). The Philippines and Thailand similarly have long histories of mussel production. The Netherlands was the world's largest mussel producer until Spain assumed that role in 1970, and abalone production was led by Taiwan from 1977 to 2002. Eventually, China overtook all other countries in these major shellfish categories.

Shellfish are important trade commodities for many countries. Of the leading shellfish-producing countries, 64% are net exporters (Table 19.1). It is interesting that China, while the leading producer of oysters, mussels, and scallops, is also a net importer of these three shellfish categories. In addition, Taiwan and Thailand are net importers of clams, Japan, China, and the US of oysters, and Chile and Thailand of mussels.

Shellfish have not been exempted from international trade conflicts. Peru, Chile, and later Canada filed a complaint with the World Trade Organization against France in 1993 related to approved names of scallops. France had developed an order that established official names for scallops in the domestic market that would have prevented scallop producers from other countries from selling their scallops as 'Coquille Saint-Jacques,' a readily recognized name in France. A mutual agreement was reached in 1996 that allows the use of 'St. Jacques' for all scallops.

In 2001, the Great Eastern Mussel Farms of Tenants Harbor, Maine, filed an antidumping petition (request to investigate whether a company is selling a product in another country at unfair market prices, referred to as 'dumping,' a practice that is not allowable under World Trade Organization guidelines) against Prince Edward Island mussel producers in 2001. Tariffs were imposed initially, but when the Prince Edward Island mussel growers raised prices the antidumping lawsuit was eventually withdrawn.

Table 19.1 Leading shellfish-producing countries, 2017.

Type of shellfish/country	Metric tons of production, 2017	Net exporter/importer	
		Exporter/importer	Net quantity (metric tons)
Abalone			
China	148,539	Net exporter	2439
Korea, Republic of	16,027	Net exporter	1681
South Africa	1122	Net exporter	731
Chile	1031	Net exporter	185
Australia	873	Net exporter	1978
Oysters			
China	4,879,422	Net importer	(1761)
Korea, Republic of	315,255	Net exporter	6314
Japan	173,800	Net importer	(2381)
United States	140,892	Net importer	(3095)
France	64,910	Net exporter	4623
Mussels			
China	997,045	Net importer	(2799)
Chile	341,447	Net importer	(80)
Spain	241,785	Net exporter	47,611
New Zealand	99,716	Net exporter	34,315
Thailand	84,923	Net importer	(2545)
Scallops			
China	2,007,529	Net importer	(12,149)
Japan	135,100	Net exporter	39,664
Peru	11,927	Net exporter	3199
Russian Federation	5463	Net exporter	4674
Chile	4706	Net exporter	714
Clams, cockles, ark shells			
China	5,414,819	Net exporter	94,473
Korea, Dem. People's Rep.	62,000	Net exporter	17
Taiwan	56,587	Net importer	(1648)
Thailand	44,832	Net importer	(1833)
Italy	36,500	Net exporter	7486

Shellfish products and consumer demand

A 'product' to a consumer is a category of something that they buy. How consumers define a given product varies over time and with different groups of consumers. Shellfish products have a long tradition of being labeled by the location of harvest, and harvest locations often are associated with distinct species that are native to those locations. For example, the Pacific or Japanese oyster (*Crassostrea gigas*) was native to the northwest Pacific Ocean. While it has been seeded for farming in many parts of the world and has become the most widely farmed oyster, it is still referred to as the Pacific oyster. Similarly, the European flat oyster (*Ostrea edulis*) that is native to Europe but now also farmed elsewhere in the world is still referred to as the European oyster.

Within these broad categories of product names, oyster companies have historically relied upon the tie of their oysters to the waterbody where grown in their marketing programs. For example, Blue Point and Cape Cod in the USA, Caraquet and Malpèque in Canada, and Marennes-Oléron in France are known as prestigious oyster-growing areas, and oyster companies use the name of the region to attempt to charge premium prices for those products.

Oysters

When early markets began to develop for oysters, they were initially shipped in the shell. However, in the early 1920s, oyster meat began to be packed in tins and later in jars for shipping because shipping costs could be reduced and oyster shells could be recycled. Oyster packing-houses emerged in many coastal areas where oysters abounded to process, shuck, and package oysters for shipping to new markets that were developing away from the coast. Shellfish dealers emerged, many with buyboats, who were licensed to purchase shellfish from watermen and farmers. Supply chains for oysters and other types of shellfish continued to develop as food markets became more concentrated and supermarkets and hypermarkets began to dominate the retail food sector. Shellfish producers in the 21st century have a variety of supply chains to choose from in terms of whether to market their products directly to consumers, supermarkets, restaurants and oyster bars, or to processing plants. Packaging can include sacks, boxes, or bushel bags.

Oysters are consumed in a variety of fresh raw (often in the shell), frozen, and in cooked forms at home and in restaurants. Processing typically involves manual shucking of shellstock oysters and grading the meat into different sizes. Value-added products are available on the market in cooked, canned, breaded, and smoked forms with a wide variety of retail packages and sizes (for example, cups, pints, quarts, and gallons).

Clams

Clam products and their names are perhaps even more varied than those for oysters. To begin with, clams are often designated as either 'hard-shelled' or 'soft-shelled' clams. Quahogs, Atlantic hard-shelled clams, are sold for chowders and stews when large, as 'cherrystones' when medium-sized, and as 'littlenecks' when small. Cherrystones and littlenecks are steamed, stuffed and baked, or eaten raw. Cockles are often prepared in soups and sauces. Razor and Manila clams are often steamed as are soft-shelled clams, those species with thin and flexible shells.

Mussels

Records of mussel farming date back to the 13th century in France and frequently are attributed to a legend of an Irishman who was shipwrecked on the coast of France. The shipwrecked sailor found mussels on nets stretched over stakes when attempting to catch birds to eat. 'Bouchot' mussels continue to be raised in France, cultured on fixed, wooden poles, and are considered a premium product as compared to rope-cultured mussels.

Mussels continue to be marketed primarily as a fresh, whole product. Mussels do not transport well as a live product over long distances because they do not close their shells when out of water as do some other types of shellfish. Thus, major market areas tend to be in relatively close proximity to the main production areas. Most exported mussel products have been processed in cooked, canned, or frozen forms due to the difficulties involved with long-distance shipping of mussels. Some mussels are sold frozen (raw or cooked), on the half shell, or canned in oil, brine, or smoked. Away from home sales of mussels have grown, and the restaurant trade has developed an increased interest in mussels.

Farmed Spanish mussels typically are intermediate-priced products with wild-caught mussels typically the lowest-priced products. There are exceptions, however, such as wild-caught mussels from the Basin of Marennes-Oléron (Girard and Mariojouls, 2003) region of France. Considered a premiere shellfish growing area, a regional trademark was developed in 1974 that has successfully resulted in higher prices for shellfish from this region. In Asia, the Asian green mussel (*Perna viridis*) is a traditional product and crop, although production of blue mussels in Asia has grown rapidly. North American markets, for many years, were slow to develop demand for mussels; however, sales and consumption of mussels have increased in North America since the 1970s.

Effective shellfish marketing

Shellfish marketing used to involve simply selling what was harvested in local areas; however, shellfish farms have become sophisticated businesses. Growth and development of shellfish farming entails investment of capital for shoreline infrastructure and facilities necessary for growing seedstock, tumbling, sorting, grading, and for processing shellfish. Increased awareness of food safety and understanding of how to minimize food safety risks has led to greater attention to handling systems that maintain proper temperatures at all times to minimize the risk of food safety problems in the shellfish produced. As greater levels of capital are invested on a shellfish farm, the need to increase sales revenue to pay the increased bills also increases.

Successful shellfish farms attract competition from others who seek to enter markets with products that have been demonstrated to be successful. Such businesses require that marketing shellfish be approached in a business-like manner in which close attention is paid to financial performance and also to marketing strategies. Managing a successful business is a complicated undertaking that involves making good decisions on a wide variety of issues from production gear and technology, location, scope and scale of the business, personnel and staffing decisions, financial management, and, of course, marketing decisions. Each decision made by the business manager is one of many other pieces that must fit together in a coherent business model for the shellfish farm to be profitable and successful.

Knowing your customers and their preferences

Effective marketing in the 21st century requires a deep and thorough understanding of which types of individuals are most likely to purchase that particular farm's product(s). For example, a clam producer in Rhode Island, in the northeastern US, may be able to tap into existing traditional markets by targeting and highlighting their clams for traditional Rhode Island clambakes. A producer attempting to develop a new market for a product like blue mussels in the US, however, would need to develop a different type of marketing strategy other than traditional clambakes for their mussels.

Consumers today have a choice among a wide variety of products, including shellfish. One of the key questions that a successful shellfish farmer must answer correctly is why should someone purchase oysters or clams from their farm and not from a different farm. Alternatively, why should a consumer purchase oysters and not some other type of seafood?

A shellfish grower must first understand clearly what he/she really is selling in the eyes of their customers. While seemingly obvious, defining and developing a clear understanding of why customers purchase their product is in fact a very important but often difficult component of strategic marketing. For example, if a shellfish farmer views his/her shellfish as simply a type of food, then their shellfish must compete with all other types of food. If shellfish are sold to tourists who visit a coastal region to experience the local food scene, history, and culture, then what is being sold is a component of an overall tourism experience. The waterman who sells an experience on-the-water (that includes sampling oysters) must design the business differently from one that is selling mostly a commodity oyster that is similar to many other oysters available in the market. Understanding what the business truly offers its customers and how that is the same or different from other similar businesses is critical for that business to position itself effectively in its targeted markets.

Marketers frequently talk about marketing not just as sales, but about meeting a customer's need or solving a customer's problem. The choice of what to eat for dinner is a need for which people in many locations have many, many options in the 21st century. To be effective in marketing a product, the farmer must meet each buyer's needs better than other farmers that produce similar crops. Are your shellfish the freshest? Are they the highest quality? What makes them the highest quality? Is it that your customers want locally grown product? Or is it because you provide much better service and support than do others?

What specific products will best meet the needs of customers in your market area? What types of shellfish do they prefer to eat? Do they prefer to eat it fresh on the half shell, or do they prefer to purchase already shucked oysters that they prepare at home? How many different products do your customers want? Do they like the variety of having different shellfish from different areas, that all have a slightly different taste (differentiated products) or do they want the consistency of one specific type that is based on a family tradition? As importantly, how many different types of products can you manage efficiently with your work force?

For shellfish consumers, key attributes are flavor, size, and meat content. Different types of consumers may prefer different sizes and flavors, and some enjoy sampling a wide variety of different flavors of shellfish. One of the keys to a successful marketing strategy is to understand the flavor profile of the shellfish produced and identify groups of customers that prefer that type of flavor profile. While differentiation is important, there are some factors that are considered essential for all food products. These include consistent quality, shape, cleanliness, safety, year-round availability, timely shipping, and attractive packaging.

Shellfish farmers can also meet buyers' needs through the type and size of packaging because preferences for how shellfish are packaged differ among consumers. Some supermarkets and restaurants may prefer shellfish to be packed in boxes while others prefer bags. Seafood restaurants with large raw bars may prefer delivery of larger package sizes whereas shoppers at a local farmers' market likely prefer smaller package sizes or the ability to choose the individual shellfish that they purchase. Understanding which is most convenient and preferred for each customer will help to position the farm to become the preferred supplier.

Such considerations need to integrate production considerations with meeting marketing preferences of targeted consumers. Normal, diploid oysters are not sold during their spawning season when the meat becomes soft and undesirable. The shellfish producer raising diploid oysters will have to plan for the cash flow deficits in months in which spawning oysters cannot be sold. On the other hand, triploid oysters are sterile, and can be harvested and sold throughout the year because there is no spawning season. Triploid oysters also have the advantage of reaching market size in 2 instead of 3 years. A farmer who raises triploid oysters can 'turn over' (or generate revenue from sales) their crop a year earlier than with normal diploid oysters that spawn. Such product turnover creates clear cash flow advantages for the producer. However, triploid oyster seed is more expensive and requires greater levels of upfront capital during stocking than do traditional, diploid oysters.

How to locate customers

Finding good customers is often a challenge for any business. Research studies can provide ideas of which types of market segments may be best to target. For example, Hanson et al. (2003), in a national survey of consumers, found that most people who ate oysters did so because they enjoyed the taste and that oysters added variety to their diet. A 2007 survey of consumers in southern California showed that 20% of their respondents consumed raw oysters from the Gulf of Mexico states, but that they had major concerns related to the health and safety associated with consuming raw oysters from the warmer waters of the Gulf of Mexico (Posadas et al., 2011). Higher-income consumers tended to be more frequent oyster consumers in southern California.

In addition to reading research reports from marketing studies, the best way to locate and identify potential customers is to spend time talking with and observing individuals considered part of the market segment considered likeliest to purchase the business's product. It is worthwhile to visit other, similar, businesses and observe what types of people buy their products. Many prospective customers are willing to talk about what they seek in a product, where they want to purchase it, when they want to eat it, and what forms they want it in. Observing the behaviors and shopping patterns of those customers targeted by the business can reveal important clues as to how to market more effectively to those types of customers. Identifying a way to better meet customer needs and desires can lead to a very profitable business.

Communicating with customers (or what used to be called 'promotion and advertising')

Traditional marketing relied on advertisements to promote a company's product(s). Advertisements were developed to communicate the advantages of a particular product over other similar products, the values of the company producing that product, or to create a tie with the lifestyle of their targeted customers. Traditional advertising has ranged from simple but yet effective word-of-mouth advertising, flyers, printed advertisements in local newspapers or seafood magazines, and booth displays in major seafood shows to radio and television advertisements.

New, innovative economic development and tourism departments in various states have begun to offer opportunities for shellfish producers to participate in tourism-related activities. For example, states such as Virginia, Maryland, and North Carolina in the US as well as various companies and regions in Australia, France, and other countries, now offer 'oyster trails'. Participation in an oyster trail program offers opportunities to include a business profile outline, list the company's name on a visitor map, or use an oyster trail logo. Other digital marketing opportunities are often made available through departments of tourism by linking with accounts on Facebook, Twitter, Instagram, and other social media platforms. The Virginia Oyster Trail, launched by the state's governor in 2015, for example, uses promotional lines such as "Oysters are the sea's version of fine wine: their taste varies with the water they grow in." In fact, many oyster trails suggest wine pairings as part of the oyster trail experience. The North Carolina Oyster Tour highlights visits to several oyster farms and eateries that feature oysters, with demonstrations of products and preparation methods. The Maryland Crab and Oyster Trail highlights more than 100 seafood restaurants, including festivals, seafood markets, and a Waterman's Heritage Tour. In Australia, Captain Sponge's Magical Oyster Tours feature tours of oyster farms in Pambula Lake, similar to visits to traditional oyster beds in the Marennes-Oléron Region of France.

The success of these initiatives stems from growing recognition that many tourists seek a 'local' experience and are attracted to activities that offer an experience and 'taste' of local areas. The Virginia Oyster Trail touts eight distinctive flavors of oysters ('water-to-table delicacies') throughout its coastal region. In addition, tourists are encouraged to rent kayaks for an 'on-the-water' experience in which they meet watermen, learn how oysters are raised, and learn about ecosystem services provided by oysters in addition to sampling local varieties.

Consumers in the 21st century expect instant information, a stream of communications, and rapid responses. With the dramatic expansion of use and diversification of social media platforms, expectations have changed and marketing and promotion decisions have become more complex. Understanding this transformation in the marketplace is critical to develop effective marketing strategies. It has become much easier for consumers to identify groups of like minded individuals on social media with whom they wish to engage in discussions, ratings, and comments about products and services offered by businesses.

While social media provides many opportunities, it also poses new challenges for business owners. One of the complexities of marketing in the age of social media is that different types of lifestyle groups frequent different types of social media sites. While Facebook initially was frequented by mostly teen-aged groups, it later became a platform used more often by older people with grandchildren than by young people who have migrated to other types of platforms. In other cases, individuals who are more visually oriented, but span many different age ranges, often prefer to spend more time on Instagram or Pinterest than on other types of social media. Posts can still be linked to Facebook and Twitter accounts. Twitter can be used for conversations about new products in nearby oyster bars, new tastings, shucking events, and other activities. Tagging can be used effectively if targeted toward the consumers the business is really trying to reach. Some types of customers, though, spend most of their time searching the web or conversing on email and do not use Twitter, Facebook, or other types of social media. Video clips can be effective to promote various product lines, especially to communicate important messages about production methods, environmental sustainability, and safety of the products being sold. YouTube, Facebook pages, or websites are all options for posting video clips, with YouTube often the medium of choice. Careful planning for the best way to communicate with customers, and costs of alternative communication strategies is important to make decisions as to the most important strategy for the company.

Customer loyalty: where does it come from?

The most successful businesses are those that are able to develop strong customer loyalty to their business and to their products. The extent to which a customer demands a particular business's product over others will determine the extent of repeat sales as well as whether there is an opportunity to charge a higher price. Consumers with the means to do so will pay more for a product that uniquely satisfies that customer's needs.

Care and attention to the consistency and quality of the product can be driving features of customer loyalty. Many businesses prioritize a high degree of customer service by developing personal relationships and providing assistance to customers when they are experiencing difficulties. In doing so, a business can create a high degree of customer loyalty if it is willing to make the effort to provide such a high level of service to its customers.

Developing a relationship of trust with customers is an important foundation for customer loyalty. Such relationships can be built through invitations to visit the farms, shucking events that include local personalities, and linking websites, Facebook, or Instagram postings between the farm and their customers' restaurant sites that reinforce and emphasize that relationship.

Customer loyalty can be developed for the product itself, for customer service, through branding, or a combination of factors. Branding is becoming more common in the food sector as a way to increase customer loyalty. A brand expresses the company's core values as well as key attributes of the product. Building a brand is a slow process that must be developed over long periods of time, but, when successful, results in strong customer loyalty. The process of developing a brand begins with developing and articulating its core values, and then ensuring that each and every product offered under that brand meets the expectations set by those core values. Messages to customers must be crafted to articulate the business's values as these relate to the key attributes of the product that are most important to the customers. These messages then need to be communicated and delivered effectively through the correct communication channels and platforms.

All shellfish marketed under the brand must be consistent in terms of quality, size, and cleanliness. Creating and maintaining a close and trusting relationship with buyers through excellent customer service, being hospitable, and taking good care of them consistently over time is critical. The process of branding requires persistent reinforcement of the key messages and core values associated with the brand.

Some of the earliest shellfish brands emerged with the European Union Protected Designation of Origin (PDO) program. This program built upon the fact that consumers tended to identify shellfish products by the geographic location where grown. With the PDO program, for example, the bouchot mussels raised in the Bay of Mont Saint-Michel in France, the Mejillón de Galicia PDO (mussel of Galicia, Spain), and the Scardovari Mussel PDO of Italy were officially identified by their brand names that contributed to company efforts to sell them as premium and luxury products.

In North America, development and marketing of branded oysters is a phenomenon that began to take hold only in the late 1990s. One of the earliest oyster brands was the Kusshi oyster. Developed in British Columbia, Canada, it was the first tumbled oyster to hit the market in North America. Its unique golfball shape, with a consistently deep cup and plump meat, allowed the company to differentiate their oysters from others, build a brand, and enabled them to charge a higher price. Hundreds of specialty oysters have since appeared on the market. In the following years, oysters with names such as Lambertini, Nordic Knute, Shigoku, Kumamoto, and Fat Bastard have become common in the market. Companies and restaurants began to offer 'samplers,' 'assortments,' and gift packages that emphasize a variety of sizes, shapes, names, and flavors of oysters.

Co-branding has been used successfully by many companies. Any product associated with the experience of producing or eating shellfish can serve as a co-branding opportunity. There are many opportunities to co-brand shellfish with wineries, craft breweries, or other innovative partnerships (for example, manufacturers of shucking knives, seaweed products, and so on). Quality, not the quantity becomes relatively more important in co-branding than in some other marketing programs.

Brands can be so successful that they can become targets for brand theft and fraud. A company working to develop a brand must recognize the need to register a trademark for its brand early on. Once the branded product is launched on the market, the company must monitor activity and be vigilant for attempts to imitate or steal it by selling product that likely does not meet its standards and violates its trademark.

Quality, consistency, and safety

Products need to consistently present the key quality attributes expected by consumers. Consistent quality of products offered is always an important factor in the successful retention of customers as well as in the process of recruiting new customers. While it may be tempting to add a few smaller oysters or shellfish with unattractive shells to a box, customers will likely notice and become more willing to talk with other suppliers who can deliver consistent quality. Consistent sizing is a component of overall shellfish product quality as is safety and cleanliness (for example, washing off salt and growth so that shellfish are clean and fresh for customers). Monitoring harvest time, date, and water quality throughout the production and processing phases and providing that information to customers demonstrates a commitment to quality.

Consistent supply of product to meet customer needs is as important as consistent quality. When a good customer has successfully helped to develop a market for shellfish from a particular farm, that market will require consistent supply for consumers to continue to seek that product. Most aquaculture businesses at some point in time will experience losses and have difficulty meeting the orders of all their customers. While product may need to be rationed, or even purchased from another farm to meet orders, it is critical to save enough product to keep at least the best, core customers supplied throughout their most important marketing seasons even if it means making less money.

Excellent customer service includes the responsibility to talk honestly with customers about problems that occur with the product. Customers are much likelier to return to a producer to purchase additional product if the supplier has been honest than if attempts were made to hide a problem.

New products

For many years, oysters were primarily sold in jars. There has been dramatic growth and expansion in sales of oysters on the half shell with many new oyster bars opening up in coastal areas around the world. This trend provides new opportunities for shellfish producers to differentiate their product by telling the history and story of their shellfish business. The shellfish producer's story, combined with a tie to the specific water body where the shellfish are grown, and the unique flavor profile that results from the salinity of that water body offer excellent opportunities to develop a brand. Within a given brand, lines of multiple products with slightly different sizes, characteristics and flavors can further be developed with good results. There are numerous examples of this in the raw oyster market. New oyster products with enticing names are being sold in many of the new oyster bars. Catchy names like 'Naked Cowboy,' 'Kumamoto' or 'First Best,' can be helpful to draw attention to one's product in a crowded market and to encourage buyers to taste or sample it for a trial period.

Saltiness often differentiates oysters raised in different water bodies and varying levels of saltiness may appeal to different consumers. 'Salt Bomb' is an oyster product developed for new oyster consumers who prefer a saltier flavor than do more experienced raw bar patrons who may prefer more subtle flavor differences among oysters.

The size of oysters can be critical in many markets and can be managed by culling. Taylor Shellfish developed their Shigoku tumbled oyster and then further differentiated that product line by size with 'Fat Bastard' as a larger Shigoku product and smaller product as small and extra small Pacific oysters. The Hama Hama Oyster Farm Sampler includes tumbled 'Blue Pools,' 'Sea Cows,' and 'Hammersley' products.

The market for 'petite' oysters has grown rapidly in North America in the early 2000s. The availability of petite oysters has led to 'Dollar Oyster' deals during happy hour in local pubs and neighborhood pizza joints. The smaller size of petite oysters allows restaurants to offer premium varieties at a lower cost. It also allows a less intimidating first-time trial by consumers who may be reluctant to attempt a larger raw oyster during a social event with their friends.

Branding, however, involves much more than just coming up with a creative name. Over the longer term, the product attributes that are most important to buyers need to be emphasized and supported consistently and absolutely by the product throughout the growing period, the processing phase, its transportation to markets, its sales staff, its customer service program, and the company itself. The reputation of the company in terms of its ability to consistently deliver the quality and volumes promised is always critical for repeat sales and to manage the brand.

Getting a foothold in markets can be challenging for new shellfish producers. Introducing one's farm and its products often is best accomplished through face-to-face meetings, especially for those restaurateurs whose store concept is based on telling the farm's story to their patrons. It is critical for the shellfish producer to introduce new products to buyers carefully and to encourage trial. High-end eating establishments, particularly those that offer a raw bar, frequently spend time educating their wait staff on lineups of shellfish products, sometimes with weekly tastings of new products.

Modified atmospheric and vacuum packaging technologies provide opportunities for adding further value to shellfish products to preserve freshness, safety, and quality of products. The new packaging technologies provide additional opportunities to add consumer convenience, especially to shellfish products, such as mussels, that cannot be transported fresh over long distances. Fresh cooked and pre-cooked dishes, vacuum-packed products, and intermediate products (as ingredients for soups, chowders, stuffings) have emerged and grown as newer packaging technologies have become available.

Understanding one's customers includes learning what their preferences are in terms of when they prefer to eat shellfish. Seasonality issues require careful planning. Meat yield can be greater in colder months than in warmer months and can affect customer satisfaction as well as the financial effects of selling less weight of product. The overall supply of shellfish will fluctuate based on the wild harvest of shellfish. Yet customers' preferences may also vary with the season of the year, and peak demand can occur in months other than those that are optimal from a production standpoint.

Setting prices

Setting a price for shellfish products requires careful thought and attention. Among other factors, decisions on pricing must reflect the value proposition for each type of product, the extent of competing products, and the expected volume of sales, given the overall business model and how the business has positioned itself in the overall market.

It is important to understand that decisions related to pricing a product must consider the volume that can realistically be sold at that price. There are many examples of markets in which there are a few individuals who are willing to pay a very high price for a product. Typically, these individuals will do so for only a small quantity and only for special occasions. Product that sells, for example, for US$77/kg (US$35/lb) would appear to be very profitable; however, if only 10 pounds can be sold at that price once a year, total revenue received is only US$350, which would be a very low level of revenue. Obviously, the total revenue received must exceed the total costs of production and marketing for the business to be profitable. Thus, it is the price multiplied by the quantity that results in the revenue that is used to pay the expenses. Expenses to the business include not only the cost of the shellfish seed and labor, but also non-cash expenses like depreciation. Even more importantly, the revenue generated must be adequate to compensate the business owner and family members for the time spent working in the business or they will not continue to raise shellfish and seek something else to do to make a living.

The price to set for one's product must also be based on the business model selected for the business. One business model is to be the largest supplier in that business category. To be the largest, dominant firm in the business typically requires that the business out-compete other businesses that sell similar products by charging lower prices; however, if the price charged is less than the cost of production, then the business is not profitable even if it is the lowest-price producer in the market. Smaller-scale businesses typically find it difficult to compete on price, particularly in industries that exhibit economies of scale. Economies of scale are common in industries that require high levels of capital investment that result in high fixed costs that must be spread over large volumes of product sold to cover costs. In such cases, the price per unit decreases the larger the company. Smaller businesses without adequate capital to expand to be able to compete on price will need to identify market opportunities to sell at higher prices. Key questions become: (1) who will pay the higher prices required for that product to be profitable; (2) what quantity will they buy at that price; and (3) for how long will they continue to pay that price?

Successful pricing strategies must match the value offer for that product. The value offer is the unique and specific marketing mix that the business can provide. Consumers tend to view low-priced product as a low-quality product; conversely, a high price implies a quality product.

Taste is, to some degree, subjective; with wines, consumer preferences have been shown to vary with price (increasing with higher prices) to some degree, regardless of taste. To successfully charge higher prices, the product must deliver the degree of quality expected and defined by the customer. Customers expect to receive the degree of value based on how high or low the price was that they paid.

Supply and value chains for shellfish

Various business models for shellfish rely on different types of supply chains (previously referred to as market channels), or the market levels that a product passes through to its end consumer. More recent approaches to supply chain management refer to the value chain instead of the supply chain, because value is added to the product as it moves from the harvester to a processor or wholesaler and on to restaurants and retailers.

Supply/value chains can be quite short and direct, or they can be extremely complicated. Shellfish producers who sell their product directly to restaurants or oyster bars have a short, fairly direct supply chain: producer to restaurant to consumer. The advantage of direct sales is that the shellfish producer, by cleaning and packaging the product appropriately and delivering it, can capture the profit margins of the various portions of the supply chain. However, there are costs associated with each marketing function (cleaning, packaging, transporting, and delivering) that the producer must be willing and able to cover. Direct sales also typically require investment in facilities for cleaning and packing, and trucks and drivers for delivery. Another disadvantage of direct sales to restaurants is that most restaurant owners prefer to minimize the number of suppliers that they work with. The more suppliers for all the various products and ingredients, the greater the time that it takes to place orders. Shellfish producers selling directly may be more successful if they are prepared to offer a wider range of shellfish products by either producing them on their lease sites or purchasing from other producers.

Specialized seafood restaurants, particularly those with raw oyster bars and with a restaurant concept that emphasizes local sourcing, may prefer working with smaller-scale producers. Seasonal availability of shellfish products can be problematic for restaurants whose business plan is based on certain signature dishes that need to be available year-round for the restaurant to be successful. On the other hand, greater numbers of farm-to-table restaurant concepts have incorporated seasonal availability into the restaurant's concept. In some cases, promotional messages such as countdowns to when specific local oysters will be available are used to alert customers to the new season's varieties.

More complex supply and value chains may involve sales from the shellfish producer to a general distributor, broker, or wholesaler who then warehouses product and delivers to various restaurants and supermarkets. Using such supply chains allows the producer to concentrate on managing only the production phases and avoids the complexities of managing sales and delivery to large numbers of individual restaurants, oyster bars, and supermarkets. One disadvantage of selling to a wholesaler is that wholesalers often sell product from a variety of sources, making it difficult for an individual producer to differentiate their farm's product and capture higher prices. For shellfish producers with a business model of being the largest, lowest-price supplier, integrating across the supply chain (referred to as vertical integration) may be a profitable option.

In such a model, the producer develops the processing, wholesaling, and distribution capacity within the company to handle these marketing functions. Clearly, a vertically integrated shellfish company would need adequate capital to build or purchase the necessary facilities for processing, storage, and transportation/distribution to ensure that product is maintained under high-quality conditions throughout and delivered in high-quality and sufficient volumes to meet customers' needs.

Regardless of the business model selected, it is important to clearly map out the distribution system that will be used to engage the business effectively with supply chain partners. Distribution systems require investment in trucks, processing, handling, storage, and shipping facilities, packaging, ice, and personnel trained to handle the product appropriately. Thus, careful planning is needed to determine the scope, scale, and cost of building and operating the facilities and trucks needed to deliver quality product through the supply chain. Quality control monitoring processes must be maintained throughout the distribution system.

The shellfish producer may need to obtain a dealer license and have a HACCP (Hazard Analysis of Critical Control Points)-certified processing facility (floor drains, stainless steel tables and sinks, and a bathroom). A refrigerated truck may be needed for interstate delivery, and packaging and labeling equipment will be necessary. Time must be spent to deliver invoices and to collect payments.

Shipping costs are an important part of the distribution system. Overnight air freight shipping is expensive and often requires large volumes to do so at a reasonable cost. Ground rates can be less expensive than air freight, but ground shipping may limit hauls to several hundred miles and prohibit long-distance deliveries. Air freight can double the costs due to the need for product to be packed with cold packs, liners, and foam. However, if customers prefer to buy directly from the farm and price points are feasible, then adding a shipping component may be worthwhile.

Partnerships across the supply chain have become increasingly common, due in part to the trend in many countries of increased concentration at the retail level. Major hypermarkets, such as Wal-Mart, Tesco, Carrefour, RT-Mart/Auchan, and Edeka, have put downward pressure on prices and have stimulated a series of mergers across wholesalers and food service distributors, to counteract the market power of the very large retailers. Alliances and partnerships across the supply chain can provide similar advantages, but only if each participant, or actor, in the supply chain receives value for the services that each provides. Important benefits that can be obtained through such supply chain partnerships include access to cold storage and distribution centers and networks of wholesalers and distributors as well as through joint advertising and promotional companies.

Whatever supply chain and distribution system is chosen by a shellfish company, developing a close relationship with all those involved is critical. Inviting buyers to visit the farm to understand the operation, the people, and the products enables the buyer to tell a better story and sell more product. Personal relationships can be critical in supply chain management, particularly in times of economic downturns and tight markets.

The seasonality of demand and of supply for many types of shellfish can create financial problems. Shellfish cannot be sold at times of the year when water temperatures are too high to ensure the safety of the food or when harvest areas are closed. In the US, EU, and other countries, closures of harvest areas may occur when algal toxins are found through routine water monitoring, when rainfall results in terrestrial runoff into coastal areas that results in *E. coli* contamination, and for other reasons. Therefore, planning for adequate cash flow along with frequent communication with customers when the business cannot meet its orders due to supply problems is vital.

Regulations that affect shellfish marketing

Shellfish producers face a wide array of overlapping and redundant regulations from many different agencies at the local, state/provincial, federal/national, and international levels, particularly in developed countries in North America, the European Union, and Australia. Many regulations that affect shellfish producers are those related to obtaining leases to farm shellfish in public waters in coastal areas. While there are a few places in the world where producers have title to tideland areas, this is an exception. Approval of leases to farm shellfish typically involves consideration of: (1) possible user conflicts (with recreational boaters, recreational and commercial fishermen, wild shellfish harvesters, aboriginal users); (2) possible environmental effects (to marine mammals, to habitat for nursery areas, to water quality); and (3) those opposed to viewing working waterfronts and shellfish gear from their homes.

In addition to such regulatory issues related to siting and leasing, there are a number of health regulations that affect shellfish. Given that many consumers prefer to consume shellfish raw, the incidence of vibriosis and other shellfish diseases can pose significant health risks. Thus, regional/state, federal/national, and international regulations exist in and between countries to reduce risk of human health problems from eating shellfish. For example, the EU requires that shellfish sold come from EU-approved establishments, premises, and approved production areas. For non-EU countries exporting to the EU, each consignment is required to have an appropriately signed health certification. In the US, a National Shellfish Sanitation Program (NSSP) certification that ensures sanitary control of shellfish production and sale for human consumption is required to market oyster products. In addition, the US Food and Drug Administration (FDA) has international agreements with foreign governments in which processing plants and dealers are inspected and certified for import and export of shellfish as well as for import and export.

What changes are coming for shellfish marketing?

Markets change constantly, and successful businesses are those that adapt to those changing conditions. Underlying the ever-changing market conditions are the changing preferences of consumers worldwide. If anything, the pace of change in market demand has increased, particularly in the 21st century, with nearly instantaneous communications through social media and digital communication formats. It is more imperative than ever that shellfish producers incorporate into their businesses mechanisms to stay informed of the many changes in the marketplace. Such changes can represent either threats or opportunities to the business.

A constant threat for any food product in the 2000s is that of a food safety scare, and consumer preferences can change quickly when illnesses or deaths occur as a result of eating shellfish. Such news stories spread rapidly over the internet and among digital social groups, and the negative responses can have far-reaching effects. Constant vigilance as to the safety of the products by all those who handle shellfish products is an essential part of a successful business. The oyster sector provides an example of this importance. The demand for shucked oysters has been fairly consistent since the 1990s. The growth in the oyster market has come primarily from increased demand for 'designer' oysters that are consumed fresh and raw in the half-shell market, primarily in raw oyster bars. Risks of food poisoning are greater when raw food products are consumed. Given that the growth in oyster markets is primarily in the consumption of raw product, the overall market risks are greater. There is a potential need to identify ways to reinvigorate markets for shucked product that would be used in prepared, cooked dishes both at home and away from home.

Threats in the marketplace can also create opportunities for entrepreneurs to take advantage of the trends in consumer preferences that create potential threats. For example, the growing concern in the minds of consumers over the safety of food eaten, while a potential threat to the shellfish and other food industries, also presents opportunities for shellfish businesses to seek ways to provide guarantees of product safety as a way to differentiate their product, and possibly command higher prices or to increase market share. New smart technologies have been emerging in the 2000s that may offer interesting applications for shellfish packaging, labeling, and marketing. For example, Smart Seafood TTI is a product developed in collaboration with FDA and state regulators in the US that monitors temperatures in packages of fresh oysters. The label uses a color-coded indicator of temperature. If proper temperature control has been maintained, the indicator remains green, but changes to red if not. Since the primary method to maintain shellfish safe for human consumption is to maintain product temperature at appropriate levels, this smart label offers consumers a way to evaluate the food safety status of shellfish products at the point of purchase in supermarkets.

The 'millennial' generation (or generation Y/Gen Y, individuals born between the early 1980s and the mid- to late 1990s) began to exert its influence in the marketplace in the early 2000s. It is a large generation, with numbers of individuals exceeding those of the 'Baby Boomer' generation (those born between the mid-1940s and mid-1960s). Research has shown that millennials place a broader emphasis on quality that goes beyond expecting food to be raised and harvested in an environmentally sustainable manner to include food that is safe and free of contaminants, and raised in a socially responsible manner. The millennial generation is not a homogeneous group; there are different segments of consumers within this generation. As in other age segments, some millennials also are 'foodies', who spend more time at home, cooking and preparing food than do other segments of millennials, and place a high value on taste and flavor. The millennial generation has driven much of the demand for local foods, preferring to connect with those who grow their food and enjoy knowing the 'story' behind the food they are eating. Locally raised food is assumed to be more environmentally sustainable because it presumably entails fewer 'food miles' than food produced and transported long distances. Such consumers tend to prefer highly differentiated and individualized products with different flavors and different presentations.

Shellfish possess many characteristics that fit the preferences of millennials (and other consumer segments) for quality food that is locally sourced. Shellfish farming is good for the environment because shellfish remove nutrients from the water, and shellfish gear provides habitat for juvenile fish. In fact, Seafood Watch lists clams, mussels, and oysters and farmed scallops and abalone as Best Choices in their sustainability ratings. Shellfish also have positive nutritional profiles that include high levels of zinc, omega-3 fatty acids, and shellfish products are low in fat and cholesterol. Oysters even have a reputation as an aphrodisiac. Thus, the millennial generation offers opportunities for increased consumption of shellfish.

However, while the characteristics of shellfish products are well aligned with core values of the millennial generation, there are challenges into turning this opportunity into effective demand and sales of shellfish. Many consumers around the world, particularly those that have grown up away from coastal areas, do not have a tradition of eating shellfish. Thus, such consumers are unlikely to be comfortable purchasing and preparing shellfish or even selecting shellfish to eat in restaurants away from home. To expand markets for shellfish in areas populated by consumers not familiar with shellfish products, programs that introduce consumers to shellfish preparation and consumption are essential to develop the level of familiarity with various types of shellfish that leads to comfort in purchasing and preparing shellfish. In some cases, youth cooking programs have been successful to introduce children to various types of shellfish and their various forms of preparation. College students often will respond favorably to the challenge and adventure of experiencing new culinary alternatives. Promotions at fraternity and sorority parties on university campuses, and tail-gating parties (pre-game parties prior to sporting events typically held in the stadium parking lots) either by individual companies or by shellfish associations, can help to develop new groups of future consumers by introducing shellfish to young people in venues where they are pre-disposed to experiment with new foods and culinary experiences.

Shellfish producers, as other businesses, must pay attention to signals that market changes are on the way and learn to adapt to those changes by seeking opportunities that can be developed from emerging threats. One constant in the business world is that successful businesses and industries will eventually attract competition. As new competitors enter any given market, the overall supply increases and price decreases. Successful businesses are those that are prepared for competition by constantly evaluating market conditions and strategically adapting to changes to stay one foot ahead of potential competitors. The shellfish business's marketing plan must include long-term strategies to adapt to such constantly changing market conditions.

Developing the marketing plan and strategy

Successful businesses are those that have been able to align decisions made on the marketing, production, financial, personnel, and other facets of the business into one coherent and effective business model. Even small-scale businesses need to address the many questions that must be answered for the business to be successful. Given the growing complexity of business management, it is essential to spend time each year thinking through the challenges and opportunities for the business and committing to paper the goals, strategies, and likely financial outcomes in the form of a business plan. The overall business plan should identify goals both for the long and the short terms and adjust these at least annually. This section focuses on the marketing portion of a business plan, but greater detail on developing a marketing strategy and plan can be found in Engle et al. (2017). The Appendix to this chapter includes an example of a marketing plan. Chapter 20 in this book provides detail on business planning for shellfish farms, and additional resources on developing a comprehensive business plan are listed in the 'References cited and further reading' section of this chapter (Center for Farm Financial Management, 2018; Engle, 2010, 2012a, 2012b, 2012c, 2012d; Engle et al., 2017, 2019).

The marketing strategy and planning process must begin with some honest soul-searching related to one's strengths and weaknesses. For example, a shellfish producer who does not enjoy interacting with people is unlikely to be an effective salesperson. Given that effective sales and marketing are vital to shellfish businesses on all scales, a producer who prefers to work alone will need to either sell to a wholesaler, form a partnership with another company that handles all sales, or bring someone into the company to handle all marketing and sales to be successful.

Careful thought related to one's strengths and weaknesses should provide the basis for establishing short- and long-term business goals. Business goals need to be specific. A goal such as 'Expand the business' is not very useful; however, if the long-term goal is to: 'Expand the business to be the largest supplier of clams to the northeast market of the USA', then it is possible to establish short-term goals that specify the percentage or dollar value of increased sales each year that will be needed to achieve that goal. With specific annual growth rates set as goals for each year, then, it becomes possible to annually monitor and evaluate whether the business is on track or not.

The marketing plan document typically begins with a brief executive summary that provides an overview of the business and a statement of where the business is planning to head over the next several years. The executive summary is followed by a description of the overall market situation, a summary of the strengths and weaknesses of the business, and the short- and long-term goals of the business.

The next section of the marketing plan should focus on an explicit statement that summarizes the overall strategy to be pursued for the marketing efforts of the business that is accompanied by specific marketing objectives. The marketing plan should discuss the following: (1) description of the product and the value offer; (2) targeted market segments; (3) new product development; (4) pricing; (5) supply and value chains; (6) processing, distribution, and storage; (7) customer service and retention; and (8) branding.

The final section of the marketing plan is the financial analysis of the marketing activities. The costs associated with each component of the marketing plan need to be assessed and estimated. The information from the marketing plan must be fed back to the overall business plan and may necessitate changes in the overall business plan because the process of developing the marketing plan may uncover changes that are needed on the farm. In some cases, farm realities may make some aspects of the marketing plan unrealistic and require re-thinking key aspects of the marketing strategy to ensure that farm production and marketing activities will mesh successfully and profitably. The best time to identify any discrepancies is before investments are made so that adequate adjustments are made before spending hard-earned capital.

Clear definitions of which market segment(s) is targeted is a critical component of the marketing plan and strategy. In-depth understanding of and familiarity with consumers in that market segment will provide the basis for identifying new products and services that will better meet the needs of the core customers of the business. Time needs to be spent especially in considering how to provide the best possible service to those customers. Specific customer service activities need to be described in the marketing plan, whether these include follow-up contact with customers for feedback on their level of satisfaction with the product, identifying new opportunities to better meet their needs, or other ways to connect and develop a relationship with customers.

The marketing plan must go beyond identifying its target market to develop an assessment of its size and to project likely sales volumes. This requires several steps that begin with an estimate of how many people are included in the targeted market segment. Clearly, the definition of the targeted market segment must include its geographic boundaries and its demographics (for example, age, income, ethnicity). A great deal of data are available online that can be used to make this first determination. The second step is then to estimate what percentage of that population segment is likely to purchase the type of product that the business sells. Those who currently purchase that particular product will need to be enticed to change to a new supplier, and the estimate of potential sales volume must include a realistic estimate of the number of people likely to switch. The third step is to estimate the quantity likely to be purchased each week of the year. This step will identify potential seasonality issues with demand that will be important for managing cash flow of the business. Finally, an assessment is made of whether this volume of sales (and the associated expected price) is adequate to meet the business's financial goals or whether additional market segments also need to be targeted.

The choice of product to be raised, while a marketing decision, must be balanced by the feasibility of the required production system on that particular farm. Racks, baskets, and flip bags require different levels of capital, experience, and management. Decisions about whether to sell to a packing house, supermarket, restaurant, or directly to end consumers are related to whether and how the company will process and distribute the product. These decisions have important implications for prices received at the farm, the volume that can be sold, and how complex the business will need to be. The overall production plan must be closely aligned with the marketing and distribution plan to avoid bottlenecks and achieve the overall efficiencies needed to offer product at competitive prices. The production plan needs to be based on production practices that are necessary to meet consumer demand and keep customers happy.

Distribution and delivery systems must be planned based on distance from lease sites to points of sale. The frequency of delivery of the volumes required, the sizes and types of vehicles required to meet delivery schedules, and the associated delivery costs must be estimated and evaluated carefully. New facilities may be needed to wash, process, and ship product of the level of quality demanded by core customers of the business.

The marketing plan, as well as the overall business plan developed, is something that should be re-visited every year. Whether or not the business achieved its short-term goals should be determined. If not achieved, then an analysis should be made as to what happened and corrective actions taken. Long-term goals should also be re-evaluated, and new short-term goals set for the coming year. The greatest value of a business plan is the annual analysis and evaluation of the business. The most successful businesses are those that make adjustments every year. Often the adjustments needed are minor, but a 1% increase in sales or decrease in production costs every year becomes a 5% change in five years. Given the changing nature of markets, annual adjustments to the business are critical for it to remain relevant and economically viable.

Conclusion

At the end of the day, the successful shellfish businessperson is one who has discovered the products that he/she can be the best at growing and selling and for which the targeted consumers are willing and able to purchase sufficient quantities at a price that is profitable. Completing a thorough business plan with a well-conceived marketing plan and strategy, evaluated and adjusted annually, is essential for a profitable shellfish business.

Appendix – hypothetical sample market plan for oyster farm

Executive summary

Our oyster farm is located in the US Chesapeake Bay area approximately 50 miles from Baltimore and 65 miles from Washington, DC. We are a family operation, farming oysters for supplemental income, but would like to develop our business into a full-time source of income for our son. We currently raise oysters using bottom culture for sale to a shucking plant, but we are considering investing in container culture systems and a tumbler to begin to target more of our product to the higher-priced half-shell market.

The overall goal long-term for our business is to develop our oyster farm into a full-time operation that provides adequate financial support for our son. In the short- and medium-term, our business goals are to: (1) initiate container culture of oysters; (2) develop several local markets for half-shell products; and (3) develop a web-based and social media-based market through which we can sell sufficient volumes at a price that allows us to meet our revenue targets by Year 5.

Overall market summary and current situation

The market for oysters on the half shell has been booming. Oyster bars are springing up in many locations, and innovative 'designer' oysters are increasingly being enjoyed by greater numbers of consumers. Thus, it appears that greater numbers of consumers are interested in experimenting with consumption of raw oysters on the half shell. This market trend likely is related to the broader growth in craft breweries and wineries and appears to have been extended to raw oyster bars that source locally raised oysters, often with pairings of local wines. Those growers who successfully can enter and compete in the half-shell market are enjoying higher prices and greater revenue. The clientele driving these sales appear to be more of the millennial generation, mostly in coastal areas, higher-income, professional, educated individuals, and those who would be considered 'foodies'. While some purchase shellfish in supermarkets, more are enjoying oysters at raw bars in restaurants or at the increasing numbers of stand-alone oyster bars. There are more than 2.8 million residents of Baltimore, Maryland, and 6 million residents in the Washington, DC area. The Atlantic coast beaches draw more than 15 million tourists a year. In addition, there has been growth in internet based sales that offers additional marketing opportunities.

Analysis of strengths and weaknesses of the business

A core strength of our business is our experience raising oysters. We have been raising oysters for more than 20 years and understand the product that we raise. Our leases include areas that could be used for container culture and we are located in an area that draws many tourists from the east coast states and elsewhere. Thus, we have the ability to use the name of our water body, that already has a positive perception among many prospective consumers.

We have no debt in our current oyster business, other than the business credit card that runs a small outstanding balance of a few hundred dollars that we pay off each month. Thus, we have a good credit rating and have the ability to take on additional debt. Our current business is profitable, generating net profits that average about US$25,000 a year. While this provides us with sufficient supplemental income, it is not adequate to support our son on a full-time basis.

We believe that there are opportunities to develop markets for our products among millennials, not so much in terms of the oysters themselves, but in two other ways: (1) connecting with millennials via social media; and (2) packaging oysters with smart labels that include meters that indicate the freshness and the safety of the product.

There are social and cultural changes under way in our society in terms of increased demand for locally grown food, increased concern and attention to the safety of the food we eat, and in terms of how we connect and communicate with each other, with the many internet and social media platforms. We hope to take advantage of these changes by providing a quality product with labeling that provides real-time food safety information.

Threats to our business would include disease outbreaks in our region that would shut down our sales. Other potential threats might include downturns in the local and national economies that would reduce the tourist trade, reduce incomes, and possibly reduce sales of shellfish.

Competition

The half-shell oyster market that is focused on 'designer' oysters is a very well-developed, likely maturing market; thus, there is quite a lot of competition, in spite of the large populations of year-round residents and tourists. We believe that our advantage in the market will be in how we relate to and communicate with the millennials who we are targeting. We intend to provide better customer service through direct connections with them via social media, an app that we intend to develop, through our food safety labeling, and a reliable delivery system. One of the challenges, if we are successful with our app and our social media efforts, will be the risk of having our app imitated by larger companies who move in to take our business. It will be important to have the app adequately protected legally, but then to invest the on-going time and financing to keep the app fresh, with new ideas, and new offers.

Marketing strategy: the game plan

Our overall long-term mission is to produce and deliver high-quality, preferred oysters for half-shell consumption via the internet. We believe that what is unique about our plan is not so much the specific products that we sell, but how we connect with millennials who are interested in quality, safe, nutritious, and healthy seafood and who prefer frequent feedback and connections with vendors that they frequent. Our principal strategy will be to employ social media to identify prospective customers and to develop them into loyal customers.

In the shorter term we plan to reduce market risk by taking advantage of our positive reputation and local networks by meeting existing requests from several local restaurants for fresh oysters on the half shell. While local sales are not expected to generate large amounts of revenue for our business given the degree of competition in this market, they will help with cash flow as we work over several years to build a solid Internet-based market.

One critical issue and key to success will be our ability to develop the necessary marketing skills. While we are confident in our ability to successfully produce oysters using container culture methods, we are not experienced marketers. Due to our positive reputations in the local area, we are also confident that we can sell at least a certain amount of our product in local raw bars because we have friends and neighbors asking us to supply their establishments with raw product; however, we suspect that the volume of local sales will be relatively small. Thus, we will need to learn how to reach prospective customers via the Internet and learn how to package and ensure timely delivery of our new products in a way that guarantees quality and increases our chances of repeat sales. This plan depends heavily on our son's ability to successfully develop an Internet-based market. He is quite active on social media and has connections with several app developers who we plan to work with.

Specific marketing and financial objectives for our business include the following

1 To expand our current experimental production of container-produced triploid oysters into reliable sales by Year 2 of this plan.
2 Develop in Year 1 a detailed plan for packaging methods and shipping methods that we will use, including projected costs. This will include additional investment in infrastructure, such as a packing shed, refrigeration, storage space for insulated shipping containers, office area with computer/printer for UPS or FedEx delivery labels, as well as packaging materials and shipping costs.
3 Develop and launch in Year 2, a modern, interactive web site that targets millennials and provides comprehensive information on:
   a environmental sustainability of our oysters
   b safety of our oyster products
   c how to determine whether oysters are safe to eat
   d nutritional information on our oysters
   e story of our farming family.
4 Develop and launch in Year 2 an app that offers engaging and informative tools attractive to millennials and others interested in experimenting with oysters as a new food.
5 Develop and launch in Year 2 social media promotional programs using: Twitter, YouTube, Facebook, Pinterest, and Reddit.

Planned expenses will include an annual salary of US$40,000 a year for our son, who will manage the sales, social media, and marketing. Our son will handle all sales, focused primarily on building our internet-based sales business. Advertising will be through social media accounts that will be developed. We are still investigating expenses to cover the costs of storage and shipping and will complete that detailed plan in Year 1. These will include the estimated investment cost of facilities for packing and shipping product. A packing facility with adequate storage for product and for shipping materials, for grading, sizing, and cleaning, a loading dock for pickup by shipping companies, and other required infrastructure will be designed. Costs of the packaging materials, smart labels, shipping boxes, ice packs, and other marketing supplies will be included in the financial plan. We plan to sell the majority of our product fresh, but there is a chance that some amount of storage might become necessary. We will make arrangements for some degree of cold storage in the form of off-farm rental storage in the early years, and possibly develop on-farm storage in later years. In Year 1, we will also begin to work on the website and app development and are costing out options for both of these.

Beginning in Year 2, with our first container-raised oysters, we will initiate a series of activities to introduce our oysters to our new targeted markets: raw oyster bars and Internet-based sales. For local raw bars, we will offer tastings in person, to talk with patrons about our farm and our oysters. For internet-based sales, we plan to offer a series of competitions that focus on university settings: clubs, fraternities and sororities, and sports events. Internet-based competitions will involve naming competitions to come up with new, attractive, and innovative names of new oysters.

For our internet-based sales, we plan to market our product as a premium-priced, high-quality product.

Our sales targets for the next 5 years are:

- Year 1: minimal additional revenue, based only on our experimental container production to date.
- Year 2: US$5,000 in sales to local raw oyster bars; US$5,000 internet sales.
- Year 3: US$5,000 in sales to local raw oyster bars; US$10,000 internet sales.
- Year 4: US$5,000 in sales to local raw oyster bars; US$25,000 internet sales.
- Year 5: US$5,000 in sales to local raw oyster bars; US$40,000 internet sales.

We are working with our accountant to develop a cash flow budget that will project our anticipated expenses and construct an operating line of credit to help us get through our first several years of raising oysters in containers. We will reconcile our actual cash flow with our projected budget for each quarter and make adjustments as needed. If we are not able to achieve our financial goals, we will adjust expenses and our overall plan accordingly.

References cited and further reading

Bardach, J.E., Ryther, J.H., and McLarney, W.O. (1972) *Aquaculture: The Farming and Husbandry of Freshwater and Marine Organisms*. John Wiley & Sons, New York.

Center for Farm Financial Management (2018) AgPlan. University of Minnesota, Minnesota, USA. Available at www.agplan@umn.edu.

Duncan, P.F. and Ghys, A. (2019) Shells as Collector's Items. In: Smaal A., Ferreira J., Grant J., Petersen J., and Strand Ø. (eds.) *Goods and Services of Marine Bivalves*. Springer, Cham, pp. 381–441.

Engle, C.R. (2010) *Aquaculture Economics and Financing: Management and Analysis*. Blackwell Scientific, Ames, IA.

Engle, C.R. (2019) Aquaculture Businesses: A Practical Guide to Economics and Marketing. 5m Publishing, Sheffield.

Engle, C.R., Quagrainie, K., and Dey, M. (2017) *Seafood and Aquaculture Marketing Handbook*. 2nd edition. Blackwell Scientific, Ames, IA.

Engle, C.R. (2012a) Introduction to financial management of aquaculture businesses. SRAC Publication No. 4400, Southern Regional Aquaculture Center, Stoneville, MS.

Engle, C.R. (2012b) Assessing the financial position of an aquaculture business: using balance sheets. SRAC Publication No. 4401, Southern Regional Aquaculture Center, Stoneville, MS.

Engle, C.R. (2012c) Determining the profitability of an aquaculture business: using income statements and enterprise budgets. SRAC Publication No. 4402, Southern Regional Aquaculture Center, Stoneville, MS.

Engle, C.R. (2012d) Evaluating the liquidity/cash position of an aquaculture business: using cash flow statements. SRAC Publication No. 4403, Southern Regional Aquaculture Center, Stoneville, MS.

Food and Agriculture Organization (n.d.) FishStatJ database. United Nations. Available at www.fao.org..

Girard, S. and Mariojouls, C. (2003) French consumption of oysters and mussels analysed within the European market. *Aquaculture Economics & Management*, 7(5–6): 319–333.

Hanson, R., Jouse, L., Sureshwaran, S., Posadas, B., and Liu, A. (2003) Opinions of U.S. consumers towards oysters: results of a 2000–2001 survey. Mississippi Agricultural & Forestry Station Bulletin 1133, Mississippi State University, MS.

Posadas, B.C., Posadas, R.A., and Andrews, L.S. (2011) Consumer preferences for post-harvest processed raw oyster products in southern California. Mississippi Agricultural & Forestry Station Bulletin 1194, Mississippi State University, MS.

CHAPTER 20

Business planning for aquaculture

Peter Cook

With the world population being expected to increase to 9.6 billion by 2050, global food production must also increase, and it is often predicted that aquaculture will play a significant role in this. Based on regional trends in seafood production, Kobayashi et al. (2015) used a multi-market model to predict that the total world fish supply will increase from 154 million tons in 2011 to 186 million tons in 2030, with the majority of that increase being produced through aquaculture. Fast-growing seafood demand in countries such as China, provides an opportunity for global aquaculture to significantly increase production. It is not surprising, therefore, that aquaculture is often portrayed as having an attractive future, and that suggestions of good future profits from investing in the industry are common. Why is it then that many prospective and existing fish farmers have found it difficult to secure financing for aquaculture operations? In a recent article on financing aquaculture Neil Sims, of Kampachi Worldwide Holdings, stated that "the cash is there, but information is lacking" (Godfrey, 2018). This suggests that part of the problem may be perceived risk, with many lenders being wary of an industry that they do not fully understand, and one in which risk is probably higher than some other investment opportunities. In addition, the failure of some aquaculture operators to produce a properly constructed business plan, must also rate highly among the major reasons for lender rejection.

A well-prepared business plan will help the operation in two important ways. First, outlining all aspects of the business on paper will help to anticipate potential problems that might be soluble in advance of any actual expenditure of funds. Exposing assumptions, plans and financial predictions to critical evaluation, early in the process, may help to distinguish between optimism and reality – not that optimism is a bad thing – but it does need to be securely anchored into the real world. Second, almost all new aquaculture ventures require some level of external investment and, without a sound business plan, investors are unlikely to be attracted to the project. Many potential lenders may be unfamiliar with aquaculture and, therefore, cautious. Lack of an adequate business plan will add to their reluctance to invest.

There are several different ways of assessing the financial viability of an aquaculture business. Engle (2012), for example, used balance sheets to estimate assets versus liabilities over specified time frames. This approach helps to assess the liquidity of the business at any point in time. There are, however, many different ways in which aquaculture can be practiced and it is necessary, therefore, to tailor a business plan to the specific characteristics of the planned operation.

Writing a business plan is not easy and usually requires a significant input of time (see Chapter 19, Appendix). It is not good enough to simply download a generic plan from the internet, and just insert a few details to give the impression that it is specific for a particular project. Investors have seen these before, and are unlikely to be impressed by them. A business plan should be designed specifically for each individual project, and should detail, on paper, how and when the project will be initiated, and how it will progress over the subsequent 10 to 15 years.

The business plan will need to be logical and structured, and will need to include, as a minimum, answers to the following questions.

- What species will be produced?
- What are the site and environmental conditions required for that species?
- How will it be produced?
- What market will it be sold into?
- How much will it cost to set up and operate the farm?
- How profitable will it be, and over what time period?
- What resources (including human resources) will be required to set up and run the farm?
- What are the major risks and how will they be mitigated?

Market

No matter how much time, effort, and funds are put into an aquaculture enterprise, the operation is doomed to fail if the product cannot be sold at a reasonable price. The time to assess the size and location of the market, and the likely price paid for the product, is before the operation is started, not when the first product is ready to sell. The types of questions that need to be addressed in a market analysis are as follows.

- Where will the product be sold – locally or exported?
- What segment(s) of the market are you aiming at (for example, wholesalers, seafood stores, supermarkets, restaurants, direct farm-gate customers)?
- What is the preferred form of the product for that market (for example, live, on-ice, frozen, canned)?
- If the product is to be processed before delivery to market, are there local facilities for this and will those facilities have sufficient capacity to handle the product?
- How, and how quickly, can the farm deliver its product to the market? This is especially important for the sale of live product – will it be delivered by local road transport or, if the product is to be exported, will there be sufficient and appropriate air freight capacity?

A thorough analysis of the potential market is essential and should be carried out in advance of any other aspect of the project. If this shows that an existing market has the capacity to absorb all of the planned production, then the business plan should reflect this, but if this is not so, then the plan will need to explore the possibilities of alternative markets or alternative forms of products. Value adding should not be discounted. Although it may be easy and convenient, for example, to sell head-on whole fish, it may ultimately be a better and more lucrative business option to partner with a processor to produce a value-added product. In the latter case, of course, processing cost must be accounted for in the business plan. In addition, the concept of risk must be built into a marketing plan (see also Chapter 19). For example, it may appear (on a price per kilogram basis) more lucrative to sell live or chilled product but, obviously, there is less risk of product spoilage if the fish are sold frozen or canned. Investors would expect a realistic level of risk to be incorporated into a marketing plan.

Site selection and production feasibility

Although most aquaculture projects start with the intention to grow one particular species, it is absolutely essential that the water and environmental conditions at the chosen site are suited to the requirements of that species. For example, if your site is selected to take advantage of the idyllic lifestyle of a tropical island, it would be useless to try to grow a cold water species, such as Atlantic salmon, at that location. Much literature has been written on criteria that should be used for aquaculture site selection (see Chapter 10) but, for the purposes of a business plan, it is at least necessary to show that the characteristics of the selected site are suited to the requirements of the selected species. Every species has a unique set of biological requirements that will influence its culture potential. To gain the information needed, it is necessary to access and critically evaluate all available scientific information on the chosen species. Justification of site suitability should include aspects such as environmental conditions, adequate supply of suitable water, proximity to processing facilities, transport arrangements, feed supplies, and access to disease testing and diagnostic laboratories. In addition, investors would want to see that the farm could be operated legally within the limitations of national, state and local laws and regulations.

For whatever species is to be grown, there are usually several different options of the technology and farm design that could be used. For example, abalone could be grown on farms in which water is pumped ashore to concrete tanks, or in nets suspended from floats at offshore locations, or on artificial structures deployed into the sea. The choice of technology will depend on many factors including land availability and cost, sea conditions, local regulations, and so on but, for the business plan, the choice of technology would need to be stated and justified. A technology that may be ideal for one location, may be completely impractical for another. A plan would need to explain, not only why a particular technology was chosen, but also why other technologies were considered inappropriate.

Financial analysis

The ultimate objective of a commercial aquaculture operation is to make a profit for the owner or for shareholders and, for this, production cost must be less than income from sales. Production managers often distinguish between two types of costs – fixed costs and variable costs. Fixed costs are based on inputs that do not vary much with the level of production and would include salaries of permanent staff, overheads, land or rental costs, insurance, and depreciation costs.

Variable costs change with the level of production and could include purchase of fingerlings, feeds, electricity, and casual labor. Because fixed costs tend to remain fairly constant throughout a range of production levels, this suggests the importance of scale, where larger operations tend to be more profitable because unavoidable overheads generally do not increase at the same rate as running costs.

For many aquaculture operations capital cost is generally large. This includes purchase or lease of land, earthworks and construction, installation costs, and commissioning costs. Generally, these expenses have to be met before the first of the stock enters the farm and the business plan must, therefore, detail how these expenses will be financed. Loans from banks or other institutions may be available but, obviously, in that case, interest on loans much be accounted for. To partially decrease the immediate impact of up-front capital costs, phased development could be considered where various parts of the farm are added, as production level increases. For example, at the very beginning, perhaps the number of grow-out structures required may be less than when the farm is in full production. There is no point in spending up-front money on structures that will not be required till later although, of course, the business plan would need to incorporate such costs later in the plan.

A financial analysis must detail all costs of running the farm, and a suggested list of the major ones is outlined below.

Source of juveniles

If the farm does not have its own hatchery, financial predictions would need to include the availability and cost of juveniles. Once juveniles have been sourced, planned stocking densities and associated costs of feed and water treatment need to be listed, and a realistic estimate of stock mortalities must be included.

Feeds

The cost of feeds, and the labor required to feed the farm stock, are expensive, and often represent up to 40–50% of total variable costs. The amount of feed required per unit of production depends on the feed conversion rate (FCR). For many species, a good indication of this can be determined from literature sources but, where this is not available, data from comparable species could be used as an initial approximation. Many companies, throughout the world, produce aquaculture feeds and the business plan must specify not only how much feed will be required and its cost, but also where it will come from and whether sufficient volumes can be guaranteed.

Management and labor costs

In addition to the labor required for feeding, labor is also required for many other farm activities such as tank or net cleaning, system maintenance, grading and harvesting. All of these should be listed as variable costs because they all vary according to the total amount of stock held at any one time. Management costs, however, are recorded as fixed costs because, no matter the size of the farm, certain levels of management input are still required. It is also true, however, that as a farm increases in size, some additional management skills, such as human resources, workplace safety, financial control, quality assurance, and worker training, may be best handled by additional management staff, skilled in these specific areas. Alternatively, these services can be bought in from consultants on a pay-as-you-use basis, or annual retainers.

Energy input

Particularly for land-based, pump-ashore farms, energy costs can be very high. The cost of pumping water to culture tanks depends on the volume of water required, the height to which it is pumped, and the unit cost of the electricity required to operate the pumps. In some cases, the cost of providing filtration for pumped water would also need to be estimated. For offshore cage farms, energy input may be less, but costs such as vehicle and boat fuel, and energy used in equipment cleaning and maintenance, still need to be estimated.

Insurance

There are many aspects of an aquaculture business that may require insurance. This whole subject is, however, extremely complex and entire books have been written about it. Perhaps the most complex area is insurance of the farm's stock and this is so variable from one species to another, and from one technology to another, that no attempt will be made in this chapter (or in the business plan) to accurately estimate specific costs. As a general example, however, one well-known international insurance company suggests that to cover a stock valued at US$400,000, a minimum premium of US$10,000 would be required (2.25%), while in an FAO publication on aquaculture it is suggested that a premium of about 4.25% of stock value is not uncommon.

Other costs

Other things that should be considered in farm running costs include compliance costs, book-keeping, and depreciation of equipment and infra-structure.

The business plan

Having gathered the types of information outlined above, it is then necessary to integrate that information into a comprehensive plan that shows projected expenses and income, preferably over a 10 to 15 year period.

The cash flow statement

This is an essential tool to predict, over a specified time period, how funds will flow out of and into the business. At the beginning, cash will flow out to meet the costs of licensing, land acquisition, earthworks, construction, and purchase of initial stock. Income will not be achieved till much later, when the first crop is sold. The cash flow statement must take this time-lag into account and must specify how it will be financed. Expenses are not necessarily continuous, but are more likely to come in pulses as various phases of development are achieved. A cash flow statement should, therefore, predict when, and how much, will be needed to finance the operation.

The time taken to generate the first income depends, to a large extent, on the species being produced, with fast-growing species perhaps being marketable after only about 1 year, while species such as abalone may take up to 4 or 5 years to reach market size. It may be acceptable, therefore, for a cash flow projection for fast-growing species to cover only about 4 or 5 years, while for slow-growing species it would be better to cover at least a 7 to 10 year period.

Cash flow statements should start as soon as the business begins, and are probably best represented on some type of spreadsheet. For the first year or two of operation, it is best to construct a cash flow on a monthly basis, but in later years, quarterly periods may be sufficient. The opening balance represents the cash available to start the business. Payments and disbursements are subtracted from this, and income is added to it, to give a closing balance at the end of each month. This closing balance is then used as an opening balance to start the next month. A cash flow sheet should include both net and cumulative cash flows, net cash flow being the difference between cash receipts and disbursements over a specified period of time (for example, each month) while cumulative cash flow is the sum of all net cash flows from start up to present.

For purposes of illustration, the following example will consider the costs of the construction of an abalone farm, and its running costs over the first few years of operation. In the Tables 20.1, 20.2 and 20.3, hypothetical values are put into Excel worksheets, purely to illustrate how a cash flow prediction might be constructed in a business plan. It should be noted that because many of the values in such spreadsheets are extremely variable from one time to another, and from one geographic location to another, they are not intended to represent actual values at the time of this publication, but are rather included to describe the categories of expenses that should be considered. In addition, some expenses such as compliance costs and insurance, have been omitted from the spreadsheets, again because of the huge geographic variations in such costs.

Table 20.1 represents a capital expenditure cash flow that might be expected during the construction phase. It assumes that an initial loan of US$1 million has been secured to initiate construction. For the purposes of illustration, it has been assumed that the interest on the loan will be repaid at a constant US$4200 per month. A lease value for the land has been set at US$1000 per month and it is assumed that, for the first year, a project manager is employed at a total cost of US$6000 per month. A construction cost of US$40,000 for the first month and US$20,000 per month thereafter is included, while costs for tanks, pumps, pumping, and vehicles are included only in months 11 and 12. This shows that, at the end of the first year, a total of US$499,400 has been spent, leaving US$500,600 available for the next phase of the project.

Assuming that construction by the end of Year 1 is sufficient for the first juveniles to be brought onto the farm at the beginning of Year 2, Tables 20.2 and 20.3 represent the running costs for Years 2 and 3. It is assumed that 100,000 spat, of 10 mm shell length (SL) are purchased at a cost of US$0.10 per individual, and that these grow at an average rate of 2 mm per month throughout Year 2, with an average mortality of 1% per month. The five tanks purchased at the end of Year 1 are sufficient to house all animals for the first 6 months of Year 2 but, from that time onwards, additional tanks are purchased each month. Water is pumped through the tanks at a rate of 2.8 m^3 per hour and, at a pumping cost of US$0.18/$m^3$/hour, this allows a calculation of the monthly cost of pumping water. Animals are fed at an average of 7% body weight per day, and feed costs US$2.75/kg. The model assumes that the farm is run by one manager and two

Table 20.1 Capital expenditure over first year (US$).

Initial loan												1,000,000
Year 1 – construction	Jan	Feb	Mar	Apr	May	Jun	Jul	Aug	Sept	Oct	Nov	Dec
Management salary	6000	6000	6000	6000	6000	6000	6000	6000	6000	6000	6000	6000
Interest	4200	4200	4200	4200	4200	4200	4200	4200	4200	4200	4200	4200
Lease	1000	1000	1000	1000	1000	1000	1000	1000	1000	1000	1000	1000
Earthworks/construction	40,000	20,000	20,000	20,000	20,000	20,000	20,000	20,000	20,000	20,000	20,000	20,000
Pumps and plumbing											25,000	25,000
First 5 grow-out tanks												5000
Equipment and vehicles												50,000
Total spent during each month	51,200	31,200	31,200	31,200	31,200	31,200	31,200	31,200	31,200	31,200	56,200	111,200
Total spent at end of first year												499,400
Funds still available at end of first year												500,600

additional workers throughout Year 2. The bottom line in Table 20.2 shows that the cumulative running cost by the end of Year 2, is US$238,027.

A similar spreadsheet is shown (Table 20.3) for Year 3 of the operation, the difference being that the animals are assumed to grow at 3 mm per month during this year, more tanks are added as total biomass on the farm increases, and an additional laborer is included in the estimates. By the end of Year 3, the total cumulative running cost has been US$678,557 and, together with the initial capital cost of US$499,400, this would mean that total cost to date would have been US$1,177,957. To finance the ongoing operation would, therefore, require further capital to be borrowed. It should be noted, however, that from that point on, animals would be ready to sell and this would, obviously, begin to repay capital. The time taken for overall cash flow to become positive would vary considerably with different species and different project scenarios, but an investor would want to see the business plan continued until at least the time of positive cash flow. As mentioned previously, in a real-life situation, it is unlikely that a single cohort of animals would be taken through the whole process independently from subsequent cohorts and this, again, could make significant improvements in cash flow because this would lessen the overall influence of fixed overheads.

The figures used in Tables 20.1, 20.2 and 20.3 are obviously hypothetical because precise values will depend on the species being cultured, the geographical location of the farm, and the time when the project is initiated. They do, however, illustrate the type of approach that an investor would expect in order to estimate capital requirement, ongoing financial commitment, expected long-term profit, and the time frame for return on investment.

As stated above, the total expenditure by the end of Year 3 in this hypothetical example, would have been US$1,117,957. These tables do not, however, record any income to the farm. If we assume (for the purposes of illustration) that the first sale occurs at the end of Year 3 when the animals measure 68 mm SL, the total weight of animals available to sell would be 4.941 metric tons, valued at a farm-gate price of US$169,435. Clearly, the first sale would not be sufficient to cover total accumulated expenditure, and it for this reason that the business plan would need to follow subsequent cohorts of production until the cash flow becomes positive.

In a recent publication, O'Shea et al. (2019) analyzed the various types of risks that apply to aquaculture operations. According to their analysis, the greatest risks to aquaculture businesses were early stage project development risk, and commodity price risk. Both of these have a high probability of occurrence, with medium-to-high severity. They suggested, however, that early stage development risk could be mitigated through proper site selection, identification of high-quality management teams, and ample contingency funding. Commodity price risk could be reduced through mitigating factors such as long-term supply agreements, product differentiation and branding, and correct species selection. The fact remains, however, that aquaculture is still seen as a high risk business by many potential investors. Even with a well-prepared business plan, the perception of high risk may persist. Without a comprehensive business plan, however, it would probably be almost impossible to attract sufficient funding to initiate a new aquaculture venture.

Table 20.2 Running costs for second year (US$).

Initial loan												1,000,000
Year 2 – monthly running costs	Jan	Feb	Mar	Apr	May	Jun	Jul	Aug	Sept	Oct	Nov	Dec
Interest	4200.00	4200.00	4200.00	4200.00	4200.00	4200.00	4200.00	4200.00	4200.00	4200.00	4200.00	4200.00
Lease	1000.00	1000.00	1000.00	1000.00	1000.00	1000.00	1000.00	1000.00	1000.00	1000.00	1000.00	1000.00
Spat cost ($)	10000.00											
Spat nos. (1%/month mortality)	100,000.00	99,000.00	98,010.00	97,030.00	96,060.00	95,100.00	94,150.00	93,210.00	92,270.00	91,350.00	90,440.00	89,530.00
Spat size (mm)	10.00	12.00	14.00	16.00	18.00	20.00	22.00	24.00	26.00	28.00	30.00	32.00
Total mass of all spat (kg)	40.00	49.50	58.80	72.77	86.45	114.12	131.81	186.42	230.67	319.73	406.98	465.56
No. additional tanks							1.00	1.00	1.00	1.00	1.00	1.00
Cost @1000 each tank							1000.00	1000.00	1000.00	1000.00	1000.00	1000.00
Total tanks used	5.00	5.00	5.00	5.00	5.00	5.00	6.00	7.00	8.00	9.00	10.00	11.00
Water required @ 2.8 m^3/hour/tank	14.00	14.00	14.00	14.00	14.00	14.00	16.80	19.60	22.40	25.20	28.00	30.80
Electricity service charge	16.50	16.50	16.50	16.50	16.50	16.50	16.50	16.50	16.50	16.50	16.50	16.50
Pumping cost (US$/month)	181.44	181.44	181.44	181.44	181.44	181.44	217.73	254.02	290.30	326.59	362.88	399.17
Feed reqd (per month) @7%BW/day	84.00	103.95	123.48	152.82	181.55	239.65	276.80	391.48	484.41	671.43	854.66	977.68
Feed cost@US$2.75/kg	231.00	285.86	339.57	420.25	499.25	659.04	761.20	1076.58	1332.12	1846.44	2350.31	2688.61
Management salary	6000	6000	6000	6000	6000	6000	6000	6000	6000	6000	6000	6000
Wages	6000	6000	6000	6000	6000	6000	6000	6000	6000	6000	6000	6000
Running cost each month	27,628.94	17,683.80	17,737.51	17,818.19	17,897.19	18,056.98	19,195.43	19,547.09	19,838.92	20,389.53	20,929.69	21,304.28
Cumulative running cost to date	27,628.94	45,312.74	63,050.25	80,868.44	98,765.63	116,822.61	136,018.04	155,565.13	175,404.06	195,793.59	216,723.28	238,027.56

Table 20.3 Running costs for third year (US$).

Initial loan												1,000,000
Year 3 – monthly running costs	Jan	Feb	Mar	Apr	May	Jun	Jul	Aug	Sept	Oct	Nov	Dec
Interest	4200.00	4200.00	4200.00	4200.00	4200.00	4200.00	4200.00	4200.00	4200.00	4200.00	4200.00	4200.00
Lease	1000.00	1000.00	1000.00	1000.00	1000.00	1000.00	1000.00	1000.00	1000.00	1000.00	1000.00	1000.00
Animal nos.	88,635	87,749	86,872	86,003	85,143	83,292	83,449	82,615	81,789	80,971	80,161	79,359
Body size (mm)	35	38	41	44	47	50	53	56	59	62	65	68
Total mass of all animals (kg)	647.04	877.49	1094.59	1419.05	1702.86	2082.3	2586.92	2561.07	3516.93	3967.58	4408.86	4840.9
No additional tanks added each month	1	1	1	1	1	1	2	2	2	2	2	2
Cost @ US$1000 each	1000.00	1000.00	1000.00	1000.00	1000.00	1000.00	2000	2000	2000	2000	2000	2000
Total no. tanks used	12	13	14	15	16	17	19	21	23	25	27	29
Water required @ 2.8 m^3/hour/tank	33.6	36.4	39.2	42	44.8	47.6	53.2	58.8	64.4	70	75.6	81.2
Electricity service charge	16.50	16.50	16.50	16.50	16.50	16.50	16.50	16.50	16.50	16.50	16.50	16.50
Pumping cost (US$/month)	435.456	471.744	508.032	544.32	580.608	616.896	689.472	762.048	834.624	907.2	979.776	1052.352
Feed reqd (per month) @7%BW/day	1,358.784	1,842.729	2,298.639	2,980.005	3,576.006	4,372.83	5,432.532	5,378.247	7,385.553	8,331.918	9,258.606	10,165.89
Feed cost@US$2.75/kg	3,736.656	5,067.5048	6,321.2573	8,195.0138	9,834.0165	12,025.283	14,939.463	14,790.179	20,310.271	22,912.775	25,461.167	27,956.198
Management salary	6000	6000	6000	6000	6000	6000	6000	6000	6000	6000	6000	6000
Wages	9000	9000	9000	9000	9000	9000	9000	9000	9000	9000	9000	9000
Running cost each month	25,388.61	26,755.75	28,045.79	29,955.83	31,631.12	33,858.68	37,845.44	37,768.73	43,361.39	46,036.47	48,657.44	51,225.05
Cumulative running cost to date	263,416.17	290,171.92	318,217.71	348,173.54	379,804.66	413,633.34	415,508.68	489,277.41	532,638.80	578,675.27	627,332.71	678,557.76

References cited and further reading

Engle, C.R. (2012) Assessing the financial position of an aquaculture business: using balance sheets. SRAC Publication No. 4401, Southern Regional Aquaculture Center, Stoneville, MS.

Engle, C.R. (2019) *Aquaculture Businesses: A Practical Guide to Economics and Marketing*. 5m Publishing, Sheffield.

Godfrey, M. (2018) Financing aquaculture: The cash is there, but information is lacking. SeaFoodSource, July 18. Available at: https://www.seafoodsource.com/features/financing-aquaculture-the-cash-is-there-but-information-is-lacking (accessed 14 January 2020).

Knapp, G. and Rubino, M.C. (2016) The political economics of marine aquaculture in the United States. *Reviews in Fisheries Science and Aquaculture*, 24(3): 213–229.

Leeds, R. (1986) Financing aquaculture projects. *Aquacultural Engineering*, 5: 109–113

Kobayashi, M., Msangi, S., Batka, M., Vannuccini, S., Dey, M.M., and Anderson, J.L. (2015) Fish to 2030: the role and opportunity for aquaculture. *Aquaculture Economics & Management*, 19(3): 282–300.

O'Shea, T., Jones, R., Markham, A., Norell, E., Scott, J., Theuerkauf, S., and Waters, T. (2019) Towards a blue revolution: catalysing private investment in sustainable aquaculture production systems. The Nature Conservancy and Encourage Capital, Arlington, VA.

Strombom, D.B. and Tweed, S.M. (1992) Business planning for aquaculture – is it feasible? NRAC Fact Sheet, No. 150, University of Massachusetts.

Webster, D., Buttner, J., and Flimlin, G. (2008) Planning for success in your aquaculture business. NRAC Publication, No. 101, University of Maryland.

Index